# Maximum Entropy and Bayesian Methods

# Fundamental Theories of Physics

*An International Book Series on The Fundamental Theories of Physics:*
*Their Clarification, Development and Application*

Volume 39

# Maximum Entropy and Bayesian Methods

*Dartmouth, U.S.A., 1989*

*edited by*

Paul F. Fougère

*Massachusetts Institute of Technology,*
*Department of Mathematics,*
*Cambridge, U.S.A.*

KLUWER ACADEMIC PUBLISHERS

DORDRECHT / BOSTON / LONDON

Library of Congress Cataloging-in-Publication Data

```
Maximum entropy and Bayesian methods, Dartmouth, U.S.A., 1989 / edited
  by P.F. Fougere.
        p.   cm. -- (Fundamental theories of physics ; v. 39)
    Proceedings of the 9th annual Workshop on Maximum Entropy and
  Bayesian Methods, Dartmouth, Mass., 1989.
    Includes index.

    1. Entropy (Information theory)--Congresses.  2. Bayesian
  statistical decision theory--Congresses.   I. Maximum Entropy
  Workshop (9th : 1989 : Dartmouth, Mass.) II. Series.
  Q370.M367  1990
  500--dc20                                              90-45264
```

ISBN-13: 978-94-010-6792-8      e-ISBN-13: 978-94-009-0683-9
DOI: 10.1007/978-94-009-0683-9

Published by Kluwer Academic Publishers,
P.O. Box 17, 3300 AA Dordrecht, The Netherlands.

Kluwer Academic Publishers incorporates
the publishing programmes of
D. Reidel, Martinus Nijhoff, Dr W. Junk and MTP Press.

Sold and distributed in the U.S.A. and Canada
by Kluwer Academic Publishers,
101 Philip Drive, Norwell, MA 02061, U.S.A.

In all other countries, sold and distributed
by Kluwer Academic Publishers Group,
P.O. Box 322, 3300 AH Dordrecht, The Netherlands.

*Printed on acid-free paper*

*To Marguerite…*
*and to my friends in the MaxEnt Community*

# PUBLICITY

It has been thirty years since Professor Edwin T. Jaynes published his fundamental papers in the physical review, on Statistical Mechanics, in which he showed that essentially all of statistical mechanics can be derived using a single, extremely general principle, that of Maximum Entropy, called here MaxEnt for short. Since that time he has worked unceasingly on the development of MaxEnt and of Bayesian probability theory, the twin foundations of the principles of inductive reasoning under the conditions of insufficient and/or inexact data. In freeing probability theory from the straight-jacket imposed by the frequentist view of probability theory, Prof. Jaynes has widened the field of application of fundamental probability theory to a host of applications where no frequency interpretation is possible and at the same time made it possible to solve those problems which do admit of a frequentist interpretation much more simply and elegantly. In making a clear and consistent separation between ontology (what really is true about the physical world) and epistemology (what is know about the world), by explaining the "mind projection fallacy", Prof. Jaynes makes another fundamental contribution to the art and science of thinking.

The applications of these methods have spread to so many fields of endeavor that ten years ago an annual MaxEnt workshop was established to bring together the growing population of researchers who were exploiting and further developing these methods. The workshop has been held at Laramie Wyoming, Seattle Washington, Calgary, British Columbia, Cambridge England and, in 1989, Dartmouth College, New Hampshire. Several volumes of proceedings have been published; this one is the ninth.

In this volume we have important papers by Ed Jaynes on probability theory as logic and by C. Ray Smith and Gary Erickson on the axiomatic basis of Cox's Theorem, relaxing one of the conditions thought necessary to establish the basic rules of inductive reasoning, if that reasoning is to be consistent. G. Larry Bretthorst, one of Jaynes' brightest students, gives a tutorial on the use of Bayes' Theorem in time series analysis and spectral estimation. Tom Loredo gives a brilliant tutorial on Bayesian analysis and applies Bayes' Theorem to the tiny quantity of data obtained on the Supernova 1987A, achieving some truly remarkable results.

In a paper of fundamental importance, David Hestenes shows how to separate the physics from the probability notions basic to quantum mechanics.

There are many papers here, exploiting Bayesian probability theory as well as MaxEnt methods. Several of these applications are based upon John Burg's fundamental advances, now known as the 'Burg Technique' and the 'Maximum Entropy Method' of spectral estimation. A paper by Al Rahbee exploits these two methods of Burg's in an important application to mass spectrometry and achieves truly remarkable mass resolutions.

For two dimensional image reconstruction, the software designed by Steve Gull, Jeff Daniells, John Skilling, Sibusiso Sibisi and others, and known as MEMSYS 3, is extremely powerful and is rapidly becoming the standard MaxEnt reconstruction algorithm. Important papers by Skilling and Sibisi continue this tradition.

# TABLE OF CONTENTS

# PREFACE

This volume represents the proceedings of the Ninth Annual MaxEnt Workshop, held at Dartmouth College in Hanover, New Hampshire, on August 14-18, 1989. These annual meetings are devoted to the theory and practice of Bayesian Probability and the Maximum Entropy Formalism. The fields of application exemplified at MaxEnt '89 are as diverse as the foundations of probability theory and atmospheric carbon variations, the 1987 Supernova and fundamental quantum mechanics. Subjects include sea floor pressures, neutron scattering, plasma equilibrium, drug absorption in man, nuclear magnetic resonance, radar and astrophysical image reconstruction, mass spectrometry, generalized parameter estimation, delay estimation, pattern recognition, heave responses in underwater sound and many others.

The first ten papers are on probability theory, and are grouped together beginning with the most abstract followed by those on applications. The tenth paper involves both Bayesian and MaxEnt methods and serves as a bridge to the remaining papers which are devoted to Maximum Entropy theory and practice. Once again, an attempt has been made to start with the more theoretical papers and to follow them with more and more practical applications. Papers number 29, 30 and 31, by Kesaven, Seth and Kapur, represent a somewhat different, perhaps even "unorthodox" viewpoint, and are included here even though the editor and, indeed many in the audience at Dartmouth, disagreed with their content. I feel that scientific disagreements are essential in any developing field, and often lead to a deeper understanding. The very last paper, by John Cyranski, owes its position to a mundane problem with typing, but curiously enough, the material on "fuzzy sets" is also at the center of another debate, and indeed was debated at MaxEnt '88 in Cambridge, England

It is unfortunate that a high point of the workshop, namely the debate between John Skilling and John Burg, which was moderated with great flair by Lee Schick, cannot be fully reported here. The debate brought into sharp focus the nature of the disagreement about the proper form of entropy to be maximized to obtain a power spectrum from a finite length time series. The last word on this subject has not yet been spoken.

The workshop owes a debt of gratitude to Professor Chi Hau Chen, who performed very diligently as treasurer, and to Ms. Linda Hathorn, Conference Coordinator, whose single-minded attention to every detail made the conference a joy to myself as well as to all other attendees. Finally, Dr. David Larner was very helpful in guiding a neophyte editor through the sometimes tortuous path of preparing a book-length typescript for publication. For this help, I am indeed grateful.

# PROBABILITY THEORY AS LOGIC

E. T. JAYNES
*Dept. of Physics*
*Washington University,*
*St. Louis MO 63130, U.S.A.*

ABSTRACT. At the 1988 workshop we called attention to the "Mind Projection Fallacy" which is present in all fields that use probability. Here we give a more complete discussion showing why probabilities need not correspond to physical *causal* influences, or "propensities" affecting mass phenomena. Probability theory is far more useful if we recognize that probabilities express fundamentally *logical inferences* pertaining to individual cases. We note several examples of the difference this makes in real applications.

## 1. Introduction

> *"Man is surely mad. He cannot make a worm;*
> *yet he makes Gods by the dozen."* — Montaigne.

It seems that mankind has always been occupied with the problem of how to deal with ignorance. Primitive man, aware of his helplessness against the forces of Nature but totally ignorant of their causes, would try to compensate for his ignorance by inventing hypotheses about them. For educated people today, the idea of directing intelligences willfully and consciously controlling every detail of events seems vastly more complicated than the idea of a machine running; but to primitive man (and even to the uneducated today) the opposite is true. For one who has no comprehension of physical law, but is aware of his own consciousness and volition, the natural question to ask is not: "*What* is causing it?", but rather: "*Who* is causing it?"

The answer was to invent Gods with the same consciousness and volition as ourselves, but with the additional power of psychokinesis; one in control of the weather, one in control of the seas, and so on. This personification of Nature must have been going on for thousands of years before it started producing permanent written records, in ancient Egypt and Greece. It appears that the adult citizens of those times really believed very literally in all their local Gods.

This oldest of all devices for dealing with one's ignorance, is the first form of what we have called the "Mind Projection Fallacy". One asserts that the creations of his own imagination are real properties of Nature, and thus in effect projects his own thoughts out onto Nature. It is still rampant today, not only in fundamentalist religion, but in every field where probability theory is used.

Of course, we are not arguing against a scientist's practice of formulating hypotheses about what is happening in Nature. Indeed, we see it as the highest form of creativity – far

1

*P. F. Fougère (ed.), Maximum Entropy and Bayesian Methods,* 1–16.

transcending mere mathematical skill – to conceive the right hypothesis in any field, out of one's educated imagination. Copernicus, Newton, Faraday, Darwin, Mendel, Pasteur, Wegener, Einstein are our heroes for having done this.

The difference between an imaginative scientist on the one hand, and primitive man and religious fundamentalists on the other, is that the scientist clearly recognizes the creations of his imagination as *tentative working hypotheses* to be tested by observation; and he is prepared to test and reject a hundred different hypotheses in order to find the right one.

## 2. The Mind Projection Fallacy

The writer became fully aware of this fallacy only recently, after many years of being vaguely troubled by the kind of logic that is used in a dozen different fields. Eventually there came that sudden flash of understanding of what was causing this.

I first learned about Bose and Fermi statistics, as an undergraduate, by this argument: "You and I cannot distinguish between the particles: *therefore*, the particles behave differently than if we could." In some vague way, the logic of this bothered me. It seemed to be claiming for man the powers of psychokinesis formerly reserved for those Gods. But a few years later, as a graduate student in Berkeley, I heard J. R. Oppenheimer expound the same argument, in the obvious conviction that this was an expression of deep new wisdom, a major advance in thinking.

Oppy proceeded to give a dozen other "physical" arguments in support of quantum theory, using the same pattern of logic. I was, of course, too cowed by his authority to raise any open objection, and made strenuous efforts to understand such arguments as lines of rational thought. But I never succeeded, and quantum theory has always seemed to have more the character of a religion than a science.

Then in studying probability theory, it was vaguely troubling to see reference to "gaussian random variables", or "stochastic processes", or "stationary time series", or "disorder", as if the property of being gaussian, random, stochastic, stationary, or disorderly is a real property, like the property of possessing mass or length, existing in Nature. Indeed, some seek to develop statistical tests to determine the presence of these properties in their data.

There was a short phase of studying philosophy, hoping that the answer to what was troubling me might be found there. But the reasoning of philosophers was far worse in the same vague way, and it was clear that they understood even less than I did about the meaning of quantum theory and probability theory (at least, I was familiar with the mathematics of the theories and could solve real problems by applying it). There were several more experiences like this in other fields.

A change started with the much appreciated opportunity to spend a year at St. John's College, Cambridge, in the quiet contemplation that is possible only when one is free of all other responsibilities. This started some new lines of thought which finally congealed a few years later. It was in late 1987 that the sudden flash of understanding came, and I saw that the things that had been troubling me vaguely for 40 years were all the *same* basic error and that this error occurs not only in science and philosophy; it is ubiquitous in every area where people try to think.

Why did it take so long to see this? We can reason clearly only in areas where we have an established symbolism for the concepts; but this error had no name. Even after sensing it intuitively, there was a struggle to find an appropriate name for this vaguely seen thing. Finally the term "Mind Projection Fallacy" seemed to be the only one that expressed the idea without calling up wrong connotations.

As soon as the error had a definite name and description, it was much easier to recognize. Once one has grasped the idea, one sees the Mind Projection Fallacy everywhere; what we have been taught as deep wisdom, is stripped of its pretensions and seen to be instead a foolish *non sequitur*. The error occurs in two complementary forms, which we might indicate thus:

(A) (My own imagination) $\longrightarrow$ (Real property of Nature)

(B) (My own ignorance) $\longrightarrow$ (Nature is indeterminate)

Form (B) arose out of quantum theory; instead of covering up our ignorance with fanciful assumptions about reality, one accepts that ignorance but attributes it to Nature. Thus in the Copenhagen interpretation of quantum theory, whatever is left undetermined in a pure state $\psi$ is held to be unknown not only to us, but also to Nature herself. That is, one claims that $\psi$ represents a physically real "propensity" to cause events in a statistical sense (a certain proportion of times on the average over many repetitions of an experiment) but denies the existence of physical causes for the individual events below the level of $\psi$. Its zealots accuse those who speculate about such causes of being "obsolete, mechanistic materialists", to borrow a favorite phrase.

Yet no experiment could possibly demonstrate that no cause exists; the most that one could ever conclude is that no cause was found. But if we ask, "Well, how hard did you *try* to find the cause?" we will be told: "I didn't try at all, because the theory assures us there is none." Then in what sense can one say that any experiments confirm such a theory? How can anyone feel confident that no causes exist at the submicroscopic level when experiments give no evidence for this? Clearly, such a claim is pure type (B) Mind Projection Fallacy.

It is evident that this pattern of thought is also present throughout orthodox statistics, whenever someone states, or implies, that his probabilities are real causative agents *en masse* for events that are not determined, individually, by anything. And we see that there can be no such thing as a statistical test for "absence of cause" or "randomness" or "disorder" for the same reason that there is no test for ugliness or foolishness; those qualities exist only in the eye of the observer. Now let us see one aspect of this in a specific example.

## 3. Example: The Poisson Distribution

At our Cambridge meeting in 1988, the fallacy of supposing that conditional probabilities must express a real physical causation in Nature (but one which operates only statistically), was illustrated by the example of drawing two balls from an Urn, comparing forward inference which may express such a causation with backward inference which cannot. Now let us give less trivial (and more useful) calculations which illustrate that in order to conduct sound reasoning we are not only permitted, but *required*, to use conditional probabilities as logical inferences, in situations where physical causation could not be operative.

The elementary Poisson distribution sampling problem provides a very nice example. The Poisson distribution is usually derived as a limiting "low counting rate" approximation to the binomial distribution, but it is instructive to derive it by using probability theory as logic, directly from the statement of independence of different time intervals, using only the primitive product and sum rules. Thus define the prior information:

$I \equiv$ "There is a positive real number $\lambda$ such that, given $\lambda$, the probability that an event $A$, or count, will occur in the time interval $(t, t + dt)$ is $p(A|\lambda I) = \lambda dt$. Furthermore, knowledge of $\lambda$ makes any information $Q$ about the occurrence or nonoccurrence of the event in any other time interval irrelevant to this probability: $p(A|\lambda Q I) = p(A|\lambda I)$."

In orthodox statistics one would not want to say it this way, but instead would claim that $\lambda$ is the sole causative agent present; the occurrence of the event in any other time interval exerts no *physical influence* on what happens in the interval $dt$. Our statement is very different.

Denote by $h(t)$ the probability there is no count in the time interval $(0, t)$. Now the proposition:

$$R \equiv \text{``No count in } (0, t + dt)\text{''} \tag{1}$$

is the conjunction of the two propositions:

$$R = [\text{``No count in } (0, t)\text{''}] \cdot [\text{``No count in } (t, t + dt)\text{''}] \tag{2}$$

and so, by the independence of different time intervals, the product rule gives:

$$h(t + dt) = h(t) \cdot [1 - \lambda dt] \tag{3}$$

or $\partial h / \partial t + \lambda h(t) = 0$. The solution, with the evident intial condition $h(0) = 1$, is

$$h(t) = e^{-\lambda t} \ . \tag{4}$$

Now consider the probability, given $\lambda$ and $I$, of the proposition

$B \equiv$ "In the interval $(0, t)$ there are exactly $n$ counts, which happen at the times $(t_1, \cdots, t_n)$ within tolerances $(dt_1, \cdots, dt_n)$, where $(0 < t_1, \cdots < t_n < t)$."

This is the conjunction of $(2n + 1)$ propositions:

$B = [\text{no count in } (0, t_1)] \cdot (\text{count in } dt_1) \cdot [\text{no count in } (t_1, t_2)] \cdot (\text{count in } dt_2) \cdots$
$\quad [\text{no count in } (t_{n-1}, t_n)] \cdot (\text{count in } dt_n) \cdot [\text{no count in } (t_n, t)].$

so by the product rule and the independence of different time intervals, the probability of this is the product of all their separate probabilities:

$$p(B|\lambda I) = [e^{-\lambda t_1}] \cdot (\lambda dt_1) \cdot [e^{-\lambda(t_2 - t_1)}] \cdots [e^{-\lambda(t_n - t_{n-1})}] \cdot (\lambda dt_n) \cdot [e^{-\lambda(t - t_n)}]$$

or, writing the proposition $B$ now more explicitly as $B = \text{`}dt_1 \cdots dt_n\text{'}$,

$$p(dt_1 \cdots dt_n | \lambda t I) = e^{-\lambda t} \lambda^n \, dt_1 \cdots dt_n \ , \qquad (0 < t_1, \cdots < t_n < t) \tag{5}$$

Then what is the probability, given $\lambda$, that in the interval $(0, t)$ there are exactly $n$ counts, whatever the times? Since different choices of the count times represent mutually exclusive propositions, the continuous form of the sum rule applies:

$$p(n|\lambda t I) = \int_0^t dt_n \cdots \int_0^{t_3} dt_2 \int_0^{t_2} dt_1 \, e^{-\lambda t} \lambda^n$$

or,

$$p(n|\lambda t I) = e^{-\lambda t}\frac{(\lambda t)^n}{n!} \tag{6}$$

the usual Poisson distribution. Without the time ordering in our definition of $B$, different choices of count times would not all be mutually exclusive events, so the sum rule would not apply in the above way.

As noted, conventional theory obtains this same formula from the premise that events in disjoint time intervals exert no physical influences on each other; the only causative agent operating is $\lambda$. Some authors have turned this around, and supposed that if we verify (6) in the frequency sense, that proves that the events were indeed causally independent!

This is an astonishing conclusion, when we note that one could design a hundred different mechanisms (or write a hundred different computer programs), which in various ways that are completely deterministic, generate the seemingly "random" data. That is, the time of the next event is completely determined by the times of the previous events by some complicated rule. Yet all of them could constrain the long-run frequencies to agree with (6) without showing any signs of correlations.

If an experimenter did not know what that complicated rule was, there is almost no chance that he could discover it merely by accumulation of more data. Then the Mind Projection Fallacy might lead him to claim that no rule exists; and we seem to be back to quantum theory. This is why "randomness" is a slippery, undefined, and unverifiable notion.

Now consider the next problem: let $0 < t_1 < t_2$ and let $n_1$ and $n_2$ be the numbers of counts in the time intervals $(0, t_1)$ and $(0, t_2)$. What is the forward conditional probability $p(n_2|n_1, \lambda, t_1, t_2, I)$? By the aforementioned logical independence, the probability that there are $(n_2 - n_1)$ counts in $(t_1, t_2)$ still has the form (1) independent of $n_1$, so

$$p(n_2|n_1, \lambda, t_1, t_2, I) = e^{-\lambda(t_2-t_1)}\frac{[\lambda(t_2 - t_1)]^{n_2-n_1}}{(n_2 - n_1)!}, \quad t_1 < t_2, \ n_1 \leq n_2 \tag{7}$$

Then what is the joint probability for $n_1$ and $n_2$? By the product rule,

$$p(n_1 n_2|\lambda t_1 t_2 I) = p(n_1|\lambda t_1 I)\, p(n_2|n_1 \lambda t_1 t_2 I)$$

$$= \left[e^{-\lambda t_1}\frac{(\lambda t_1)^{n_1}}{n_1!}\right] \cdot \left[e^{-\lambda(t_2-t_1)}\frac{[\lambda(t_2 - t_1)]^{n_2-n_1}}{(n_2 - n_1)!}\right] \tag{8}$$

Now this can be rearranged into

$$\left[e^{-\lambda t_2}\frac{(\lambda t_2)^{n_2}}{n_2!}\right] \cdot \left[\binom{n_2}{n_1}\left(\frac{t_1}{t_2}\right)^{n_1}\left(1 - \frac{t_1}{t_2}\right)^{n_2-n_1}\right] \tag{9}$$

and we recognize the first factor as the unconditional probability $p(n_2|\lambda t_2 I)$, so by the other way of writing the product rule,

$$p(n_1 n_1|\lambda t_1 t_2 I) = p(n_2|\lambda t_2 I)\, p(n_1|n_2 \lambda t_1 t_2 I) \tag{10}$$

the backward conditional distribution must be given by the binomial:

$$p(n_1|n_2 \lambda t_1 t_2 I) = \binom{n_2}{n_1}\left(\frac{t_1}{t_2}\right)^{n_1}\left(1 - \frac{t_1}{t_2}\right)^{n_2-n_1}, \qquad (0 \leq n_1 \leq n_2) \tag{11}$$

But this is totally different from $p(n_2|n_1 \lambda t_1 t_2)$; it does not even contain $\lambda$!

When we reason forward from given $n_1$ to inferred $n_2$, knowledge of $\lambda$ makes $n_1$ irrelevant for predicting the number of counts after $t_1$. In conventional "random variable" probability theory one might think that $\lambda$ is always the sole relevant quantity because it is the sole physical causative agent; and therefore $p(n_1|n_2 \lambda t_1 t_2) \equiv p(n_1|\lambda t_1)$. But our analysis shows that when we reason backward from $n_2$ to $n_1$, knowledge of $\lambda$ does not make $n_2$ irrelevant; on the contrary, knowledge of $n_2$ makes $\lambda$ irrelevant!

We could hardly make the point more strongly that *physical* dependence and *logical* dependence are very different things. Some may find this result so disconcerting that their first reaction is to doubt the correctness of (11). If you find yourself with such feelings, please consider: If you already knew the actual number $n_2$ of events in the long interval, how would you then use knowledge of $\lambda$ to improve your estimate of $n_1$ beyond what is given by (11)?

The point is that knowledge of $\lambda$ does not determine $n_2$; it gives us only probabilities for different values of $n_2$. But if we know the *actual value* of $n_2$ over an interval that includes $(0, t_1)$, common sense surely tells us that this takes precedence over anything that we could infer from $\lambda$. That is, possession of the datum $n_2$ makes the original sampling probabilities (those conditional only on $\lambda$) irrelevant to the question we are asking.

In the above we considered $\lambda$ known in advance (*i.e.* specified in the statement of the problem). More realistically, we will not know $\lambda$ exactly, and therefore information about either $n_1$ or $n_2$ will enable us to improve our knowledge of $\lambda$ and take this into account to improve our estimates of other things. How will this change our results?

Consider the case that $\lambda$ is unknown, but suppose for simplicity that it is known not to be varying with time. Then we are to replace (6) – (10) by extra integrations over $\lambda$. Thus in place of (6) we have

$$p(n|I) = \int p(n\lambda|I)d\lambda = \int p(n|\lambda I)\, p(\lambda|I)\, d\lambda \tag{6a}$$

where $p(n|\lambda I)$ is given by (6), and $p(\lambda|I)$ is the prior probability density function (*pdf*) for $\lambda$. In a similar way, (7) is replaced by a probabiity mixture of the original distributions:

$$p(n_2|n_1, t_1, t_2, I) = \int p(n_2|n_1, \lambda, t_1, t_2, I)\, p(\lambda|I)d\lambda\,, \qquad t_1 < t_2, \ n_1 \leq n_2 \tag{7a}$$

and the joint probability for $n_1$ and $n_2$ becomes

$$p(n_1 n_2|t_1 t_2 I) = \int p(n_1 n_2|\lambda t_1 t_2 I)\, p(\lambda|I)\, d\lambda \tag{8a}$$

where the integrand is given by (8); again it can be rearranged as in (9), yielding

$$p(n_1 n_2|t_1 t_2 I) \int \left[e^{-\lambda t_2} \frac{(\lambda t_2)^{n_2}}{n_2!}\right] \left[\binom{n_2}{n_1}\left(\frac{t_1}{t_2}\right)^{n_1}\left(1 - \frac{t_1}{t_2}\right)^{n_2 - n_1}\right] p(\lambda|I)\, d\lambda$$

but now we recognize in this the factor

$$p(n_2|t_1 t_2 I) = \int p(n_2|\lambda t_1 t_2 I)\, p(\lambda|I)\, d\lambda \tag{9a}$$

and so from the product rule in the form $p(n_1 n_2 | t_1 t_2 I) = p(n_2 | t_1 t_2 I) \, p(n_1 | n_2 t_1 t_2 I)$ we conclude that

$$p(n_1 | n_2 t_1 t_2 I) = \binom{n_2}{n_1} \left(\frac{t_1}{t_2}\right)^{n_1} \left(1 - \frac{t_1}{t_2}\right)^{n_2 - n_1}, \qquad (0 \le n_1 \le n_2) \qquad (11a)$$

in agreement with (11); this result is the same whether $\lambda$ is known or unknown. Since this derivation allowed full opportunity for updated knowledge of $\lambda$ to be taken into account, (11a) is a Bayesian predictive distribution.

In reasoning from $n_1$ to $n_2$, the difference between (7) and (7a) represents the penalty we pay for not knowing $\lambda$ exactly; but in reasoning from $n_2$ to $n_1$ there is no penalty. Indeed, if the probability (11) is independent of $\lambda$ when $\lambda$ is known, it is hard to see how it could matter if $\lambda$ was unknown; and probability theory so indicates. Possession of the datum $n_2$ makes the original sampling probabilities – whether $\lambda$ is known or unknown – irrelevant to the question we are asking.

## 4. Discussion

The phenomena we have just demonstrated are true much more generally. Conventional sampling probabilities like $p(n | \lambda t)$ are relevant only for "pre-data" considerations; making predictions before we have seen the data. But as soon as we start accumulating data, our state of knowledge is different. This new information necessarily modifies our probabilities in a way that is incomprehensible to one who tries to interpret probabilities as expressing physical causation or long-run relative frequencies; but as (11) and (11a) illustrate, this updating appears automatically when we use probability theory as logic.

For example, what is the probability that in 10 binomial trials we shall find 8 or more successes? The binomial sampling distribution might assign to this event a probability of 0.46. But if the first 6 trials yield only 3 successes, then we know with certainty that we shall *not* get 8 or more successes in those 10 trials; the sampling probability 0.46 becomes irrelevant to the question we are asking.

How would a conventional probabilist respond to this example? He can hardly deny our conclusion, but he will get out of it by saying that conventional probability theory does not refer to the individual case as we were trying to do; it makes statements only about long-run relative frequencies, and we agree.

But then we observe that probability theory as logic *does* apply to the individual case, and it is just that individual case that concerns us in virtually all our problems of inference (*i.e.*, reasoning as best we can when our information is incomplete). The binomial distribution (11) will yield a more reliable estimate of $n_1$ than will the Poisson distribution (6) *in each individual case* because it contains cogent information, pertaining to that individual case, that is not in (6).

Orthodox probabilists, who use only sampling probability distributions and do not associate them with the individual case at all, are obliged to judge any estimation method by its performance "in the long run"; *i.e.* by the sampling distribution of the estimates when the procedure is repeated many times. That is of no particular concern to a Bayesian, for the same reason that a person with a ten-digit hand calculator has no need for a slide rule. The real job before us is to make the best estimates possible from the information we have *in each individual case*; and since Bayesians already have the solution to that problem, we have no need to discuss a lesser problem.

To see that long-run performance is indeed a lesser problem, note that even if we had found a procedure whose long-run performance is proved to be as good as can be obtained (for example, which achieves the minimum possible mean-square error), that would not imply that this procedure is best – or even tolerably good – in any particular individual case. One can trade off better performance for one class of samples against worse performance for another in a way that has no effect on long-run performance, but has a very large effect on performance in the individual case.

We do, of course, want to know how accurate we can expect our estimates to be; but the proper criterion of this is not the sampling distribution, but the width of the Bayesian posterior probability distribution for the parameter. This gives an indication of the accuracy of our estimate *in each individual case*, not merely a smeared-out average over all cases. In this sense also, the sampling distribution is the answer to a lesser problem, and the sampling distribution criterion of performance is not the one an informed scientist really wants.

The relevant question for a scientist is not: "How accurately would the estimate agree with the true value of the parameter in the long run over all possible data sets?" but rather: "How accurately does the one data set that I actually have determine the true value of the parameter?" This is the question that Bayes' theorem answers. When no sufficient statistic exists (or if one uses an estimator which is not a sufficient statistic, even though one does exist) the answers to these questions can be very different, and the sampling distribution criterion can be misleading.

This was perceived already by R. A. Fisher in the 1930's. He noted that different data sets, even though they may lead to the same estimate, may still justify very different claims of accuracy because they have different spreads, and he sought to correct this by his conditioning on "ancillary statistics". But ancillary statistics do not always exist, and when they do, as noted by Bretthorst (1988), this procedure is mathematically equivalent to applying Bayes' theorem. Indeed Bayes' theorem automatically generates a log-likelihood that is spread over a range corresponding to that of the data, whether or not ancillary statistics exist. This is one of its built-in safety features; it protects us against being misled about the accuracy of our estimates.

If we choose estimates by sampling distribution criteria, the conclusions we draw will depend, not just on the data that we actually have, but on what other data sets one thinks might have been observed, but were not. Thus an estimation procedure that works well for the data set that we have observed can be rejected on the grounds that it would have worked poorly for some other data set that we have not observed! We return to this point in "Optional Stopping" below, and see the can of worms it opens up.

But if anyone insists on seeing it, the Bayesian can of course calculate the sampling distribution of his estimates. Needless to say, Bayesian procedures look very good by sampling theory criteria, even though they were not derived with such criteria in mind. In fact, analysis of many cases has shown that the Bayesian point and interval estimates based on noninformative priors (say, the mean $\pm$ standard deviation of the posterior distribution) are usually optimal by sampling theory criteria. With informative priors, Bayesian methods advance into an area where sampling theory cannot follow at all.

For example, the frequency distribution of errors in estimating $n_1$ that result from using the Bayesian predictive distribution (11a) will be better (more accurate in the long run) than the distribution that one gets from using the direct sampling distribution (6). To a Bayesian, this is obvious without any calculation; for if a method is better in each individual case, how could it fail to be better also in the long run?

Orthodox probabilists would not accept that argument, but they would be convinced by comparing the two sampling distributions of the estimates, either analytically or experimentally. A simple hand-waving argument leads us to predict that the mean-square error with (11a) will be less than that with (6) by a factor $[1 - (t_1/t_2)]$. Although we could demonstrate this by an explicit sampling theory calculation, it would be more interesting to conduct "Monte Carlo" computer experiments to check our claim.

Of course, probability theory as logic need not be applied only to the individual case. It can be applied equally well to prediction of long-run relative frequencies, if that happens to be the thing of interest. Indeed, it can sometimes make better predictions, because by using the notion of probability of an hypothesis it has the means for taking into account relevant information that "random variable" theory cannot use.

The philosophical difference between conventional probability theory and probability theory as logic is that the former allows only sampling distributions, interprets them as physically real frequencies of "random variables", and rejects the notion of probability of an hypothesis as being meaningless. We take just the opposite position: that the probability of an hypothesis is the fundamental, necessary ingredient in all inference, and the notion of "randomness" is a red herring, at best irrelevant.

But although there is a very great philosophical difference, where is the functional difference? As illustrated above, by "probability theory as logic" we mean nothing more than applying the standard product and sum rules of probability theory to whatever propositions are of interest in our problem. The first reaction of some will be: "What difference can this make? You are still using the same old equations!" To see why it makes a difference, consider some case histories from Statistical Mechanics and Artificial Intelligence.

## 5. Statistical Mechanics

Mark Kac (1956) considered it a major unsolved problem to clarify how probability considerations can be introduced into physics. He confessed that he could not understand how one can justify the use of probability theory as Boltzmann used it, in writing down a simultaneous probability distribution $f(x, v; t) d^3x\, d^3v$ over position and velocity of a particle, because:

"In every probabilistic model in physics and in all other sciences there must be some lack of specification over which you can average. - - - That's the whole problem as to how probability can be introduced in kinetic theories of mechanics. - - - I am unable to find a probabilistic model which will lead to the full Boltzmann equation. I will show you how one can very easily be led to the equation in velocity space, however. - - - Once we have spatial homogeneity, then we have a lack of specification of position. And consequently we have wide freedom to average over all possible positions. If you don't have spatial homogeneity, then the problem becomes over-determined. There's absolutely no room, or at least I can't find any room, to introduce a stochastic element. I don't know what's random anymore, and so I cannot find a stochastic model which will lead to the full Boltzmann equation."

Mark Kac was a fine mathematician, but he had this mental hangup which prevented him from comprehending the notion of a probability referring to an individual case, rather than an "ensemble" of cases. So he was reduced to inventing clever mathematical models, instead of realistic physical situations, for his probability analyses. In a very idealized model he found that he could get a Boltzmann-like equation only if the $n$-particle probability distribution factored into a product of single-particle distributions. This led him to say of the Boltzmann equation:

" - - - it is philosophically rather peculiar. Because if you believe in it you must ask yourself why nature prepares for you at time zero such a strange factorized distribution. Because otherwise you can't get Boltzmann's equation."

We see here the Mind Projection Fallacy, in the form of a belief that his $n$-particle probability distributions were real things existing in Nature. The answer which I gave Mark Kac at the time was: "Nature does not prepare distributions, factorized or otherwise; she prepares *states*." But his thinking was so far removed from this viewpoint that he thought I was joking.

Today we could explain the point a little better: "The probability distributions in phase space used by Maxwell, Boltzmann, and Gibbs are not realities existing in Nature; they are descriptions of incomplete human information about Nature. They yield the best predictions possible from the information contained in them. Probability distributions are not "right" or "wrong" in a factual sense; rather, some distributions make better predictions than others because they contain more relevant information. With a factorized distribution, getting knowledge of the position of one particle would tell us nothing about the position of any other. But at soon as the particles interact, knowing the position of one *does* tell us something about where the others are. Therefore the probability distributions which lead to the best physical predictions for interacting particles are not factorized."

But we think that Kac's mathematical conclusion is quite correct; the Boltzmann equation does, in effect, suppose factorization. But then it follows that it cannot take full account of the effect of interactions. Indeed, when the particles interact, a factorized distribution cannot predict correctly either the equation of state or the hydrodynamic equations of motion. One can do much better by using the nonfactorized distribution of Gibbs, which contains more relevant information than does the Boltzmann distribution, in just the same sense that (11) contains more relevant information than does (6). Mark Kac had this important fact in his grasp but could not see it, and so he never appreciated the superiority of Gibbs' methods, and continued trying to justify Boltzmann's methods.

We could give other recent case histories of workers (Feller, Uhlenbeck, Montroll) who were highly competent mathematically, but were conceptually such captives of the Mind Projection Fallacy that it prevented them from seeing important results that were already present in their equations. For others, this fallacy prevents them from finding any useful results at all. For example, Pool (1989) quotes a current worker in statistical mechanics stating as one of the long-standing problems of statistical mechanics:

"Where does the randomness necessary for statistical behavior come from if the universe is at heart an orderly, deterministic place?"

Statements like this are spread uniformly and densely throughout the literature of quantum theory and statistical mechanics. People who believe that probabilities are physically real, are thereby led to doubt the reality of mechanical causes; eventually they come to doubt the reality of physical objects like atoms.

Once we have learned how to use probability theory as logic, we are free of this mental hangup and able at least to perceive, if not always solve, the real problems of science. Most of those long-standing problems of statistical mechanics are seen as non-problems.

We do not seek to explain "statistical behavior" because there is no such thing; what we see in Nature is *physical* behavior, which does not conflict in any way with deterministic physical law. Quite the contrary, probability theory as logic easily explains what we see, as a *consequence* of deterministic physical law.

We are not puzzled by "irreversibility" because (one of those important results which has been in our equations for over a Century, but is still invisible to some), given the present macrostate, the overwhelming majority of *all possible* microstates lead, via just those evil, deterministic mechanical equations of motion, to the *same* macroscopic behavior; just the reproducible behavior that we observe in the laboratory. So what else is there to explain? There would be a major mystery to be explained if this behavior were *not* observed.

The Maximum Entropy formulation makes these things obvious from the start, because it sees Statistical Mechanics not as a physical theory of "random behavior", but as a process of inference: predicting macroscopic behavior as best we can from our incomplete information about microstates. In this endeavor, as in any other problem of inference, we never ask, "Which quantities are random?" The relevant question is: "Which quantities are known, and which are unknown?" Indeed, it appears to us that whenever we get down to a specific calculation, all of us are obliged to use the term "random" merely as a synonym for "unknown".

## 6. Artificial Intelligence

This field provides examples differing in detail, but not in the basic situation. Recently, its stagnation has been noted by many, leading to the appearance of articles of the genre: "*Whatever happened to AI?*" We can tell you what has happened by noting two recent references.

Peter Cheeseman (1988) in an eight-page article, tried to point out the need for Bayesian inference in AI, only to be buried under an avalanche of criticism (a total of 62 pages by 26 authors), which prompted a 14-page reply by Cheeseman. To elicit such a response must mean that Cheeseman's needle struck a rather sensitive nerve. Most of the critics simply refused to take note of his message (which was that Bayesian methods solve some currently important AI problems) and attacked his work on other grounds. Obviously, we cannot go into all the counter-arguments here, but we can indicate their general flavor.

The first critic objected to Bayesian methods on the grounds that they do not tell us how to create hypotheses (although neither do any other methods). This is like criticizing a computer because it will not push its own button; of course, it is up to us to tell Bayesian theory *which* problem we want it to solve. Would anybody want it otherwise?

The second critic complained that Bayesian inference "seems to be ruled out as a candidate for representing commonsense reasoning" on the grounds that people often reason differently. Indeed they do, particularly in AI. As we have been pointing out for many years, people who lack Bayesian training commit all kinds of errors of reasoning, which Bayesian methods would have corrected.

The third critic wrote so confusingly that I have no idea what he was trying to say. The fourth was concerned with Cheeseman's failure to recite the long history of the subject. The fifth and sixth rejected Cox's theorems on the grounds that he assumed differentiability, although Aczél (1966) removed that assumption long ago. Another critic resorted to name-calling, accusing Cheeseman of being a "necessarian"; and even worse, a physicist.

[Indeed, one with some training in physics is in a good position to perceive the logical situation here, because we are familiar with it in other contexts. The present context is that we learn about real physical phenomena via probability theory, although probability is not itself a physical phenomenon. We also learn about the size and shape of objects via light, although light does not itself possess the properties of size and shape.]

And so it went on and on, critics calling up every conceivable subterfuge in order to avoid having to recognize what Bayesian methods *actually do* in real problems. It was like the rantings of a desperately ill patient who refuses to take the one medicine that could save him, and accuses the doctors of trying to poison him.

Our second example is explicit enough so that we can indicate one thing that Bayesian methods can do for AI. Dan Shafer (1989) tries to explain what is called in AI a "certainty factor" or "confidence factor". He states that (on a scale of 0 to 100) this "expresses the degree of confidence the user or the system has in a response or a conclusion" and warns us that this is something very different from a probability. In his words [italics mine]:

"Probability—which predicts or describes the likelihood that *in a given group of items* any single item will have a particular attribute—does not enter into the issue"

Again, the mental hangup which cannot comprehend the notion of probability applied to an individual case.

Next he considers two propositions: $A \equiv$ "The patient's temperature is $> 101$." and $B \equiv$ "The patient has been resting for an hour." Then we have the technical problem: suppose we have the confidence factors $c(A), c(B)$ for propositions $A$ and $B$; what is the confidence factor for their conjunction $AB =$ "The patient has a fever"? He notes "three popular methods" for calculating this: (1) the minimum; (2) the product; (3) the average. But then he notes that we now have a computer program named GURU, which is much superior because it offers the user his choice of not just three, but *seven* different rules for calculating this confidence factor! This certainly reveals the poverty of Bayesian analysis, which can offer only one solution to this problem.

A "confidence factor" is a very explicit attempt to represent degrees of plausibility by real numbers; and so the user of it is automatically at the mercy of Cox's theorems. Cox's first functional equation, expressing the associativity of Boolean algebra, shows that for the conjunction of propositions, any AI algorithm that is not mathematically equivalent to the product rule of probability theory, will contain demonstrable inconsistencies when we try to apply it to more than two propositions.

But these inconsistencies never appear in Shafer's discussion because it never reaches even the first level of comprehension, where one sees that the problem requires the notion of a conditonal confidence factor $c(A|B)$. That is, knowing that a patient is not resting but exercising vigorously ought to increase one's "degree of confidence" that he has an elevated temperature, etc. The confidence factor for the conjunction $AB$ cannot be assessed rationally if one fails to take this correlation into account. A potential inconsistency for three propositions hardly matters if we have not yet achieved consistency for two.

If we can judge from Shafer's exposition, the AI theory of confidence factors is stumbling about in a condition more primitive than the probability theory of Cardano, 400 years ago. Yet the problems they are trying to solve are just the ones that *were* solved long ago in the Bayesian literature, as Cheeseman tried to point out.

One disadvantage of having a little intelligence is that one can invent myths out of his own imagination, and come to believe them. Wild animals, lacking imagination, almost never do disastrously stupid things out of false perceptions of the world about them. But humans create artificial disasters for themselves when their ideology makes them unable to perceive where their own self-interest lies. We predict that AI will continue to stumble about, producing results that a Bayesian considers always trivial, usually quantitatively wrong, and often qualitatively wrong, until it recogizes that to dissociate itself from Bayesian probability theory was a disastrous error.

Almost everything we have noted here applies as well to the field of fuzzy sets; but it would be repetitious.

## 7. Optional Stopping

A different kind of comparison appears in the issue of optional stopping, which has been a point of controversy between Bayesians and Orthodoxians for 30 years. This is another case where the Mind Projection Fallacy is doing serious damage, leading researchers to erroneous conclusions and wasted effort in the area of medical testing. The issue is: can an overzealous experimenter produce evidence supporting a foregone false conclusion, by deciding when to stop taking data?

Here orthodox theory as expounded by Armitage (1975) holds that the conclusions we should draw from an experiment depend not only on the experimental procedure and the resulting data, but also on the private thoughts that went through the experimenter's mind when he took the data! How can this be?

To see how, consider again binomial sampling, observing $r$ successes in $n$ trials. Two medical researchers use the same treatment independently, in different hospitals. Neither would stoop to falsifying the data, but one had decided beforehand that because of finite resources he would stop after treating $n = 100$ patients, however many cures were observed by then. The other had staked his reputation on the efficacy of the treatment, and decided that he would not stop until he had data indicating a rate of cures definitely greater than 60%, however many patients that might require. But in fact, both stopped with exactly the same data: $n = 100$, $r = 70$. Should we then draw different conclusions from their experiments?

One who thinks that the important question is: "Which quantities are random?" is then in this situation. For the first researcher, $n$ was a fixed constant, $r$ a random variable with a certain sampling distribution. For the second researcher, $r/n$ was a fixed constant (approximately), and $n$ was the random variable, with a very different sampling distribution. Orthodox practice will then analyze the two experiments in different ways, and will in general draw different conclusions about the efficacy of the treatment from them.

This would really cause trouble in a high-energy physics laboratory, where a dozen researchers may collaborate on carrying out a big experiment. Suppose that by mutual consent they agree to stop at a certain point; but they had a dozen different private reasons for doing so. According to the principles expounded by Armitage, one ought then to analyze the data in a dozen different ways and publish a dozen different conclusions, from what is in fact only one experiment!

Bayesian inference will not get us into this absurd situation, because it perceives automatically what common sense demands; that what is relevant for this inference is not the relative probabilities of imaginary data sets which were not observed, but the relative likelihoods of different parameter values, based on the one real data set which *was* observed; and this is the same for all the experimenters.

Actually, as Jimmie Savage (1962) explained long ago, we need not worry about being misled by zealots because, contrary to what many assume, it is not posible to sample deliberately to a foregone conclusion that is appreciably false. From the above data, most statisticians would estimate the true cure rate to be about $f \pm \sqrt{f(1-f)/n}$, or 70% ± 5% at one standard deviation, where $f = r/n$ is the observed frequency of cures. If the true incidence of cures in the whole population is only 50%, then the probability that a zealot can ever produce data like this (*i.e.* data which lead to an estimated interval that strongly excludes the true value) is extremely small in an honestly conducted experiment, even if he samples to the end of time.

Thus to produce data strongly supporting a false conclusion it is not enough merely to be zealous; one must be actively dishonest. Today it is not the zealots, but the medical testers who follow Armitage, who mislead themselves and others.

Orthodox writers love to charge Bayesians with "subjectivity" because we use probability as a way of describing our information. But we have seen a few examples of their idea of objectivity and some of its consequences, including inability to take prior information into account, inability to get rid of nuisance parameters, using inefficient criteria of performance, and inability to see important facts. Strangest of all, orthodox teaching can lead one to draw conclusions that depend on whether an experimenter subjectively imagined data sets which were not observed! A person who does that is in no position to charge anybody with "subjectivity".

Invariably, those who attack Bayesian methods only reveal their ignorance of what Bayesian methods are. The blame for this lies with those educators who continue to teach only orthodox methods and deprecate Bayesian methods without even fully defining them, much less examining their performance. The fact is that Bayesian methods of inference easily solve technical problems on which orthodox methods break down, and they protect us automatically against the absurd errors in orthodox methods. Thus they achieve scientific "objectivity" in the true sense of that word.

## 8. Recapitulation

In our simplest everyday inferences, in or out of science, it has always been clear that two events may be physically independent without being logically independent; or put differently, they may be logically dependent without being physically dependent. From the sound of raindrops striking my window pane, I infer the likely existence of clouds overhead, $p(\text{clouds}|\text{sound}) \simeq 1$, although the sound of raindrops is not a physical causative agent producing clouds. From the unearthing of bones in Wyoming we infer the existence of dinosaurs long ago: $p(\text{dinosaurs}|\text{bones}) \simeq 1$, although the digging of the bones is not the physical cause of the dinosaurs.

Yet conventional probability theory cannot account for such simple inferences, which we all make constantly and which are obviously justified. As noted, it rationalizes this failure by claiming that probability theory expresses partial physical causation and does not apply to the individual case.

But if we are to be denied the use of probability theory not only for problems of reasoning about the individual case; but also for problems where the cogent information does not happen to be about a physical cause or a frequency, we shall be obliged to invent arbitrary *ad hockeries* for dealing with virtually all real problems of inference; as indeed the orthodox school of thought has done.

Therefore, if it should turn out that probability theory used as logic *is*, after all, the unique, consistent tool for dealing with such problems, a viewpoint which denies this applicability on ideological grounds would represent a disastrous error of judgment, which deprives probability theory of virtually all its real value – and even worse, deprives science of the proper means to deal with its problems.

As an analogy, small children start counting on their fingers and toes. Then suppose that the teaching of arithmetic had fallen under control of an Establishment ideology which proclaims that the rules of elementary arithmetic apply *only* to fingers and toes; and then invents a different *ad hoc* rule for counting apples, still another for counting dollars, and so on. Imagine the effect this would have on our civilization.

Our position is that this is exactly what *has* happened in probability theory, and its effects are visible all about us. In theoretical physics we see the stagnation of quantum theory, an astonishing number of physicists having left it for other fields such as biophysics, and the wheel-spinning concentration on non-problems in statistical mechanics. In experimental science we see the absurdity of orthodox methods of data analysis. In Artificial Intelligence we see the consequences of militant refusal to adopt the only methods that can deal with their unsolved problems.

Fortunately, children quickly reach a level of maturity where they can perceive the number 13 as an abstract notion in its own right, not standing necessarily for ten fingers and three toes. It is high time that science reached the level of maturity where we can grasp the idea of probability of an hypothesis as an abstract notion in its own right, necessary for organizing our reasoning in a consistent way. Of course, just as the rules of arithmetic remain valid when applied to fingers and toes, the rules of probability theory remain valid when applied to calculation of frequencies.

Note that probability theory as logic is more general than just Bayesian inference; it automatically includes calculation of sampling distributions, as in our derivation of (6), and Maximum Entropy calculations, in the situations where they are appropriate. Bayesian analysis requires that we have a model in addition to the data. If we have only an hypothesis space but no model, then MAXENT is the best we can do without more information.

It may appear that in our recent concentration on Bayesian methods we have abandoned MAXENT. Not at all; it is an accomplished fact and we are using it constantly for its original purpose: to assign our priors. It is just for that reason that, for some of us, the focus of attention has now shifted to the Bayesian sequel. The exciting "new" fact (although it would not have surprised Harold Jeffreys in the least fifty years ago) is the flexibility of Bayesian analysis. As Bretthorst (1988) demonstrates by many specific examples, it can accommodate itself easily to all kinds of complicating circumstances that would bring orthodox methods to a standstill.

Scientists, engineers, economists and statisticians who are ignorant of Bayesian methods are handicapped in the performance of their work. In physics and astronomy the greatest experts in instrumentation may conduct multi-million-dollar data gathering operations – and then present their final conclusions in the form of confidence intervals which ignore not only some highly cogent prior information, but usually part of the information in the data. It is as if the best chefs in Paris had spared no effort or expense to prepare the finest food a restaurant can offer – and then spilled half of it down the drain and served the rest on paper plates.

## Appendix: A Basic Blunder

Finally, we must comment on a curious article entitled "The Basic Bayesian Blunder", by the philosopher H. E. Kyburg (1987), which amused us at this meeting. He formulates a problem of little interest except that the mathematical issue reduces to this: It is our basic blundering Bayesian belief that a weighted average of real numbers may take on all those, and only those, values in the range spanned by them. Thus any number in (.887, .917) can be written as a weighted average of (.887, .907, .917).

Since the average 0.900 specified by Kyburg lies in that interval, Bayesians do indeed, just as he charges, believe that we can assign prior probabilities $(\alpha, \beta, \gamma)$ leading to that average. He devotes several pages to arguing, by reasoning that we are completely unable to follow, that there is no solution. For answer it should suffice to exhibit an infinity of solutions. The system of equations to be solved is

$$\alpha + \beta + \gamma = 1$$

$$.887\alpha + .907\beta + .917\gamma = .900$$

and one verifies at once that the exact general solution of this system is

$$(\alpha, \beta, \gamma) = \left( \frac{65 - r}{140}, \frac{43 + 3r}{140}, \frac{32 - 2r}{140} \right)$$

where r is arbitrary. To meet the further requirement of nonnegativity $(\alpha, \beta, \gamma) \geq 0$, we see by inspection that r is confined by $(-43 \leq 3r \leq 48)$. There is a continuum of prior probability assignments which meet all the specified conditions, and which therefore enable a Bayesian to incorporate additional information.

## References

P. Armitage (1975), *Sequential Medical Trials*, Thomas, Springfield, Illinois. Second edition: Blackwell, Oxford.

G. Larry Bretthorst (1988), *Bayesian Spectrum Analysis and Parameter Estimation*, Springer Lecture Notes in Statistics, Vol. 48.

Peter Cheeseman (1988), "An inquiry into computer understanding", Comput. Intell. **4**, 58-66. See also the following 76 pages of discussion.

H. E. Kyburg (1987), "The Basic Bayesian Blunder", in *Foundations of Statistical Inference*, Vol II, I. B. MacNeill & G. J. Umphrey, Editors, Reidel Publishing Company, Holland.

Mark Kac (1956), it Some Stochastic Problems in Physics and Mathematics; Colloquium Lectures in Pure and Applied Science #2, Socony-Mobil Oil Company, Dallas, Texas.

R. Pool (1989), "Chaos Theory: How Big an Advance?", Science, **245**, 26-28.

L. J. Savage (1962) *The Foundations of Statistical Inference*, G. A. Barnard & D. R. Cox, Editors, Methuen & Co., Ltd., London

Dan Shafer (1989), "Ask the Expert", PC AI Magazine, May/June; p. 70.

# PROBABILITY THEORY
# AND THE ASSOCIATIVITY EQUATION

C. Ray Smith

Research, Development and engineering Center
U.S. Army Missile Command
Redstone Arsenal, Alabama, U.S.A.

AND

Gary J. Erickson
Department of Electrical Engineering
Seattle University, Seattle, Washington, U.S.A.

## ABSTRACT

At recent MaxEnt Workshops, several tutorials on Bayesian probability theory based on a Rationality desideratum, a Consistency desideratum and Boolean algebra were presented. The associativity equation is an important component of this approach to probability theory, and past tutorials could not devote the space to prove the uniqueness of or solve this functional equation. We demonstrate the unique relation of the associativity equation with rationality and consistency; then, we determine the solution of this equation by a method that does not assume differentiability.

## 1. INTRODUCTION

Our tutorial "From Rationality and Consistency to Bayesian Probability" in the 1988 Cambridge Proceedings [Smith and Erickson, 1989] concentrated on background, concepts and rationale — this paper is hereinafter referred to as RC. The presentation at Dartmouth was another tutorial, similar in spirit to the Cambridge talk, but with emphasis on different points. The present paper, representing our contribution to the Dartmouth Proceedings, will actually be a supplement to RC. Specifically, we want to address two problems identified to be of interest to a number of participants and omitted from RC: (1) establishing the functional composition of $u(AB|E)$ and (2) solving the associativity equation with methods from the theory of functional equations. As a supplement to RC, this paper tends to be tutorial; accordingly, several technical matters are sidestepped in the interest of clarity and brevity. Our apologies to the reader, but he or she may find it necessary to refer to RC for context, notation and specific equations.

P. F. Fougère (ed.), Maximum Entropy and Bayesian Methods, 17–30.

## 2. THE FUNCTIONAL COMPOSITION OF $u(AB|E)$

The plausibility for the compound proposition $AB$, given our prior information $E$, is symbolized by $u(AB|E)$. We want to explore the possible functional dependencies of $u(AB|E)$ on the plausibilities for $A$ and $B$ that are allowed by the desiderata. The plausibilities for $A$ and $B$ that we must consider are

$$u(A|E), \qquad u(B|E), \qquad u(A|BE), \qquad u(B|AE). \qquad (2\text{-}1)$$

Also, we will need the three desiderata stated and discussed in RC. The first desideratum is: The numerical measures of plausibilities are real numbers. We summarize the remaining desiderata in a form that will be most useful to us here:

**Rationality.** As new information supporting the truth of a proposition is supplied, the number which represents the plausibility is to increase monotonically and continuously. The deductive limit must   (2-2)
obtain where appropriate (e.g., for modus ponens and modus tollens).

**Consistency.** The truth values of the propositions in a plausibility can be analyzed via any legitimate operations, theorems and identities from Boolean algebra without altering the value of the plausi   (2-3)
bility.

The plausibility $u(AB|E)$ can depend upon two or more of the plausibilities in Eq. (2-1). In fact, the possible functional dependencies are exhausted by

$$u(AB|E) = F_1[u(A|E), u(B|E)] \qquad (2\text{-}4\text{a})$$
$$u(AB|E) = F_2[u(A|E), u(A|BE)] \qquad (2\text{-}4\text{b})$$
$$u(AB|E) = F_3[u(A|E), u(B|AE)] \qquad (2\text{-}4\text{c})$$
$$u(AB|E) = F_4[u(A|BE), u(B|AE)] \qquad (2\text{-}4\text{d})$$
$$u(AB|E) = F_5[u(A|E), u(B|AE), u(B|E)] \qquad (2\text{-}4\text{e})$$
$$u(AB|E) = F_6[u(A|E), u(B|AE), u(A|BE)] \qquad (2\text{-}4\text{f})$$
$$u(AB|E) = F_7[u(A|E), u(B|AE), u(B|E), u(A|BE)]. \qquad (2\text{-}4\text{g})$$

$$u(AB|E) = F_8[u(B|E), u(B|AE)] \qquad (2\text{-}5\text{a})$$
$$u(AB|E) = F_9[u(B|E), u(A|BE)] \qquad (2\text{-}5\text{b})$$
$$u(AB|E) = F_{10}[u(B|E), u(A|BE), u(A|E)] \qquad (2\text{-}5\text{c})$$
$$u(AB|E) = F_{11}[u(B|E), u(A|BE), u(B|AE)]. \qquad (2\text{-}5\text{d})$$

Here, it is understood that the functional dependencies are generic in the sense that only the potential independent variables are of interest — the order in which they appear in the functions is not relevant to this analysis. With this convention, we can immediately shrink the list of candidate functions by using the commutativity of the logical product $(AB = BA)$. By consistency, Eq. (2-3), we must have

$$u(AB|E) = u(BA|E). \tag{2-6}$$

Upon interchanging $A$ and $B$ in Eqs. (2-4b), (2-4c), (2-4e) and (2-4f), we obtain $F_8$, $F_9$, $F_{10}$ and $F_{11}$, respectively. Hence, the functions in Eq. (2-5) are redundant, and there remain the seven possible classes of functions in Eq. (2-4) relating $u(AB|E)$ and the plausibilities in Eq. (2-1). (The commutativity of $AB$ is understood to be always in effect.) Next, we inspect each class of functions for compliance with Rationality and Consistency in Eqs. (2-2) and (2-3). The criterion for rejection of a candidate function is simple: If a function violates Rationality or Consistency for *any* legitimate problem, then it is summarily removed from further consideration. So, our strategy is to examine the behavior of a given function in a variety of problem types.

As a baseline, we consider a situation in which the truth value of $AB$ is determined serially by examining first $A$, then $B$, as described by the following tree diagram:

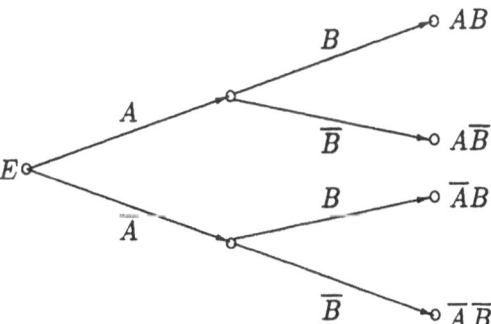

The proposition $AB$ can be reached only via the upper branch of the diagram, and the plausibility $u(AB|E)$ can depend only on the plausibilities $u(A|E)$ and $u(B|AE)$ as expressed in Eq. (2-4c). Now, Eq. (2-4c) is clearly descriptive of a problem with the structure of the tree diagram and, by the Consistency desideratum, must apply to all other problems. We hope, by the experience gained in the subsequent study, the reader will "insist" that Eq. (2-4c) is the only functional form allowed by the desiderata and that this entire study was unnecessary — in fact, the tree diagram was designed to assist in this process. For future reference, we write Eq. (2-4c) in the notation used in RC:

$$u(AB|E) = F[u(A|E), u(B|AE)]. \tag{2-7}$$

But what about the other functional dependencies in Eq. (2-4): Are any of them compatible with Rationality and Consistency? An affirmative reply to this query would be unsettling, to say the least, so it is important that one demonstrate that $F_3$ (or $F_9$) is the only functional dependency in Eq. (2-4) that is compatible with the desiderata. We turn next to this task. This line of inquiry has been carried out by Dr. Myron Tribus [1969], pp. 14–18; unfortunately, this book is out of print. The details of our analysis differ from Tribus', but we attempt to capture the pedagogical nature of his presentation.

**2a.**     $u(AB|E) = F_1[u(A|E), u(B|E)]$.

In our first encounter with testing the compatibility of a proposed function with the desiderata, we take a somewhat exploratory approach. One might be tempted to devise a systematic approach, using Boolean algebra to introduce all conditional relations between $A$ and $B$ and then checking how the function fares with respect to the desiderata. Such a formal approach is unnecessary, but we admit that we tried that approach.

Because $F_1$ cannot reflect any interdependence of $A$ and $B$, we have a clue for bringing $F_1$ into conflict with Rationality. We consider a problem in which our prior information $E$ reveals that $A$ and $B$ satisfy the conditional relation

$$AB = A, \qquad\qquad (2\text{-}8)$$

in which case Eq. (2-4a) becomes

$$u(AB|E) = u(A|E) = F_1[u(A|E), u(B|E)].$$

This equation is self–consistent only if $F_1$ does not depend upon $u(B|E)$; the problem is that there is no way to "inform" the plausibilities in $F_1$ that $AB = A$. For $A = \mathbb{1}$, we are dealing with the deductive limit of modus ponens, but $u(B|E)$ is unable to absorb the information $B = \mathbb{1}$ (from modus ponens).

Another way to analyze $F_1$ is by considering two problems with common features and with differences. In problem 1, we know, via $E_1$, that $A_1$ and $\overline{B}_1$ have the same truth values, so that $A_1 B_1 = \mathbb{0}$ (more qualitatively, we could say that $A_1 B_1$ is very implausible). In this case,

$$F_1[u(A_1|E_1), u(B_1|E_1)] = u(\mathbb{0}|E_1). \qquad\qquad (2\text{-}9)$$

In problem 2, no such conditional exists, but it is supposed that

$$u(A_2|E_2) = u(A_1|E_1) \quad \text{and} \quad u(B_2|E_2) = u(B_1|E_1).$$

Therefore,

$$\begin{aligned} u(A_2 B_2|E_2) &= F_1[u(A_2|E_2), u(B_2|E_2)], \\ &= F_1[u(A_1|E_1), u(B_1|E_1)]. \end{aligned}$$

In view of Eq. (2-9), it follows that

$$u(A_2 B_2 | E_2) = u(Q | E_1). \qquad (2\text{-}10)$$

Now, if $A_2 B_2$ is very plausible, Rationality requires

$$u(A_2 B_2 | E_2) > u(A_1 B_1 | E_1),$$

contradicting Eq. (2-10). Thus, $F_1$ is not compatible with the desiderata and must be rejected.

**2b.** $\quad u(AB|E) = F_2[u(A|E), u(A|BE)].$

Suppose $A$ is independent of $B$, in the sense that $u(A|BE) = u(A|E)$. In this case, Eq. (2-4b) becomes

$$u(AB|E) = F_2[u(A|E), u(A|E)].$$

This asserts that the plausibility for $AB$ is independent of that for $B$, whether the plausibility for $B$ is large or small. Therefore, $F_2$ violates Rationality and must be rejected. Problems with $F_2$ persist even when $A$ and $B$ are not independent. In particular for $AB = B$, $u(A|BE) = u(1|E)$ and Eq. (2-4b) gives

$$u(AB|E) = F_2[u(A|E), u(1|E)].$$

This relates the plausibility for $B$ to that for $A$ only; because $AB = B$, the right–hand side ought to collapse to $u(B|E)$. This provides a second reason for rejecting $F_2$ — but one reason is enough.

**2c.** $\quad u(AB|E) = F_3[u(A|E), u(B|AE)].$

We have indicated already that this expression satisfies all of the desiderata. One may wish to subject $F_3$ to the same analyses we use in eliminating the other candidate functions. One should recall that $F_3$ and $F_9$ are equivalent and that $F_3$ will usually be denoted by $F$ — cf. Eq. (2-7).

**2d.** $\quad u(AB|E) = F_4[u(A|BE), u(B|AE)].$

This functional dependence is suspicious from the beginning, because it is incompatible with the tree diagram — this observation supplies the clue for the dismissal of $F_4$. Suppose the initial information $E$ tells us that the propositions $A$ and $B$ always have the same truth value: $A = B$. For this problem, Eq. (2-4d) reduces to

$$u(A|E) = F_4[u(1|E), u(1, E)].$$

So, whether $A$ is plausible or implausible, its value is fixed by $F_4$ once and for all (in the deductive limit, $F_4$ can lead us to conclusions known to be false). This behavior is not consistent with the desideratum of Rationality: $F_4$ is unacceptable.

**2e.**     $u(AB|E) = F_5[u(A|E), u(B|AE), u(B|E)].$

Suppose the information $E$ includes the conditional relation $AB = A$, so that $u(B|AE) = u(\mathbf{1}|E)$. In this case, Eq. (2-4e) becomes

$$u(A|E) = F_5[u(A|E), u(\mathbf{1}|E), u(B|E)]. \qquad (2\text{-}11a)$$

The right–hand side of this expression can reduce to the left–hand side only if $F_5$ is independent of $u(B|E)$ — this would bring us back to $F_3$, making $F_5$ redundant.

Let us show in another way that it is the last variable in $F_5$ that is inappropriate. Consider another problem in which $E'$ informs us that $A'$ and $B'$ have the same truth value $(A' = B')$, so that

$$u(A'|E') = F_5[u(A'|E'), u(\mathbf{1}|E'), u(A'|E')]. \qquad (2\text{-}11b)$$

Suppose further that $u(A'|E') = u(A|E)$. According to Eqs. (2-11a) and (2-11b),

$$F_5[u(A|E), u(\mathbf{1}), u(B|E)] = F_5[u(A|E), u(\mathbf{1}), u(A|E)].$$

[Note: The *value* of $u(\mathbf{1}|E)$ is not conditional on $E$, so we can use $u(\mathbf{1})$ when desired.] Thus, $F_5$ exhibits unacceptable behavior and must be excised from our list.

**2f.**     $u(AB|E) = F_6[u(A|E), u(B|AE), u(A|BE)].$

Comparing $F_6$ with $F_5$ leads us to focus on the third variable in $F_6$. As in the preceding discussion, we consider problem 1 with $A_1 B_1 = A_1$ and problem 2 with $A_2 = B_2$ and assume further that $u(A_1|E_1) = u(A_2|E_2)$. Then, Eq. (2-4f) tell us

$$F_6[u(A_1|E_1), u(\mathbf{1}), u(A_1|B_1 E_1)] = F_6[u(A_1|E_1), u(\mathbf{1}), u(\mathbf{1})].$$

The conclusion is that $F_6$ has unacceptable behavior.

**2g.**     $u(AB|E) = F_7[u(A|E), u(B|AE), u(B|E), u(A|BE)].$

The functional dependence in $F_7$ is rather peculiar: it duplicates that in $F_3$ and $F_9$. In fact, we can write

$$F_7 = \sqrt{F_3 F_9}. \qquad (2\text{-}12)$$

Because $F_3$ and $F_9$ are consistent with the desiderata, so is the form in Eq. (2-12). Hence, $F_7$ is redundant, and we can concentrate our attention on $F_3 = F$ (and thereupon $F_9$).

The preceding arguments may not be formal enough to suit everyone. But the method ought to be clear, so that anyone can develop his or her own arguments.

So far, $F$ is unknown except for its dependent variables. In the next section, we turn to the problem of ascertaining explicit functional forms of $F$.

## 3. SOLVING THE ASSOCIATIVITY EQUATION

Rationality and Consistency have fixed the functional composition of $u(AB|E)$, namely $u(AB|E) = F[u(A|E), u(B|AE)]$. Although the functional form of $F$ is not known, it is definitely not arbitrary—for example, we know already that it is continuous and monotonic. Moreover, Cox [1946, 1961] has shown how the Consistency desideratum can be used to impose a rather strong constraint on $F$. In fact, this constraint is a manifestation of the associativity of the logical product, as we show next.

Writing

$$ABC = (AB)C = A(BC) \tag{3-1}$$

and treating $(AB)$ as one proposition and $C$ as another, then $A$ and $(BC)$ similarly, we obtain from Eq. (2-7)

$$u(ABC|E) = F[u(AB|E), u(C|ABE)]$$
$$= F[u(A|E), u(BC|AE)].$$

Applying Eq. (2-7) to $u(AB|E)$ and $u(BC|AE)$ in these expressions leads to

$$F\{F[u(A|E), u(B|AE)], u(C|ABE)\} = F\{u(A|E), F[u(B|AE), u(C|ABE)]\}. \tag{3-2}$$

Finally, the notation

$$x = u(A|E), \quad y = u(B|AE), \quad z = u(C|ABE) \tag{3-3}$$

allows us to write Eq. (3-2) in the form

$$F[F(x, y), z] = F[x, F(y, z)], \tag{3-4}$$

which is a functional equation known as the *associativity equation*. It is said to be a composite functional equation because the unknown function is a function of itself. In the subsequent analysis, symbols like $x, y, s, t$ will be used to denote arbitrary variables —they do not represent specific plausibilities as in Eq. (3-3).

We cannot really judge whether the associativity equation is merely a neat result or represents real progress until we possess its explicit solutions. But when it comes to solving functional equations, just as in the case of differential equations, there is no general omnipotent method that works for all equations. In the case of the associativity equation, we are in luck: This equation has been studied extensively, beginning more than 150 years ago, so several techniques for solving it are already in existence—see Aczél [1966] for an excellent bibliography on this equation. One way to proceed is to derive from Eq. (3-4) a second–order partial differential equation for $F(x, y)$ and to solve that equation. This is the well–known approach used by Cox [1946, 1961]—see also Tribus [1969]. Another approach is to derive from Eq. (3-4) a first–order partial differential equation that has the structure of a vanishing Jacobian, thereby gaining access to a relevant theorem from calculus. This approach is the basis of an analysis of the generalized associativity equation by Aczél [1966, Sec. 7.2.2 and p. 325] and is used also in an interesting paper by Dr. Yoel Tikochinsky [1988]. The former approach requires $F(x, y)$ to be twice differentiable, and the latter requires $F(x, y)$ to have continuous first derivatives. Finally, a third method employs an iterative technique to solve the associativity equation directly. This is perhaps the most widely used technique in the theory of functional equations. One of its assets is that it does not require differentiability of the functions.

The requirement of differentiability is disturbing to some people. For this reason alone, a brief account of this method would appear timely; additionally, we believe the method is intrinsically interesting and instructive. The major source of the material in our presentation is Sec. 6.2 of Aczél [1966] and pp. 106–122 of Aczél [1987]. The analysis by Aczél is thorough, well–written, and rigorous; however, it takes a fairly long time to work through this material, perhaps because it comprises an area that is not part of our usual mathematical training. Nevertheless, because our presentation omits several technical points, the reader may wish to consult the references just cited to fill in the gaps.

Ordinarily, the range and domain of a function are provided as part of the definition of the function. But in the problem under study, the range and domain of $F(x, y)$ are not known in advance, except that because $x, y$ and $F(x, y)$ represent plausibilities, they are to be defined over the same interval $J$. So, part of the analysis of the associativity equation must be concerned with finding the intervals for which solutions exist—this has been done carefully and accounts for several pages of work in Aczél. This is of more than academic interest, because for some functional equations the nature of the solution will change with the ranges and domains of the functions involved [Smital, 1988]; fortunately, this is not the case here.

We do not give any of Aczél's analysis establishing the allowed intervals, but we will have to appeal to some of the results. In our discussion below, it will be useful to have notation for the end points of $J$. The lower end of $J$ corresponds to the plausibility of a proposition known by the evidence $E$ to be impossible: $u(Q \,|E) = \omega$; the upper end of $J$ corresponds to the certain proposition: $u(1 \,|E) = \epsilon$. Although, we will write

$$J = [\omega, \epsilon], \tag{3-6}$$

one should be warned that the solution at the end points must be examined carefully (the use of a closed interval requires some justification) — see Aczél [1963].

In order to solve the associativity equation using the iterative technique given below, it is necessary to have one condition beyond the continuity and increasing monotonicity of $F(x, y)$. The additional condition we consider is either of the two: cancellativity or *strictly* increasing monotonicity (these do not exhaust the possibilities). In fact, in this problem one condition implies the other (see below). Therefore, we can concentrate on either condition; we focus on cancellativity. The function $F(x, y)$ is said to be cancellative (or reducible) if both

$$F(x, s) = F(x, t) \quad \text{and} \quad F(s, y) = F(t, y) \tag{3-7}$$

imply $s = t$. A violation of Rationality can occur if cancellativity is not required; for example, two values of $x$ could lead to the same value for $u = F(x, y)$. It is easy to see that if $F(x, y)$ is continuous and strictly increasing, it is also cancellative; the converse is also true, but the associativity equation is required in the proof. In what follows, we treat $F(x, y)$ as continuous, strictly increasing and cancellative.

We finally turn to the task of solving the associativity equation, using the iterative technique mentioned earlier. The solution is developed from the sequence of functions $\{g_n(x)\}$, defined as follows:

$$g_1(x) = x, \quad g_{n+1}(x) = F[x, g_n(x)], \quad n = 1, 2 \dots . \tag{3-8}$$

Below, we state two theorems involving the functions $g_n(x)$. The proof of these theorems makes use of the lemma

$$F[g_n(x), x] = F[x, g_n(x)]. \tag{3-9}$$

To prove the lemma, note first for $n = 1$,

$$F[g_1(x), x] = F(x, x) = F[x, g_1(x)]. \tag{3-10}$$

For $n = 2$, we use the definition of $g_2(x)$ to write

$$F[g_2(x), x] = F\{F[x, g_1(x)], x\}.$$

Then, we use the associativity equation, Eq. (3-4), Eq. (3-10) and the definition of $g_2(x)$ to write

$$\begin{aligned} F[g_2(x), x] &= F\{x, F[g_1(x), x]\} \\ &= F\{x, F[x, g_1(x)]\} \\ &= F[x, g_2(x)]. \end{aligned}$$

Now, using mathematical induction, one verifies Eq. (3-9) in a few lines.

The two theorems mentioned earlier are

$$g_{m+n}(x) = F[g_m(x), g_n(x)] = F[g_n(x), g_m(x)], \tag{3-11}$$

$$g_{mn}(x) = g_m[g_n(x)] = g_n[g_m(x)], \tag{3-12}$$

where $m = 1, 2, \ldots, n = 1, 2, \ldots$ . The proof of Eq. (3-11) follows the pattern used in proving the lemma; specifically, one uses the definition in Eq. (3-8), the lemma and Eq. (3-4). The proof of Eq. (3-12) is similar to that of Eq. (3-11), except that Eq. (3-11) itself is needed.

Next, let

$$t = g_m(x). \tag{3-13}$$

Because $F(x,y)$ is continuous and strictly increasing (in both $x$ and $y$), so is $g_m(x)$; thus, $g_m(x)$ has a continuous, strictly increasing inverse. If this inverse, denoted by

$$x = g_m^{-1}(t), \tag{3-14}$$

is substituted into Eq. (3-12), there results

$$g_m\{g_n[g_m^{-1}(t)]\} = g_n\{g_m[g_m^{-1}(t)]\} = g_n(t),$$

which leads immediately to

$$g_n[g_m^{-1}(t)] = g_m^{-1}[g_n(t)]. \tag{3-15}$$

Now, let $t = a \in J$ be some fixed number. Then, upon setting $t = a$ in Eq. (3-15) and defining the function $f(n/m)$ by the resulting expression, we obtain

$$f(n/m) = g_n[g_m^{-1}(a)] = g_m^{-1}[g_n(a)]. \tag{3-16}$$

The variables $x$ and $y$ have been superseded by the rational numbers $n/m$; it is therefore important to determine the nature of the dependence of $f(n/m)$ on $n/m$. Because of the convention chosen below [corresponding to the deductive limit leading to Eq. (3-27)], the value of $a \in J$ in Eq. (3-16) will be selected to satisfy

$$F(a, a) < a. \tag{3-17}$$

But, by Eqs. (3-8) and (3-4)

$$\begin{aligned} g_{n+1}(a) &= F[a, g_n(a)] \\ &= F\{a, F[a, g_{n-1}(a)]\} \\ &= F[F(a, a), g_{n-1}(a)]. \end{aligned}$$

By Eq. (3-17) and because $F(x,y)$ is strictly increasing

$$F[F(a, a), g_{n-1}(a)] < F[a, g_{n-1}(a)].$$

The right–hand side of this inequality is seen to be $g_n(a)$; therefore

$$g_{n+1}(a) < g_n(a). \tag{3-18}$$

Next, by the definition in Eq. (3-16), we have

$$f(\frac{n+1}{m}) = g_m^{-1}[g_{n+1}(a)].$$

Because $g_m^{-1}(x)$ is strictly increasing and because of Eq. (3-18)

$$g_m^{-1}[g_{n+1}(a)] < g_m^{-1}[g_n(a)].$$

Next, noting the right–hand side of this expression is $f(n/m)$, we obtain the desired result:

$$f(\frac{n+1}{m}) < f(\frac{n}{m}). \tag{3-19}$$

That is, $f(r)$ is a strictly decreasing function of the rationals $r$.

We can now complete the construction of the solutions of the associativity equation. Using Eq. (3-16), we write

$$F[f(\frac{n_1}{m}), f(\frac{n_2}{m})] = F\{g_{n_1}[g_m^{-1}(a)], g_{n_2}[g_m^{-1}(a)]\}$$
$$= g_{n_1+n_2}[g_m^{-1}(a)]$$
$$= f(\frac{n_1}{m} + \frac{n_2}{m}), \tag{3-20}$$

where the second line follows from the theorem in Eq. (3-11) and the last line follows by the definition of $f(n/m)$ in Eq. (3-16). By considering two sequences of rationals $r_k$ and $r_k'$ with

$$\lim_{k\to\infty} r_k = s \qquad \text{and} \qquad \lim_{k\to\infty} r_k' = t,$$

one obtains from Eq. (3-20)

$$F[f(s), f(t)] = f(s+t). \tag{3-21}$$

(Here, we recall that two fractions can always be expressed with common denominators. Also, to be more rigorous, one should define the end points $s$ and $t$ by a Dedekind section.) Note that $s$ and $t$ are intermediate variables and need not (indeed, do not) belong to $J$. Finally, with $x = f(s)$ and $y = f(t)$, we obtain from Eq. (3-21) the solution of the associativity equation:

$$F(x,y) = f[f^{-1}(x) + f^{-1}(y)], \tag{3-22}$$

where $f(x)$ is any continuous, strictly decreasing function of $x$.

To obtain the form of the solution that is familiar to most of us, we introduce [cf., Aczél, 1963]

$$G(x) = \exp[-f^{-1}(x)]. \tag{3-23}$$

The result in Eq. (3-22) becomes

$$F(x,y) = G^{-1}[G(x)G(y)], \tag{3-24}$$

where $G(x)$ is continuous and strictly increasing in $x$. Reinstating our original variables in Eqs. (2-7) and (3-3), we obtain from Eq. (3-24)

$$G[u(AB|E)] = G[u(A|E)]G[u(B|AE)].\tag{3-25}$$

This is the expression quoted in Eq. (31) of RC.

For the next stage of the development of Bayesian probability theory that emerges from the solution in Eq. (3-25), see Cox [1946, 1961], Jaynes [1957], Tribus [1969, 1988] and Smith and Erickson [1989]. By considering deductive limits, one arrives at

$$G(\epsilon) = 1\tag{3-26}$$

and

$$G(\omega) = 0.\tag{3-27}$$

Note that at no point in this analysis is it necessary to assign specific numerical values to $\omega$ and $\epsilon$, so plausibilities have escaped explicit quantification.

The preceding analysis has been brief and has omitted some detail, but we have retained the structure of Aczél's treatment and hope to motivate many readers to take a careful look at his monographs.

## APPENDIX

An examination of the uniqueness of the function $G(x)$ in Eq. (3-24) leads to a rather interesting result, as we discuss next.

Suppose another function $H(x)$ is proposed for defining $F(x, y)$ in the manner of Eq. (3-24):

$$F(x,y) = H^{-1}[H(x)H(y)].\tag{A-1}$$

Equating Eqs. (3-24) and (A-1), we obtain

$$G^{-1}[G(x)G(y)] = H^{-1}[H(x)H(y)].\tag{A-2}$$

Letting

$$u = G(x), \qquad v = G(y)\tag{A-3}$$

in Eq. (A-2) gives

$$G^{-1}(uv) = H^{-1}\{H[G^{-1}(u)]H[G^{-1}(v)]\}.\tag{A-4}$$

Defining

$$k(u) = H[G^{-1}(u)],\tag{A-5}$$

we obtain from Eq. (A-4)

$$k(uv) = k(u)k(v),\tag{A-6}$$

which is one of Cauchy's functional equations. The solving of the Cauchy functional equations is the subject of Secs. 2.1.1 and 2.1.2 of Aczél [1966]. The solution of Eq. (A-6) that applies for $0 < u \leq 1$ and our other conventions is

$$k(u) = u^\alpha, \qquad (A\text{-}7)$$

where $\alpha$ is an arbitrary positive number [note that $k(u)$ in Eq. (A-5) is strictly monotonic]. Reinstating the original variables in Eqs. (A-3) and (A-5), we find

$$H(x) = [G(x)]^\alpha, \qquad \alpha > 0. \qquad (A\text{-}8)$$

That is, if $G(x)$ generates a given solution of the associativity equation, so does $[G(x)]^\alpha$. The study of the plausibility $u(\overline{A}|E)$ [cf. Eqs. (43) – (45) of RC] leads us to consider the condition in Eq. (A-8) from another point of view.

## REFERENCES

Aczél, J. (1963), 'Remarks on Probable Inference,' *Ann. Univ. Sci. Budapest. R. Eötvös Sect. Math.* **6**, pp. 3– 11.

Aczél, J. (1966), *Lectures on Functional Equations and Their Applications*, Academic Press, New York.

Aczél, J. (1987), *A Short Course on Functional Equations*, D. Reidel, Dordrecht.

Cox, R.T.(1946), 'Probability, Frequency and Reasonable Expectation,' *American Journal of Physics*, **14**, pp. 1–13.

Cox, R.T. (1961), *The Algebra of Probable Inference*, The Johns Hopkins Press, Baltimore, Maryland.

Jaynes, E.T. (1957), 'How Does the Brain Do Plausible Reasoning?' Stanford University Microwave Laboratory Report 421. Reprinted in G.J. Erickson and C.R. Smith, eds. (1988), *Maximum–Entropy and Bayesian Methods in Science and Engineering*. I. *Foundations*, Kluwer Academic Publishers, Dordrecht, pp. 1–23.

Smital, J. (1988), *On Functions and Functional Equations*, Adam Hilger, Bristol, England.

Smith, C. Ray and G. J. Erickson, (1988), 'From Rationality and Consistency to Bayesian Probability,' in *Maximum–Entropy and Bayesian Methods*, J. Skilling, ed., Kluwer Academic Publishers, Dordrecht. pp. 29–44.

Tikochinsky, Y. (1988), 'On the Generalized Multiplication and Addition of Complex Numbers,' *J. Math. Phys.* **29**, pp. 398–399.

Tribus, Myron (1969), *Rational Descriptions, Decisions and Designs*, Pergamon Press, New York.

Tribus, M. (1988),'An Engineer Looks at Bayes,' in G.J. Erickson and C. R. Smith, eds. (1988), *Maximum–Entropy and Bayesian Methods in Science and Engineering*. I.
*Foundations*, Kluwer Academic Publishers, Dordrecht, pp. 31–52.

## Objective Bayesianism and Geometry

Carlos C. Rodriguez
State University of New York at Albany
Department of Mathematics and Statistics
Albany, New York 12222

This research was supported by PHS grant
number 5-R01-CA41171-03 awarded by the
National Cancer Institute, DHHS.

**Abstract**. We suggest in this paper that the concepts of
utility, prior probability and entropy are not independent
but must be related through the following formula: "*The
expected utility of a Theory is an increasing function of
its entropy.*" It follows that associated to each regular
class of theories (i.e. parametric statistical model) there
is a unique one parameter family of densities able to act as
prior distributions. These entropic priors form the
exponential family generated by the invariant measure in the
class, that has the entropy of each theory as sufficient
statistic.

## Introduction.

We show in this paper that agreement of the principle of
maximum entropy with decision theory implies the form of the
prior density up to a parameter. We are dealing here with the
general problem of inference. We assume that we are uncertain
about which member $\mu$ of a class $\mathfrak{M}$ of probability measures
generated the observed data D.

To fix the notation and the ideas we summarize the
remarkable fact that maximum entropy is the unique (up to an
irrelevant monotone transformation) method for ranking the
elements of $\mathfrak{M}$. We assume that all the probabilities in $\mathfrak{M}$ are
defined over the same measurable space $(X, \mathcal{B})$. The problem is to
choose the best hypothesis in $\mathfrak{M}$ relative to an initial estimate $\nu$
(not necessarily in $\mathfrak{M}$). We assume that $\nu$ is a $\sigma$-finite positive
measure on $(X, \mathcal{B})$ that dominates $\mathfrak{M}$ (i.e. $\nu \gg \mu \; \forall \mu \in \mathfrak{M}$). In other
words, we want to find a ranking procedure specified as a
functional over $\mathfrak{M}$ and written for a given $\nu$ as $S(\mu : \nu)$ i.e.
$S(\bullet : \nu) : \mathfrak{M} \to \mathbf{R}$. We then modify our initial estimate $\nu$ with the
$\mu \in \mathfrak{M}$ that maximizes the ranking $S(\mu : \nu)$.

The novel idea of Shore and Johnson, extended and clarified
by Skilling was to seek relevant criteria to restrict the class
of possible ranking procedures S. Surprisingly, only four axioms
are needed to reduce S to the entropy functional. These axioms
are based on two fundamental principles: 1) The arbitrary way in
which we label the elements of X (i.e., the coordinate system in
X) should be immaterial for the ranking methods and 2) the
ranking procedure should explicitly reject knowledge of unassumed
correlations. Remarkably, 1), 2) and a boundary condition (Axiom
IV) are sufficient to specify the form of S up to an irrelevant
monotone transformation. Following Skilling 1987 we enunciate

*P. F. Fougère (ed.), Maximum Entropy and Bayesian Methods*, 31–39.
© 1990 *Kluwer Academic Publishers.*

each axiom followed by its consequence (see Shore and Johnson, 1980 or Skilling 1987 for proofs).

Axiom I:  (Subset Independence)
    It should not matter whether one treats pieces of X separately or jointly.

Consequence:  $S(\mu:\nu) = \int U(x, \frac{d\mu}{d\lambda}(x)) \lambda(dx)$ i.e., S must be the

integral w/r to some $\mu$-dominating measure $\lambda$ of some arbitrary function U.  In this axiom spatial correlations are explicitly rejected.

Axiom II:  (Coordinate invariance)
    The selected $\mu \in \mathfrak{M}$ should be invariant under coordinate transformations of X.

Consequence:  $S(\mu:\nu) = \int U\left(\frac{d\mu}{d\lambda}(x)\right) \lambda(dx)$.

Axiom III:  (System Independence)
    If only the marginal distributions of $\mu$ are known then the procedure should select the product of the marginals.

Consequence:  $U(y) = -y \log y$

Axiom IV:  (Boundary Condition)
    If $\nu \in \mathfrak{M}$ best $\mu$ should be $\nu$ itself.

Consequence:  $\lambda = \nu$.

    Hence, the only ranking procedure compatible with these axioms is the one that selects $\mu \in \mathfrak{M}$ to maximize

$$S(\mu:\nu) = -\int \frac{d\mu}{d\nu}(x) \log \frac{d\mu}{d\nu}(x) \ \nu(dx)$$

i.e., the entropy of $\mu$ relative to the intial measure $\nu$.  For example in image processing we can identify the image $f(x)$ with

the density $\frac{d\mu}{d\lambda}(x)$ for some measure $\lambda$.

2.  The Entropic Prior
    The axiomatic derivation of the previous section have led us to choose the best (image) hypothesis $\mu$ as the solution of the variational problem

$$\max_{\mu \in \mathfrak{M}} \ S(\mu:\nu) \ . \tag{2.1}$$

But if a utility function $u(\mu,\xi)$ is assumed then classical Bayesian decision theory tells us that the best $\mu$ is the one that maximizes the expected utility (reward) i.e., $\mu^*$ satisfying

$$\mathcal{R}(\mu^*) = \max_{\mu \in \mathfrak{M}} \ \mathcal{R}(\mu) = \max_{\mu \in \mathfrak{M}} \ \int u(\mu,\xi)\Pi(d\xi) \tag{2.2}$$

where $\Pi$ denotes the prior probability measure on $\mathfrak{M}$. The idea is to make these two different ways of ranking $\mu$'s in $\mathfrak{M}$ (i.e. (2.1) and (2.2) to agree. In this way an objective relation between entropy, utility, and the prior distribution is obtained

$$\int u(\mu,\xi)\Pi(d\xi) = T(S(\mu:\nu)) \tag{2.3}$$

where T is an arbitrary monotone increasing transformation.
    In order to avoid integration over the functional space $\mathfrak{M}$ we shall assume that $\mathfrak{M}$ is a regular model (see Rodriguez, 1988) i.e., it admits a smooth k dimensional (k<∞) parametrization. Let $\theta \subset \mathbb{R}^k$ be the parameter space. Hence, we can identify $\mathfrak{M}$ with $\theta$ and translate the utility u and the prior distribution $\Pi$ as if they were defined on $\theta$. However, the parameters are only arbitrary labels for the "real things" that are the measures $\mu$ (or equivalently the images f) and therefore we must always demand invariance over smooth reparametrizations. In particular this implies that the prior densities must be computed not with respect to the Lebesgue measure in $\mathbb{R}^k$ but with respect to the invariant measure (unique upto a constant) in $\theta$; otherwise, a relabeling of the elements in $\mathfrak{M}$ would change the prior probabilities which is absurd. The assumed regularity of $\mathfrak{M}$ makes it a Riemannian manifold with metric tensor $g(\theta)$, which coincides with the Fisher information matrix of $\mathfrak{M}$ (see Amari, 1985). We may also assume that $\mathfrak{M}$ is compact and use "max" instead of "sup" in the formulas. It is well known, and easy to show, that the invariant measure in a metric space with metric $g(\theta)$ is given by

$$\eta(d\theta) \ \alpha\sqrt{\det g(\theta)} \ \lambda(d\theta)$$

where $\lambda$ is the lebesgue measure in $\mathbb{R}^k$.
    We can then re-write (4.2) using the parametrization as

$$\mathcal{R}(\theta^*) = \max_{t \in \theta} \ \mathcal{R}(t) = \max_{t \in \theta} \ \int u(t,\theta)\pi(\theta)\eta(d\theta) \tag{2.4}$$

where $\pi(\theta) = \dfrac{d\Pi}{d\eta}(\theta)$. In order for the solutions of (2.1) and (2.4) to agree $\pi$ must satisfy a Fredholm integral equation:

$$\int u(\theta,t)\,\pi(t)\,\eta(dt) = T(S(\mu_\theta:\nu)) \tag{2.5}$$

this is just (2.3) in terms of the assumed parametrization. Notice that the prior $\pi$ and the utility u are logically independent concepts and therefore the functional form of $\pi$ must be independent of u. In particular when $u(\theta,t) = \delta(t-\theta)$ i.e., all-nothing utility, we must have

$$\pi(\theta) = U(S(\mu_\theta:\nu)) \tag{2.6}$$

where U is the increasing function T obtained when $u(\theta,t) = \delta(t-\theta)$. To find U consider the following hypothetical setting

$$\theta = \theta_1 \times \theta_2 , \quad X = X_1 \times X_2$$

$$\mu_\theta(A \times B) = \mu_{\theta_1}(A)\,\mu_{\theta_2}(B) \text{ when } \theta=(\theta_1,\theta_2) \text{ and}$$

$$\nu(A \times B) = \nu_1(A)\,\nu_2(B), \quad \pi(\theta) = \pi_1(\theta_1)\pi_2(\theta_2)$$

where

$$\mu_{\theta_1}(A) = \mu_\theta(A \times X_2), \quad \mu_{\theta_2}(B) = \mu_\theta(X_1 \times B)$$

$$\nu_1(A) = \nu(A \times X_2) , \quad \nu_2(B) = \nu(X_1 \times B) .$$

Hence, replacing in (2.6) we obtain that

$$U(y_1)\, U(y_2) = U(y_1 + y_2) \tag{2.7}$$

where

$$y_i = S(\mu_{\theta_i}:\nu) \quad i=1,2 .$$

Therefore, in the above setting U must transform sums into products. The only increasing functions that satisfy (2.7) are

$$U(y) \propto \exp\{\alpha y\} \text{ for some } \alpha > 0 . \tag{2.8}$$

Thus, we arrive at the universal family of entropic priors associated to a regular model $\mathfrak{M}$

$$\pi(\theta \mid \alpha) = \exp\{\alpha S(\mu_\theta:\nu) - \beta(\alpha,\eta)\} \tag{2.9}$$

i.e., the exponential family with sufficient statistic $S(\mu_\theta:\nu)$ generated by the invariant measure $\eta$ in the regular space $\mathfrak{M}$. Equation (2.9) provides another interpretation for the maxent

distribution (i.e., the solution of (2.1)). The maximum entropy
distribution is the Bayes action for the all-nothing utility.
However, since (2.9) is a density with respect to the invariant
measure $\eta$, we have

$$Pr(d\theta|\alpha) \propto \exp\{\alpha S(\mu_\theta:\nu)\}\sqrt{detg(\theta)} \ \lambda(d\theta)$$

and the region with the highest concentration of probability mass
is around the $\theta^*$ solving

$$\max_{\theta \in \Theta}\{\alpha S(\mu_\theta:\nu) + \log\sqrt{detg(\theta)}\} . \qquad (2.10)$$

The most likely distribution $\mu_{\theta*}$ is then a compromise between the
initial guess $\nu$ and the non-informative (Jeffreys) prior
$\sqrt{detg(\theta)}$ (see Jeffreys, 1939).


3. **When Data are Available**
     Let us now assume that a vector $D \in R^N$ of data is observed.
Assume further that given $\theta \in \Theta$ the functional form of the
distribution of D is known.  i.e.,

$$Pr(dD|\theta) = p(D|\theta) \ \lambda(dD) \qquad (3.1)$$

where  $p(D|\theta)$ is the joint pdf of the data D with respect to the
Lebesgue measure $\lambda$ on $R^N$. The universal entropic prior (2.9) is
then updated by conditioning on the observed data D.  i.e.,

$$\pi(d\theta|D,\alpha) \propto p(D|\theta)\pi(\theta|\alpha) \ \eta(d\theta). \qquad (3.2)$$

Hence, the Bayes' rule under the all-nothing utility function is
given by the solution of

$$\max_{\theta \in \Theta}\{\log p(D|\theta) + \alpha \ S(\mu_\theta : \nu) + \log\sqrt{det \ g(\theta)}\} . \qquad (3.3)$$

i.e., the mode of the posterior. We can then interpret the
solution of (3.3) as a **regularized** maximum likelihood estimate.
The entropy plays the role of the regularizing function and $\alpha$ is
the regularizing parameter.  In general, the Bayes rule for an
arbitrary utility function $u(\theta,t)$ is given by

$$\max_{t \in \Theta} \int u(\theta,t)p(D|\theta) \ \exp\{\alpha \ S(\mu_\theta:\nu)\} \ \eta(d\theta) . \qquad (3.4)$$

Apart from the all-nothing utility function considered above we
can use $u(\theta,t)$ proportional to an invariant distance on $\mathfrak{M}$ between
$\mu_\theta$ and $\mu_t$, i.e.,

$$u(\theta,t) \propto -d(\mu_\theta, \mu_t) \ . \tag{3.5}$$

Examples of this kind of distances are given by the $\iota$-geodesic distances $d_\iota$ introduced in Rodriguez, 1988. However, the requirement of invariance under smooth reparametrizations of $\mathfrak{M}$ is, I think, not as important for utility functions as it is for prior distributions. For, we can regard some parametrizations as more (or less) expensive (e.g. in terms of computations) than others. Thus, it may be justifiable to use a non-invariant utility like

$$u(\theta,t) \propto - <\theta-t, \ \theta-t> \tag{3.6}$$

where $<\cdot,\cdot>$ is a given constant scalar product in $\theta$. When this is the case we obtain (adding and substracting $t^*(D,\alpha) = E(\theta|D,\alpha)$ inside the inner product in (3.6)) that (3.4) is equivalent to

$$\max_{t\in\theta}\left\{-\int<\theta-t^*, \ \theta-t^*>\pi(\theta|D,\alpha)\,\eta(d\theta) - <t-t^*, \ t-t^*>\right\}$$

and the Bayes rule is just the posterior mean of $\theta$. i.e.,

$$t^*(D,\alpha) = \frac{\int \theta p(D|\theta)\exp\{\alpha S(\mu_\theta:\nu)\}\eta(d\theta)}{\int p(D|\theta)\exp\{\alpha S(\mu_\theta:\nu)\}\eta(d\theta)} \ . \tag{3.7}$$

Even though, the posterior mean is a popular Bayes rule among practitioners it seems to me that in most cases the only justification for its use is easiness with computations. It is much more natural to use invariant utility functions as in (3.5).

4.  **A Rose is a Rose is a Rose is...**
      An entropic family of prior distributions is associated to each regular statistical model (or hypothesis space). In particular when the starting model is the one dimensional family of distributions (2.9) its associated set of entropic priors relative to the initial prior m on the $\theta$'s is

$$\tilde\pi(\alpha|\alpha') = \exp\{\alpha' \ S(\pi_\alpha:m) - \beta(\alpha',\tilde\eta)\} \tag{4.1}$$

where $\tilde\pi$ is a density with respect to the invariant measure on the space of distributions of the form (2.9) i.e. $\tilde\eta$ with

$$\tilde\eta(d\alpha) \propto \sqrt{g(\alpha)} \ \lambda(d\alpha) \ . \tag{4.2}$$

In (4.2) $\lambda$ represents the Lebesgue measure on $R$ and $g(\alpha)$ is the Fisher information amount in (2.9) i.e.

$$g(\alpha) = E_\alpha\left[\left(\frac{\partial}{\partial\alpha} \log \Pi(\theta|\alpha)\right)^2\right] = var_\alpha(S(\mu_\theta:\nu)) \ . \qquad (4.3)$$

Hence, for given $\alpha'>0$ we can write the joint distribution of $\theta$ and $\alpha$ as

$$Pr(d\theta,d\alpha|\alpha') = Pr(d\theta|\alpha) \ Pr(d\alpha|\alpha')$$

$$\propto \exp\{\alpha \ S(\mu_\theta:\nu)\}\eta(d\theta) \ \exp\{\alpha'S(\pi_\alpha:m)\}\tilde{\eta}(d\alpha) \ .$$

Replacing $\eta$ and $\tilde{\eta}$ and taking logs we obtain that the solution of

$$\max_{\theta,\alpha}\{\alpha \ S(\mu_\theta:\nu) + \log\sqrt{\det g(\theta)} + \alpha' \ S(\pi_\alpha:m) + \log\sigma_\alpha(S(\mu_\theta:\nu))\}$$

gives the most likely a priori pair $(\theta,\alpha)$. A posteriori, after observing the data D, we modify the a priori pair by solving

$$\max_{\theta,\alpha}\left\{\log p(D|\theta) + \alpha S(\mu_\theta:\nu)+\log\sqrt{\det g(\theta)}+\alpha'S(\pi_\alpha:m)+\log \sigma_\alpha(S(\mu_\theta:\nu))\right\}$$

$$(4.4)$$

which is the analogous of (3.3). We can compute the entropic prior family of $\alpha'$ provided that (4.1) is regular (i.e. $g(\alpha')>0$). This entropic prior of the entropic prior will depend on another parameter $\alpha^2$ on which we may put another entropic prior depending on $\alpha^3$ etc. i.e., a Rose is a... This infinite regret stops at level $i$ if for example $g(\alpha^i)=0$ for all $\alpha^i>0$. In such a case we have $\alpha^i=0$ since whatever the prior $\pi$ on $\alpha^i$ and for all $\epsilon>0$ we have

$$Pr(\alpha^i\geq\epsilon) = \int_\epsilon^\infty \pi(\alpha^i)\eta(d\alpha^i) = \int_\epsilon^\infty \pi(\alpha^i)\sqrt{g(\alpha^i)} \ d\alpha^i = 0 \ .$$

Let us denote by $m^i$ the initial measure at level i we can write

$$Pr(d\alpha^i|\alpha^{i+1}) \propto \exp\{\alpha^{i+1}S(\pi_\alpha:m^i)\}\sqrt{g(\alpha^i)} \ d\alpha^i$$

where

$$g(\alpha^i) = E_{\alpha^i}\left(\frac{\partial}{\partial\alpha^i} \log \pi(\alpha^{i+1}|\alpha^i)\right)^2 = var_{\alpha^i}(S(\pi_{\alpha^{i-1}}:m^{i-1})) \equiv \sigma^2_{\alpha^i}.$$

Therefore the expression analogous to (4.4) that considers all levels is given by

$$\max_{\theta,\alpha^0,\alpha^1\dots} \left\{ \log p(D|\theta) + \log\sqrt{\det g(\theta)} + \sum_{i=0}^{\infty}\left[\alpha^i S(\pi_{\alpha^{i-1}}:m^{i-1})\log\sigma_{\alpha^i}\right]\right\}$$

$$(4.5)$$

where $\alpha^{(-1)} \equiv \theta$, $\alpha^0 \equiv \alpha$, $m^{(-1)} \equiv \nu$.

<u>Example</u>:  Take $\mu_\theta = N(\theta,1)$, $\theta\in R$, $\nu=N(0,1)$ we have $g(\theta)=1$ and
$S(\mu_\theta:\nu) = -\frac{1}{2}\theta^2$.

Hence the entropic prior family is defined for all $\alpha>0$ by

$$\pi(\theta|\alpha) \equiv N(0, 1/\alpha) \text{ w/r to the lebesgue measure .}$$

Without loss of generality let's reparametrize by
$\sigma = 1/\sqrt{\alpha}$ and (abusing the notation) keep the same name $\pi$.  We
have

$$\log\pi(\theta|\sigma) = -\frac{x^2}{2\sigma^2} - \log\sigma + \text{cnt. and } g(\sigma) = \frac{2}{\sigma^2}$$

from where $\eta(d\sigma) \propto \frac{d\sigma}{\sigma}$ (the Classic Jeffreys prior).  Take as the
initial distribution on $\theta$, $m^0=N(\mu_0,\sigma_0^2)$.  Hence,

$$S(N(0,\sigma^2):N(\mu_0,\sigma_0^2)) = -\left\{\log\frac{\sigma_0}{\sigma} - \frac{1}{2} + \frac{1}{2\sigma_0^2}(\sigma^2+\mu_0^2)\right\}$$

obtaining

$$\pi(\sigma|\alpha') \propto \sigma^{\alpha'}\exp\left\{-\frac{\alpha^1}{2\sigma_0^2}\sigma^2\right\} \text{ w/r to } \eta$$

i.e.,
$$\Pr(d\alpha|\alpha') \propto \sigma^{\alpha'-1}\exp\left\{-\frac{\alpha^1}{2\sigma_0^2}\sigma^2\right\}d\sigma \equiv \Gamma\left(\alpha^1, \frac{\alpha^1}{2\sigma_0^2}\right)$$

$$\max_{\theta,\alpha}\left\{\log p(D|\theta) - \frac{\alpha}{2}\theta^2 - \log\alpha\right\}$$

$$(4.6)$$

obtaining $\alpha=0$ and $\theta=\hat\theta$, the MLE!
      (Of course, the objective function is singular at $\alpha=0$ but
for all $\iota>0$ if we take $\alpha>\iota$ (a.e., take $\iota$ the smallest number that
can be handled by the modern computer) then the solution is $\alpha=0$
and $\theta=\hat\theta$ for all practical purposes.

# References

Amari, S., 1985, <u>Differential Geometrical Methods in Statistics</u>.
Lecture Notes in Statistics, <u>28</u> Springer-Verlag.

Jeffreys, H., 1939, <u>Theory of Probability</u>.  Clarendon Press,
Oxford.

Rodriguez, C., 1988.  The metrics induced by the Kullback
Number.  <u>Maximum Entropy and Bayesian Methods</u>, 415-422.
(J. Skilling, Ed.)  Kluwer Academic Publishers.

Shore, J. and Johnson, R., 1980.  Axiomatic derivations of the
principle of maximum entropy and the principle of minimum
cross-entropy, <u>IEEE Trans. on Information Theory</u>, IT-<u>26</u>,
26-37.

Skilling, J., 1987.  The axioms of maximum entropy.  <u>Proceedings
of the 7th Maximum Entropy Workshop</u>.  G. Erikson and
R. Smith (Eds.) Kluwer Academic Publishers.

Skilling, J., 1988.  Classical MaxEnt data analysis.  <u>Maximum
Entropy and Bayesian Methods</u>.  (J. Skilling, Ed.) Kluwer
Academic Publishers.

# CONSISTENCY PRINCIPLE FOR DATA-BASED PROBABILISTIC INFERENCE

V. SOLANA
Instituto de Matemáticas
National Research Council of Spain
Serrano 123, Madrid, Spain, E-28006.

**ABSTRACT.** Probabilistic methods to make inference from a random sample of observations should satisfy certain requirements for consistency. A principle of invariance under conditioning of the random variable domain is stated. Such a principle is formalized using the paradigm of a two-step inference method based on data. Invariance requirements are expressed both in the data encoding and probability distribution selection steps.

Distributions may be assigned by minimization of cross-entropy using fractile constraints encoding data and any reference distribution. The paper shows that this version of entropy method satisfies the data encoding and probability selection invariance requirements.

## 1. INTRODUCTION

A central problem in probabilistic inference is the assignment of a probability distribution to a random variable on the basis of only a random sample of observations of the variable on a given domain. All probabilistic methods to make inference about probabilities from data of observations should satisfy certain requirements for consistency. The paper states a necessary principle for consistency of any data-based inference method. This principle is that the inference method must be invariant under conditioning of the random variable domain.

The existence of a set of principles for consistency of any inductive inference method has been conjectured by Jeffreys (1939) and Jaynes (1957). Several principles were formulated by these and other authors, but certain principles of this set could be still unknown. Moreover, some of these principles could be only stated for a particular class of inference characterized by the use of only one type of available information, for instance, the data of observations of a random variable or the observed values of any set of ordered moments or functional expectations. This paper considers the most common case of inference, i.e. the inductive inference about probabilities from data of observations of a random variable on a given domain. It assumes the existence in this case of a set of consistency principles namely the data-based inference rationale.

Probabilistic inference problem has been aimed to the axiomatic derivation of an inference method from a set of a few axioms taken from an inference rationale, by Shore and Johnson (1980), Tikochinsky, Tishby and Levine (1984) and Skilling (1988). These authors claimed that to minimize the cross-entropy functional is the only consistent variational method when some expectation values are known.

41

*P. F. Fougère (ed.), Maximum Entropy and Bayesian Methods, 41–52.*
© 1990 *Kluwer Academic Publishers.*

An alternative approach to deal with probabilistic inference is here applied when the available information is a given sample of data of observations of a random variable. The approach is just to construct a extended data-based inference rationale with new principles and to find an inference procedure that satisfy all principles of the extended rationale. There should be a problem if no one among the available methods satisfies the desirable rationale.

Two possible principles of the data-based inference rationale are the data monotonicity and the invariance under any non-singular transformation of random variable. The data monotonicity principle prescribes that the assigned probability distributions must be a non-increasing function of all data element of a sample (Solana and Lind, 1989). The invariance principle (Jeffreys, 1939; Shore and Johnson, 1980) requires that if an inference method based on data is invariant under a set of transformations then the assigned probabilities must be invariant under the same set of transformations. Other invariance requirements were proposed for consistency, viz. the system independence and the subset independence principles of invariance (Shore and Johnson, 1980; Skilling 1988).

All these invariance principles together with other possible consistency requirements can be interpreted as being particular cases of one of the first principles for a data-based inference rationale. This first principle, namely the strong invariance principle (Solana and Lind, 1989) has been so restated:

*" In any probability assignment, alternative ways of using the same data, should lead to the same result ".*

Another particular case of the strong invariance principle is the new consistency principle of invariance under conditioning of the random variable domain. The principle is formalized in section 2 for a data-based inference method having two steps: the encoding of data of observations as a set of constraints and, the probability distribution selection among all possible distributions satisfying the constraints. This principle stipulates different requirements for each step of the inference method.

It is shown in section 4 of this paper that, fortunately, there is one inference method that does not violate the principle of invariance under conditioning of the random variable domain. The method is the cross-entropy minimization of distributions with fractile constraints encoding data (Lind and Solana, 1988) which is here presented in section 3. This method also satisfies the data-monotonicity principle and the invariance under admissible transformations of random variables.

## 2. THE INVARIANCE PRINCIPLE OF RANDOM VARIABLE DOMAIN CONDITIONING

This principle refers to the invariance of an inference method under conditioning of the random variable domain. A general method of inference about probability based on observations data of a random variable consists of two distinct steps (Solana and Lind, 1989). In this section first the paradigm of a two-step inference method based on data is described using operators for encoding data and for choosing probability distribution.

Second, the principle of invariance under random variable domain conditioning is stated and formalized for each method step.

Let the data be (i) the set $S = \{x_j\}, j = 1, 2, ..., r$, of observations data of the random variable $X$ ordered as increasing values $x_j$; and (ii) a given domain $E$ of $X$. For the sake of simplicity, a continous domain $E = [x_0, x_{r+1}]$ described by only the end points data $x = x_0$ and $x = x_{r+1}$ is here considered. Extension of the invariance principle statements to the case of a multiple domain $E$ is not difficult.

In the first step, or *data encoding*, the inference method creates a set of constraints from data. Let $H$ be an operator that encodes the data (i) and (ii) into constraints, this step may be simbolically represented by

$$I_X(S) = H(S, x_0 \leq X \leq x_{r+1}), \qquad (1)$$

where $I_X(S)$ denotes the encoded information about $X$ from the random sample $S$ and the given domain $E$.

In the second step, or *probability selection*, a probability distribution being assigned to the random variable $X$ is selected from among all possible distributions subject to the only available information of the set of constraints (1) encoding the data. Let $F$ be an inference operator that selects the probability distributions. Then this step is represented by

$$Q_X(x \mid S) = F[I_X(S), P_X(x)], \qquad (2)$$

where $P_X(x)$ is a reference probability distribution.

Next, the values of random variable $X$ are conditioned to lie into a subdomain $E_c = [x_{c1}, x_{c2}]$ , $E_c \subset E$. Let the set $S_c = \{x_j\}$ , $j = s+1, s+2, ..., l-1$, and $0 \leq s \leq l \leq r+1$, be a censored sample of observations data such that $x_{c1} \leq x_j \leq x_{c2}$ , where $x_{s+1}$ and $x_{l-1}$ are respectively the first value of $S$ above $x = x_{c1}$ and the last value of $S$ below $x = x_{c2}$.

There are different ways to encode information about the censored data sample $S_c$ and the subdomain $E_c$. The strong invariance principle expresed in section 1 prescribes that the same result should be obtained by using the same data in alternative ways. Then, application of this principle to the conditioning of random variable domain must be examined for each step of the inference method in all possible alternative conditioning ways.

## Data encoding

There are two alternative ways to encode information about constraints:

(I) First, the information $I_X(S)$ is conditioned by $x_{c1} \leq X \leq x_{c2}$. The resulting information in this way is a set of conditioned constraints which may be represented by $I_X(S \mid x_{c1} \leq X \leq x_{c2})$.

(II) Second, the new set of constraints is obtained by applying the encoding operator $H$ to the censored sample $S_c$ and subdomain $E_c$. The information encoded in this way may be represented in accordance with (1) by $I_X(S_c) = H[S_c, x_{c1} \leq X \leq x_{c2}]$.

Then, it is most important when applying the strong invariance principle to ensure that alternative conditioning ways (I) and (II) convey exactly the same data.

Consider the case of data encoding on a data-bounded subdomain $E_c = [x_s, x_l]$ , $0 \leq s \leq l \leq r$. Here the subdomain boundaries $x_{c1} = x_s$ and $x_{c2} = x_l$ correspond to certain observations data of $S$ or one of the end points of $E$. Hence, all data used by the encoding ways (I) and (II) belong to the initial data (i) and (ii). Since no new data points have been added throughout the encoding processes in the case of a data-bounded subdomain, the data encoded in the ways (I) and (II) are equivalent.

Otherwise, when data encoding is done on a subdomain $E_c$ such that $[x_{c1}, x_{c2}] \neq [x_s, x_l]$ then the encoding ways (I) and (II) take additional data points $x = x_{c1}$ and $x = x_{c2}$ into account. The data encoding processes involve the knowledge of new data, the subdomain end points $x = x_{c1}$ and $x = c_{c2}$, which are of the same nature as the initial data (i) and (ii). Since alternative conditioning processes are different, the equivalence of data being really conveyed by different encoding ways (I) and (II) can not be ensured in this case.

Then the strong invariance principle requires that the same encoded information should be obtained by using the same data, i.e. the observations on a given domain, in the ways (I) and (II).

Consequently the application of the strong invariance principle in this phase must be only claimed when the random variable domain is conditioned by a data-bounded subdomain.

## Probability selection

In this step there are two possibilities of alternative conditioning ways.

The **first possibility of alternative conditioning ways** follows on the encoding way (I) biffurcation. It considers that encoded information $I_X(S)$ is conditioned by $x_{c1} \leq X \leq x_{c2}$, alternatively before or after the probability selection step.

There are two alternative ways to select a probability distribution subject to the conditioned information:

(I.1)    First, the probability distribution is obtained from the distribution $Q_X(x \mid S)$ given by (2) when it is directly conditioned by $x_{c1} \leq X \leq x_{c2}$, using the classical probability conditioning rule. The conditioned distribution is so represented by $Q_X(x \mid S, x_{c1} \leq X \leq x_{c2})$.

(I.2)    Second, the probability distribution is selected from among all possible distributions subject to the conditioned constraints $I_X(S \mid x_{c1} \leq X \leq x_{c2})$ by employing a conditioned reference distribution. The resulting distribution may be expressed according to (2), by

$$Q_X[x \mid (S \mid x_{c1} \leq X \leq x_{c2})] = F\left[I_X(S \mid x_{c1} \leq X \leq x_{c2}), P_X(x \mid x_{c1} \leq X \leq x_{c2})\right].$$
$$(3)$$

Both conditioning ways involve the knowledge of the new data points $x = x_{c1}$ and $x = x_{c2}$, but these data are essentially different from the initial information constraints (1).

Therefore the information used in the two ways (I.1) and (I.2) is the same set of constraints, and the strong invariance principle must be here applied whatever is the condition stated by the subdomain $E_c$. Then this principle requires that the same probability distribution should be obtained by using the same data, i.e. the initially encoded constraints, in the ways (I.1) and (I.2) for any subdomain.

The second possibility of alternative conditioning ways considers the data encoding ways (I) and (II). Then, there are two ways to select a probability distribution:

(I)  First, the probability distribution is given by the same distribution obtained in the first possibility of alternative conditioning ways, denoted by $Q_X(x \mid S, x_{c1} \leq X \leq x_{c2})$ or $Q_X[x \mid (S \mid x_{c1} \leq X \leq x_{c2})]$.

(II)  Second, the probability distribution is chosen from among all possible distributions subject to the censored constraints $I_X(S_c)$ using a conditioned reference distribution. The selected distribution is

$$Q_X(x \mid S_c) = F\left[I_X(S_c), P_X(x \mid x_{c1} \leq X \leq x_{c2})\right] .  \tag{4}$$

In this possibility the application of the strong invariance principle must be only claimed when the random variable domain is conditioned by a data-bounded subdomain. In this case no new data points have been added throughout the alternative conditioning ways (I) and (II).

By summarizing, the application of the strong invariance principle to a two-step data-based inference method states, in the case of random variable domain conditioning, the following requeriments:

In the data encoding step:

- *Identical set of constraints should be obtained whether the constraints encoded from a sample of observations data in a given domain, are conditioned by any data-bounded subdomain, or the set of constraints is directly encoded from the censored sample of data in the same data-bounded subdomain;*

In the probability selection step:

- *Identical probability distribution should be obtained whether probabilities of the assigned distribution for a given domain are conditioned on any random variable subdomain, or the probability distribution is directly assigned on that subdomain taking the conditioned set of constraints into account.*

- *Identical probability distribution should be obtained whether probabilities of the assigned distribution for a given domain are conditioned by any data-bounded subdomain, or the distribution is directly assigned either from the constraints of censored data in the same data-bounded subdomain, or taking the conditioned set of constraints on the same subdomain into account.*

This particular application of the strong invariance principle determines the principle of invariance under random variable domain conditioning. Thus, the new invariance principle for consistency of a two-step inference method stipulates two requirements: first, that the encoded information is invariant in case of a data-bounded subdomain, and second, that the assigned probability distribution is invariant for any subdomain. These requirements may be formally expresed:

In the data encoding step, by

$$I_X(S_c) = I_X(S \mid x_s \leq X \leq x_l), 0 \leq s \leq l \leq r+1; \tag{5}$$

In the probability selection step, by

$$Q_X \left[ x \mid (S \mid x_{c1} \leq X \leq x_{c2}) \right] = Q_X(x \mid S, x_{c1} \leq X \leq x_{c2}). \tag{6}$$

and

$$Q_X(x \mid S, x_s \leq X \leq x_l) = Q_X(x \mid S_c), 0 \leq s \leq l \leq r+1. \tag{7}$$

## 3. CROSS-ENTROPY METHOD USING FRACTILE CONSTRAINTS

A data-based inference method using entropy, namely the cross-entropy method with fractile constraints, has been developed by Lind and Solana (1988). This method is presented in this section for only the case of a scalar random variable; the method consists of two steps, data encoding and probability selection.

### Data encoding

Let the data be the set $S = \{x_j\}, j = 1, 2, ..., r$, of observations data of $X$ ordered as increasing values $x_j$, and a finite or infinite domain $E = [x_0, x_{r+1}]$ of the random variable $X$. The basis of the data encoding process is the well-known symmetry (exchangeability) property of a random sample of independent realizations of $X$. Let $x$ be a new observation. Then $x$ has equal probability of falling in the $r + 1$ intervals in which the elements of the sample divides the domain $E$. This may be expressed as the following sample rule:

*The elements of an ordered random sample of size r of a random variable are the $j/(r+1)$-fractiles $(j = 1, 2, ..., r)$ of the probability of any future observation of the random variable on the only basis of the observed data and the given domain.* (Lind and Solana, 1988).

Let $Q_X(x \mid S)$ be the distribution of the new observation $x$ inferred from the sample $S$ such that $x \in E$, having the density function $q_X(x \mid S)$. The sample rule prescribes the fractiles at points $x = x_j$ $(j = 1, 2, \ldots, r)$ while conventionally it assigns the values 0 and 1 at the domain end points. This give the following set of fractile-pair constraints,

$$[x; Q_X(x \mid S)]_j = (x_j, Q_j), j = 0, 1 \ldots, r + 1 \,, \tag{8}$$

where probability fractiles $Q_j$ are

$$Prob\,[X \leq x_j \mid S] = j/(r+1), j = 0, 1, \ldots, r+1 \,. \tag{9}$$

## Probability selection

Given is a reference distribution $P_X(x)$ having density $p_X(x)$ on $E$. The value of $P_X(x)$ at the fractile point $x = x_j$ is denoted $P_j$. Consider that $E$ is partitioned into $r + 1$ intervals $E_j = [x_j, x_{j+1})$. The probability selection process is to find a posterior distribution $Q_X(x \mid S)$ that minimizes the cross-entropy functional (Kullback)

$$D(q, p) = \int_E q_X(x \mid S) \left[\log q_X(x \mid S) - \log p_X(x)\right] dx \,, \tag{10}$$

and satisfies the fractile-pair constraints (8, 9).

The general solution is determined using Lagange's multiplier method (Lind and Solana, 1988). It may be expressed in terms of the interval multipliers

$$\mu_j = (Q_{j+1} - Q_j)/(P_{j+1} - P_j), j = 1, 2, \ldots, r \,, \tag{11}$$

this solution is

$$Q_X(x \mid S) = Q_j + \mu_j \left[P_X(x) - P_j\right], x \in E_j, j = 1, 2, \ldots, r \,, \tag{12}$$

with density

$$q_X(x \mid S) = \mu_j\, p_X(x), x \in E_j, j = 1, 2, \ldots, r \,. \tag{13}$$

The posterior density $q_X(x \mid S)$ has piecewise the form of the reference density function $p_X(x)$ scaled over each interval $E_j$ by the constant factor $\mu_j$.

## 4. CONSISTENCY OF THE CROSS-ENTROPY METHOD WITH FRACTILE CONSTRAINTS

It is showed in this section that the fractile constraints version of the cross-entropy method satisfies the invariance requirements for consistency of a data-based inference method stipulated by the principle of invariance under random variable domain conditioning. These requirements stated in section 2 are the invariance of encoded information for a data-bounded subdomain, and the invariance of the assigned probability distribution for any subdomain or a data-bounded subdomain, according to the possibilities of alternative conditioning ways.

### 4.1 Invariance of the encoded information

In the cross-entropy method the sample rule is the operator $H$ that encodes the fractile-pair constraints (8, 9) from data.

Consider the two ways to encode information described in section 2.

In the way (I), the conditioned constraints are obtained from the fractile-pair constraints (8, 9) which are conditioned by the data-bounded subdomain $E_c = [x_s, x_l]$. Thus the encoded information is so given

$$I_X(S \mid x_s \leq X \leq x_l): \quad (x, Q_X(x \mid S, x_s \leq X \leq x_l))_j = x_j, Q_X(x_j \mid S, x_s \leq X \leq x_l),$$

$$j = s, s+1, \ldots, l. \tag{14}$$

In the alternative way (II), constraints are directly obtained by applying the sample rule to the censored sample $S_c$ and the data-bounded subdomain $E_c = [x_s, x_l]$. Then the sample rule operator states that the elements of the censored sample $S_c$ of size $(l - s - 1)$ are the $(j - s)/(l - s)$-fractiles $(j = s+1, s+2, \ldots, l - 1,$ and $0 \leq s \leq l \leq r + 1)$ of the probability of any future observation of the random variable $X$ contained into $E_c$. Hence the encoded information, including conventional probability values 0 and 1 at the end points of $E_c$, are the following fractile-pair constraints

$$I_X(S_c): \quad (x, Q_X(x \mid S_c))_j = x_j, Q_X(x_j \mid S_c); \quad j = s, s+1, \ldots, l, \tag{15}$$

where the probability fractiles $Q_X(x_j \mid S_c)$ are

$$Prob \ [X \leq x_j \mid S_c] = (j - s)/(l - s), j = s, s+1, \ldots, l. \tag{16}$$

The invariance of the encoded information (14) and (15) prescribed by the expression (5), is easily proved as follows. First, application of probability conditioning rule to (14) gives the following probability fractiles $Q_X(x_j \mid S, x_s \leq X \leq x_l)$

$$Prob \ [X \leq x_j \mid S, x_s \leq X \leq x_l] = \frac{Prob \ [X \leq x_j, x_s \leq X \leq x_l \mid S]}{Prob \ [x_s \leq X \leq x_l \mid S]},$$

$$j = s, s+1, \ldots, l. \tag{17}$$

Next, by substituting the probability fractiles $(j - s)/(r + 1)$ and $(l - s)/(r + 1)$ given by (9) into (17)

$$Prob \ [X \leq x_j \mid S, x_s \leq X \leq x_l] = \frac{j - s}{l - s}, \quad j = s, s+1, \ldots, l. \tag{18}$$

Therefore the probability fractiles $Q_X(x_j \mid S, x_s \leq X \leq x_l)$ in the way (I) are exactly the same as the probability fractiles $Q_X(x_j \mid S_c)$ obtained in the alternative way (II) given by (16).

## 4.2 Invariance of the assigned probability distribution

Here the cross-entropy minimization of distributions subject to fractile constraints is the operator $F$ that selects the distribution (12) given as a function of the interval multiplier (11).

Consider the **first possibility of alternative ways** and the random variable conditioning by any subdomain $E_c = [x_{c1}, x_{c2}]$. The two ways (I.1) and (I.2) to choose a probability distribution are subjected to the same conditioned probability fractiles

$$I_X(S \mid x_{c1} \leq X \leq x_{c2}) :$$

$$[x_j, Q_X [x_j \mid (S \mid x_{c1} \leq X \leq x_{c2})]] = [x_j, Q_X(x_j \mid S, x_{c1} \leq X \leq x_{c2})],$$

$$j = s, s+1, \ldots, l \text{ and } 0 \leq s \leq l \leq r+1. \tag{19}$$

In the way (I.1), the probability distribution $Q_X(x \mid S, x_{c1} \leq X \leq x_{c2})$ is directed obtained from $Q_X(x \mid S)$ given by (12), when this is conditioned by $x_{c1} \leq X \leq x_{c2}$. Application of probability conditioning rule results in

$$Prob\ [X \leq x \mid S, x_{c1} \leq X \leq x_{c2}] = \frac{Prob\ [X \leq x, x_{c1} \leq X \leq x_{c2} \mid S]}{Prob\ [x_{c1} \leq X \leq x_{c2} \mid S]}, \tag{20}$$

which may be so written in terms of probability values of $Q_X(x \mid S)$

$$Q_X(x \mid S, x_{c1} \leq X \leq x_{c2}) = \frac{Q_X(x \mid S) - Q_X(x_{c1} \mid S)}{Q_X(x_{c2} \mid S) - Q_X(x_{c1} \mid S)}. \tag{21}$$

Specialization of (21) at the $x_j$ points gives the following probability fractiles of the conditioned constraints pairs

$$Q_X(x_j \mid S, x_{c1} \leq X \leq x_{c2}) = \frac{Q_X(x_j \mid S) - Q_X(x_{c1} \mid S)}{Q_X(x_{c2} \mid S) - Q_X(x_{c1} \mid S)}, $$

$$j = s, s+1, \ldots, l \text{ and } 0 \leq s \leq l \leq r+1. \tag{22}$$

In the alternative way (I.2), the probability distribution $Q_X [x \mid (S \mid x_{c1} \leq X \leq x_{c2})]$ is selected from all distributions subject to the conditioned constraints (19, 22). First, the conditioned reference distribution is calculated using the probability conditioning rule

$$P_X(x \mid x_{c1} \leq X \leq x_{c2}) = \frac{P_X(x) - P_X(x_{c1})}{P_X(x_{c2}) - P_X(x_{c1})}. \tag{23}$$

Specialization of (23) at the points $x_j$ gives the conditioned probability values

$$P_X(x_j \mid x_{c1} \leq X \leq x_{c2}) = \frac{P_j - P_X(x_{c1})}{P_X(x_{c2}) - P_X(x_{c1})}, $$

$$j = s, s+1, \ldots, \text{ and } 0 \leq s \leq l \leq r+1. \tag{24}$$

The application of the operator $F$ that minimizes cross-entropy to the conditioned constraints (22) and the reference distribution (23), selects the following posterior distribution

$$Q_X\left[x \mid (S \mid x_{c1} \le X \le x_{c2})\right] =$$

$$= Q_X(x_j \mid S, x_{c1} \le X \le x_{c2}) + \mu_j^*\left[P_X(x \mid x_{c1} \le X \le x_{c2}) - P_X(x_j \mid x_{c1} \le X \le x_{c2})\right],$$

$$\text{for } x \in E_j \,, \ j = s, s+1, \ldots, l \text{ and } 0 \le s \le l \le r+1, \tag{25}$$

where the new interval multipliers are

$$\mu_j^* = \frac{Q_X(x_{j+1} \mid S, x_{c1} \le X \le x_{c2}) - Q_X(x_j \mid S, x_{c1} \le X \le x_{c2})}{P_X(x_{j+1} \mid x_{c1} \le X \le x_{c2}) - P_X(x_j \mid x_{c1} \le X \le x_{c2})}. \tag{26}$$

Next, the invariance of the assigned probability distributions stipulated by the expression (6) is showed.

By introducing the conditioned probability fractiles (22) into (26) and taking into account (8)

$$\mu_j^* = \left[\frac{Q_{j+1} - Q_j}{P_{j+1} - P_j}\right]\left[\frac{P_X(x_{c2}) - P_X(x_{c1})}{Q_X(x_{c2} \mid S) - Q_X(x_{c1} \mid S)}\right],$$

$$j = s, s+1, \ldots, l \text{ and } 0 \le s \le l \le r+1. \tag{27}$$

Now, by substituting (22), (23), (24) and (27) into (25)

$$Q_X\left[x \mid (S \mid x_{c1} \le X \le x_{c2})\right] =$$

$$= \frac{Q_X(x_j \mid S) - Q_X(x_{c1} \mid S) + \left[(Q_{j+1} - Q_j)/(P_{j+1} - P_j)\right]\left[P_X(x) - P_X(x_j)\right]}{Q_X(x_{c2} \mid S) - Q_X(x_{c1} \mid S)}$$

$$x \in E_j \,, \ j = s, s+1, \ldots, l \text{ and } 0 \le s \le l \le r+1. \tag{28}$$

Since equation (28) contains partially the expression of $Q_X(x \mid S)$ given by (12) when substituting the coefficients $\mu_j$ by (11), the posterior distribution may be so written

$$Q_X\left[x \mid (S \mid x_{c1} \le X \le x_{c2})\right] = \frac{Q_X(x \mid S) - Q_X(x_{c1} \mid S)}{Q_X(x_{c2} \mid S) - Q_X(x_{c1} \mid S)}. \tag{29}$$

Hence the posterior distribution (29) in the way (I.2), is exactly the same posterior probability distribution obtained in the way (I.1) given by (21).

Now, consider the second possibility of alternative ways and the case of conditioning by a data-bounded subdomain $E_c = [x_s, x_l]$.

In this case, the cross-entropy method selects in the ways (I) and (II) a probability distribution subject to identical set of constraints (14, 15).

Identical conditioned distribution is assigned by the ways (I.1) and (I.2) for any subdomain, and consequently for a data-bounded subdomain. In this case, by substituting (9) and (12) into (21) specialized for a data-bounded subdomain it is obtained

$$Q_X(x \mid S, x_s \leq X \leq x_l) = \frac{j - s + [(P_X(x) - P_j)/(P_{j+1} - P_j)]}{l - s}$$

$$x \in E_j, j = s, s + 1, \ldots, l; 0 \leq s \leq l \leq r + 1. \tag{30}$$

In the way (II), the cross-entropy minimization applies to the censored sample constraints (15) using the reference distribution (23) specialized for a data-bounded subdomain. It gives the following conditioned distribution

$$Q_X(x \mid S_c) =$$

$$= Q_X(x_j \mid S_c) + \mu_j^{**}[P_X(x \mid x_s \leq X \leq x_l) - P_X(x_j \mid x_s \leq X \leq x_l)]$$

$$x \in E_j, j = s, s + 1, \ldots, l; \text{ and } 0 \leq s \leq l \leq r + 1, \tag{31}$$

and the new interval multipliers

$$\mu_j^{**} = \frac{Q_X(x_{j+1} \mid S_c) - Q_X(x_j \mid S_c)}{P_X(x_{j+1} \mid x_s \leq X \leq x_l) - P_X(x_j \mid x_s \leq X \leq x_l)}$$

$$j = s, s + 1, \ldots, l; \text{ and } 0 \leq s \leq l \leq r + 1. \tag{32}$$

Next, the invariance of the assigned probability distribution prescribed by the expression (7) is showed.

By substituting the probability fractiles of the censored sample constraints (16) into (32)

$$\mu_j^{**} = \frac{1}{l - s} \left[ \frac{1}{P_X(x_{j+1} \mid x_s \leq X \leq x_l) - P_X(x_j \mid x_s \leq X \leq x_l)} \right]$$

$$j = s, s + 1, \ldots, l \text{ and } 0 \leq s \leq l \leq r + 1. \tag{33}$$

Finally the substitutions of (33), (16) and (23, 24) specialized for data-bounded subdomains, into (31) give

$$Q_X(x \mid S_c) = \frac{j - s}{l - s} + \frac{1}{l - s} \left[ \frac{P_X(x) - P_j}{P_{j+1} - P_j} \right]$$

$$x \in E_j, j = s, s + 1, \ldots, l; \text{ and } 0 \leq s \leq l \leq r + 1 \tag{34}$$

Therefore the conditioned probability distribution (34) assigned in the way (II) is exactly the same conditioned distribution as obtained in the ways (I.1) and (I.2) given by (30).

## 5. CONCLUSIONS

Probabilistic methods assigning probability distributions on the basis of only a random sample of observed data on a given domain are considered in this paper. The approach is to construct a data-based inference rationale and to search an inference procedure that fulfil all the requirements derived from the principles of this rationale.

A new consistency principle of invariance under conditioning of the random variable domain is stated as a particular of the strong invariance principle. Such a consistency principle is formalized for a data-based inference method having the data encoding and the probability selection steps. The principle imposes different requirements for each step of the inference method.

The cross-entropy method using fractile constraints satisfies all the requirements of the principle of invariance under random variable domain conditioning. Apparently, there is no other method available that satisfy these invariance requirements.

Furthermore, the cross-entropy method with fractile constraints can claim the advantages over existing methods of satisfying the data-monotonicity principle and the invariance under monotonic transformations of random variables.

**ACKNOWLEDGMENT.** The work of this paper was carried out as a part of the research project PB 87-0458 supported financially by the DGICYT of Spain.

# References

[1] JEFFREYS, H. (1939, 2nd and 3th ed. 1948, 1961) *Theory of probability*, Oxford University Press, London.

[2] JAYNES, E.T. (1957) *"How Does the Brain Do Plausible Reasoning?"*, Microwave Laboratory Report No. 421. University of Stanford; and G.J. Erickson and C.R. Smith (eds.). Maximum-Entropy and Bayesian Methods in Science and Engineering. Vol. I, 1-24, Kluwer Academic Publications, 1988.

[3] SHORE J.E. and JOHNSON, R.W. (1980) *"Axiomatic Derivation of Principle of Maximum Entropy and the Principle of Minimum Cross-Entropy"*, IEEE Transactions on Information Theory, IT-26, 1, 26-37.

[4] TIKOCHINSKY, Y., TISHBY, N.Z. and LEVINE R.D. (1984) *"Consistent Inference Probabilities for Reproducible Experiments"*, Physics Review Letters, V. 51, 1357-1360.

[5] SKILLING, J. (1988) *"The axioms of Maximum Entropy"*, G.J. Erickson and C.R. Smith (eds.). Maximum-Entropy and Bayesian Methods in Science and Engineering, Vol. 1, 173-187, Kluwer Academic Publications.

[6] SOLANA, V. and LIND, N.C. (1989) *"A Monotonic Property of Distributions Based on Entropy with Fractile Constraints"*, 8th. International Maximum Entropy Conference, St. John's College, Cambridge, England 1988; John Skilling (ed.), Kluwer Academic Publications, Dordrecht, Netherland.

[7] SOLANA, V. AND LIND, N.C. (1989) *"Two Principles for Data Based Probabilistic System Analysis"*, Proceedings 5th. International Conference on Structural Safety and Reliability, San Francisco, August 1989. Shinozuka (ed.), ASCE American Society of Civil Engineers.

[8] LIND, N.C. and SOLANA, V. (1988) *"Estimation of Random Variables with Fractile Constraints"*, IRR No. 11, Institute for Risk Research, University of Waterloo, Waterloo, Ontario, Canada.

# AN INTRODUCTION TO PARAMETER ESTIMATION USING BAYESIAN PROBABILITY THEORY

G. LARRY BRETTHORST
*Washington University,*
*Department of Chemistry*
*1 Brookings Drive,*
*St. Louis, Missouri 63130*

ABSTRACT. Bayesian probability theory does not define a probability as a frequency of occurrence; rather it defines it as a reasonable degree of belief. Because it does not define a probability as a frequency of occurrence, it is possible to assign probabilities to propositions such as "The probability that the frequency had value $\omega$ when the data were taken," or "The probability that hypothesis $x$ is a better description of the data than hypothesis $y$." Problems of the first type are parameter estimation problems, they implicitly assume the correct model. Problems of the second type are more general, they are model selections problems and do not assume the model. Both types of problems are straight forward applications of the rules of Bayesian probability theory. This paper is a tutorial on parameter estimation. The basic rules for manipulating and assigning probabilities are given and an example, the estimation of a single stationary sinusoidal frequency, is worked in detail. This example is sufficiently complex as to illustrate all of the points of principle that must be faced in more realistic problems, yet sufficiently simple that anyone with a background in calculus can follow it. Additionally, the model selection problem is discussed and it is shown that parameter estimation calculation is essentially the first step in the more general model selection calculation.

# 1 Introduction

The problem of estimating the value of a parameter is basic to science and engineering, yet the procedures needed to solve this problem are rarely taught to scientists and engineers. This is mostly historical and has to do with the fact that probability theory, as traditionally interpreted, treats all probabilities as frequencies (here the word frequency is used in the sense of the number of times an event occurs). This interpretation of probability theory, called sampling theory, has no way of addressing the parameter estimation problem when only a single data set is available. To address the problem within sampling theory, one must imagine one is estimating the distribution of a random parameter within an ensemble of data sets. One then tries to determine the mean and standard deviation of this parameter within the ensemble. This is not the problem faced by scientist and engineers; there is typically only a single data set, and one is trying to determine the value the parameter had at the time the data was taken.

To solve this problem within probability theory, a wider interpretation is needed. This interpretation was first given by Laplace [1] and then essentially ignored for the next century.

53

*P. F. Fougère (ed.), Maximum Entropy and Bayesian Methods, 53–79.*
© 1990 *Kluwer Academic Publishers.*

In Laplace's interpretation, probability theory is just common sense reduced to numbers, and a probability represents a reasonable degree of belief; not a frequency of occurrence. In the mid 1930, Jeffreys [2] rediscovered the works of Laplace and derived probability theory as an axiomatic theory of inference. Then in 1957, E. T. Jaynes [3] using the methodology of Shannon [4], the mathematics of Abel [5] and Cox [6], and the qualitative principles of Laplace [1], proved that if one represents a reasonable degree of belief as a real number, then the only consistent rules for manipulating probabilities are those given by Laplace. In this wider interpretation of probability theory, called Bayesian probability theory, problems of the form "What is the best estimate of a parameter one can make from the data and one's prior information?" make perfect sense. In this paper the basic principles of Bayesian probability theory are outlined and an example, the estimation of a single stationary sinusoidal frequency, is worked with each step in the calculation explained in detail. For an introduction to Bayesian probability theory see the works of Tribus [7] and Zellner [8] and for a derivation of the rules of probability theory see Jaynes [3].

## 2   The Rules of Probability Theory

There are only two basic rules for manipulating probabilities, the product rule and the sum rule; all other rules may be derived from these. If $A$, $B$, and $C$ stand for three arbitrary propositions then the product rule is

$$P(A, B|C) = P(A|C)P(B|A, C) \tag{1}$$

where $P(A, B|C)$ is the joint probability that "$A$ and $B$ is true given that $C$ is true," $P(A|C)$ is the probability that "$A$ is true given $C$ is true," and $P(B|A, C)$ is the probability that "$B$ is true given that both $A$ and $C$ are true." The notation "$|C)$" means conditional on the truth of proposition $C$. In Bayesian probability theory *all* probabilities are conditional. To use a notation such as $P(A)$ to stand for the probability of $A$ does not make sense until the evidence on which it is based is given. Anyone using such notation either does not understand this or is being extremely careless in their notation. In either case one should be careful when interpreting such material.

In Aristotelian logic the proposition "$A$ and $B$" is the same as "$B$ and $A$" so the truth value of the propositions must be the same in the product rule. That is the probability of "$A$ and $B$ given $C$" must be equal to the probability of "$B$ and $A$ given $C$," so the order may be rearranged in Eq. (1) to obtain

$$P(B, A|C) = P(B|C)P(A|B, C).$$

This equation may be combined with Eq. (1) to obtain a seemingly trivial result

$$P(A|B, C) = \frac{P(A|C)P(B|A, C)}{P(B|C)}. \tag{2}$$

This is Bayes' theorem. It is named after Rev. Thomas Bayes, an 18th century mathematician who derived a special case of the theorem. Bayes' calculations [9] were published in 1763 after his death. Exactly what Bayes intended to do with the calculation, if anything, remains a mystery today. Nevertheless, today in Bayesian inference, this theorem is the starting point for all Bayesian calculations.

The sum rule of probability theory may be stated as

$$P(A + B|C) = P(A|C) + P(B|C) - P(A, B|C) \tag{3}$$

where $P(A + B|C)$ means the probability that "$A$ or $B$ is true given that $C$ is true." If the propositions $A$ and $B$ are mutually exclusive, that is the probability $P(A, B|C)$ is zero, then the sum rule becomes

$$P(A + B|C) = P(A|C) + P(B|C).$$

The sum rule will prove to be useful in parameter estimation problems, because it allows one to investigate an interesting parameter while removing an uninteresting parameter from consideration. To see this, suppose one wished to compute the probability of $A$ given $C$, but there is a third proposition $B$ which must be taken into account. To make this more concrete, suppose $C$ stands for the data, $A$ the frequency of a sinusoidal oscillation, and $B$ the amplitude of the sinusoid. Now suppose one wishes to compute the probability of the frequency given the data, $P(A|C)$. But there is this other parameter $B$; the way to proceed is to compute the joint probability of the frequency and the amplitude given the data, $P(A, B|C)$, and then use the sum rule to eliminate the amplitude $B$. Suppose, for arguments sake, the amplitude $B$ could take on one of two mutually exclusive values $B = \{B_1, B_2\}$. If one computes the probability of the frequency and (amplitude 1 or amplitude 2) given the data then

$$P(A|C) \equiv P(A, [B_1 + B_2]|C) = P(A, B_1|C) + P(A, B_2|C).$$

This probability distribution summarizes all the information in the data relevant to the proposition $A$, and $P(A|C)$ is called the marginal probability of $A$ given $C$. The marginal probability does not depend on the amplitude at all. To see this, suppose that the proposition $A$ had not been present, then $P(B_1 + B_2|C) = P(B_1|C) + P(B_2|C) = 1$, thus when a parameter has been marginalized from a joint probability distribution the reference to that parameter is dropped because the probability distribution no longer depends on the parameter. Of course the amplitude could take on more than two values, for example $B_j = \{B_1, \cdots, B_m\}$, in which case the marginal probability distribution would become

$$P(A|C) = \sum_{j=1}^{m} P(A, B_j|C)$$

provided the amplitudes are mutually exclusive. In real problems, the parameter $B$ would take on a continuum of values; but *as long as $B$ is a constant through the run of the data* the sum rule becomes

$$P(A|C) = \int dB P(A, B|C), \tag{4}$$

where the integral is over all possible values of the parameter $B$. It is in this form that the sum rule will be used in the example to remove uninteresting or *nuisance* parameters. Note that $dB$ refers to a number, while $B$ appearing inside $P(A, B|C)$ refers to a proposition. A notation could be invented to reflect this distinction, but it is unnecessary; provided that one realizes that when a capital letter appears outside of a probability symbol it refers to a number associated with a proposition, while inside the probability symbol it refers to the proposition.

## 3   Assigning Probabilities

The product rule Eq. (1), Bayes theorem Eq. (2), and the sum rule in the form of Eq. (4) are the only rules of probability theory needed to solve most problems of inference. But these rules tell one how to manipulate probabilities after they have been assigned; *there is nothing within the theory to tell one how to assign probabilities.* This must come from outside probability theory and is one of the major reasons why probability theory, as formulated by Laplace, was rejected. If probabilities are frequencies there is no problem in assigning their values; it is only when probabilities are interpreted as a reasonable degree of belief, that their assignment becomes a real question. There are many ways to address the assignment of probabilities – see Ref. [8,10,11,12] for more on this. Here, the principle of maximum entropy will be used to assign all probabilities – see Ref. [13] for an extended discussion of maximum entropy.

Suppose one has a discrete probability distribution $P(i|I)$, where $i$ stands for some proposition (for example, the probability of one of the faces of a die) and $I$ represents the information on which the probability distribution is based, then Shannon's $H$ theorem [4] states that

$$H \equiv -\sum_{i=1}^{N} P(i|I) \log P(i|I) \tag{5}$$

is a measure of the amount of ignorance in the probability distribution. Shannon's theorem is derived based on a qualitative requirement plus the requirement that the measure be consistent. The principle of maximum entropy then states that if one has some testable information $I$, one can assign a probability distribution to a proposition $i$ such that $P(i|I)$ contains only the information $I$. This assignment is done by maximizing $H$ subject to the constraints represented by the information $I$.

To demonstrate its use, suppose one has a die and wishes to assign a probability to each of the six faces. If nothing is known about the die except that the probabilities must total 1, then

$$1 - \sum_{j=1}^{6} P(j|I) = 0$$

must be satisfied. If this constraint is satisfied, then one can multiply it by a constant $\beta$ (called a Lagrange multiplier) and because the constraint is zero it can be add it to the entropy without changing the value of $H$:

$$H = -\sum_{j=1}^{6} P(j|I) \log P(j|I) + \beta \left[ 1 - \sum_{j=1}^{6} P(j|I) \right].$$

But the probabilities and $\beta$ are not known; they must be assigned. To assign the probabilities, $H$ is constrained to be a maximum with respect to variations in the unknowns:

the probabilities, and $\beta$. Because $H$ measures the amount of ignorance in the probability distribution, constraining $H$ to be a maximum is asking for the least informative (highest entropy) probability distribution subject to the known prior information. This maximum is located by differentiating $H$ with respect to $P(k|I)$ and $\beta$, and setting derivatives equal to zero. Here there are six unknown probabilities and one unknown Lagrange multiplier. But when the derivatives are taken, there will be seven equations; thus all of the unknowns are completely determined. This system of equations is then solved for the values of $P(j|I)$ and for $\beta$. Taking the derivatives one obtains

$$-[\log P(k|I) + 1] + \beta = 0,$$
$$1 - \sum_{j=1}^{6} P(j|I) = 0$$

from which one finds

$$P(j|I) = \frac{1}{6} \quad \text{and} \quad \beta = 1 - \log 6.$$

When nothing is known, except that the probability distribution should be normalized, the principle of maximum entropy reduces to the uniform prior. This is Laplace's principle of insufficient reason [1]. But the principle maximum entropy is much more general because it allows one to assign probabilities that are maximally uninformative, while still incorporating the known information. If information were available that indicated that the die was not honest, then this information could be used in a maximum entropy calculation to obtain a nonuniform probability distribution – see Ref. [14] for this calculation and much more on maximum entropy.

The principle of maximum entropy represents a way of assigning probabilities based only on the information that one actually possesses. In the previous example, that information was that the probabilities should total one. Almost any information can be incorporated into a maximum entropy calculation. In the example that follows, the joint probability $P(e|I)$ of a set of noise values will be needed, where $e \equiv \{e_1, \cdots, e_N\}$. One should read $P(e|I)$ as the probability that the noise should have value $e_1$ at time $t_1$, and value $e_2$ at time $t_2$, etc. This probability density will be derived as a second example of the principle of maximum entropy.

In real experiments, not much is actually known about the noise. Typically when an experiment is performed, any systematic component in the data is defined to be the "signal" and the random part is defined to be noise. With this definition of "signal" the noise must have zero mean value

$$\frac{1}{N} \sum_{i=1}^{N} e_i = 0$$

where $N$ is the total number of data values. But even though the mean value of the noise is zero, it will have a mean-square value:

$$\frac{1}{N} \sum_{i=1}^{N} e_i^2 = \sigma^2. \tag{6}$$

Assuming the second moment of the noise exists, implies only that the noise is present; it does not assume that $\sigma^2$ is actually known. In addition to the second moment, there could be higher moments

$$\frac{1}{N}\sum_{i=1}^{N} e_i^x = \gamma_x \tag{7}$$

where $x$ is an integer and $\gamma_x$ represents the average value of the moment. Additionally, there could be correlations such that

$$\frac{1}{N-m}\sum_{i=1}^{N-m} e_i e_{i+m} = \rho_m \qquad (1 < m < N) \tag{8}$$

where $\rho_m$ is a correlation coefficient. However without seeing several samples of the noise it is not known if correlations exist or if higher moments exist, except of course what is implied by Eq. (6). All that is actually known is that Eq. (6) must be satisfied.

Maximum entropy can be used to assign a probability density function that incorporates only what is actually known. If the noise has mean-square $\sigma^2$, then the expected value of $e_j^2$ is given by

$$\gamma_j \left[\sigma^2 - \int d\mathbf{e}\, e_j^2 P(\mathbf{e}|I)\right] = 0$$

where $\gamma_j$ is a Lagrange multiplier. This equation must be true for every $e_j^2$. All proper probability density functions must be normalized so

$$\beta \left[1 - \int d\mathbf{e}\, P(\mathbf{e}|I)\right] = 0.$$

If there are $N$ data values, there are $N+1$ constraints and the entropy functional to be maximized is given by

$$H = -\int d\mathbf{e}\, P(\mathbf{e}|\sigma, I) \log P(\mathbf{e}|\sigma, I) + \beta \left[1 - \int d\mathbf{e}\, P(\mathbf{e}|\sigma, I)\right] + \sum_{j=1}^{N} \gamma_j \left[\sigma^2 - \int d\mathbf{e}\, e_j^2 P(\mathbf{e}|\sigma, I)\right]$$

where the notation $P(\mathbf{e}|\sigma, I)$ has been adopted to indicate that it is the probability of the noise given $\sigma$ and the information $I$.

This case is fundamentally different from the previous example, because $P(\mathbf{e}|\sigma, I)$ is a function of $N$ continuous variables. Instead of having to determine a fixed number of unknown probabilities there are infinitely many probabilities corresponding to the continuous variables. These must be determined from the $N+1$ constraints. The principle of maximum entropy may still be used to assign the *probability density function* by taking the derivatives and solving the resulting system of equations. Taking the derivatives of $H$ with respect to $P(\mathbf{e}|\sigma, I)$ gives

$$-[\log P(\mathbf{e}|\sigma, I) + 1] - \beta - \sum_{i=1}^{N} \gamma_i e_i^2 = 0$$

as one of the equations, and taking the derivatives with respect to the Lagrange multipliers just returns the constraint equations:

$$1 - \int d\mathbf{e}\, P(\mathbf{e}|\sigma, I) = 0$$

and

$$\sigma^2 - \int dee_j^2 P(\mathbf{e}|\sigma, I) = 0.$$

Solving this system of equations one finds the probability of the noise to be

$$P(\mathbf{e}|\sigma, I) = (2\pi\sigma^2)^{-\frac{N}{2}} \exp\left\{ -\sum_{i=1}^{N} \frac{e_i^2}{2\sigma^2} \right\} \tag{9}$$

where

$$\gamma_j = \frac{1}{2\sigma^2}$$

and

$$\beta = \log\left[\int d\mathbf{e} \exp\left\{ -\sum_{i=1}^{N} \frac{e_i^2}{2\sigma^2} \right\}\right] - 1.$$

There are several interesting points to note about this probability density function. First, it was not assumed that the noise was correlated and the resulting maximum entropy probability density function does not contain correlations. Had Eq. (8) been used as a constraint, the resulting probability density function would have been very different. Second, no assumptions were made about the odd moments, and the resulting probability density function has no odd moments. Third, if one computes the expected value of the even moments one finds

$$(2\pi\sigma^2)^{-\frac{N}{2}} \int_{-\infty}^{+\infty} de_1 \cdots de_N e_j^{2n} \exp\left\{ -\sum_{i=1}^{N} \frac{e_i^2}{2\sigma^2} \right\} = 1 \cdot 3 \cdot 5(2n-1)\sigma^{2n},$$

which for $n = 1$ reduces to

$$(2\pi\sigma^2)^{-\frac{N}{2}} \int_{-\infty}^{+\infty} de_1 \cdots de_N e_j^{2n} \exp\left\{ -\sum_{i=1}^{N} \frac{e_i^2}{2\sigma^2} \right\} = \sigma^2,$$

just what one would expect to find. Thus maximum entropy has not introduced any spurious correlations or effects into the probability density function that were not already implicit in the constraints. Equation (9) is the least informative probability density function that is consistent with the given second moment. If one has information that correlations exist or that the odd moments are not zero, that information can always be used in a maximum entropy calculation to assign a probability density function to the noise, that probability density will always have more compact support (lower entropy) for a given value of $\sigma$ than Eq. (9). Therefore, this new noise probability density function will always make more precise estimates of the parameters.

## 4  Example – Estimating a Frequency

In the second section the rules of probability theory were outlined, and in the third section the procedures for assigning probability distributions were given, in this section a nontrivial parameter estimation problem is worked. Each step in the calculation explained in detail. The example is sufficiently complex to illustrate all of the points of principle that must be faced in more complex problems, yet sufficiently simple that anyone with a background in calculus should be able to follow the mathematics, and additionally it gives an important and surprising result.

In this example the probability of a frequency of oscillation $\omega$ is computed conditional on the data $D$ and the prior information $I$. This probability density will be computed from Bayes' theorem, but first, exactly what problem is being solved must be explained. In this example, there is a time series $y(t)$ that is being considered. The time series is postulated to contain a single stationary sinusoidal signal $f(t)$ plus noise. The basic model is: a discrete data set $D = \{d_1, \cdots, d_N\}$ has been recorded; it is sampled from $y(t)$ at discrete times $\{t_1, \cdots, t_N\}$; with model equation

$$d_i = y(t_i) = f(t_i) + e_i, \qquad (1 \leq i \leq N) \tag{10}$$

where $e_i$ represents the numerical value of the noise at time $t_i$. It is possible in different problems to have other types of noise. For example, the noise could be multiplicative, or it could be simple digitizing noise. The signal $f(t)$ for a single sinusoidal frequency may be written

$$f(t) = B_1 \cos(\omega t) + B_2 \sin(\omega t) \tag{11}$$

which has three parameters $(B_1, B_2, \omega)$ that may be estimated from the data.

The problem to be solved is to compute the probability of the frequency $\omega$ conditional on the data and the prior information, this is abbreviated as $P(\omega|D, I)$. But Eq. (11) has two other parameters, effectively the amplitude and phase of the sinusoid. In this problem the two parameters $B_1$ and $B_2$ are referred to as *nuisance parameters,* because the probability distribution that is to be calculated does not depend on these parameters. To perform this calculation one applies Bayes' theorem to compute the joint probability of all of the parameters and then uses the sum rule, Eq. (4), to eliminate the nuisance parameters. Applying Bayes' theorem gives

$$P(\omega, B_1, B_2|D, I) = \frac{P(\omega, B_1, B_2|I)P(D|\omega, B_1, B_2, I)}{P(D|I)} \tag{12}$$

which indicates that to compute the joint probability density one must obtain three terms. The first term, $P(\omega, B_1, B_2|I)$, is the probability of the parameters given only the information $I$. This term is referred to as a prior probability density, or simply as a prior. The second term, $P(D|\omega, B_1, B_2, I)$, is the probability of the data given the parameters and the information $I$. This term is called the direct probability density of the data, and it is often referred to as a likelihood function when one considers a single data set. The third term, $P(D|I)$, is the probability of the data given only the information $I$, this term will be shown to be a normalization constant.

Equation (12) is the joint probability density of all the parameters including the amplitudes. The sum rule, Eq. (4), can be applied to remove the dependence on the amplitudes:

$$P(\omega|D, I) = \int dB_1 dB_2 \frac{P(\omega, B_1, B_2|I)P(D|\omega, B_1, B_2, I)}{P(D|I)}. \tag{13}$$

The calculation of the posterior probability density of a single stationary sinusoidal frequency is at least a four step problem: assign the three probability density functions, and then remove the dependence on the amplitudes by integration.

## 4.1  Assignment of the direct probability

The calculation of each of the three terms proceeds by applying the rules of probability theory and by supplying additional information $I$ when necessary. Each of the terms will be taken separately starting with $P(D|\omega, B_1, B_2, I)$, and followed by $P(\omega, B_1, B_2|I)$. The probability density of the data given the value of the parameters is essentially asking how unlikely is the data, but the name of this term is a little misleading; it is not the probability of the data that is needed here but the probability of the noise. What is taken to be noise depends critically on what one takes to be signal. Equation (10) is a definition of what is meant by the noise. The probability of the noise was assigned in the previous section, Eq. (9). One can take the difference between the data and the model,

$$e_i = d_i - f(t_i)$$

and substitute this into Eq. (9) to obtain

$$P(D|f, \sigma, I) = (2\pi\sigma)^{-\frac{N}{2}} \exp\left\{ -\frac{1}{2\sigma^2} \sum_{i=1}^{N} [d_i - f(t_i)]^2 \right\}$$

as the probability of the data given the model $f$, where a new parameter $\sigma^2$ (the variance of the noise) has been added. Inserting the single stationary sinusoidal frequency model, Eq. (11), for $f$ and changing notation $f \rightarrow \omega, B_1, B_2$ to indicate that it is the parameters that interests us, one obtains

$$P(D|\omega, B_1, B_2, \sigma, I) = (2\pi\sigma)^{-\frac{N}{2}} \exp\left\{ -\frac{1}{2\sigma^2} \sum_{i=1}^{N} [d_i - B_1 \cos(\omega t_i) - B_2 \sin(\omega t_i)]^2 \right\}. \quad (14)$$

The usual way to proceed is to fit the sum in the exponent. Finding the parameter values which minimize this sum is called "least-squares." The equivalent procedure, in this case, of finding parameter values that maximize $P(D|\omega, B_1, B_2, \sigma, I)$ is called "maximum-likelihood." The maximum-likelihood procedure is more general than least-squares: it has theoretical justification when the likelihood is not Gaussian.

In this calculation uniform sampling will be assumed, because it simplifies some of the analytic details. Expanding the exponent in Eq. (14) one obtains

$$P(D|B_1, B_2, \omega, \sigma, I) = (2\pi\sigma)^{-\frac{N}{2}} \exp\left\{ -\frac{Q}{2\sigma^2} \right\} \quad (15)$$

where

$$Q \equiv N\overline{d^2} - 2[B_1 R(\omega) + B_2 I(\omega)] + B_1^2 c + B_2^2 s, \quad (16)$$

and

$$R(\omega) = \sum_{i=1}^{N} d_i \cos(\omega t_i) \quad (17)$$

$$I(\omega) = \sum_{i=1}^{N} d_i \sin(\omega t_i) \quad (18)$$

are the cosine and sine transforms of the data,

$$\overline{d^2} = \frac{1}{N} \sum_{i=1}^{N} d_i^2$$

is the observed mean-square data value, and

$$c \equiv \sum_{i=1}^{N} \cos^2(\omega t_i) = \frac{N}{2} + \frac{\sin(N\omega)}{2\sin(\omega)} \qquad (19)$$

$$s \equiv \sum_{i=1}^{N} \sin^2(\omega t_i) = \frac{N}{2} - \frac{\sin(N\omega)}{2\sin(\omega)} \qquad (20)$$

where the convention $t_i = \{-T, -T+1, \cdots, T-1, T\}$, and $2T+1 = N$ has been adopted. Use of this convention means that frequencies are measured in radians and may take on values between zero and $2\pi$. The cross term, $\sum_{i=1}^{N} \cos(\omega t_i) \sin(\omega t_i)$, is zero for uniform sampling.

The assignment of the direct probability of the data is now complete. However, the use of Eq. (9) has modified the problem to include a fourth parameter, the variance of the noise $\sigma^2$. Inserting Eq. (15) into Eq. (13) one obtains

$$P(\omega|\sigma, D, I) = (2\pi\sigma^2)^{-\frac{N}{2}} \int dB_1 dB_2 \frac{P(\omega, B_1, B_2|I) \exp\{-Q/2\sigma^2\}}{P(D|I)} \qquad (21)$$

as the posterior probability of the frequency, where the parameter $\sigma$ is assumed known. If the variance is not known, at the end of the calculation the product and sum rules may be used to remove the variance from the problem, just as the amplitudes will be removed.

## 4.2 Assignment of the prior probability

Assigning the prior probability is one of the most controversial areas in Bayesian probability. Yet, to a Bayesian it is the most natural of things. No one would think of trying to solve any problem in everyday life without bringing to bear all of his prior experiences. In traditional frequency interpretation of probability theory assigning a prior probability makes no sense, because prior information has no frequency interpretation. The sampling theorist finds himself in the uncomfortable position of trying to solve a problem in which he is not allowed to use what he knows, while trying to justify his use of a particular model. Most of the controversy arises when one is trying to solve a problem in which one has little prior information. If one has highly informative prior information, such as a prior measurement, there is little discussion on how to assign priors: one simply uses the posterior probability derived in analyzing the previous measurement as the prior probability for the current measurement. But this type of argument simply delays the problem of how to assign a probability to represent knowing little. In this tutorial the prior probabilities will be assigned to indicate that little, effectively no, prior information is available.

In assigning the prior probability $P(B_1, B_2, \omega|I)$ exactly what is known about the parameters will have to be stated. Before doing this note that the prior probability is the joint probability of the amplitudes and the frequency. But from the product rule, Eq. (1), this can be factored:

$$P(B_1, B_2, \omega | I) = P(B_1, B_2 | \omega, I) P(\omega | I). \tag{22}$$

The prior probability of the frequency may be assigned completely independent of the values of the amplitudes.

Here the only thing known about the frequency is that the data has been sampled uniformly, thus frequency values greater than the Nyquist frequency are aliased. So if the experimenter was even reasonably competent, the frequency must be bounded between 0 and $2\pi$ radians. Using this bound and the normalization constraint in a maximum entropy calculation results in the assignment of

$$P(\omega | I) = \frac{1}{2\pi}$$

as the prior probability of the frequency. Of course this is not the only prior probability that could be assigned. Indeed in [11] very convincing arguments are given that demonstrate a frequency to be scale parameter for which a Jeffreys' prior [2] is appropriate. There is no contradiction in arriving at different prior probability assignments. The two different assignments correspond to being in different states of knowledge, and people with different prior information $I$ will naturally make different assignments. But this probability assignment represent knowing little, effectively nothing, and regardless of what functional form one assigns to the prior; *if the prior is slowly varying compared to the direct probability, the prior will look like a constant over the range of values where the direct probability is sharply peaked* and its behavior outside of this region will make little effectively no difference in the results. It is only when the width of the prior is comparable to the width of the direct probability that it can have any significant effect.

Having assigned the prior probability of the frequency, Eq. (22) becomes

$$P(B_1, B_2, \omega | I) = \frac{P(B_1, B_2 | \omega, I)}{2\pi}.$$

The probability of the amplitudes explicitly depends on a given value of the frequency. In this calculation it will be assumed that knowing the frequency tells one nothing about the amplitudes. This is not true in general, for example if the experiment is repeatable and a previous measurement is available, knowledge of the frequency would imply a great deal about the value of the amplitude. But if knowledge of the frequency does not tell one anything about the amplitudes then $P(B_1, B_2 | \omega, I) = P(B_1, B_2 | I)$ and the joint prior probability of all of the parameters may be written as

$$P(B_1, B_2, \omega | I) = \frac{P(B_1, B_2 | I)}{2\pi}.$$

To proceed one must state exactly what is known about the amplitudes. Suppose for arguments sake one repeated this experiment a great number of times. The signal is a stationary sinusoid, when the experiment is repeated each of the amplitudes will take on both positive and negative values, (the phase will be different in each run of the data). Thus the average value of the amplitudes over many runs of the data will be zero, but the mean-square value of the amplitudes will be nonzero. *If one knows nothing else about the amplitudes, then applying the principle of maximum entropy will result in assigning a Gaussian prior probability density to the amplitudes:*

$$P(B_1, B_2|\delta, I) = (2\pi\delta^2)^{-1} \exp\left\{-\frac{B_1^2 + B_2^2}{2\delta^2}\right\}, \qquad (23)$$

where $\delta^2$ represents the uncertainty in the amplitudes. If this prior probability is to represent knowing little, then $\delta$ must be very large. But if $\delta$ is very large *this prior is effectively a uniform prior probability* over the range of values where the direct probability is peaked. Because this prior is to represent knowing little; $\delta$ will be taken to be very large, effectively infinite, and the Gaussian prior will be replaced by its limiting value: the uniform prior. This is the first example of what is called an improper prior probability density. An improper prior is a prior probability density that is not normalizable – and is not, strictly speaking, a probability density function at all. When performing a Bayesian probability calculation, one should always use a proper probability density, and then pass to the limit of an improper probability density at the end of the calculation – see Ref. [12] for more on this. But because the Gaussian is so strongly convergent the order of these operations may be interchanged. This is not always true and when in doubt the only safe course is to use a proper prior and then pass to the limit of an improper prior at the end of the calculation.

With the assignment of the uniform prior for the amplitudes the joint prior probability density of all of the parameters is given by

$$P(\omega, B_1, B_2|\delta, I) = \frac{1}{2\pi}.$$

This prior probability density may now be substituted into Eq. (21) to obtain

$$P(\omega|\sigma, D, I) = \frac{(2\pi\sigma^2)^{-\frac{N}{2}}}{2\pi} \int dB_1 dB_2 \frac{\exp\left\{-Q/2\sigma^2\right\}}{P(D|I)}. \qquad (24)$$

### 4.3   Assignment of the prior probability of the data

The prior probability of the data, $P(D|I)$, is essentially a normalization constant in the parameter estimation problems. But to someone unfamiliar with Bayesian calculations this is not obvious, nor is it obvious how to calculate it. The way to proceed is to calculate the joint probability of the data and the parameters, $P(\omega, B_1, B_2, D|I)$. This can be factored using the product rule, Eq. (1), to obtain

$$P(\omega, B_1, B_2, D|I) = P(\omega, B_1, B_2|I)P(D|\omega, B_1, B_2, I).$$

The sum rule may now be applied to remove the dependence on the parameters,

$$P(D|I) = \int d\omega dB_1 dB_2 P(\omega, B_1, B_2|I)P(D|\omega, B_1, B_2, I). \qquad (25)$$

Comparing this with Eq. (13) we see that this is just the constant needed to ensure that the total probability is one. So *in parameter estimation problems $P(D|I)$ is a normalization constant,* and in Eq. (24) the irrelevant constants can be dropped provided this probability density function is normalized at the end of the calculation. Dropping these constants, Eq. (24) becomes

$$P(\omega|\sigma, D, I) \propto \sigma^{-N} \int dB_1 dB_2 \exp\left\{-\frac{Q}{2\sigma^2}\right\}. \tag{26}$$

If $\sigma$ is actually known there are some other constants which may be dropped. However, in real data, the variance of the noise is frequently not known and must be estimated, so the retained terms will be needed later.

Notice that this equation is essentially Eq. (15); the steps in assigning the uninformative priors were unnecessary, they simply cancel from the problem. People familiar with parameter estimation problems using uninformative priors simply skip the intermediate steps and go straight to Eq. (26). That could not be done here because until one has seen the intermediate steps, one simply does not know that the uninformative prior probabilities cancel. Indeed, in model selection problems even uninformative prior probabilities do not cancel, and improper priors simply cannot be used.

## 4.4 Elimination of the nuisance parameters

Now that all terms in the posterior probability density function have been assigned, Eq. (26) must be integrated with respect to $B_1$ and $B_2$. These are Gaussian integrals and any multivariant Gaussian integral can be done – see Ref. [11] for examples of this. In this problem the integrals are particularly simple because the term involving the product $B_1 B_2$ canceled. This cancellation occurred because uniform sampling was assumed. Had nonuniform sampling been assumed, the integrals would be tractable, but the results would be much more complex analytically – see Ref. [11] for this calculation. To do these integrals, the value of $Q$ in Eq. (26) is replaced by its definition and the order of the terms is rearranged to obtain

$$P(\omega|\sigma, D, I) \propto \sigma^{-N} \exp\left\{-\frac{N\overline{d^2}}{2\sigma^2}\right\}$$

$$\times \int_{-\infty}^{+\infty} dB_1 \exp\left\{-\frac{c}{2\sigma^2}\left[B_1^2 - 2B_1 R(\omega)/c\right]\right\}$$

$$\times \int_{-\infty}^{+\infty} dB_2 \exp\left\{-\frac{s}{2\sigma^2}\left[B_2^2 - 2B_2 I(\omega)/s\right]\right\}.$$

The squares of each of these Gaussian integrals can be completed by adding and subtracting a constant from the exponent:

$$P(\omega|\sigma, D, I) \propto \sigma^{-N} \exp\left\{-\frac{N\overline{d^2} - R(\omega)^2/c - I(\omega)^2/s}{2\sigma^2}\right\}$$

$$\times \int_{-\infty}^{+\infty} dB_1 \exp\left\{-\frac{c}{2\sigma^2}\left[B_1 - R(\omega)/c\right]^2\right\}$$

$$\times \int_{-\infty}^{+\infty} dB_2 \exp\left\{-\frac{s}{2\sigma^2}\left[B_2 - I(\omega)/s\right]^2\right\}.$$

A change of variables

$$x = \sqrt{\frac{c}{2\sigma^2}} \left[ B_1 - \frac{R(\omega)}{c} \right] \qquad \text{and} \qquad dx = \sqrt{\frac{c}{2\sigma^2}} dB_1,$$

and

$$y = \sqrt{\frac{s}{2\sigma^2}} \left[ B_2 - \frac{I(\omega)}{s} \right] \qquad \text{and} \qquad dy = \sqrt{\frac{s}{2\sigma^2}} dB_2$$

reduces the integrals to standard form

$$P(\omega|\sigma, D, I) \propto \frac{\sigma^{-N+2}}{\sqrt{cs}} \exp\left\{ -\frac{N\overline{d^2} - R(\omega)^2/c - I(\omega)^2/s}{2\sigma^2} \right\} \int_{-\infty}^{+\infty} dx e^{-x^2} \int_{-\infty}^{+\infty} dy e^{-y^2}$$

where a factor of 2 was dropped. This numerical factor would be absorbed when the probability density is normalized. These integrals can now be done trivially; each contributing $\sqrt{\pi}$, which may also be dropped. This gives

$$P(\omega|\sigma, D, I) \propto \frac{\sigma^{-N+2}}{\sqrt{cs}} \exp\left\{ -\frac{N\overline{d^2} - R(\omega)^2/c - I(\omega)^2/s}{2\sigma^2} \right\} \tag{27}$$

as the posterior probability of the frequency independent of the amplitudes.

If the variance of the noise is actually known, then there are several constants that will be absorbed when this probability density function is normalized. Assuming the variance of the noise known and dropping these constants, the posterior probability density is

$$P(\omega|\sigma, D, I) \propto \frac{1}{\sqrt{cs}} \exp\left\{ \frac{R(\omega)^2/c + I(\omega)^2/s}{2\sigma^2} \right\}. \tag{28}$$

This is an exact result, and is the posterior probability of a single stationary sinusoidal frequency independent of the amplitude and phase of the sinusoid given the uniformly sampled data, the variance of the noise, and the prior information $I$.

In this form the sufficient statistic, $[R(\omega)^2/c + I(\omega)^2/s]$, is not very recognizable. There is an approximate result that is simpler and worth investigating. The functions $c$ and $s$ were defined earlier, Eqs. (19) and Eq.(20), and these equations explicitly depend on the frequency. Unless the frequency is near zero, the functions $c$ and $s$ are slowly varying and may be approximated by

$$c \approx \frac{N}{2} \qquad \text{and} \qquad s \approx \frac{N}{2}.$$

The posterior probability is approximately

$$P(\omega|\sigma, D, I) \propto \exp\left\{ \frac{C(\omega)}{\sigma^2} \right\}, \tag{29}$$

where $C(\omega)$ is the Schuster periodogram [15] and may be defined as

$$C(\omega) = \frac{2R(\omega)^2 + 2I(\omega)^2}{N} = \frac{2}{N} \left| \sum_{k=1}^{N} d_k \exp\{-i\omega t_k\} \right|^2. \tag{30}$$

The Schuster periodogram is traditionally referred to as a discrete Fourier transform power spectrum. From the standpoint of Bayesian probability, the discrete Fourier transform power spectrum answers a very specific question about single stationary sinusoidal frequency estimation.

These simple results, Eqs. (28,29,30), show why the discrete Fourier transform tends to peak at the location of a frequency when the data are noisy. Namely, the discrete Fourier transform power spectrum is directly related to the probability that a single stationary sinusoidal frequency is present in the data. Additionally, zero padding a time series (i.e. adding zeros at its end to make a longer time series) and then performing the discrete Fourier transform of the padded series, is equivalent to calculating the Schuster periodogram at smaller frequency intervals. If the signal one is analyzing is a single stationary sinusoidal frequency plus noise, then the maximum of the periodogram will be the "best" estimate of the frequency one can make in the absence of additional prior information about it.

The discrete Fourier transform and the Schuster periodogram can now be seen in a entirely new light: the highest peak in the discrete Fourier transform is an optimal frequency estimator for a data set which contains a single stationary sinusoidal frequency in the presence of Gaussian white noise. Stated more carefully, the discrete Fourier transform will give optimal frequency estimates if six conditions are met:

1. The number of data values $N$ is large,
2. There is no constant component in the data,
3. There is no evidence of a low frequency,
4. The data contain only one frequency,
5. The frequency is be stationary
   (i.e. the amplitude and phase are constant), and
6. The noise is white.

If any of these six conditions is not met, the discrete Fourier transform may give misleading or simply incorrect results in light of the more realistic models. Not because the discrete Fourier transform is wrong, but because it is answering what should be regarded as the wrong question. The discrete Fourier transform will always interpret the data in terms of a single stationary sinusoidal frequency model! The effects of violating one or more of these assumptions are shown in [11] and it is demonstrated that when the assumptions are violated, the range of parameter values that are consistent with the data is larger than when these conditions are met.

## 4.5  Eliminating the variance of the noise

Equation (28) is an exact result and is valid for any uniformly sampled data set while Eq. (29) is valid provided there is no evidence of a low frequency, and $N \gg 1$. Both results depend on knowing the variance of the noise. Frequently one has no independent knowledge of the noise. The noise variance $\sigma^2$ then becomes a nuisance parameter. It can be eliminated by first applying the product rule, to Eq. (27) followed by the sum rule Eq. (4). This gives

$$P(\omega|D, I) \propto \int_0^{+\infty} d\sigma\, P(\sigma|I) \frac{\sigma^{-N+2}}{\sqrt{cs}} \exp\left\{ -\frac{N\overline{d^2} - R(\omega)^2/c - I(\omega)^2/s}{2\sigma^2} \right\}$$

as the posterior probability of the frequency independent of the amplitudes and the variance of the noise.

To perform this integral one must supply $P(\sigma|I)$, the prior probability of the variance of the noise $\sigma^2$. The derivation of the prior which indicates "complete ignorance" of a scale parameter was first given by Jeffreys [2] using invariance arguments. It has since been derived by Jaynes [12] using what is called the marginalization paradox, and by Zellner [8] using the principle of maximum entropy. This prior, $1/\sigma$, is called a Jeffreys prior and is the second example of an improper prior. The first example was the uniform prior when extended to an infinite region – see Ref. [11] for more on improper priors. As was mentioned earlier, when using improper priors one should begin by using a proper prior and then pass to the limit of an improper prior; only then can one be absolutely sure that the limits are well-behaved. But in parameter estimation problems using a Gaussian noise prior this limit is uneventful. Multiplying Eq. (27) by the Jeffreys prior gives

$$P(\omega|D, I) \propto \frac{1}{\sqrt{cs}} \int_0^{+\infty} d\sigma \sigma^{-N+1} \exp\left\{-\frac{Q}{\sigma^2}\right\} \tag{31}$$

as the posterior probability of the frequency, where $Q$ is now taken to be

$$Q \equiv \frac{N\overline{d^2} - R(\omega)^2/c - I(\omega)^2/s}{2}.$$

The integral may be transformed into a gamma integral by the following change of variables:

$$\sigma = \sqrt{\frac{Q}{x}} \quad \text{and} \quad d\sigma = -\frac{\sqrt{Q}x^{-\frac{3}{2}}}{2} dx.$$

Using this change of variables Eq. (31) becomes

$$P(\omega|D, I) \propto \frac{Q^{-\frac{N-2}{2}}}{\sqrt{cs}} \int_0^{+\infty} dx\, x^{\frac{N-2}{2}-1} e^{-x}$$

in standard form for a gamma integral and another factor of 2 was dropped. Performing the integral gives

$$P(\omega|D, I) \propto \frac{1}{\sqrt{cs}} \Gamma\left(\frac{N-2}{2}\right) \left[\frac{N\overline{d^2} - R(\omega)^2/c - I(\omega)^2/s}{2}\right]^{\frac{2-N}{2}}$$

where $\Gamma(x)$ is a gamma function of argument $x$. Numerous constants have already been dropped, these constants will cancel when this distribution is normalized. Here the gamma function and the factor of 2 may be dropped to obtain

$$P(\omega|D, I) \propto \frac{1}{\sqrt{cs}} \left[N\overline{d^2} - R(\omega)^2/c + I(\omega)^2/s\right]^{\frac{2-N}{2}} \approx \left[N\overline{d^2} - 2C(\omega)\right]^{\frac{2-N}{2}}. \tag{32}$$

This is called a "Student t-distribution" for historical reasons, although it is expressed here in very nonstandard notation. In this case it is the posterior probability density that a single stationary sinusoidal frequency $\omega$ is present in the data when no prior information about the variance of the noise is available.

## 4.6   Resolving Power

At this point in the calculation the formal derivation of the posterior probability of the frequency is completed. But the problem of estimating the frequency of oscillation is essentially only half complete. In addition to estimating the value of the frequency, one needs to determine the accuracy of the estimate. Of course, this is given by the width of the probability density function, Eqs. (28-29), when the variance of the noise is known and Eq. (32) when the variance of the noise is unknown. But these equations do not aid understanding just how accurately the frequency has been determined; nor do they indicate what is to be gained by using Bayesian probability theory.

To understand what is to be gained from using Bayesian probability theory, an estimate of the width of the posterior probability is derived. The technique used has proven useful in a number of other examples [16,17,18]. To determine the precision of the frequency estimate, $C(\omega)$ is Taylor expanded about the maximum as a function of $\omega$. Equation (29) is used in this derivation so that the results may be directly compared to the discrete Fourier transform power spectrum. This calculation results in a Gaussian approximation of the posterior probability from which the (mean) $\pm$ (standard deviation) estimates of the frequency $\omega$ is obtained. Expanding $C(\omega)$, about the maximum $\hat{\omega}$, one obtains

$$C(\omega) = C(\hat{\omega}) - \frac{b}{2}(\hat{\omega} - \omega)^2 + \cdots$$

where

$$b \equiv - \left. \frac{\partial^2 C(\omega)}{\partial \omega^2} \right|_{\hat{\omega}} > 0. \tag{33}$$

The Gaussian approximation is

$$P(\omega|D,\sigma,I) \simeq \left[ \frac{2b}{\pi\sigma^2} \right]^{\frac{1}{2}} \exp\left\{ -\frac{b(\hat{\omega} - \omega)^2}{2\sigma^2} \right\}$$

from which the (mean) $\pm$ (standard deviation) estimate of the frequency is

$$\omega_{\text{est}} = \hat{\omega} \pm \frac{\sigma}{\sqrt{b}}.$$

The accuracy depends on the curvature of $C(\omega)$ at its peak, not on the height of $C(\omega)$. For example, if the data are composed of a sinusoid plus noise, $e_i$ of standard deviation $\sigma$,

$$d_i = \hat{B}_1 \cos(\hat{\omega} t_i) + \hat{B}_2 \sin(\hat{\omega} t_i) + e_i$$

where $\hat{B}_1$ and $\hat{B}_2$ are the true amplitudes of the signal and $\hat{\omega}$ is the true frequency. A closed form for $C(\omega)$ may be obtained by noting

$$
\begin{aligned}
R(\omega) &= \sum_{i=-T}^{+T} d_i \cos(\omega i) \\
&= \sum_{i=-T}^{+T} \hat{B}_1 \cos(\hat{\omega} i) \cos(\omega i) + \hat{B}_2 \sin(\hat{\omega} i) \cos(\omega i) + e_i \cos(\omega i) \\
&\approx \sum_{i=-T}^{+T} \hat{B}_1 \cos(\hat{\omega} i) \cos(\omega i) + \hat{B}_2 \sin(\hat{\omega} i) \cos(\omega i) \\
&= \frac{\hat{B}_1}{2} \left[ \frac{\sin \frac{N}{2}(\hat{\omega} - \omega)}{\sin \frac{1}{2}(\hat{\omega} - \omega)} + \frac{\sin \frac{N}{2}(\hat{\omega} + \omega)}{\sin \frac{1}{2}(\hat{\omega} + \omega)} \right],
\end{aligned}
$$

$$
\begin{aligned}
I(\omega) &= \sum_{i=-T}^{+T} d_i \sin(\omega i) \\
&= \sum_{i=-T}^{+T} \hat{B}_1 \cos(\hat{\omega} i) \sin(\omega i) + \hat{B}_2 \sin(\hat{\omega} i) \sin(\omega i) + e_i \sin(\omega i) \\
&\approx \sum_{i=-T}^{+T} \hat{B}_1 \cos(\hat{\omega} i) \sin(\omega i) + \hat{B}_2 \sin(\hat{\omega} i) \sin(\omega i) \\
&= \frac{\hat{B}_2}{2} \left[ \frac{\sin \frac{N}{2}(\hat{\omega} - \omega)}{\sin \frac{1}{2}(\hat{\omega} - \omega)} - \frac{\sin \frac{N}{2}(\hat{\omega} + \omega)}{\sin \frac{1}{2}(\hat{\omega} + \omega)} \right].
\end{aligned}
$$

The sums appearing in $R(\omega)$ and $I(\omega)$ were done by first changing the sines and cosines to exponential notation. In this notation the sums are of the form $\sum_{i=1}^{N} x^i$, which may be summed using the fact that $1/(1 - x) = 1 + x + x^2 + \cdots$. In the above it is only the terms involving $\hat{\omega} - \omega$ that are large, so $C(\omega)$ may be approximated by

$$
C(\omega) \approx \frac{(\hat{B}_1^2 + \hat{B}_2^2)}{4N} \left[ \frac{\sin \frac{N}{2}(\hat{\omega} - \omega)}{\sin \frac{1}{2}(\hat{\omega} - \omega)} \right]^2.
$$

So, as found by Jaynes [19], the estimate of the frequency is given by

$$
C(\hat{\omega}) \simeq \frac{N}{4}(\hat{B}_1^2 + \hat{B}_2)
$$

$$
b \simeq \frac{(\hat{B}_1^2 + \hat{B}_2^2)N^3}{48}
$$

$$
(\omega)_{\text{est}} = \hat{\omega} \pm \sigma \sqrt{\frac{48}{N^3(\hat{B}_1^2 + \hat{B}_2^2)}} \tag{34}
$$

which indicates, as intuition would lead us to expect, that the accuracy depends on the signal-to-noise ratio, and quite strongly on how much data are available.

These results can be further compared with experience, but first note that dimensionless units have been used. To convert to ordinary physical units, let the sampling interval be $\Delta t$ seconds, and denote by $f$ the frequency in Hz. Then the total number of cycles in the data record is

$$\frac{\hat{\omega}(N-1)}{2\pi} = (N-1)\hat{f}\Delta t = \hat{f}T$$

where $T = (N-1)\Delta t$ seconds is the duration of our data run. So the conversion of dimensionless $\omega$ to $f$ in physical units is

$$f = \frac{\omega}{2\pi\Delta t} \text{ Hz.}$$

The frequency estimate Eq. (34) becomes

$$f_{\text{est}} = \hat{f} \pm \delta f \text{ Hz}$$

where now, not distinguishing between $N$ and $(N-1)$,

$$\delta f = \frac{\sigma}{2\pi T}\sqrt{\frac{48}{N(\hat{B}_1^2 + \hat{B}_2^2)}} = \frac{1.1\sigma}{T\sqrt{N(\hat{B}_1^2 + \hat{B}_2^2)}} \text{ Hz.} \tag{35}$$

From this one can see that the two most important factors for improving resolution are how long one samples (the $T$ dependence) and the signal-to-noise ratio. The number of data values may be doubled in one of two ways, by doubling the total sampling time or by doubling the sampling rate. However, Eq. (35) clearly indicates that doubling the sampling time is to be preferred. This indicates that data values near the beginning and end of a record are most important for frequency estimation, which is in agreement with intuitive common sense.

Consider the following example: Suppose the RMS signal-to-noise ratio (i.e. ratio of RMS signal to RMS noise $\equiv$ S/N) of the data is S/N $= [(\hat{B}_1^2 + \hat{B}_2^2)/2\sigma^2]^{\frac{1}{2}} = 1$, and data is taken every $\Delta t = 10^{-3}$ sec. for $T = 1$ second, thus getting $N = 1000$ data points, then the theoretical accuracy for determining the frequency of a single, steady sinusoid is

$$\delta f = \frac{1.1}{\sqrt{2000}} = 0.025 \text{ Hz} \tag{36}$$

while the Nyquist frequency for the onset of aliasing is $f_N = (2\Delta t)^{-1} = 500\text{Hz}$ – greater by a factor of 20,000.

To some, this result will be quite startling. Indeed, had the periodogram itself been considered to be a spectrum estimator, one would instead calculated the width of its central peak. A noiseless sinusoid of frequency $\hat{\omega}$ would have a periodogram proportional to

$$C(\omega) \propto \frac{\sin^2\{N(\omega - \hat{\omega})/2\}}{\sin^2\{(\omega - \hat{\omega})/2\}}.$$

The half-width at half amplitude is given by $|N(\hat{\omega}-\omega)/2| = \pi/4$ or $\delta\omega = \pi/2N$. Converting to physical units, the periodogram will have a width of about

$$\delta f = \frac{1}{4N\Delta t} = \frac{1}{4T} = 0.25 \text{ Hz} \tag{37}$$

just ten times greater than the value Eq. (36) indicated by probability theory. This factor of ten is the amount of narrowing produced by the exponential peaking of the periodogram in Eq. (29), even for peak signal-to-RMS noise ratio of one.

But some would consider even the result of Eq. (37) to be a little overoptimistic. The famous Rayleigh criterion [20] for resolving power of an optical instrument supposes that the minimum resolvable frequency difference corresponds to the peak of the periodogram of one sinusoid coming at the first zero of the periodogram of the second. This is twice Eq. (37):

$$\delta f_{\text{Rayleigh}} = \frac{1}{2T} = 0.5 \text{ Hz.} \tag{38}$$

There is a widely believed "folk-theorem" among theoreticians without laboratory experience which seems to confuse the Rayleigh limit with the Heisenberg uncertainty principle, and holds that Eq. (38) is a fundamental irreducible limit of resolution. Of course there is no such theorem, and workers in high resolution NMR have been routinely determining line positions to an accuracy that surpasses the Rayleigh limit by an order of magnitude for thirty years.

The misconception is perhaps strengthened by the curious coincidence that Eq. (38) is also the minimum half-width that can be achieved by a Blackman-Tukey spectrum analysis [21] (even at infinite signal-to-noise ratio) because the "Hanning window" tapering function that is applied to the data to suppress side-lobes (the secondary maxima of $[\sin(x)/x]^2$) just doubles the width of the periodogram. Since the Blackman-Tukey method has been used widely by economists, oceanographers, geophysicists, and engineers for many years, it has taken on the appearance of an optimum procedure.

According to E.T. Jaynes [22], Tukey himself acknowledged that his method fails to give optimum resolution, but held this to be of no importance because "real time series do not have sharp lines." Nevertheless, this misconception is so strongly held that there have been attacks on the claims of Bayesian/Maximum Entropy spectrum analysts to be able to achieve results like Eq. (36) when the assumed conditions are met. Some have tried to put such results in the same category with circle squaring and perpetual motion machines. Therefore, we want to digress to explain in very elementary physical terms why it is the Bayesian result, Eq. (35), that does correspond to what a skilled experimentalist can achieve.

Suppose first that our only data analysis tool is our own eyes looking at a plot of the raw data of duration $T = 1$ sec., and that the unknown frequency $f$ in Eq. (36) is 100Hz. Now anyone who has looked at a record of a sinusoid and equal amplitude wide-band noise, knows that the cycles are quite visible to the eye. One can count the total number of cycles in the record confidently (using interpolation to help us over the doubtful regions) and will feel quite sure that the count is not in error by even one cycle. Therefore by raw "eyeballing" of the data and counting the cycles, one can achieve an accuracy of

$$\delta f \simeq \frac{1}{T} = 1 \text{ Hz.}$$

But in fact, if one draws the sine wave that seems to fit the data best, one can make a quite reliable estimate of how many quarter-cycles were in the data, and thus achieve

$$\delta f \simeq \frac{1}{4T} = 0.25 \text{ Hz}$$

corresponding just to the periodogram width, Eq. (37).

Then the use of probability theory needs to surpass the naked eye by another factor of ten to achieve the Bayesian width, Eq. (36). What probability theory does is essentially average out the noise in a way that the naked eye cannot do. If some measurement is repeated $N$ times, any randomly varying component of the data will be suppressed relative to the systematic component by a factor of $N^{-\frac{1}{2}}$, the standard rule.

In the case considered, there were $N = 1000$ data points. If they were all independent measurements of the same quantity with the same accuracy, this would suppress the noise by about a factor of 30. But in this case; not all measurements are equally cogent for estimating the frequency. Data points in the middle of the record contribute very little to the result; data points near the ends are highly relevant for determining the frequency, so the effective number of observations is less than 1000. The probability analysis leading to Eq. (36) indicates that the "effective number of observations" is only about $N/10 = 100$; thus the Bayesian width Eq. (36) that results from the exponential peaking of the periodogram now appears to be, if anything, somewhat conservative.

Indeed, that is what Bayesian analysis always does when smooth, uninformative priors for the parameters are used, because then probability theory makes allowance for all possible values that they might have. As noted before, if cogent prior information about $\omega$ was available and it was expressed in a narrower prior, still better results would be obtained; but they would not be much better unless the prior range became comparable to the width of the probability of the data.

## 4.7  Wolf's Relative Sunspot Numbers

Wolf's relative sunspot numbers [24] are, perhaps, the most analyzed set of data in all of spectrum analysis. As Marple [23] explains in more detail, these numbers (defined as: $W = k[10g + f]$, where $g$ is the number of sunspot groups, $f$ is the number of individual sunspots, and $k$ is used to reduce different telescopes to a common scale) have been collected on a yearly basis since 1700 and on a monthly basis since 1748. The exact physical mechanism which generates the sunspots is unknown, and no complete theory exists. Different analyses of these numbers have been published more or less regularly since their tabulation began. Here the sunspot numbers will be analyzed with the stationary sinusoidal frequency model, even though this model is too simple to be realistic for these numbers.

The time series from 1700 to 1985 has been plotted in Fig. 1(A). A cursory examination of this time series does indeed show a cyclic variation with a period of about 11 years. The square of the discrete Fourier transform is a continuous function of frequency and is proportional to the Schuster periodogram of the data [Fig. 1(B), continuous curve]. The frequencies could be restricted to the Nyquist [25,26] steps [Fig.1(B) open circles]; it is a theorem that the discrete Fourier transform on those points contains all the information that is in the periodogram, but one sees that the information is much more apparent to

Figure 1: Wolf's Relative Sunspot Numbers

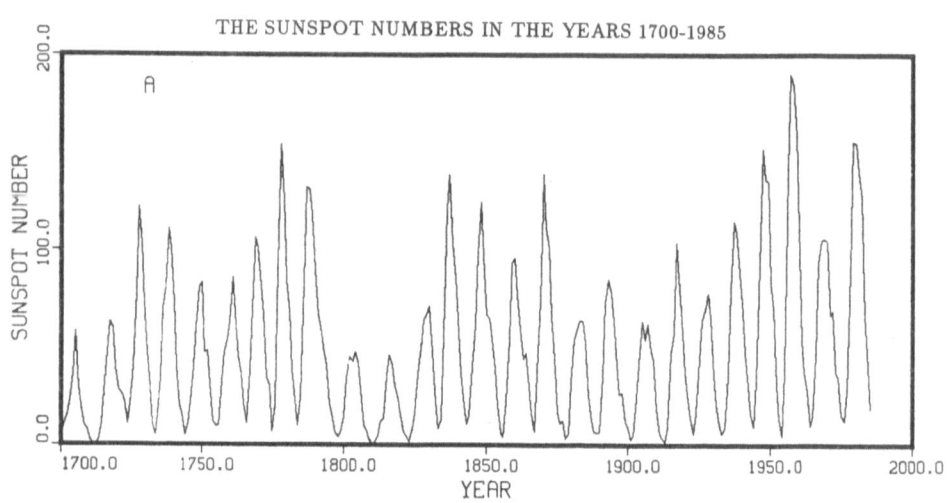

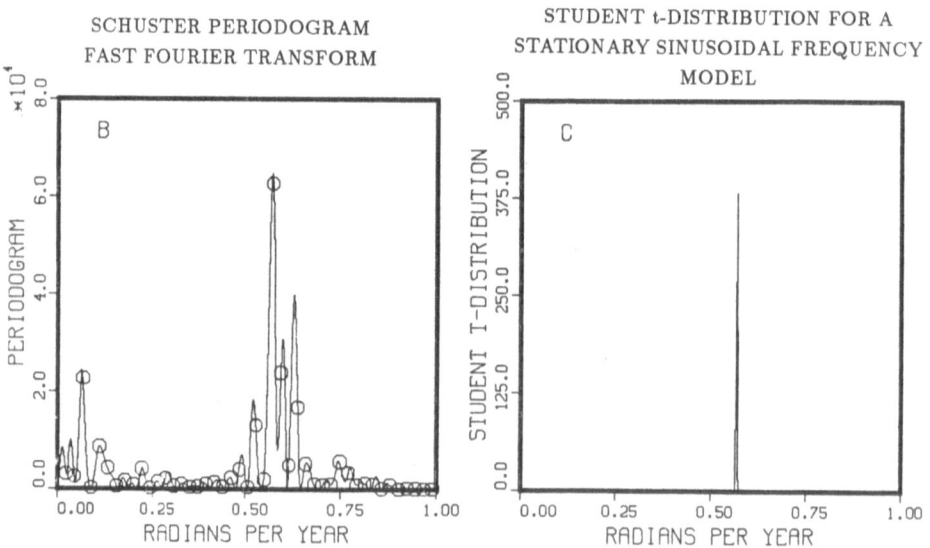

Wolf's relative sunspot numbers (A) have been collected on a yearly basis since 1700. The periodogram (B) contains evidence of several complex phenomena. In spite of this the single frequency model posterior probability density (C) picks out the 11.04 year cycle to an estimated accuracy of ±10 days.

the eye in the continuous periodogram. The Schuster periodogram or the discrete Fourier transform clearly shows a maximum with period near 11 years.

The "Student t-distribution," Eq. (32), was computed and displayed in Fig. 1(C). Now, because of the processing in Eq. (32), all details in the periodogram have been suppressed and only the peak at 11 years remains.

The accuracy of the frequency estimate was determined as follows: the maximum of the "Student t-distribution" was located, an integral was performed about this maximum in a symmetric interval, and the enclosed probability was recorded at a number of points. This gives a period of 11.04 years with the accuracy shown as follows:

| period in years | | accuracy in years | probability enclosed |
|---|---|---|---|
| 11.04 | ± | 0.015 | 0.62 |
| | ± | 0.020 | 0.75 |
| | ± | 0.026 | 0.90 |

According to this, there is not one chance in 10 that the true period differs from 11.04 years by more than 10 days. At first glance, this appears too good to be true. But what does raw "eyeballing" of the data give? In 285 years, there are about $285/11 \approx 26$ cycles. If cycles are counted to an accuracy of $\pm 1/4$ cycle, the period estimate would be about

$$(f)_{est} = 11 \text{ years} \pm 39 \text{ days}.$$

Probability averaging of the noise, as discussed above Eq. (34), would reduce this uncertainty by about a factor of $\sqrt{285/10} = 5.3$, giving

$$(f)_{est} = 11 \text{ years} \pm 7.3 \text{ days}, \quad \text{or} \quad (f)_{est} = 11 \pm 0.02 \text{ years}$$

which corresponds nicely with the result of the probability analysis.

# 5 Model Selection

But these results come from analyzing the data by a model which assumes there is nothing present but a single sinusoid plus noise. Probability theory, given this model, is obliged to consider everything in the data that cannot be fit to a single sinusoid to be noise. But a glance at the data shows clearly that there is more present than our model assumed. Therefore, probability theory must estimate the noise to be quite large.

This suggests that a more realistic model, which allows the "signal" to have more structure, might do better. Such a model can be fit to the data more accurately; therefore it will estimate the noise to be smaller. This should permit a still better period estimate! But caution forces itself upon us; by adding more and more components to the model the data can always be fit more and more accurately. It is absurd to suppose that by mere proliferation of a model arbitrarily accurate estimates of a parameter can be extracted. There must be a point of diminishing returns – or indeed of negative returns – beyond which we are deceiving ourselves.

It is very important to understand the following point. In parameter estimation problems, probability theory always gives us the estimates that are justified by the information *that was actually used* in the calculation. The parameter estimation procedures outlined in this tutorial assume that information to be absolutely true! If one puts false information into a parameter estimation calculation, then probability theory will give optimal estimates based on false information. These could be very misleading.

In real experimental data analysis, one is hardly ever sure of the true model; at best there may be a number of competing candidates. When one has a series of candidate models, probability theory can be used to rank various candidates. Therefore, it can answer questions of the form "Given a set $S \equiv \{f_1, \cdots, f_s\}$ of $s$ possible models, which model is most probable in view of the data and all of one's prior information?" Therefore, probability theory can warn one that the hypothesis may not be true; but one must ask, probability theory will not volunteer the information. To understand how to ask, consider the data $D \equiv \{d_1, \cdots, d_N\}$ sampled at discrete times $\{t_1, \cdots, t_N\}$. If the true signal in the data is $f_j$, then the data may be modeled as

$$d_i = f_j(t_i) + e_i$$

where $f_j(t_i)$ is the $j$th member of the set $S$ at time $t_i$. Suppose the model signal may be written

$$f_j(t, \Theta) = \sum_{k=1}^{m} B_k G_k(t, \Theta)$$

where $G_k(t, \Theta)$ is one of $m$ signal functions with amplitude $B_k$, $\mathbf{B} \equiv \{B_1, \cdots, B_m\}$, and $\Theta$ is a set of $r$ nonlinear parameters defined as $\Theta \equiv \{\Theta_1, \cdots, \Theta_r\}$. The parameters $\mathbf{B}$ and $\Theta$ are assumed different in every model $f_j$.

From Bayes' theorem, Eq. (2), the probability of model $f_j$, conditional on the data $D$ and the prior information $I$ is given by

$$P(f_j|D, I) = \frac{P(f_j|I)P(D|f_j, I)}{P(D|I)} \qquad (1 \le j \le s) \tag{39}$$

where $P(f_j|D, I)$ is the posterior probability of model $f_j$ given the data $D$ and the prior information $I$. It is this term that one wants to calculate for the set of models $S$. To calculate it, Bayes' theorem indicates that three terms must be computed. The first, $P(f_j|I)$, is the probability of model $f_j$ given only the information $I$. This term represents ones state of knowledge about the models before obtaining the data $D$. The second, $P(D|f_j, I)$, is the global-likelihood of the data given model $f_j$ and the prior information $I$. This term represents how well the data fit the model. The third, $P(D|I)$, is the probability of the data given only the prior information $I$ and is a normalization constant given by

$$P(D|I) = \sum_{k=1}^{s} P(f_k|I)P(D|f_k, I).$$

The global-likelihood of the data, $P(D|f_j, I)$, is obtained from the joint posterior probability of the data and the parameters,

$$P(D|f_j, I) = \int d\mathbf{B} d\Theta \, P(D, \mathbf{B}, \Theta|f_j, I).$$

In the previous example, only the amplitudes $B_1$ and $B_2$ were eliminated from consideration. In model selection problems, both the amplitudes $\mathbf{B}$ and the nonlinear $\Theta$ parameters must be eliminated from consideration.

The product rule of conditional probability theory may be used to factor $P(D, \mathbf{B}, \Theta|f_j, I)$ to obtain

$$P(D|f_j, I) = \int d\Theta\, P(\Theta|f_j, I) \left[ \int d\mathbf{B}\, P(\mathbf{B}|\Theta, f_j, I) P(D|\mathbf{B}, \Theta, f_j, I) \right].$$

The term in brackets is just a generalized version of the parameter estimation problem with three changes: 1) all numerical factors must be kept, 2) all parameters are to be considered nuisance parameters, and 3) fully normalized prior probability density functions must be used. In the previous example improper priors were used (prior probabilities that cannot be normalized) because little prior information about the parameters was assumed. Any uninformative prior will look like a constant over the range of values where the likelihood of the parameters is sharply peaked. The uninformative prior would then cancel when the distributions were normalized. The parameters estimated by this procedure are the maximum-likelihood estimates and, in the case of a Gaussian noise prior, the least-squares estimates. However, improper priors cannot be used in Eq. (39) because they do not cancel. This is easily seen. For example, if the prior is a bounded uniform prior, then as the bounds are allowed to go to infinity, the bounds will not cancel in Eq. (39) – unless every model contains exactly the same prior. Thus, the model with the larger number of parameters would automatically be excluded. To perform the indicated calculation see Ref. [11] for details and Ref.[27] for several examples of its use.

# 6 Conclusions

In this tutorial the rules of Bayesian probability theory, the procedures for assigning probabilities, and a nontrivial example have been given. This example demonstrates the power of Bayesian probability theory and illustrates how to apply the procedures to real problems in parameter estimation. The relation between parameter and model selection has been discussed and it has been shown that the parameter estimation procedures are just one step in the more general model selection problem.

# Acknowledgments

This work supported by NIH grant GM-30331, J. J. H. Ackerman principal investigator. The encouragement of Professor J. J. H. Ackerman is greatly appreciated as are the editorial comments of Dr. C. R. Smith and extensive conversations with Professor E. T. Jaynes.

# References

[1] Laplace, P. S., *A Philosophical Essay on Probabilities*, unabridged and unaltered reprint of Truscott and Emory translation, Dover Publications, Inc., New York, 1951, original publication data 1814.

[2] Jeffreys, H., *Theory of Probability*, Oxford University Press, London, 1939; Later editions, 1948, 1961.

[3] Jaynes, E. T., "How Does the Brain do Plausible Reasoning?" unpublished Stanford University Microwave Laboratory Report No. 421 (1957); reprinted in *Maximum-Entropy and Bayesian Methods in Science and Engineering* 1, pp. 1-24, G. J. Erickson and C. R. Smith *Eds.*, 1988.

[4] Shannon, C. E., "A Mathematical Theory of Communication," *Bell Syst. Tech. J.* 27, pp. 379-423 (1948).

[5] Abel, N. H., *Crelle's Jour.*, Bd. 1 (1826).

[6] Cox, R. T., "Probability, Frequency, and Reasonable Expectations," *Amer. J. Phys.* 14, pp. 1-13 (1946).

[7] Tribus, M., *Rational Descriptions, Decisions and Designs*, Pergamon Press, Oxford, 1969.

[8] Zellner, A., *An Introduction to Bayesian Inference in Econometrics*, John Wiley and Sons, New York, 1971; Second edition 1987.

[9] Bayes, Rev. T., "An Essay Toward Solving a Problem in the Doctrine of Chances," *Philos. Trans. R. Soc. London* 53, pp. 370-418 (1763); reprinted in *Biometrika* 45, pp. 293-315 (1958), and *Facsimiles of Two Papers by Bayes*, with commentary by W. Edwards Deming, New York, Hafner, 1963.

[10] Jaynes, E. T., "Prior Probabilities," *IEEE Transactions on Systems Science and Cybernetics*, SSC-4, pp. 227-241 (1968); reprinted in [13].

[11] Bretthorst, G. Larry, "Bayesian Spectrum Analysis and Parameter Estimation," in *Lecture Notes in Statistics* 48, Springer-Verlag, New York, New York, 1988.

[12] Jaynes, E. T., "Marginalization and Prior Probabilities," in *Bayesian Analysis in Econometrics and Statistics*, A. Zellner, *ed.*, North-Holland Publishing Company, Amsterdam, 1980; reprinted in [13].

[13] Jaynes, E. T., *Papers on Probability, Statistics and Statistical Physics*, a reprint collection, D. Reidel, Dordrecht the Netherlands, 1983; second edition Kluwer Academic Publishers, Dordrecht the Netherlands, 1989.

[14] Jaynes, E. T., "Where Do We Stand On Maximum Entropy?" in *The Maximum Entropy Formalism*, R. D. Levine and M. Tribus *Eds.*, pp. 15-118, Cambridge: MIT Press, 1978; Reprinted in [13].

[15] Schuster, A., "The Periodogram and its Optical Analogy," *Proc. R. Soc. London* 77, pp. 136 (1905).

[16] Bretthorst, G. Larry, *Bayesian Spectrum Analysis and Parameter Estimation*, Ph.D. thesis, Washington University, St. Louis, MO., available from University Microfilms Inc., Ann Arbor Mich. 1987; an "Excerpts from Bayesian Spectrum Analysis and Parameter Estimation," is printed in *Maximum-Entropy and Bayesian Methods in Science and Engineering* 1, pp. 75-145, G. J. Erickson and C. R. Smith *Eds.*, Kluwer Academic Publishers, Dordrecht the Netherlands, 1988.

[17] Bretthorst, G. Larry, "Bayesian Spectrum Analysis on Quadrature NMR Data with Noise Correlations," *Maximum Entropy and Bayesian Methods*, pp. 261-274, J. Skilling *ed.*, Kluwer Academic Publishers, Dordrecht the Netherlands, 1989.

[18] Bretthorst, G. Larry and C. Ray Smith, "Bayesian Analysis of Signals from Closely-Spaced Objects," *Infrared Systems and Components III*, pp 93.104, Robert L. Caswell *ed.*, *SPIE* Vol. **1050,** 1989.

[19] Jaynes, E. T., "Bayesian Spectrum and Chirp Analysis," in *Maximum Entropy and Bayesian Spectral Analysis and Estimation Problems*, pp. 1-37, C. Ray Smith and G. J. Erickson, *Eds.*, Kluwer Academic Publishers, Dordrecht the Netherlands, 1987.

[20] Lord Rayleigh, *Philos. Mag.* **5,** p. 261 (1879).

[21] Blackman, R. B. and J. W. Tukey, *The Measurement of Power Spectra*, Dover Publications, Inc., New York, 1959.

[22] Tukey, J. W., several conversations with E. T. Jaynes, in the period 1980-1983.

[23] Marple, S. L., *Digital Spectral Analysis with Applications*, Prentice-Hall, New Jersey, 1987.

[24] Waldmeier, M., *The Sunspot Activity in the Years 1610-1960*, Schulthes, Zurich, 1961.

[25] Nyquist, H., "Certain Topics in Telegraph Transmission Theory," *Trans. AIEE*, pp. 617 (1928).

[26] Nyquist, H., "Certain Factors Affecting Telegraph Speed," *Bell Sys. Tech. J.* **3,** pp. 324 (1924).

[27] Bretthorst, G. Larry, "Bayesian Model Selection: Examples Relevant to NMR," *Maximum Entropy and Bayesian Methods*, J. Skilling *ed.*, pp. 377-388, Kluwer Academic Publishers, Dordrecht the Netherlands, 1989.

# FROM LAPLACE TO SUPERNOVA SN 1987A: BAYESIAN INFERENCE IN ASTROPHYSICS

T. J. LOREDO

*Dept. of Astronomy and Astrophysics*
*University of Chicago*
*5640 South Ellis Ave.*
*Chicago, IL 60637*

ABSTRACT. The Bayesian approach to probability theory is presented as an alternative to the currently used long-run relative frequency approach, which does not offer clear, compelling criteria for the design of statistical methods. Bayesian probability theory offers unique and demonstrably optimal solutions to well-posed statistical problems, and is historically the original approach to statistics. The reasons for earlier rejection of Bayesian methods are discussed, and it is noted that the work of Cox, Jaynes, and others answers earlier objections, giving Bayesian inference a firm logical and mathematical foundation as the correct mathematical language for quantifying uncertainty. The Bayesian approaches to parameter estimation and model comparison are outlined and illustrated by application to a simple problem based on the gaussian distribution. As further illustrations of the Bayesian paradigm, Bayesian solutions to two interesting astrophysical problems are outlined: the measurement of weak signals in a strong background, and the analysis of the neutrinos detected from supernova SN 1987A. A brief bibliography of astrophysically interesting applications of Bayesian inference is provided.

## Contents

*P. F. Fougère (ed.), Maximum Entropy and Bayesian Methods*, 81–142.

---

## 1. Introduction

Few astrophysicists have expertise in the use of advanced statistical methods. The reason for this is not difficult to find, for examination of the use of statistics in the astrophysical literature reveals the lack of a clear rationale for the choice and use of advanced methods.

Unfortunately, this problem is not intrinsic to astrophysics, but has been inherited from statistics itself. To an outsider, statistics can have the appearance of being merely an "industry" where statistical methods are invented without a clear design rationale, and then evaluated by mass-producing simulated data sets and analyzing the average, long-run behavior of the methods. As a result, there often are several methods available for addressing a particular statistical question, each giving a somewhat different answer from the others, with no compelling criteria for choosing among them. Further, the reliance on long-run behavior for the evaluation of statistical methods makes the connection between textbook statistical inferences and the real life problems of scientists seem rather tenuous. This problem can be particularly acute in astrophysics, where the notion of a statistical ensemble is often extremely contrived and can hence seem irrelevant. The gamma-ray astronomer does not want to know how an observation of a gamma-ray burst would compare with thousands of other observations of that burst; the burst is a unique event which can be observed only once, and the astronomer wants to know what confidence should be placed in conclusions drawn from the one data set that actually exists. Similarly, the cosmologist is not comforted to learn that his statement about the large scale structure of the Universe would be correct 95% of the time were he to make similar observations in each of thousands of universes "like" our own. He wants to know how much confidence should be placed in his statement about our particular Universe, the only one we know exists.

Given these difficulties, it is no wonder that many scientists are dubious about results obtained using any but the simplest statistical methods, and no wonder that some openly assert, "If it takes statistics to show it, I don't believe it." It is no wonder, but it is unfortunate. Among all scientists, it is perhaps most unfortunate for the astronomer, who studies objects inaccessible to direct manipulation in a laboratory, and whose inferences are thus fraught with uncertainty, uncertainty crying out for quantification.

It is the thesis of this paper that this situation is unnecessary, that there exists a simple mathematical language for the quantification of uncertainty, that this language produces *unique* answers to well-posed problems, and that its answers are demonstrably optimal by rather simple, compelling desiderata. This language is *Bayesian Probability Theory* (BPT), and far from being a new approach to statistics, it is the *original* approach to statistics, predating the current long-run performance approach by a century. Ironically, it was originally developed by an astrophysicist: Laplace used such methods to analyze astronomical observations for comparison with his famous calculations in celestial mechanics, and developed them at length in his *Théorie Analytique des Probabilités* (Laplace 1812). Heightening the irony, many later developments of Laplace's theory also came from mathematicians and physicists analyzing astronomical problems (see Feigelson 1989 for a brief review). More recently, a full development of Laplace's theory, including the solutions to dozens of practical statistical problems, was published by Sir Harold Jeffreys while a professor of astronomy at Cambridge University in the chair previously held by Eddington (Jeffreys 1939).*

The Bayesian approach to probable inference is remarkably straightforward and intuitive. In fact, it is most likely what the reader already believes probability theory is, since the

---

* This work remains little known among astronomers. A recent obituary of Jeffreys (Runcorn 1989) fails even to mention this important work, described by the prominent statistician I. J. Good as being "of greater importance for the philosophy of science, and obviously of greater immediate practical importance, than nearly all the books on probability written by professional philosophers *lumped together*" (Good 1980).

intuitive understanding physicists have of the more common statistical notions (such as $1\sigma$ error bars) is often identical to the Bayesian interpretation of the notion, and far from the rigorous "classical" or "orthodox" interpretation. But the precise quantification of such intuitive notions in Bayesian inference allows one to extend them into the realm where subtleties often leave our intuition—and classical statistics—at a loss. In such cases, the Bayesian solution often appears beautifully intuitive *a posteriori*, our intuition having been trained and sharpened by probability theory.

The plan of this paper is as follows. First, we will have to discuss exactly what one means by the word "probability." This may sound like a topic for philosophers, but the whole course of probability theory is set by what one decides the conceptual playground of the theory is, so the discussion is crucial. Next we will see that the Bayesian notion of probability, which appears at first to be too vague for quantitative analysis, in fact allows one to develop a complete mathematical language for dealing with uncertainty that is both simpler than standard statistics and more general than it, including much of it as a special case. Following this, we will learn how to use the theory to address two classes of problems of particular interest to scientists: the estimation of parameters in a model, and the assessment of competing models. The basic ideas will be illustrated by comparing the Bayesian approach to measuring a signal in Gaussian noise with the standard long term performance approach.

Once the general theory is set up, we will outline its application to two real astrophysical problems: the measurement of a weak photon counting signal in a (possibly strong) background, and the analysis of the neutrinos detected from the supernova SN 1987A. The failure of orthodox methods to guide astronomers to a single, optimal solution to a problem as simple and fundamental as the measurement of a weak signal is a powerful indication of the poverty of such methods. The Bayesian solution to this problem is so simple that it is reduced from a research problem (Hearn 1969; O'Mongain 1973; Cherry *et al.* 1980) to an undergraduate homework problem.

This is a lot of ground to cover in the pages of a single paper, and much of it will be covered unevenly and incompletely. Hopefully, the reader will be induced to study the cited references where the theory is developed both more eloquently and more fully. To this end, the concluding section not only summarizes the contents of this work, but also points the reader to Bayesian literature covering several topics of particular interest to astrophysicists, including Bayesian spectrum analysis and the Bayesian approach to inverse problems.

## 2. What is Probability?

### 2.1 TWO DEFINITIONS OF PROBABILITY

Traditionally, probability is identified with the *long-run relative frequency of occurrence of an event*, either in a sequence of repeated experiments or in an ensemble of "identically prepared" systems. We will refer to this view of probability as the "frequentist" view; it is also called the "classical," "orthodox," or "sampling theory" view. It is the basis for the statistical procedures currently in use in the physical sciences.

Bayesian probability theory is founded on a much more general definition of probability. In BPT, probability is regarded as a real-number-valued measure of the plausibility of a proposition when incomplete knowledge does not allow us to establish its truth or falsehood with certainty. The measure is taken on a scale where 1 represents certainty of the truth of the proposition, and 0 represents certainty of its falsehood. This definition has an obvious connection with the colloquial use of the word "probability." In fact, Laplace viewed probability theory as simply "common sense reduced to calculation" (Laplace 1812, 1951). For Bayesians, then, probability theory is a kind of "quantitative epistemology", a numerical encoding of one's state of knowledge.

Few works on statistics for the physical sciences bother to note that there is controversy over so fundamental a notion as the definition of probability. In fact, two of the most influential works introducing statistical methods to physical scientists neither define probability nor discuss the complicated frequentist derivation and interpretation of concepts as simple and as widely used as the $1\sigma$ confidence region (Bevington 1969; Press *et al.* 1986). Other texts, noting that there is some controversy over the definition, adopt the frequency definition, asserting that there is little practical difference between the approaches (Eadie *et al.* 1971; Martin 1971; Mendenhall *et al.* 1981).

Of course, it is futile to argue over which is the "correct" definition of probability. The different definitions merely reflect different choices for the types of problems the theory can address, and it seems possible that either definition could lead to a consistent mathematical theory. But though this is true, it leaves open the question of which approach is more useful or appropriate, or which approach addresses the types of problems actually encountered by scientists in the most straightforward manner.

In fact, it will not take much deep thought for us to see that the Bayesian approach to probability theory is both more general than the frequentist approach, and much more closely related to how we intuitively reason in the presence of uncertainty. We will also find that Bayesian solutions of many important statistical problems are significantly simpler to derive than their frequentist counterparts. But if this is true, and if, as noted earlier, the Bayesian approach is the historically older approach, why was the frequentist definition adopted, and why has it dominated statistics throughout this century? To address these questions, our discussion of the contrast between Bayesian and frequentist reasoning will be quasi-historical. More extensive discussions of the history of probability theory and the Bayesian/frequentist controversy are available in Rényi (1972, Appendices III and IV), Jaynes (1978, 1986a), and Grandy (1987, Ch. 2).

## 2.2  SOME EARLY HISTORY: BERNOULLI, BAYES, AND LAPLACE

2.2.1. *Frequency from Probability.* Though statistical problems, particularly those related to gambling and games of chance, have entertained the minds of thinkers since ancient times, the first formal account of the calculation of probabilities is Bernoulli's *Ars Conjectandi* ("The Art of Conjecture", Bernoulli 1713). Bernoulli was what we would today term a Bayesian, holding that probability is "the degree of certainty, which is to the certainty as the part to the whole." He clearly recognized the distinction between probability and frequency, deriving the relationship between probability of occurrence in a single trial and frequency of occurrence in a large number of independent trials now known as Bernoulli's theorem, or the law of large numbers.

Bernoulli's theorem tells us that, if the probability of obtaining a particular outcome in a single trial is known to be $p$, the relative frequency of occurrence of that outcome in a large number of trials converges to $p$.

Also of interest to Bernoulli was the inverted version of this problem: supposing the probability of occurrence in a single trial is unknown, what does the observation of the outcome $n$ times in $N$ repeated, independent trials tell us about the value of the probability? Bernoulli never solved this problem, but his interest in it further emphasizes the distinction made by him and his contemporaries between probability ("degree of certainty") and frequency.

2.2.2. *Probability from Frequency: Bayes' Theorem.* A solution to Bernoulli's problem was published posthumously by the Rev. Thomas Bayes (1763). It was soon rediscovered by Laplace, in a much more general form, and this general form is known as *Bayes' Theorem* (BT). It can be derived very simply as follows.

The mathematical content of the probability theory of Bernoulli, Bayes, and Laplace was specified by taking as *axioms* the familiar sum rule,

$$p(A \mid C) + p(\overline{A} \mid C) = 1, \tag{1}$$

and product rule,

$$p(AB \mid C) = p(A \mid BC)p(B \mid C). \tag{2}$$

Here the symbols, $A, B, C$, represent propositions, $\overline{A}$ represents the denial of $A$ (read "not $A$"), and $AB$ means "$A$ and $B$," a proposition that is true only if $A$ and $B$ are both true. The vertical bar is the conditional symbol, indicating what information is assumed for the assignment of a probability. We must always assume something about the phenomenon in question, and it is good practice to put these assumptions out in the open, to the right of the bar. Failure to do this can lead to apparent paradoxes when two problems with different background assumptions are compared; see Jaynes (1980a) for an educational example.

All legitimate relationships between probabilities can be derived from equations (1) and (2). For example, we may want to know the probability that either or both of two propositions is true. Denoting this by $p(A + B \mid C)$, it can be easily shown (Jaynes 1958, 1990b; Grandy 1987) that the axioms imply

$$p(A + B \mid C) = p(A \mid C) + p(B \mid C) - p(AB \mid C). \tag{3}$$

In fact, we can take this in place of (1) as one of our axioms if we wish. If $A$ and $B$ are exclusive propositions, so that only one of them may be true, $p(AB \mid C) = 0$, and equation (3) becomes the familiar sum rule for exclusive propositions: $p(A + B \mid C) = p(A \mid C) + p(B \mid C)$.

It is important to keep in mind that the arguments for a probability symbol are propositions, not numbers, and that the operations inside the parentheses are logical operations. The symbols for logical operations are here chosen to make the axioms mnemonic. Thus logical "and," represented by juxtaposition in the argument list, leads to multiplication of probabilities. Similarly, logical "or," indicated by a "+" in the argument list, leads to sums of probabilities. But the meanings of juxtaposition and the "+" sign differ inside and outside of the probability symbols.

The propositions $AB$ and $BA$ are obviously identical: the ordering of the logical "and" operation is irrelevant. Thus equation (2) implies that $p(A \mid BC)p(B \mid C) = p(B \mid AC)p(A \mid C)$. Solving for $p(A \mid BC)$, we find

$$p(A \mid BC) = p(A \mid C)\frac{p(B \mid AC)}{p(B \mid C)}. \tag{4}$$

This is Bayes' theorem; it is a trivial consequence of axiom (2).

Bayesian probability theory is so-called because of its wide use of BT to assess hypotheses, though of course Bayesians use all of probability theory, not just BT. To see how BT can help us assess an hypothesis, make the following choices for the propositions $A$, $B$, and $C$. Let $A = H$, an hypothesis we want to assess. Let $B = D$, some data we have that is relevant to the hypothesis. Let $C = I$, some background information we have indicating the way in which $H$ and $D$ are related, and also specifying any alternatives we may have to $H$.* With these propositions, BT reads

$$p(H \mid DI) = p(H \mid I)\frac{p(D \mid HI)}{p(D \mid I)}. \tag{5}$$

Thinking about this a little, we see that BT represents *learning*. Specifically, it tells us how to adjust our plausibility assessments when our state of knowledge regarding an hypothesis changes through the acquisition of data. It tells us that our "after data" or *posterior probability* of $H$ is obtained by multiplying our "before data" or *prior probability* $p(H \mid I)$ by the probability of the data assuming the truth of the hypothesis, $p(D \mid HI)$, and dividing it by the probability that we would have seen the data anyway, $p(D \mid I)$. The factor $p(D \mid HI)$ is called the *sampling distribution* when considered as a function of the data, or the *likelihood function*, $\mathcal{L}(H)$, when considered as a function of the hypothesis. For reasons that will become clear below, $p(D \mid I)$ is sometimes called the *global likelihood*, and usually plays the role of an ignorable normalization constant.

Two points are worth emphasizing immediately about BT. First, there is nothing about the passage of time built into probability theory. Thus, our use of the terms "after data," "before data," "prior probability," and "posterior probability" do not refer to times before or after data is available. They refer to *logical* connections, not temporal ones. Thus, to be precise, a prior probability is the probability assigned before *consideration* of the data, and similarly for the other terms.

Second, for those who may have been exposed to BT before and heard some ill-informed criticisms of it, the $I$ that is always to the right of the bar in equation (5) is not some major premise about nature that must be true to make our calculation valid. Nor is it some strange, vague proposition defining some universal state of ignorance. It simply is the background information that defines the problem we wish to address at the moment. It may specify information about $H$ that we are content to assume true, or it may simply specify some alternative hypotheses we wish to compare with $H$. We will have the chance to elaborate on this point below, when we see how to use (5) to solve concrete problems.

To solve Bernoulli's problem, Bayes used a special case of BT to evaluate different propositions about the the value of the single trial probability of an outcome, given its relative frequency of occurrence in some finite number of trials (Bayes 1763; Jaynes 1978). Later, independently, Laplace greatly developed probability theory, with BT playing a key role. He used it to address many concrete problems in astrophysics. For example, he used BT to estimate the masses of the planets from astronomical data, and to quantify the uncertainty of the masses due to observational errors. Such calculations helped him choose which problems in celestial mechanics to study by allowing him to identify significant perturbations and to make predictions that would be testable by observers.

---

*        To be precise, $H$ is a proposition asserting the truth of the hypothesis in question ("The plasma temperature is $T$."), $D$ is a proposition asserting the values of the data ("The observed photon energy is $\epsilon$."), etc., but we will usually be a bit free with our language in this regard.

## 2.3  FREQUENTIST PROBABILITY

Despite the success of Laplace's development of probability theory, his approach was soon rejected by mathematicians seeking to further develop the theory. This rejection was due to a lack of a compelling rationale for some of the practices of Bernoulli, Bayes, Laplace, and their contemporaries.

First, the idea that probability should represent a degree of plausibility seemed too vague to be the foundation for a mathematical theory. The mathematical aspect of the theory followed from the axioms (1) and (2), but it was certainly not obvious that calculations with degrees of plausibility had to be governed by those axioms and no others. The axioms seemed arbitrary.

Second, there were problems associated with how prior probabilities should be assigned. The probability axioms described how to manipulate probabilities, but they did not specify how to assign the probabilities that were being manipulated. In most problems, it seemed clear how to assign the sampling probability, given some model for the phenomenon being studied. But finding compelling assignments of prior probabilities proved more difficult. In a certain class of problems, Bernoulli and his successors found an intuitively reasonable principle for such an assignment that we will call the *Principle of Indifference* (PI; it is also known as the Principle of Insufficient Reason). It is a rule for assignment of probabilities to a finite, discrete set of propositions that are mutually exclusive and exhaustive (*i.e.*, one proposition, and only one, must be true). The PI asserts that if the available evidence does not provide any reason for considering proposition $A_1$ to be more or less likely than proposition $A_2$, then this state of knowledge should be described by assigning the propositions equal probabilities. It follows that in a problem with $N$ mutually exclusive and exhaustive propositions, and no evidence distinguishing them, each proposition should be assigned probability $1/N$.

While the PI seemed compelling for dealing with probability assignments on discrete finite sets of propositions, it was not clear how to extend it to cases where there were infinitely many propositions of interest. Such cases arise frequently in science, whenever one wants to estimate the value of a continuous parameter, $\theta$. In this case, $\theta$ is a label for a continuous infinity of propositions about the true value of the parameter, and we need to assign a prior probability (density) to all values of $\theta$ in order to use BT. We might specify indifference about the value of $\theta$ by assigning a flat probability density, with each value of $\theta$ having the same prior probability as any other. Unfortunately, it seems that we could make the same statement about prior probabilities for the value of $\theta' \equiv \theta^2$. But a flat density for $\theta'$ does not correspond to a flat density for $\theta$. For this reason, inferences about continuous parameters seem to have a disturbing subjectivity, since different investigators choosing to label hypotheses differently by using different parameters could come to different conclusions.

The mathematicians of the late nineteenth and early twentieth centuries dealt with these legitimate problems by surgical removal. To eliminate the arbitrariness of the probability axioms, they drastically restricted the domain of the theory by asserting that probability had to be interpreted as relative frequency of occurrence in an ensemble or in repeated random experiments. The algebra of relative frequencies obviously satisfied the axioms, so their arbitrariness was removed.

As a byproduct, the second problem with the Laplace theory disappeared, because the frequency definition of probability made the concept of the probability of an hypothesis illegitimate. This is because the frequency definition can only describe the probability of a *random variable:* a quantity that can meaningfully be considered to take on various values throughout an ensemble or a series of repeated experiments. An hypothesis, being either true or false for every element of an ensemble or every repetition of an experiment, is not a random variable; its "relative frequency of occurrence" throughout the ensemble or sequence of experiments is either 0 or 1. For example, were we to attempt to measure the radius of a planet by repeated observation, the observed radius would vary from repetition to repetition, but the actual radius of the planet would be constant, and hence not amenable to frequentist description. Put another way, were we to analyze the observations with BT, we would be attempting to find a posterior distribution for the radius; but if this posterior distribution is a frequency distribution, there is an obvious problem: how can the frequency distribution of a parameter become known from data that were taken with only one value of the parameter actually present?

For these reasons, the concept of the probability of an hypothesis is held as meaningless in frequentist theory. A consequence is that scientists are denied the ability to use BT to assess hypotheses, so the problem of assigning prior probabilities disappears. The resulting theory was originally deemed superior to BPT, especially because it seemed more objective. The apparent subjectivity of prior probability assignments was avoided, and the frequency definition of probability, by its reference to observation of repeated experiments, seemed to make probability an objective property of "random" phenomena, and not a subjective description of the state of knowledge of a statistician.

## 2.4  CRITICISM OF THE FREQUENTIST APPROACH

2.4.1. *Arbitrariness and Subjectivity.* Unfortunately, assessing hypotheses was one of the principle aims of probability theory. Denied the use of BT for this task, frequentist theory had to develop ways to accomplish it without actually calculating probabilities of hypotheses. The frequentist solution to this problem was the creation of the discipline of *statistics.* Basically, one constructs some function of observable random variables that is somehow related to what one wishes to measure; such a function is called a *statistic.* Familiar statistics include the sample mean and variance, the $\chi^2$ statistic, and the $F$ statistic. Since a statistic is a function of random variables, its probability distribution, assuming the truth of the hypothesis of interest, can be calculated. A hypothesis is assessed by comparing the observed value of the statistic with the long-run frequency distribution of the values of the statistic in hypothetical repetitions of the experiment.

Intuition was a clear guide for the construction of statistics for simple problems (the familiar statistics mentioned above refer to the rather simple gaussian distribution). But for complicated problems, there is seldom a compelling "natural" choice for a statistic. Several statistical procedures may be available to address a particular problem, each giving a different answer. For example, to estimate the value of a parameter, one can use the method of moments, the maximum likelihood method, or a more specialized *ad hoc* method. Or, to compare unbinned data with an hypothesized continuous distribution, one could use one of the three Kolmogorov-Smirnov tests, the Smirnov-Cramer-von Mises test, or any of a number of obvious generalizations of them.

To provide a rationale for statistic selection, many principles and criteria have been added to frequentist theory, including unbiasedness, efficiency, consistency, coherence, the conditionality principle, sufficiency, and the likelihood principle. Unfortunately, there is an arbitrariness to these principles, and none of them have been proved to be of universal validity (for example, there is currently a growing literature endorsing the use of biased statistics in some situations; see Efron 1975 and Zellner 1986). Further, with the exception of the concept of sufficiency (which applies to only a limited family of distributions), none of these criteria alone leads to a unique choice for a statistic. Thus in practice more than one criterion must be invoked; but there are no principles specifying the relative importance of the criteria.

Once a statistic is selected, it must be decided how its frequency distribution will be used to assess an hypothesis. To replace the Bayesian notion of the probability of an hypothesis, other real number measures of the plausibility of an hypothesis are introduced, including confidence regions, significance levels, type I and II error probabilities, test size and power, and so on. These all require the consideration of hypothetical data for their definitions.

The resulting frequentist theory is far from unified, and the proliferation of principles and criteria in the theory and the availability of a plurality of methods for answering a single question place the objectivity of the theory in question. This situation is ironic. The frequency definition was introduced to eliminate apparent arbitrariness and subjectivity in the Laplace theory. Yet a large degree of arbitrariness must enter the frequency theory to allow it to address the problems Laplace could address directly.

### 2.4.2. *Comparison with Intuition.*

Once a statistical procedure is chosen in frequentist theory, it is used to assess an hypothesis by calculating its long-term behavior, imagining that the hypothesis is true and that the procedure is applied to each of many hypothetical data sets. But this is strongly at variance with how we intuitively reason in the presence of uncertainty. We do not want a rule that will give good long term behavior; rather, we want to make the best inference possible given the one set of evidence actually available.

Consider the following three examples of everyday plausible inference. When we come to an intersection and must decide whether to cross, or wait for oncoming traffic to pass, we consider whether we will make it across safely or be hit, given the current traffic situation at the intersection. When a doctor diagnoses an illness, he or she considers the plausibility of each of a variety of diseases in the light of the current symptoms of the patient. When a juror attempts to decide the guilt or innocence of a defendant, the juror considers the plausibility of guilt or innocence in light of the evidence actually presented at the trial.

These three examples have a common structure: in the presence of uncertainty, we assess a *variety of hypotheses* (safe crossing or a collision; cold or flu or bronchitis; guilty or innocent) in the light of the *single set of evidence* actually presented to us. In addition, we may have strong, rational prior prejudices in favor of one or more hypotheses. The doctor may know that there is a flu epidemic in progress, or that the patient has had a recurrent viral infection in the past.

Bayes' theorem has just this structure. A variety of hypotheses, specified in $I$, are each assessed by calculating their posterior probability, which depends both on the prior probability of the hypothesis, and on the probability of the one data set actually observed.

In contrast, the roles of hypothesis and data are reversed in frequentist reasoning. Forbidden the concept of the probability of an hypothesis, the frequentist must assume the truth of a *single* hypothesis, and then invent ways to assess this decision. The assessment is made considering not only the data actually observed, but also *many hypothetical data sets* predicted by the hypothesis but not seen. It is as if the juror tried to decide guilt or innocence by taking into consideration a mass of evidence that might possibly have been presented at the trial but which was not.

In a word, frequentist reasoning assesses decisions to assume the truth of an hypothesis by considering hypothetical data, while the Bayesian approach assesses hypotheses directly by calculating their probabilities using only the data actually observed, the only hypothetical elements of the calculation being the hypotheses themselves.

2.4.3. *Randomness vs. Uncertainty.* Frequentist theory is forced to base inferences on hypothetical data because data, and not hypotheses, are considered to be "random variables." The concept of randomness is at the heart of the theory. But a close inspection of the notion of randomness reveals further difficulties with the frequentist viewpoint.

In frequentist theory, a quantity is random if it unpredictably takes on different values in otherwise identical repetitions of an experiment or among identically prepared members of an ensemble. To explore this concept, we will consider as an example the prototypical random experiment: the flip of a coin. Imagine an experiment specified by the statement, "A fair (*i.e.*, symetrical) coin is flipped." Since either heads or tails can come up in a flip, and since we cannot predict with certainty which will come up, the outcome of a flip is considered random. The probability of a particular outcome—heads, say—is defined as the limiting frequency with which heads comes up in an indefinitely large number of flips. This definition seems to refer to an observable property of the coin. For this reason, frequentist probability appears more objective than Bayesian probability; the latter describes a state of knowledge, while the former seems to describe an observable property of nature.

But certainly the motion of a coin is adequately described by classical mechanics; if we knew the physical properties of the coin (mass, inertia tensor, etc.), the initial conditions of the flip, and exactly how it was flipped, we could predict the outcome with certainty. If the same coin was flipped under precisely the same conditions, the outcome would be the same for each flip. What, then, gives rise to the "randomness" of the outcomes of repeated flips?

If "identical" repetitions of a coin flip experiment produce different outcomes, something must have changed from experiment to experiment. The experiments could not have been precisely identical. Hidden in the adjective, "identical", describing repetitions of an experiment or elements of an ensemble in frequentist theory is the true source of "randomness": the repeated experiments must be identical only in the sense that in each of them we are in the same state of knowledge in regard to the detailed conditions of the flip. Put another way, the description of the experiment is incomplete, so that repetitions of the experiment that agree with our description vary in details which, though not specified in our description, nevertheless affect the outcome. In the coin example, we have specified only that the same (*i.e.*, physically identical) coin be flipped in repeated experiments. But this leaves the initial conditions of the flips, and the precise manner of flipping, completely unspecified. Since the outcome of a flip depends as much on these unspecified details as on the physical properties of the coin, it is unpredictable.

There is variability in the outcome of "random" experiments only because our incomplete knowledge of the details of the experiment permit variations that can alter the outcome. In some cases, our knowledge may not constrain the outcome at all. This could be the case in a coin flipping experiment, where merely specifying that the same coin be flipped leaves so much room for variation that the outcome is totally uncertain, heads and tails being equally probable outcomes for a particular flip. But often our knowledge, though incomplete, sufficiently constrains the experiment so that some general features of the outcome can be predicted, if not with certainty, than at least with high plausibility. The best example is statistical mechanics. There, measurement of the temperature of an equilibrium system provides us with knowledge about its total energy. Though many, many microstates are compatible with the measurement, our limited knowledge of the precise microstate of the system still permits us to make very accurate predictions of, say, its pressure. This is because the vast majority of microstates compatible with our limited knowledge have very nearly identical pressures.

Thus even in frequency theory, situations are described with probability, not because they are intrinsically random or unpredictable, but because we want to make the most precise statements or predictions possible given the variations permitted by the uncertainty and incompleteness of our state of knowledge (Jaynes 1985d). "Randomness", far from being an objective property of an object or phenomenon, is the result of uncertainty and incompleteness in one's state of knowledge. Once this is realized, the frequentist distinction between the uncertainty one may have about the value of a "random variable" and the uncertainty one may have about the truth of an hypothesis appears highly contrived. Randomness, like any uncertainty, is seen to be "subjective" in the sense of resulting from an incomplete state of knowledge.*

Two operational difficulties with frequentist theory clearly indicate that it is as subjective as BPT, and in some contexts even more subjective. First, though probability is defined as long-term frequency, frequency data is seldom available for assignment of probabilities in real problems. In fact, the infinite amount of frequency data required to satisfy the frequentist definition of probability is *never* available. As a result, the frequentist must appeal to an imaginary infinite set of repeated experiments or an imaginary infinite ensemble. Often, which imaginary reference set to choose will not be obvious, as the single data set we wish to analyze can often be considered to be a member of many reasonable reference sets. This subjectivity of frequentist theory has led to statistical paradoxes where simple, apparently well-posed problems have no obvious solution. In the Bayesian approach, where probability assignments describe the state of knowledge defined by the problem statement, such paradoxes disappear (see Jaynes 1973 for an instructive example).

The second operational difficulty arises in the analysis of data consisting of multiple samples of a random quantity. Since frequentist theory requires the consideration of hypothetical data to assess an hypothesis, analysis requires the specification, not only of the phenomenon being sampled and the results of the sample, but also the specification of what other samples might have been seen. These hypothetical samples are needed to specify the reference set for the observed sample, but unfortunately their specification can depend on the thoughts of the experimenter in disturbing ways. This complicated phenomenon is best described by an example (Berger and Berry 1988).

Consider again the flip of a coin, and imagine that a coin has been flipped $N = 17$ times, giving $n_H = 13$ heads and $n_T = 4$ tails. Is this evidence that the coin is biased? Strangely, a frequentist cannot even begin to address this question with the data provided, because it is not clear from these data what the reference set for the data is. If the frequentist is told

---

*        A reader may object at this point, arguing that the success of quantum theory "proves" that phenomena can be intrinsically random. But the successes of quantum theory no more prove the randomness of nature than the success of statistical description of coin flipping proves that coin flipping is intrinsically random, or the fact that a random number algorithm passes statistical tests proves that the numbers it produces (in a purely deterministic fashion!) are random. Indeed, BPT offers much hope in helping us to unravel inference from physics in quantum theory; see Jaynes (1989a,b) for preliminary analyses.

that the experimenter planned beforehand on flipping the coin 17 times, then analysis can proceed, with probabilities determined by embedding the data in a reference set consisting of many sets of 17 flips. But this is not the only way the data could have been obtained. For example, the experimenter may have planned to flip the coin until he saw 4 tails. In that case, the reference set will be many sets of flips differing in their total number, but with each set containing 4 tails.

In the first case, the number of heads (or tails) is the random quantity, and in the second, the total number of flips is the random quantity. Depending on which quantity is identified as random, a different reference set will be used, and different probabilities will result. The results of the analysis thus depend on the stopping rule used by the experimenter. Experiments must therefore be carefully planned beforehand to be amenable to frequentist analysis, and if the plan is altered during execution for any reason (for example, if the experimenter runs out of funds or subjects), the data is worthless and cannot be analyzed. An example is worked out in Section 4.3.1, where it is shown that this so-called *optional stopping problem* can lead to dispute over whether or not an hypothesis is rejected by a given data set.

Intuition rebells against this strange behavior. Surely my conclusions, given the one data set observed, should not depend on what I or anyone else might have done if different data were obtained. And surely, if my plan for an experiment has to be altered (as is often the case in astronomy, where observations can be cut short due to bad weather or fickle satellite electronics), I should still be able to analyze the resulting data. In Bayesian probability theory, the stopping rule plays no role in the analysis, and this has been an important factor in bringing many statisticians over to the Bayesian school of thought (Berger and Berry 1988). There is no ambiquity over which quantity is to be considered a "random variable", because the notion of a random variable and the consequent need for a reference set of hypothetical data is absent from the theory. All that is required is a specification of the state of knowledge that makes the outcome of each element of the data set uncertain.

2.4.4. *The Frequentist Failure.* The frequentist approach to probability theory was motivated by important deficiencies in the Bayesian theory that it replaced. Unfortunately, frequentist theory addressed these deficiencies only by burying them under a superficially more objective facade. When examined more deeply, we see that frequentist theory only exacerbates the ambiguity and subjectivity of the Bayesian theory.

One motivation for frequentist theory was the apparent arbitrariness of the probability axioms. To make the axioms compelling, the frequency definition of probability was introduced. But this definition forbade the use of Bayes' Theorem for the analysis of hypotheses, and the resulting frequentist theory cannot by itself produce unique solutions to well-posed problems. A wide variety of principles and criteria must be added to the theory, each at least as arbitrary as the probability axioms seemed to be.

Another important motivation for frequentist theory was the subjective nature of Bayesian probability assignments, particularly in regard to prior probabilities for hypotheses. Frequentist theory replaces the subjective probability assignments of BPT with relative frequencies of occurence of random variables. But the notion of randomness is itself subjective, dependent on one's state of knowledge in a manner very similar to that of Bayesian probability. In many problems, it is substantially *more* subjective, since the identification of random variables and their probability assignments can depend in disturbing ways on the thoughts of the experimenter. Such is the case in the optional stopping problems just described.

Finally, frequentist theory is badly at odds with the manner in which we intuitively reason in the presence of uncertainty. Rather than evaluate a variety of hypotheses in the light of the available evidence, the theory attempts to evaluate a single hypothesis by considering a variety of hypothetical data. It also ignores any prior information one may have regarding the possible hypotheses.

Frequentist theory has thus failed to address the problems that motivated it, and in fact has exacerbated them. Though it has been used with great success for the analysis of many problems, it is far from unified, and can give anti-intuitive and paradoxical results. These problems signal a deep flaw in the theory, and indicate the need to find a better theory. This new theory should duplicate the successes of frequentist theory, and eliminate its defects.

In the remainder of this paper, we will see that such a better theory exists, and is in fact identical to the original probability theory of Bernoulli, Bayes, Laplace, and their contemporaries, though with a sounder rationale.

## 2.5  SOME RECENT HISTORY

Sir Harold Jeffreys was one of the earliest critics of the frequentist statistics of his day. But he did more than criticize; he offered an alternative. In his book (Jeffreys 1939) he presented Bayesian solutions to dozens of practical statistical problems. He also tried to provide a compelling rationale for Bayesian probability theory, and although he was not completely successful in this, his mass of intuitively appealing results, many of them inaccessible to frequentists, should have been a clear indication that "something is *right* here." But his work was rejected on philosophical grounds, and has remained largely unnoticed until recently.

In the 1940's and 1950's, R. T. Cox, E. T. Jaynes, and others began to provide the missing rationale for Bayesian probability theory. Their work was little appreciated at first, but others rediscovered some of this rationale, and over the past few decades there has been a slow but steady "Bayesian revolution" in statistics. Astrophysicists have been slow to reap the benefits of this revolution. But in the last 15 years Gull, Skilling, Bretthorst and others have begun working out astrophysical applications of BPT. In the remainder of this paper, we will examine the rationale and foundations of BPT, learn how it is used to address well-posed statistical problems, and then briefly review some of the recent astrophysical applications of BPT.

## 3. Bayesian Probability Theory: A Mathematical Language for Inference

The difficulties with frequentist theory, particularly its clash with common sense reasoning, lead us to conclude that it is not generally appropriate for the analysis of scientific data. The intuitive appeal of BPT and the mass of successful results from it lead us to suspect that it may be the correct theory. But can a compelling, rigorous mathematical theory be erected upon a concept as apparently vague as the notion that probability is a measure of degree of plausibility?

Happily, the answer is yes. In this section we will see that a small set of compelling qualitative desiderata for a measure of plausibility will be sufficient to completely specify a quantitative theory for inference that is identical to the probability theory used by Laplace and Jeffreys. Specifically, these qualitative desiderata will allow us to *derive* the "axioms" of probability theory, giving them an unassailable status as the correct rules for the manipulation of real number valued degrees of plausibility. We will also recognize that these rules for combination and manipulation—a "grammar" for plausible inference—are only half of the required theory. The other half of the theory is the problem of assigning initial probabilities to be manipulated—the "vocabulary" of the mathematical language—and the desiderata will provide us with rules for unambiguous assignment of probabilities in well-posed problems.

The desiderata make no reference to frequencies, random variables, ensembles, or imaginary experiments. They refer only to the plausibility of propositions. Deductive reasoning, by which we reason from true propositions to other true propositions, will be a limiting case of the theory, and will guide its development. Thus, the theory can be viewed as *the extension of deductive logic to cases where there is uncertainty* (Jaynes 1990a,b). Of course, we are free to use the resulting theory to consider propositions about frequencies in repeated experiments. In this way, connections between probability and frequency, including Bernoulli's theorem and its generalizations, will be derived consequences of the theory, and all the useful results of frequentist theory will be included in the new theory as special cases.

The missing rationale for BPT was first provided by Cox (1946, 1961) and Jaynes (1957, 1958). Similar results were soon found in other forms by other statisticians (see Lindley 1972 for a terse review). We will only be able to describe briefly this profound and beautiful aspect of BPT here. More detailed, highly readable developments of these ideas may be found in Jaynes (1957, 1958), Tribus (1969), Grandy (1987), and Smith and Erickson (1989). A particularly eloquent and complete development will be available in the upcoming book by Jaynes (1990b).

## 3.1 THE DESIDERATA

Our first desideratum for a theory of plausibility is simple:

(I) Degrees of plausibility are represented by real numbers.

Perhaps there are useful generalizations of the theory to different number systems. But if our theory is to represent something similar to the way we reason, or if we wish to consider it possible to design a computer or robot that follows our quantitative rules, at some point we will have to associate plausibility with some physical quantity, meaning we will have to associate it with real numbers.

Not yet identifying degrees of plausibility with probability, we will indicate the plausibility of proposition $A$ given the truth of proposition $C$ by the symbol $A \mid C$. We will take it as a convention that greater plausibility will correspond to a greater number.

Our second desideratum will be

(II) Qualitative consistency with common sense.

There are several specific ways we will use this; they are noted below. For example, if the plausibility of $A$ increases as we update our background information from $C$ to $C'$ (that is, $A \mid C' > A \mid C$), but our plausibility of $B$ is unaffected ($B \mid C' = B \mid C$), then we expect that the new information can only increase the plausibility that $A$ and $B$ are both true, and never decrease it ($AB \mid C' > AB \mid C$). Effectively, this desideratum will ensure that the resulting theory is consistent with deductive logic in the limit that propositions are certainly true or certainly false.

Our final desideratum is

(III) Consistency.

More explicitly, we want our theory to be consistent in 3 ways.

(IIIa) *Internal Consistency:* If a conclusion can be reasoned out in more than one way, every possible way must lead to the same result.

(IIIb) *Propriety:* We demand that the theory take into account all information provided that is relevant to a question.

(IIIc) *Jaynes Consistency:* Equivalent states of knowledge must be represented by equivalent plausibility assignments. This desideratum, a generalization of the Principle of Indifference, is the key to the problem of assigning prior probabilities. Though it seems obvious once stated, its importance has only been appreciated beginning with the work of Jaynes (1968).

Amazingly, these few compelling desiderata will be sufficient to completely specify the form of Bayesian probability theory.

## 3.2  THE GRAMMAR OF INFERENCE: THE PROBABILITY AXIOMS

Given two or more propositions, we can build other, more complicated propositions out of them by considering them together. We would like to have rules to tell us how the plausibilities of these new, compound propositions can be calculated from the plausibilities of the original propositions. We will assume for the moment that the original plausibilities are given. The rules we seek will play the role of a "grammar" for our theory.

Some of the ways we can build new propositions out of a set of propositions $\{A, B, C \ldots\}$ include logical negation ($\overline{A}$, "not $A$"), logical conjunction ($AB$, "$A$ and $B$"), and logical disjunction ($A + B$, "$A$ or $B$"), mentioned above. An example of another important operation is implication: $A \Rightarrow B$ is the proposition, "If $A$ is true, then $B$ follows." The symbolic system governing the combination of propositions like this is *Boolean Algebra.* We want our plausibility calculus to enable us to calculate the plausibility of any proposition built from other propositions using Boolean algebra.

It will come as no surprise to students of computer science that only a subset of the logical operations we have listed is needed to generate all possible propositions. For example, the proposition $A + B$ is identical to the proposition $\overline{\overline{A}\,\overline{B}}$; that is, $A + B$ is true unless both $A$ and $B$ are false. One adequate subset of Boolean operations that will be convenient for us to consider is conjunction and negation. If we can determine how to calculate the plausibility of the negation of a proposition, given the plausibility of the original proposition, and if we can determine how to calculate the plausibility of the conjunction of two propositions from their separate plausibilities, then we will be able to calculate the plausibilities of all possible propositions that can be built from one or more "elementary" propositions.

Our desiderata are sufficient to specify the desired rules for calculation of the plausibility of a negated proposition and of the conjunction of two propositions; not surprisingly, they are the sum rule and product rule, equations (1) and (2) above. We do not have the space to discuss the derivation of these rules fully here. But since the resulting rules are the foundation for probability theory, and since the kind of reasoning by which such quantitative rules are derived from qualitative desiderata is prevalent in Bayesian probability theory, it is important to have an understanding of the derivation. Therefore, we will outline here the derivation of the product rule, and only present the results of the similar derivation of the sum rule; the above mentioned references may be consulted for further details.

3.2.1 *The Product Rule.* We will first look for a rule relating the plausibility of $AB$ to the plausibilities of $A$ and $B$ separately. That is, we want to find $AB \mid C$ given information about the plausibilities of $A$ and $B$. The separate plausibilities of $A$ and $B$ that may be known to us include the four quantities $u \equiv (A \mid C)$, $x \equiv (B \mid C)$, $y \equiv (A \mid BC)$, and $v \equiv (B \mid AC)$. By desideratum (IIIb), we should use all of these, if they are relevant.

Now we invoke desideratum (II) to try to determine if only a subset of these four quantities is actually relevant. Common sense tells us right away, for example, that $(AB \mid C)$ cannot depend on only one of $x, y, u$, or $v$. This leaves eleven combinations of two or more of these plausibilities. A little deeper thought reveals that most of these combinations are at variance with common sense. For example, if $(AB \mid C)$ depended only on $u$ and $x$ we would have no way of taking into account the possibility that $A$ and $B$ are exclusive.

Tribus (1969) goes through all eleven possibilities, and shows that all but two of them exhibit qualitative violations of common sense. The only possible relevant combinations are $x$ and $y$, or $u$ and $v$. We can understand this by noting that there are two ways a decision about the truth of $AB$ can be broken down into decisions about $A$ and $B$. Either we first decide that $A$ is true, and then, accepting the truth of $A$, decide that $B$ is true. Or, we first decide that $B$ is true, and then make our decision about $A$ given the truth of $B$. Finally, we note that since the proposition $AB$ is the same as the proposition $BA$, we can exchange $A$ and $B$ in all the quantities. Doing so, we see that the different pairs, $x, y$ and $u, v$, merely reflect the ordering of $A$ and $B$, so we may focus on one pair, the other being taken care of by the commutativity of logical conjunction.

Denoting $(AB \mid C)$ by $z$, we can summarize our progress so far by stating that we seek a function $F$ such that

$$z = F(x, y). \tag{6}$$

Now we impose desideratum (IIIa) requiring internal consistency. We set up a problem that can be solved two different ways, and demand that both solutions be identical. One such problem is finding the plausibility that *three* propositions, $A, B, C$, are simultaneously true. The joint proposition $ABC$ can be built two different ways: $ABC = (AB)C = A(BC)$. The first of these equations and equation (6) tell us that $(ABC \mid D) = F[(BC \mid D), (A \mid BCD)]$, where we treat the proposition $AB$ as a single proposition. A similar equation follows from the second equality. Internal consistency then requires that the function $F$ obey the equation,

$$F[F(x, y), z] = F[x, F(y, z)], \tag{7}$$

for all real values of $x$, $y$, and $z$. Crudely, the function $F$ is "associative." The general solution of this functional equation is $F(x, y) = w^{-1}[w(x)w(y)]$, where $w(x)$ is any positive, continuous, monotonic function of plausibility. Thus equation (7) does not uniquely specify $F$, but only constrains its form. Using this solution in equation (6), our consistency requirement tells us that

$$w(AB \mid C) = w(A \mid BC)w(B \mid C). \tag{8}$$

This looks like the product rule of probability theory. But at this point we cannot identify probability with plausibility, because equation (8) involves the arbitrary function $w$.

3.2.2 *The Sum Rule.* We may apply similar reasoning to determine how to calculate $w(\overline{A} \mid B)$ from $w(A \mid B)$. Consistency with common sense and internal consistency again lead to a functional equation whose solution implies that

$$w^m(A \mid B) + w^m(\overline{A} \mid B) = 1. \tag{9}$$

Here $w(x)$ is the same function as in (8), and $m$ is an arbitrary positive number. Along the way, it is found that certainty of the truth of a proposition must be represented by $w = 1$, and by convention, $w = 0$ is chosen to represent impossibility.

A new arbitrary element—the number $m$—has appeared; but since the function $w$ is itself arbitrary, we are free to make a simple change of variables from $w(x)$ to the different monotonic function $p(x) \equiv w^m(x)$, so that we may always write

$$p(A\ B) + p(\overline{A} \mid B) = 1, \tag{10}$$

and

$$p(AB \mid C) = p(A \mid BC)p(B \mid C), \tag{11}$$

in place of equations (8) and (9). Thus the choice of $m$ is irrelevant, and does not offer us any degree of freedom we did not already have in our choice of $w(x)$.

The arbitrary function $p(x)$ indicates that our desiderata do not lead to unique rules for the manipulation of plausibilities. There are thus an infinite number of ways to use real numbers to represent plausibility. But what we *have* shown is that for any such plausibility theory that is consistent with our desiderata, there must be a function $p(x)$ such that the theory can be cast in the form of equations (10) and (11). These equations thus contain the content of all allowed plausibility theories.

Equations (10) and (11) are the "axioms" of probability theory, so we identify the quantity $p(A \mid B)$ as the probability of $A$ given $B$. That is, probability is here taken to be a technical term referring to a monotonic function of plausibility obeying equations (10) and (11). We have shown that *every allowed plausibility theory is isomorphic to probability theory*. The various allowed plausibility theories may differ in form from probability theory, but not in content. Put another way, since $p(x)$ is a monotonic function of the plausibility $x$, $x$ is a monotonic function of $p$. Therefore all allowed plausibility theories can be created by considering all possible functions $x(p)$ and the corresponding transformations of (10) and (11). Of all these theories, differing in form but not in content, we are choosing to use the one specified by $x(p) = p$, since this leads to the simplest rules of combination, equations (10) and (11).

An analogy can be made with the concept of temperature in thermodynamics, a real number encoding of the qualitative notion of hot and cold (Jaynes 1957, 1990b). Different temperature scales can be consistently adopted, each monotonically related to the others, but the Kelvin scale is chosen for the formulation of thermodynamics, because it leads to the simplest expression of physical laws.

## 3.3 THE VOCABULARY OF INFERENCE: ASSIGNING PROBABILITIES

We have found the rules for combining probabilities, a kind of "grammar" for inference. Now we ask how to assign numerical values to the probabilities to be so combined: we want to define a "vocabulary" for inference. Probabilities that are assigned directly, rather than derived from other probabilities using equations (10) and (11), are called *direct probabilities*. We seek rules for converting information about propositions into numerical assignments of direct probabilities. Such rules will play a role in probability theory analogous to deciding the truth of a proposition in deductive logic. Deductive logic tells us that certain propositions will be true or false *given* the truth or falseness of other assumed propositions, but' the rules of deductive logic do not determine the truth of the assumed propositions; their truth must be decided in some other manner, and provided as input to the theory. Direct probabilities are the analogous "input" for probability theory.

It is worth emphasizing that probabilities are *assigned*, not *measured*. This is because probabilities are measures of the plausibilities of propositions; they thus reflect whatever information one may have bearing on the truth of propositions, and are not properties of the propositions themselves. This is reflected in our nomenclature, in that all probability symbols have a vertical bar and a conditioning proposition indicating exactly what was assumed in the assignment of a probability. In this sense, BPT is "subjective," it describes states of knowledge, not states of nature. But it is "objective" in that we insist that equivalent states of knowledge be represented by equal probabilities, and that problems be well-posed: enough information must be provided to allow unique, unambiguous probability assignments.

We thus seek rules for assigning a numerical value to $p(A \mid B)$ that expresses the plausibility of $A$ given the information $B$. Of course, there are many different kinds of information one may have regarding a proposition, so we do not expect there to be a universal method of assignment. In fact, only recently has it been recognized that finding rules for converting information $B$ into a probability assignment $p(A \mid B)$ is fully half of probability theory. Finding such rules is a subject of much current research.

Rules currently exist for several common types of information; we will outline some of the most useful here. The simplest kind of information we can have about some proposition $A_1$ is a specification of alternatives to it. That is, we can only be uncertain of $A_1$ if there are alternatives $A_2, A_3 \ldots$ that may be true instead of $A_1$; and the nature of the alternatives will have a bearing on the plausibility of $A_1$. Probability assignments that make use of only this minimal amount of information are important in BPT as objective representations of initial ignorance, and they deserve a special name. We will refer to them as *least informative probabilities* (LIPs).* Probability assignments that make use of information beyond the specification of alternatives we will call *informative probabilities*.

3.3.1 *Least Informative Probabilities.* For many problems, our desiderata are sufficient to specify assignment of a LIP. Consider a problem where probabilities must be asigned to two propositions, $A_1$ and $A_2$. Suppose we know from the very nature of the alternatives that they form an exclusive, exhaustive set (one of them, and only one, must be true), but that this is all we know. We might indicate this symbolically by writing our conditioning information as $B = A_1 + A_2$. Since the propositions are exclusive, $p(A_1 A_2 \mid B) = 0$, so the sum rule (3) implies that $p(A_2 \mid B) = 1 - p(A_1 \mid B)$. But this does not specify numbers for the probabilities.

Now imagine someone else addressing this problem, but labeling the propositions differently, writing $A_1' = A_2$ and $A_2' = A_1$. This person's conditioning information is $B' = A_1' + A_2' = A_1 + A_2 = B$. Obviously, $p(A_1' \mid B) = p(A_2 \mid B)$, and $p(A_2' \mid B) = p(A_1 \mid B)$. But now note that since $B$ is indifferent to $A_1$ and $A_2$, the state of knowledge of this second

---

*         Such probabilities are also referred to as *uninformative probabilities* in the literature.

person regarding $A_1'$ and $A_2'$, including their labeling, is the same as that in the original problem. By desideratum (IIIc), equivalent states of knowledge must be represented by equivalent probability assignments, so $p(A_1' \mid B) = p(A_1 \mid B)$. But this means that $p(A_2 \mid B) = p(A_1 \mid B)$ which, through the sum rule, implies $p(A_1 \mid B) = p(A_2 \mid B) = 1/2$. We finally have a numerical assignment!

This line of thought can ge generalized to a set of $N$ exclusive, exhaustive propositions $A_i$ ($i = 1$ to $N$), leading to the LIP assignments $p(A_i \mid B) = 1/N$ (Jaynes 1957, 1990b). This is just Bernoulli's principle of indifference mentioned earlier, now seen to be a consequence of consistency when all the information we have is an enumeration of an exclusive exhaustive set of possibilities, with no information leading us to prefer some possibilities over the others.

Note that other information could lead to the same assignment. For example, if we are tossing a coin, and we know only that it has head and tail sides, we would assign least informative probabilities of 1/2 to the possibilities that heads or tails would come up on a single toss. Alternatively, we may have made careful measurements of the shape and inertia tensor of the coin, compelling us to conclude that both outcomes are equally likely and hence to assign *informative* probabilities of 1/2 to both heads and tails. The difference between these assignments would show up once we flipped the coin a few times and then reassessed our probabilities. If three flips gave three heads, in the first state of knowledge this would constitute evidence that the coin was biased and lead us to alter our probability assignment for the next toss, but in the informative state of knowledge it would not, since our information leads us to believe very strongly that the two sides are equally probable.

When the set of possibilities is infinite, as when we want to assign probabilities to the possible values of continuous parameters, the analysis becomes more complicated. This is because it may not be obvious how to transform the original problem to an equivalent one that will help us determine the probability assignment. In the finite discrete case, the only transformation that preserves the identity of the possibilities is permutation, leading to the PI. But in the continuous case, there is an infinite number of possible reparametrizations.

The key to resolving this dilemma is to realize that specifying the possibilities not only provides labels for them, but tells you about their nature. For example, the finite discrete problem we solved assumed that the nature of the possibilities indicated they formed an exhaustive, exclusive set (this implied $p(A_1 A_2 \mid B) = 0$, which we used in the sum rule). In problems with continuous parameters, transformations that lead to equivalent problems that can help one assign a LIP can often be identified by the nature of the parameters themselves. Information unspecified in the problem statement can be as important for this identification as the specified information itself, for problems that differ with respect to unspecified details are equivalent.

For example, suppose we want to find the probability that a marble dropped at random (*e.g.*, by a blindfolded person) will land in a particular region of a small target on the floor. Intuition tells us that the probability is proportional to the area of the region. How could we have established this by logical analysis? Draw an $(x, y)$ coordinate system on the target, so the possibilities are specified by intervals in $x$ and $y$. Write the probability that the ball will fall in the small area $dx\,dy$ about $(x, y)$ as $p(x, y, dx\,dy \mid I) = f(x, y)dx\,dy$; here $I$ specifies the target region. But nothing in the problem specified an origin for the coordinate system, so our assignments to $p(x, y, dx\,dy \mid I)$ and $p(x' = x + a, y' = y + b, dx'dy' \mid I)$ must be the same for any choice of $a$ or $b$. It follows that $f(x, y) = f(x + a, y + b)$ for any $(a, b)$, so $f(x, y) = const$ (the constant is determined by normalization to be 1/[target area]), and the probability is proportional to the area $dx\,dy$.* Such arguments can produce LIPs for many interesting and useful problems (Jaynes 1968, 1973, 1980; Rosenkrantz 1977; Bretthorst 1989). This tells us that mere specification of the possibilities we are considering, *including their physical meaning*, is a well-defined state of knowledge that can be associated with an unambiguous probability assignment.

### 3.3.2 *Informative Probabilities, Bayes' Theorem, and Maximum Entropy.*

Besides the specification of possibilities, $I$, we may have some additional information $I_A$ that should lead us to probability assignments different from least informative assignments. Rather than $p(A_i \mid I)$, we seek $p(A_i \mid II_A)$, an informative probability assignment.

One way to find $p(A_i \mid II_A)$ is to use Bayes' Theorem, equation (5), to update our assignments for each of the $A_i$ one at a time. To do this, the additional information $D \equiv I_A$ must be able to play the role of data, that is, it must be meaningful to consider for each $A_i$ the "sampling probability" $p(D \mid IA_i)$ that occurs on the right hand side of BT. Specifically, $D$ has to be a possible consequence of one or more of the $A_i$ considered individually, since each application of BT will require us to assume that one of the $A_i$ is true to calculate the likelihood of the additional information. If the information $D$ is of this type, we do not need any new rules for probability assignment; our rules of combination tell us how to account for the additional information by using BT.**

But data—observation of one of the possible consequences of the $A_i$—is not the only kind of information we may have about the various possibilities. Our information may refer directly to the possibilities themselves, rather than to their consequences. In our coin example above, the evidence $E$ provided by the measurements took the form of the proposition, "the probability of heads is the same as that of tails." Information like this cannot be used in Bayes' theorem because it does not refer to a consequence of one of the possibilities being considered. For example, here our possibilities are $A_1$ = heads on the next toss, $A_2$ = tails. To use BT, we need $p(E \mid IA_1)$, which in words is "the probability that heads and tails are equally probable if heads comes up on the next toss." But since either heads or tails *must* come up on the next toss, asserting that one or the other will come up tells us nothing about their relative probabilities. Put another way, a statement about the relative probabilities of $A_1$ or $A_2$ is not a possible logical implication of the truth of either of them, so it cannot be used in BT. Yet such information is clearly relevant for assessing the plausibility of the propositions. We must therefore find rules that will allow us to use information of this kind to make probability assignments.

Such a rule exists for converting certain types of information called *testable information* to a probability assignment. The information $E$ is testable if, given a probability distribution over the $A_i$, we can determine unambiguously if the distribution is consistent with the information $E$. In the example above, this was trivially true; $E$ asserted all probabilities were equal, and only one distribution is consistent with this. But in general, there may

---

\*       We expect the result to also be invariant with respect to rotations and scale changes. Since the area element is already invariant to these operations, considering them does not alter the result.

\*\*       Of course, we must now address the problem of assigning $p(D \mid IA_i)$. This is no different in principle than assigning $p(A_i \mid I)$, and is treated analogously. We start by specifying what other consequences of $A_i$ are possible, assign a LIP, and then account for any other information we have about the possible consequences. In this sense, the distinction between prior probabilities and sampling probabilities is somewhat artificial; both are direct probabilities, and the same rules are used for their assignment.

be many distributions consistent with testable information $E$. For example, we may know that the mean value of many roles of a die was 4.5 (rather than 3.5 expected for a fair die), and want to use this knowledge to assign probabilities to the six possible outcomes of the next role of the die. This information is testable—we can calculate the mean value of any probability distribution for the six possible outcomes of a roll and see if it is 4.5 or not—but it does not single out one distribution. But despite the multiplicity of distributions consistent with this information, our common sense seems to tell us something about the distribution which represents knowledge of the mean value, *and nothing else*, beyond the fact that it must be one of the distributions with the indicated mean value. For example, we would reject the assignment $\{p_1 = 0.3,\ p_6 = 0.7,$ all other $p_i = 0\}$ as unreasonable, despite the fact that it agrees with the mean value constraint. This is because this particular distribution, by excluding several of the possibilities that the evidence does not compel us to exclude, violates our propriety desideratum (IIIb).

Denote the operation of altering a LIP distribution to reflect testable information $E$ by $\mathcal{O}$, writing $p(H \mid IE) = \mathcal{O}[p(H \mid I); E)]$. Shore and Johnson (1980) have shown that our desiderata are sufficient to specify the operation $\mathcal{O}$. They consider three general types of transformations of a problem into an equivalent one, and show that the requirement that the solutions of these equivalent problems be consistent uniquely specifies $\mathcal{O}$: It selects from among all the possible normalized distributions satisfying the constraints imposed by $E$, the one with maximum entropy, where the entropy of a finite discrete distribution over exclusive, exhaustive alternatives is defined by

$$H = -\sum_{i=1}^{N} p_i \log p_i, \tag{12}$$

and that of a continuous distribution is defined analogously by

$$H = -\int p(\theta) \log \left( \frac{p(\theta)}{m(\theta)} \right) d\theta, \tag{13}$$

with $m(\theta)$ the LIP assignment for the parameter $\theta$. (Actually any monotonic function of entropy will suffice.) This rule is of enormous practical and theoretical importance; it is called *the maximum entropy principle* (MAXENT).

MAXENT assignments have a number of intuitively appealing interpretations, and were in fact introduced long before the work of Shore and Johnson, based on just such interpretations (Jaynes 1957a,b, 1958; Tribus 1969). For example, we can seek a measure of the amount of uncertainty expressed in a distribution. Arguments originating with Shannon (1948) show that a few compelling desiderata lead to entropy as the appropriate measure of uncertainty. It then seems reasonable to choose from among all distributions satisfying the constraints imposed by $E$ that which is otherwise the most uncertain (*i.e.*, assuming the least in addition to $E$); this leads to MAXENT (Jaynes 1957, 1958). An example of the use of MAXENT to assign a direct probability distribution will be mentioned in Section 5.1 below; instructive worked examples can be found in Jaynes (1958, 1963, 1978), Tribus (1962, 1969), Fougere (1988, 1989), and Bretthorst (1990).

## 3.4  THE FREQUENCY CONNECTION

Since there is presumably no end to the types of information one may want to incorporate into a probability assignment, BPT will never be a finished theory. Yet the existing rules are already sufficient for the analysis of uncertainty in many problems in the physical sciences, and in this sense the theory is complete.

Note that the entire theory has been developed without ever even mentioning relative frequencies or random variables. Yet the success of some frequentist methods indicates that there must be a connection between frequency and probability. There is, and such connections arise naturally in the theory, as *derived consequences* of the rules, when one calculates the probabilities of propositions referring to frequencies.

For example, given that the probability of a particular outcome in a single trial of an experiment is $p$, we can calculate the probability that $N$ repetitions of the experiment will give this outcome $n$ times. This calculation is just what Bernoulli did to prove his large number theorem—the equality of long-term relative frequency and probability in a single trial—mentioned above. Note, however, that the theorem is restricted to the case where each trial is independent of all the others; BPT is not so restricted.

Bernoulli's theorem is an example of reasoning from probability to frequency. But BPT, through Bayes' Theorem, also allows us to reason from observed frequency to probability. The observed frequency constitutes data which we can use to estimate the value of the single trial probability. Such a calculation can be done for any number of trials; it is not restricted to the infinite case. This is of immense importance. In frequentist theory, there is no way to reason from an observed frequency in a finite number of trials to the value of the probability (identified as long-term frequency). This is an awkward situation, because the theory by definition deals with long-term frequencies, but has no way of inferring their values from actual data.

Other connections between frequency and probability can also be derived within BPT by considering other propositions about frequencies. One connection of particular interest is a kind of consistency relationship between Bayes' Theorem and MAXENT. Testable information—information that refers directly to the relative probabilities of the events or hypotheses under consideration—cannot be used in Bayes' Theorem because such information does not refer to possible consequences in a single trial. But if we consider many repeated trials, and reinterpret the testable information as referring to relationships between relative frequencies in many trials rather than probabilities in single trials, we *can* use the information in BT to infer probabilities. For any finite number of trials, precise values of the probabilities will not be specified; rather, BT will provide a distribution for the values. But as the number of trials becomes infinite, the assignment from BT converges to the MAXENT assignment for a single trial (Jaynes 1978, 1982, 1988a; van Campenout and Cover 1981). This result has been used as a justification for MAXENT when the notion of repeated independent trials is meaningful. But MAXENT is not restricted to such cases.

## 4. Some Well-Posed Problems

Our theory so far is rather abstract; now we take a step toward concreteness by illustrating how two common types of statistical problems are addressed using BPT. We begin by noting that any problem we wish to address with BPT must be *well-posed*, in the sense that enough information must be provided to allow unambiguous assignment of all probabilities required in a calculation. As a bare minimum, this means that an exhaustive set of possibilities must be specified at the start of every problem.* We will call this set the *sample space* if it

---

* Readers familiar with Kolmogorov's measure theory approach to probability theory will

refers to possible outcomes of an experiment, or the *hypothesis space* if it specifies possible hypotheses we wish to assess.

Using experimental data to analyze parametrized models is an important task for physical scientists. The two classes of well-posed problems we will focus on here are designed for such analysis. They are called *estimation* and *model comparison.*** Estimation explores the consequences of assuming the truth of a particular model, and model comparison assesses a model by comparing it to one or more alternatives. These problems thus differ in regard to the specification of an hypothesis space. We discuss them in turn. Further details may be found in the excellent review of Bretthorst (1990).

## 4.1  BAYESIAN PARAMETER ESTIMATION

4.1.1 *Parametrized Models.* A parametrized model is just a set of exclusive hypotheses, each labeled by the value of one or more parameters. The parameters may be either continuous or discrete. For simplicity, we will focus attention on a model with a single parameter, $\theta$.

In an estimation problem one assumes that the model is true for *some* (unknown) value of its parameter, and explores the constraints imposed on the parameter by the data using BT. The hypothesis space for an estimation problem is thus the set of possible values of the parameter, $\mathcal{H} = \{\theta_i\}$. The data consist of one or more samples; to make the problem well-posed, the space of possible samples, $\mathcal{S} = \{s_i\}$, must also be specified. The hypothesis space, the sample space, or both can be either discrete or continuous.

Writing the unknown true value of the parameter as $\Theta$, we can use BT to address an estimation problem by calculating the probability that each of the possible parameter values is the true value. To do this, make the following identifications in equation (5). Let $D$ represent a proposition asserting the values of the data actually observed. Let $H$ be the proposition $\Theta = \theta$ asserting that one of the possible parameter values, $\theta$, is the true value (we will abbreviate this by just using the proposed value, $\theta$, as $H$ in BT). The background information $I$ will define our problem by specifying the hypothesis space, the sample space, how the hypotheses (parameter values) and sample values are related, and any additional information we may have about the hypotheses or the possible data. Symbolically, we might write $I$ as the proposition asserting (1) that the true value of the parameter is in $\mathcal{H}$; (2) that the observed data consisting of $N$ samples is in the space $\mathcal{S}^N$; (3) the manner in which the parameter value relates to the data, $I_r$; and (4) any additional information $I_A$; that is, $I = (\Theta \in \mathcal{H})(D \in \mathcal{S}^N)I_r I_A$. Of course, the physical nature of the model parameters and the data is implicit in the specification of $\mathcal{H}$, $\mathcal{S}$, and $I_r$.

Bayes' Theorem now reads*

$$p(\theta \mid DI) = p(\theta \mid I)\frac{p(D \mid \theta I)}{p(D \mid I)}. \tag{14}$$

recognize this as similar to the requirement that probabilities refer to elements of a $\sigma$-field. The close connection of BPT with Kolmogorov's theory is elaborated on in Jaynes (1990b).

**          Some model comparison problems are also called *significance tests* in the literature.

*          Bayes' Theorem refers to probabilities, not probability densities. Thus when considering continuous parameters, we technically should write $p(\theta \mid DI)d\theta = p(\theta \mid I)d\theta p(D \mid \theta I)dD/p(D \mid I)dD$, where the $p$'s are here understood to be densities. But the differentials cancel, so equation (14) is correct for densities as well as probabilities.

To use it, we need to know the three probabilities on the right hand side. The prior $p(\theta \mid I)$ and the likelihood $p(D \mid \theta I)$ are both direct probabilities and must be assigned *a priori* using the methods described previously; concrete examples are given below. The term in the denominator is independent of $\theta$. Given the prior and the likelihood, its value can be calculated using the probability axioms as follows.

First, recall that we are assuming the model to be true for *some* value of its parameter(s). Thus the proposition, "$\Theta = \theta_1$ or $\Theta = \theta_2$ or ..." is true, and so has a probability of 1, given $I$. Writing this proposition symbolically as $(\theta_1 + \theta_2 + \ldots)$, we thus have from axiom (2),

$$p(D[\theta_1 + \theta_2 + \ldots] \mid I) = p(D \mid I)p(\theta_1 + \theta_2 + \ldots \mid I)$$
$$= p(D \mid I). \tag{15}$$

But by expanding the logical product on the left, and again using (2), we also have

$$p(D[\theta_1 + \theta_2 + \ldots] \mid I) = p(D\theta_1 \mid I) + p(D\theta_2 \mid I) + \ldots$$
$$= \sum_i p(D\theta_i \mid I)$$
$$= \sum_i p(\theta_i \mid I)p(D \mid \theta_i I). \tag{16}$$

Equations (14) and (15) together imply that

$$p(D \mid I) = \sum_i p(\theta_i \mid I)p(D \mid \theta_i I). \tag{17}$$

This expresses $p(D \mid I)$ in terms of the prior and the likelihood, as promised. Each term in the sum is just the numerator of the posterior probability for each $\theta_i$. Thus in an estimation problem, the denominator of BT is just the normalization constant for the posterior. The probability, $p(D \mid I)$, is sometimes called the *prior predictive distribution*, since it is the probability with which one would predict the data, given only the prior information about the model. Though here it is just a normalization constant, it plays an important role in model comparison, as will be shown below.

The trick we just used to calculate $p(D \mid I)$—inserting a true compound proposition and expanding—arises frequently in BPT. It is just like expanding a function in a complete orthogonal basis; here we are expanding a *proposition* in a complete "orthogonal" basis. This trick is important enough to deserve a name: it is called *marginalization*.* The quantity $p(D \mid I)$ is sometimes called the *marginal likelihood* or the *global likelihood*. Of course, when dealing with continuous parameters, the sum becomes an integral, and (17) reads

$$p(D \mid I) = \int p(\theta \mid I)p(D \mid \theta I)d\theta. \tag{18}$$

Inserting the various probabilities into BT, we can calculate the posterior probabilities for all values of the parameters. The resulting probability distribution represents our inference about the parameters completely. As an important matter of interpretation, note that in this and any Bayesian distribution, it is the *probability* that is distributed, not the parameter

---

* This name is historical, and refers to the practice of listing joint probabilities of two discrete variables in a table, and listing in the *margins* the sums across the rows and columns.

(Jaynes 1986a). Stating an inference by saying something like "the parameter is distributed as a gaussian..." is misleading. The parameter had a single value during the experiment, and we want to infer something about this single value. We do not know it precisely, but the data tell us something about it. We express this incomplete knowledge by spreading our belief regarding the true value among the possible values according to the posterior distribution.

4.1.2 *Summarizing Inferences.* We can present the full posterior graphically or in a table. But usually we will want to summarize it with a few numbers. This will be especially true for multiparameter problems, where graphical or tabular display of the full posterior may be impossible because of the dimension of the parameter space. There are various ways to summarize a distribution, depending on what is going to be done with the information.

One summarizing item is a "best fit" value for the parameter. Which value to choose will depend on what is meant by "best". One obvious choice is the most probable parameter value, the *mode*. It is the best in the sense of being the single value one has greatest confidence in. But its selection does not reflect how our confidence is spread among other values at all. For example, if a distribution is very broad and flat with a small "bump" to one side, the mode will not be a good summarizing item, since most of the probability will be to one side of it. In this case, the *mean* of the distribution would be a better "best" value. On the other hand, if the distribution has two narrow peaks, the mean could lie between them at a place where the probability is small or even zero. So some common sense has to be used in choosing a best fit value.

There is a formal theory for making decisions about best fit values; it is called *decision theory* (Eadie *et al.* 1971; Berger 1985). Decision theory is very important in business and economics where one frequently must make a decision about a best value and then act as if it were true. But in the physical sciences, best values are usually just a convenient way to summarize a distribution. For this, common sense is usually a good enough guide, and a formal decision theory is not needed.

Besides a best value, it is useful to have a simple measure of how certain one is of this value. Again, decision theory can be brought to bear on this problem, but the traditional practice of quoting either the standard deviation (second moment about the mean) or the size of intervals containing specified fractions of the posterior probability is usually adequate. Of course, since probability is a measure of the plausibility of the parameter values, when we quote an interval, we should choose its boundaries so that all values inside it have higher probability than those outside. Such an interval is called a *credible region*, or a *highest posterior density interval* (HPD interval) when it is used to summarize the posterior distribution of one or more continuous parameters.*

In multiparameter problems, we may be interested only in certain subsets of the parameters. Depending on how many parameters are of interest, the distribution may be summarized in different ways. If the values of all of the parameters are of interest, a best fit point can be found straightforwardly by locating the mean or mode in the full parameter space. To quantify the uncertainty in the best fit point, all of the second moments can be calculated and presented as an $N \times N$ matrix; but off-diagonal moments are not an intuitively appealing measure of the width of the distribution. Alternatively, one can calculate

---

*        Sometimes the name *confidence interval* is given to credible intervals, and indeed it reflects well the intuitive meaning of a credible interval. But "confidence interval" has a technical meaning in frequentist theory that is different from its meaning here, and so we avoid this term.

an HPD region in the full parameter space, and present it by plotting its *projection* onto one, two, or three dimensional subspaces of the full parameter space. Some information is lost in such projections—the HPD region cannot be uniquely reconstructed from them — but they conservatively summarize the HPD region in the sense that they will show the full range of parameter values permitted in the region. They will also probably indicate the nature of any correlations among parameters, though two dimensional *cross sections* of the HPD better reveal correlations.

If only a subset of the parameters is of interest, the other parameters are called *nuisance parameters* and can be eliminated from consideration by marginalization. For example, if a problem has two parameters, $\theta$ and $\phi$, but we are interested only in $\theta$, then we can calculate $p(\theta \mid DI)$ from the full posterior $p(\theta\phi \mid DI)$ by using the trick we used to calculate $p(D \mid I)$. The result is $p(\theta \mid DI) = \int d\phi \, p(\theta\phi \mid DI)$; this is called the marginal distribution for $\theta$. Using BT and the product rule, the marginal distribution can be written

$$p(\theta \mid DI) = \frac{1}{p(D \mid I)} \int p(\phi \mid I) p(\theta \mid \phi I) p(D \mid \theta \phi I) d\phi. \tag{19}$$

Marginalization is of great practical and theoretical importance, because it can often be used to significantly reduce the dimensionality of a problem by eliminating nuisance parameters, making numerical calculations and graphical presentation much more tractable. Denied the concept of the probability of a parameter value, frequentist theory is unable to deal with nuisance parameters, except in special cases where intuition has led to results equivalent to marginalization (Lampton, Margon, and Bowyer 1976; Dawid 1980). Marginalization is thus an important technical advantage of BPT. It is a quantitative way of saying, in regard to the uninteresting parameters, "I don't know, and I don't care."

As useful and necessary as summaries of distributions are, we must always remember that the entire distribution is the full inference, not the summary.

## 4.2 BAYESIAN MODEL COMPARISON

Estimation problems assume the truth of the model under consideration. We often would like to test this assumption, calling into question the adequacy of a model. If the model is inadequate, then some alternative model must be better, and so BPT assesses a model by comparing it to one or more alternatives. This is done by assuming that some member of a set of competing models is true, and calculating the probability of each model, given the observed data, with BT. As we will see in Section 5, the Bayesian solution to this problem provides a beautiful quantification of Ockham's razor: simpler models are automatically preferred unless a more complicated model provides a significantly better fit to the data.

To use BT for model comparison, $I$ asserts that one of a set of models is true. This means that $I$ will have all the information needed to address an estimation problem for each model, plus any additional information $I_0$ that may lead us to prefer certain models over others *a priori*. Denote the information needed to address an estimation problem with model number $k$ as $I_k$ ($k = 1$ to $M$). Then symbolically we may write $I = (I_1 + I_2 + \ldots + I_M)I_0$. Let $D$ stand for the data, and let $k$ stand for the hypothesis, "Model number $k$ is true." BT can now be used to calculate the *probability of a model*:

$$p(k \mid DI) = p(k \mid I)\frac{p(D \mid kI)}{p(D \mid I)} \tag{20}$$

To use this, we must calculate the various probabilities. Here we will consider the case where we have no prior information preferring some models over the other, so the prior is $p(k \mid I) = 1/M$.

To calculate $p(D \mid kI)$, note that since $k$ asserts the truth of model number $k$, only the information $I_k$ in $I$ is relevant: $kI = k(I_1 + I_2 + \ldots)I_0 = I_k$. Thus, $p(D \mid kI) = p(D \mid I_k)$, the marginal likelihood for model $k$, described above. Labeling the parameters of model $k$ by $\theta_k$, this can be calculated from

$$p(D \mid kI) = \int d\theta_k p(\theta_k \mid I_k) p(D \mid \theta_k I_k). \qquad (21)$$

To calculate $p(D \mid I)$, we marginalize by inserting the true proposition ($k = 1 + k = 2 + \ldots$). This gives

$$p(D \mid I) = \sum_k p(k \mid I) p(D \mid kI). \qquad (22)$$

As in an estimation problem, $p(D \mid I)$ is simply a normalization constant. In model comparison problems, we can avoid having to calculate it by focusing attention on the ratios of the probabilities of the models, rather than the probabilities themselves. Such ratios are called *odds*, and the odds in favor of model $k$ over model $j$ we will write as $O_{kj} \equiv p(k \mid DI)/p(j \mid DI)$. From the above equations, the odds can be calculated from

$$O_{kj} = \left[ \frac{p(k \mid I)}{p(j \mid I)} \right] \frac{\int d\theta_k p(\theta_k \mid I_k) p(D \mid \theta_k I_k)}{\int d\theta_j p(\theta_j \mid I_j) p(D \mid \theta_j I_j)}$$

$$\equiv \left[ \frac{p(k \mid I)}{p(j \mid I)} \right] B_{kj}, \qquad (23)$$

where the factor in brackets is called the *prior odds* (and is here equal to 1), and $B_{kj}$ is called the *Bayes factor*. The Bayes' factor is just the ratio of the prior predictive probabilities, $B_{kj} = p(D \mid I_k)/p(D \mid I_j)$.

Equation (22) is the solution to the model comparison problem. In principle, such problems are little different from estimation problems; Bayes' theorem is used similarly, with an enlarged hypothesis space. In practice, more care must be exercised in calculating probabilities for models than for model parameters when there is little prior knowledge of the values of the parameters of the models under consideration. This is illustrated by way of an example in Section 5 below.

## 4.3 PROBLEMS WITH FREQUENTIST MODEL ASSESSMENT

As a basic principle for the design of well-posed problems, we have demanded that an exhaustive set of possibilities be specified at the beginning of any problem. In an estimation problem, this is accomplished by asserting the truth of a model, so that the hypotheses labeled by values of model parameters form an exhaustive set of alternatives. In model comparison, we satisfied this principle by explicitly specifying a set of competing models. How does this compare with frequentist methods for estimation and model assessment?

One of the most important frequentist statistics in the physical sciences is the $\chi^2$ statistic. It is used both for parameter estimation and for assessing the adequacy of a model (see, e.g., Lampton, Margon, and Bowyer 1976). The use of $\chi^2$ for obtaining best fit parameters and confidence regions is mathematically identical to Bayesian parameter estimation for models with gaussian "noise" probabilities and with flat priors for the parameters. This is because $\chi^2$ is proportional to the log of the likelihood when there is gaussian noise, and BT tells us that the posterior is proportional to the likelihood when the priors are flat.

Besides being used for estimation, frequentist theory also uses the $\chi^2$ statistic to assess an hypothesis by calculating the tail area above the minimum $\chi^2$ value in the $\chi^2$ distribution— the probability of seeing a $\chi^2$ value as large or larger than the best fit value if the model is true with its best fit parameters. This is very different in character from the Bayesian approach to model assessment. In particular, in this $\chi^2$ goodness-of-fit (GOF) test and other GOF tests (*e.g.*, the Kolmogorov-Smirnov test, the Smirnov-Cramer-von Mises test, etc.) *no explicit alternatives are specified*. At first sight, this seems to be an important advantage of frequentist theory, because it may be difficult to specify concrete alternatives to a model, and because it appears restrictive and subjective to have to specify an explicit set of alternatives to assess a model.

Deeper thought reveals this apparent advantage of frequentist GOF tests to be a defect, a defect that can be all the more insidious because its manifestations can be subtle and hidden. The resulting problems with GOF tests and other frequentist procedures that rely on tail areas began to be discussed openly in the statistics literature at least as early as the late 1930s (Jeffreys 1939), and continue to be expounded today (see Berger and Berry 1988 and references therein). Disturbingly, they are seldom mentioned in even the most recent frequentist texts. We will briefly note some of these important problems here.

4.3.1. *Reliance on Many Hypothetical Data Sets.* The $\chi^2$ GOF test is based on the calculation of the probability $P$ that $\chi^2$ values equal to or larger than that actually observed would be seen. If $P$ is too small (the critical value is usually 5%), the model is rejected. The earliest objections to the use of tests like $\chi^2$ focused on the reliance of such tests, not only on the probability of the observed value of the statistic, but on the probability of values that have not been observed as well. Jeffreys (1939) raised the issue with particular eloquence:

> What the use of $P$ implies, therefore, is that a hypothesis that may be true may be rejected because it has not predicted observable results that have not occurred. This seems a remarkable procedure. On the face of it the fact that such results have not occurred might more reasonably be taken as evidence for the law, not against it.

Indeed, many students of statistics find that the unusual logic of $P$-value reasoning takes some time to "get used to."

Later critics strengthened and quantified Jeffreys' criticism by showing how $P$-value reasoning can lead to surprising and anti-intuitive results. This is because the reliance of $P$-values on unobserved data makes them dependent on what one believes such data might have been. The intent of the experimenter can thus influence statistical inferences in disturbing ways, a phenomenon alluded to in Section 2.4.3 above, and known in the literature under the name *optional stopping*. Here is a simple example (after Iverson 1984, and Berger and Berry 1988).

Suppose a theorist predicts that the number of A stars in an open cluster should be a fraction $a = 0.1$ times the total number of stars in that cluster. An observer who wants to test this hypothesis studies the cluster and reports that his observations of 5 A stars out of 96 stars observed rejects the hypothesis at the 95% level, giving a $\chi^2$ $P$-value of 0.03. To check the observer's claim, the theorist calculates $\chi^2$ from the reported data, only to find that his hypothesis is acceptable, giving a $P$-value of 0.12. The observer checks his result, and insists he is correct. What is going on?

The theorist calculated $\chi^2$ as follows. If the total number of stars is $N = 96$, his prediction is $n_A = 9.6$ A stars and $n_X = 86.4$ other stars. Pearson invented the $\chi^2$ test for just such a problem; $\chi^2$ is calculated by squaring the difference between the observed and expected numbers for each group, dividing by the expected numbers, and summing (Eadie *et al.* 1971). From the predictions and the observations, the theorist calculates $\chi^2 = 2.45$, which has a $P$-value of 0.12, using the $\chi^2$ distribution for one degree of freedom (given $N$, $n_X$ is determined by $n_A$, so there is only one degree of freedom).

Unknown to the theorist, the observer planned his observations by deciding beforehand that he would observe until he found 5 A stars, and then stop. So instead of the number of A and non-A stars being random variables, with the sample size $N$ being fixed, the observer considers $n_{A,obs} = 5$ to be fixed, and the sample size as being the random variable. From the negative binomial distribution, the expected value of $N$ is $5/a = 50$, and the variance of the distribution for $N$ is $5(1-a)/a^2 = 450$. Using the observed $N = 96$ and the asymptotic normality of the negative binomial distribution, these give $\chi^2 = 4.70$ with one degree of freedom, giving a $P$-value of 0.03 as claimed.

The reason for the difference between the two analyses is due to different ideas of what other data sets might have been observed, resulting in different conclusions regarding what observed quantities should be treated as "random." But why should the plans of the observer regarding when to stop observing affect the inference made from the data? If, because of poor weather, his observing run had been cut short before he observed 5 A stars, how then should his analysis proceed? Should he include the probability of poor weather shortening the observations? If so, shouldn't he then include the probability of poor weather in the calculation when he *is* able to complete the observations?

Because of problems like this, some statisticians have adopted the *conditionality principle* as a guide for the design of statistical procedures. This principle asserts that only the data actually observed should be considered in a statistical procedure. Birnbaum (1962) gave this principle an intuitively compelling rationale through a *reductio ad absurdum* as follows. Suppose there are two experiments that may be performed to assess an hypothesis, but that only one can be performed with existing resources. A coin is flipped to determine which experiment to perform, and the data is obtained. If the data are analyzed with any method relying on $P$-values, we have to consider what other data might have been observed. But in doing so, should we consider the possibility that the coin could have landed with its other face up, and therefore consider all the data that might have come from the *other* experiment in our analysis? Most people's intuition compels them to assert that only data from the experiment actually performed should be relevant. Birnbaum argued that if this is accepted, the conditionality principle follows, and only the one data set actually obtained should be considered. Of course, BT obeys the conditionality principle, since it uses only the probability of the actually observed data in the likelihood and the marginal likelihood.

In the same work, Birnbaum shows that another technical criterion (sufficiency) already widely employed by statisticians implies with the conditionality principle that all the evidence of the data is contained in the likelihood function. This *likelihood principle* is also adhered to in BPT. Though widely discussed in the literature (see Berger and Wolpert 1984, and references therein), the likelihood principle has so far had little effect on statistics in the physical sciences.

As Jeffreys himself noted (Jeffreys 1939), the fundamental idea behind the use of $P$-values—that the observation of data that depart from the predictions of a model call the model into question—is natural. It is the expression of this principle in terms of $P$ that is unacceptable. The reason we would want to reject a model with large $\chi^2$ is not that $\chi^2$ is large, but that large values of $\chi^2$ are less probable than values near the expected value. But very small values, with $P$ near 1, are similarly unexpected, a fact not expressed by $P$-values.*

We have argued that only the probability of the actually observed $\chi^2$ value is relevant. But this probability is usually negligible even for the expected value of $\chi^2$ or any other GOF statistic. $P$-values adjust for this by considering hypothetical data. Bayes' Theorem remedies the problem by dividing this small probability by another small probability, the marginal likelihood. But the use of BT requires the specification of alternative hypotheses. The apparent absence of such alternatives in frequentist tests is the basis for the next two criticisms of such tests.

### 4.3.2. *Reliance on a Single Hypothesis.*
GOF tests require one to assume the truth of a single hypothesis, without reference to any alternatives. But this is clearly a weakness when such tests are used to evaluate parameterized models, because they require one to assume, not only that the model under consideration is true, but also that the best fit parameter values are the true values. This raises two questions regarding the logic of GOF tests.

First, if we decide to reject the hypothesis, then certainly we must reject probabilities calculated conditional on the truth of the hypothesis. But the $P$-value itself is such a probability! Thus when an hypothesis is rejected, tail area reasoning seems to invalidate itself. Bayes' theorem avoids this problem because rather than calculating probabilities of hypothetical data conditional on a single hypothesis, it calculates the probabilities of various hypotheses conditional on the observed data (Jaynes 1985c, 1986a).

Second, even if the model is true or adequate, it is almost certain that the best fit parameter values are not the true values. This again seems to put the logical status of the test in question, since its probabilities must always be calculated conditional on an hypothesis we are virtually certain is false. One might appeal to intuition and argue that if the model is rejected with its best fit parameter values, then surely the model as a whole must be rejected. But if the best fit model is acceptable, the acceptability of the model as a whole does not necessarily follow. For example, we feel that a model that produces a good fit over a wide range of its parameter space is to be preferred to a model with the same number of parameters but which requires parameter values to be carefully "fine-tuned" to explain the data; the data are a more natural consequence of the former model. Frequentist GOF tests have no way to account for such characteristics of a model, since they consider only the best fit parameter values.

Bayesian methods account for our uncertainty regarding the model parameters naturally and easily through marginalization. The probability of model $k$ is proportional to its marginal likelihood $p(D \mid I_k)$, which takes into account all possible parameter values. Bayesian methods also take this uncertainty into account when making predictions about future data. The probability of seeing data $D'$, given the observation of data $D$ and the

---

\* Many astronomers seem to consider a fit with, say, $\chi^2 = 16$ with 25 degrees of freedom ($P = 0.915$) to be better than one with, say, $\chi^2 = 27$ ($P = 0.356$); in fact, the former value of $\chi^2$ is 40% *less* probable than the latter, despite the fact that its $P$-value is over 2.5 times greater. To account for this, Lindley (1965) has advocated a 2-tailed $\chi^2$ test, in which $P$ is calculated by integrating over all less probable values of $\chi^2$, not just greater values.

truth of model $k$, is easily shown to be

$$p(D' \mid DI_k) = \int d\theta_k p(\theta_k \mid DI_k) p(D' \mid \theta_k I_k). \qquad (24)$$

This is called the *posterior predictive distribution*, and it is derived by marginalizing with respect to $\theta_k$. It says that the probability of $D'$ is just its average likelihood, taking the average over the posterior distribution for $\theta_k$ based on the observed data, $D$. Surely any model assessment based on how unexpected the data are should rely on the marginal likelihood or the predictive distribution, and not on distributions assuming the truth of particular parameter values.

### 4.3.3. *Implicit Alternatives: The Myth of Alternative-Free Tests.*

Though GOF tests appear to make no assumptions about alternatives, in fact the selection of a test statistic corresponds to an implicit selection of a class of alternatives. For example, the $\chi^2$ statistic is the sum of the squares of the residuals, and thus contains none of the information present in the order of the data points. The $\chi^2$ test is thus insensitive to patterns in the residuals, and will not account for small but statistically significant trends or features in the residuals in its assessment of an hypothesis. Thus the $\chi^2$ test implicitly assumes a class of alternatives for which the data are exchangeable, so that their order is irrelevant (Jaynes 1985c).

This characteristic of test statistics has long been recognized, beginning with the work of Neyman and Pearson (the inventor of $\chi^2$) in 1938. It has led to the characterization of statistical tests, not only by $P$-values, but also by their *power*, the probability that they correctly identify a true model against a particular alternative. But though modern statistical theory insists that tests be characterized both by $P$-values and by their power, few statistics texts for the physical sciences even mention the concept of power (Eadie *et al.* 1971 is a notable exception), and as a rule, the power of a test is never considered by physical scientists.

It is a far from trivial asset of Bayesian probability theory that by its very structure it forces us to specify a set of alternative hypotheses explicitly, in $I$, rather than implicitly in the choice of a test statistic.

### 4.3.4. *Violation of Consistency and Rationality.*

The many problems of alternative free GOF tests and tail area reasoning should come as no surprise in the light of the Cox-Jaynes derivation of the probability axioms. This is because the $P$-value is an attempt to find a real number measure of the plausibility of an hypothesis. But in Section 3 we saw that any such measure that is consistent with common sense and is internally consistent must be a monotonic function of the probability of the hypothesis. In general, a $P$-value will not be a monotonic function of the probability of the hypothesis under consideration, and so anti-intuitive and paradoxical behavior of $P$-value tests should be expected.

It also comes as no surprise that some of the most useful tail area tests have been shown to lead to $P$-values that *are* monotonic functions of Bayesian posterior probabilities with specific classes of alternatives, thus explaining the historical success of these tests (see, *e.g.*, Zellner and Siow 1980; Bernardo 1980). Of course, the Bayesian counterparts to these tests are superior to the originals because they reveal the assumed alternatives explicitly, showing how the test can be generalized; and because they produce a probability for the hypothesis being assessed, a more direct measure of the plausibility of the the the hypothesis than the $P$-value.

## 5. Bayesian and Frequentist Gaussian Inference

We will now apply BPT to a common and useful statistical problem: infering the amplitude of a signal in the presence of gaussian noise of known standard deviation $\sigma$, given the values $x_i$ of $N$ independent measurements. We will solve this problem with both frequentist and Bayesian methods. The Bayesian result is mathematically identical to the familiar frequentist result, but it is derived very differently and has a different interpretation. Bayesian and frequentist results will *not* be identical in general; our study will tell us about the conditions when identity may be expected.

### 5.1 THE STATUS OF THE GAUSSIAN DISTRIBUTION

We begin by first discussing the model: in what situations is a "gaussian noise" model appropriate?

In frequentist theory, the noise model should be the frequency distribution of the noise in an infinitely large number of repetitions of the experiment. But there is seldom even a moderate finite number of repetitions available to provide us with frequency data, so some other justification for the gaussian distribution must be offered. Sometimes it is used simply because it has convenient analytical properties. Often it is justified by appealing to the central limit theorem (CLT), which states that if the noise in a single sample is the the result of a number of independent random effects, the gaussian distribution will be a good approximation to the actual frequency distribution of the noise in many trials regardless of the distributions for each of.the effects, if the number of independent effects is large. But in general noise is not the result of a large number of independent effects; and even when it is, there is no way to be sure that the gaussian distribution is an adequate approximation for a finite number of effects without knowing the distributions describing each effect.

Bayesians interpret a noise distribution as an expression of our state of knowledge about the size of the noise contribution in the single data set actually being considered. Of course, if frequency data from many independent repetitions of an experiment are available, they will be relevant for assigning a noise distribution. But such data is typically not available, and the methods for assigning direct probabilities described in Section 3 must be used to find the quantitative expression of our state of knowledge about the noise.

Usually by noise we mean effects from unknown causes that we expect would "average out": positive and negative values are equally likely. Thus we expect the mean of the noise distribution to be zero. Additionally, we usually expect there to be a "typical scale" to the noise; we do not expect very large noise contributions to be as probable as smaller ones. Thus we expect the noise distribution to have some finite standard deviation, though we may not have a good idea what its value should be.

The information that a distribution have zero mean and standard deviation $\sigma$ is testable: given any distribution, we can see if its mean vanishes and if its second moment is $\sigma^2$. Thus we can use MAXENT to assign the noise distribution, using the zero mean and $\sigma$ as constraints. The resulting distribution is the gaussian distribution! Thus in BPT the gaussian distribution is appropriate whenever we know or consider it reasonable to assume that the noise has zero mean and finite standard deviation, but we do not have further details about it (Jaynes 1985c, 1987; Bretthorst 1988b, 1990). Additionally, we often need not specify the actual value of $\sigma$ if it is not known. We can consider it a parameter of our model, and estimate it from the data or marginalize it away.

The status of the gaussian distribution in BPT is thus very different from its status in frequentist theory. In BPT it simply represents the most conservative distribution consistent with minimal information about the noise phenomena, and it will be appropriate whenever such information is all we know about the noise, regardless of whether or not the CLT applies. This accounts for the great practical success of models assuming gaussian noise.

The reasoning used in BPT to assign the gaussian distribution can be easily generalized to other situations. For example, there is no single distribution for directional data on a circle or on a sphere that has all of the properties of the gaussian distribution on a line, and so there is some controversy over what distributions are the counterparts of the gaussian distribution for directional data (Mardia 1972). But if our knowledge is restricted to specification of a mean direction and an expected angular scale for deviations, then MAXENT identifies the correct distributions as the von Mises distribution for circular data and the Fisher distribution for spherical data (these distributions are discussed in Mardia 1972).

Having justified our model, we now describe the development of frequentist and Bayesian procedures for estimating the amplitude $\mu$ of a signal for which there are $N$ measurements $x_i$ contaminated with noise with standard deviation $\sigma$.

## 5.2  ESTIMATING THE SIGNAL

5.2.1.  *The Frequentist Approach.* In frequentist theory, since the signal strength $\mu$ is not a random variable taking on values according to a distribution, we are forbidden to speak of a probability distribution for $\mu$. But the $x_i$ *are* considered random variables, and their distribution is just gaussian,

$$p(x_i) = \frac{1}{\sigma\sqrt{2\pi}} \exp\left[-\frac{1}{2}\frac{x_i - \mu}{\sigma}\right]^2. \tag{25}$$

To estimate $\mu$, the frequentist must choose a statistic—a function of the random variables $x_i$—and calculate its distribution, connecting it with $\mu$. A few of the many possible statistics for estimating $\mu$ include $x_3$ (the value of the third sample); $(x_1 + x_N)/2$ (the mean of the first and last samples); the median of the observations; or their mean, $\bar{x} = \sum_i x_i/N$.

To choose from among these or other statistics, some criteria defining a "best" statistic must be invoked. For example, it is often required that a statistic be *unbiased*, that is, that the average value of the statistic in many repeated measurements converges to the true value of $\mu$. But the distributions for all of the above mentioned statistics can be calculated and reveal them *all* to be unbiased, so additional criteria must be specified. Unfortunately, all such criteria have a certain arbitrariness to them. For example, the criterion of unbiasedness focuses on the long-term mean value of the statistic. But the long-term median or most probable value would also reflect the intuitive notion behind the idea of bias, and in general would lead to a different choice of "best" statistic.

Of course, intuition suggests that to estimate the mean of a distribution, one should take the mean of the sample.* Various criteria of frequentist theory are chosen with this in mind, and eventually identify the mean, $\bar{x}$, as the "best" estimate of $\mu$.

Now we would like to know how certain we are that $\bar{x}$ is near the unknown true value of $\mu$. Interestingly, frequentist theory treats this problem as logically distinct from estimating

---

\*        Such intuitive reasoning does not always lead to good statistics; see Section 8.2.

best values, and in general completely different statistics and procedures can be used for these problems. In this simple gaussian problem, intuition again compels us to focus our attention on $\bar{x}$, and a *confidence region* for $\mu$ is found from $\bar{x}$ as follows.

Suppose $\mu$ were known. Then the distribution for $\bar{x}$ can be calculated from equation (25); a somewhat tedious calculation gives

$$p(\bar{x} \mid \mu) = \left[\frac{N}{2\pi\sigma^2}\right]^{1/2} \exp\left[-\frac{N}{2\sigma^2}(\bar{x} - \mu)^2\right]. \tag{26}$$

This distribution is a gaussian about $\mu$ with standard deviation $\sigma/\sqrt{N}$. With $\mu$ known, we can calculate the probability that $\bar{x}$ is in any interval $[a, b]$ by integrating (26) over this region with respect to $\bar{x}$. But when $\mu$ is unknown, this is not possible. However, since $p(\bar{x} \mid \mu)$ is a function only of the difference between $\bar{x}$ and $\mu$, we can always calculate the probability $\beta$ that $\bar{x}$ lies in some interval *relative to the unknown mean*, such as the interval $[\mu + c, \mu + d]$, and the result will be independent of $\mu$. Using equation (26), we find

$$\beta \equiv p(\mu + c < \bar{x} < \mu + d) = \frac{1}{2}\left[\text{erf}\left(\frac{d}{\sigma\sqrt{2/N}}\right) - \text{erf}\left(\frac{c}{\sigma\sqrt{2/N}}\right)\right]. \tag{27}$$

For a given $\beta$ of interest, there are many choices of $c$ and $d$ that satisfy (27). For example, for the "$1\sigma$" value $\beta = 68\%$, we may choose any of $[c, d] = [-\infty, x]$, $[-\sigma/\sqrt{N}, \sigma/\sqrt{N}]$, or $[-x, \infty]$. A priori, there is no reason to prefer any one of these to the others in frequentist theory, and again some criterion must be invoked to select one as "best" (Lampton, Margon, and Bowyer 1976). Popular criteria are to choose the smallest interval satisfying (27), or the symmetric one. For this problem, both criteria lead to the choice $[-\sigma/\sqrt{N}, \sigma/\sqrt{N}]$

In summary, the frequentist inference about $\mu$ might be stated by estimating $\mu$ with $\bar{x}$, and giving a "$1\sigma$" confidence interval of $\bar{x} \pm \sigma/\sqrt{N}$, the familiar "root $N$" rule.

### 5.2.2. The Bayesian Approach.

The Bayesian solution to this problem is to simply calculate the posterior distribution for $\mu$ using BT. We begin by specifying the background information $I$. $I$ will contain the information leading to the MAXENT assignment of a gaussian distribution for a single datum, equation (25). $I$ will also specify the hypothesis space, a range of possible values for $\mu$. We will assume we know $\mu$ to be in the range $[\mu_{min}, \mu_{max}]$; we discuss this assumption further below.

With this $I$, we must assign the prior and the likelihood. A simple consistency argument (Jaynes 1968) shows that the LIP assignment for $\mu$ is the uniform density,

$$p(\mu \mid I) = \frac{1}{\mu_{max} - \mu_{min}}. \tag{28}$$

The likelihood follows from (25) using the product rule: the joint probability of the $N$ independent observations is the product of their individual probabilities,

$$p(\{x_i\} \mid \mu I) = \frac{1}{\sigma^N(2\pi)^{N/2}} \exp\left[-\frac{1}{2\sigma^2}\sum_i(x_i - \mu)^2\right]$$

$$= \frac{1}{\sigma^N(2\pi)^{N/2}} \exp\left[-\frac{Ns^2}{2\sigma^2}\right] \exp\left[-\frac{N}{2\sigma^2}(\bar{x} - \mu)^2\right], \tag{29}$$

where we have separated out the dependence on $\mu$ by expanding the argument of the exponential and completing the square. Here $s^2$ is the sample variance, $s^2 = \sum_i (x_i - \bar{x})^2 / N$. Together, the prior and the likelihood determine the marginal likelihood to be

$$p(\{x_i\} \mid I) = \frac{1}{\sqrt{N}} (\sigma \sqrt{2\pi})^{1-N} \exp\left[-\frac{Ns^2}{2\sigma^2}\right] \frac{\operatorname{erf}\left(\frac{\bar{x} - \mu_{\max}}{\sigma\sqrt{2/N}}\right) - \operatorname{erf}\left(\frac{\bar{x} - \mu_{\min}}{\sigma\sqrt{2/N}}\right)}{2(\mu_{\max} - \mu_{\min})}, \qquad (30)$$

where the error functions arise from integrating (29) with respect to $\mu$ over the interval $[\mu_{\min}, \mu_{\max}]$. Equation (30) is constant with respect to $\mu$.

With these probabilities, BT gives our complete inference regarding $\mu$ as

$$p(\mu \mid \{x_i\}I) = \left[\frac{\operatorname{erf}\left(\frac{\bar{x} - \mu_{\max}}{\sigma\sqrt{2/N}}\right) - \operatorname{erf}\left(\frac{\bar{x} - \mu_{\min}}{\sigma\sqrt{2/N}}\right)}{2}\right] \left(\frac{N}{2\pi\sigma^2}\right)^{1/2} \exp\left[-\frac{N}{2\sigma^2}(\bar{x} - \mu)^2\right]. \quad (31)$$

This is just a gaussian about $\bar{x}$ with standard deviation $\sigma/\sqrt{N}$, truncated at $\mu_{\min}$ and $\mu_{\max}$. The factor in brackets is the part of the normalization constant due to the truncation.

As a best fit value, we might take the mode of the distribution, $\mu = \bar{x}$ (assuming that $\bar{x}$ is in the allowed range for $\mu$). Alternatively, we might take the mean. The mean value, and the limits of any HPD region, will depend on our prior range for $\mu$. But as long as the prior range is large compared to $\sigma/\sqrt{N}$, the effect of the prior range will be negligible. In fact, if we are initially completely ignorant of $\mu$, we can consider the limit $[\mu_{\min}, \mu_{\max}] \to [-\infty, \infty]$, for which the term in brackets becomes equal to 1. The mean is then the same as the mode, and the "$1\sigma$" HPD region is $\bar{x} \pm \sigma/\sqrt{N}$, the same as in the frequentist case.

5.2.3. *Comparison of Approaches.* Despite the mathematical identity of the Bayesian and frequentist solutions to this simple problem, the meaning of the results and their methods of derivation could hardly be more different.

First, the interpretations of the results are drastically different. To a Bayesian, $\bar{x}$ is the most plausible value of $\mu$ given the one set of data at hand, and there is a plausibility of 0.68 that $\mu$ is in the range $\bar{x} \pm \sigma/\sqrt{N}$. In contrast, the frequentist interpretation of the result is a statement about the long term performance of adopting the procedure of estimating $\mu$ with $\bar{x}$ and stating that the true value of $\mu$ is in the interval $\bar{x} \pm \sigma/\sqrt{N}$. Specifically, if one adopts this procedure, the average of the $\mu$ estimates after many observations will converge to the true value of $\mu$, and the statement about the interval containing $\mu$ will be true 68% of the time. Note that this is not a statement about the plausibility of the single value of $\bar{x}$ or the single confidence region actually calculated. Frequency theory can only make statements about the long-term performance of the adopted procedure, not about the confidence one can place in the results of the procedure for the one available data set.

Mathematically, these conceptual differences are reflected in the choice of the interesting variable in the final gaussian distributions, equations (26) and (31). The frequentist approach estimates $\mu$ and finds the probability content of a confidence region by integrating over possible values of $\bar{x}$, thus taking into consideration hypothetical data sets with different sample means than that observed. The Bayesian calculation finds the estimate and the probability content of an HPD region by integrating over $\mu$, that is, by considering different hypotheses about the unknown true value of $\mu$. The symbolic expression of frequentist and Bayesian interval probabilities expresses this difference precisely: The frequentist calculates $p(\mu - \sigma/\sqrt{N} < \bar{x} < \mu + \sigma/\sqrt{N})$, the fraction of the time that the sample mean will be

within $\sigma/\sqrt{N}$ of $\mu$ in many repetitions of the experiment. In contrast, the Bayesian calcu-
lates $p(\bar{x} - \sigma/\sqrt{N} < \mu < \bar{x} + \sigma/\sqrt{N} \mid DI)$, the probability that $\mu$ is within $\sigma/\sqrt{N}$ of the
sample mean of the one data set at hand.

The second important difference between the frequentist and Bayesian calculations is the
uniqueness and directness of the Bayesian approach. Frequentist theory could only produce
a unique procedure by appealing to *ad hoc* criteria such as unbiasedness and shortest confidence intervals. Yet such criteria are not generally valid (Jaynes 1976; Zellner 1986). For
example, there is a growing literature on biased estimators, because prior information or
evidence in the sample can identify a procedure that is appropriate for the case in consideration, but that would not have optimal long term behavior (Efron 1975; Zellner 1986).
In contrast, BPT provides a unique solution to any well posed problem, and this solution
is guaranteed by our desiderata to be the best one possible given the information actually
available, by rather inescapable criteria of rationality and consistency.

As a third important difference, we note that the frequentist calculation of the "covering
probability" of the confidence region depended on special properties of the distribution for
the statistic that was chosen. First, the statistic—the sample mean, $\bar{x}$—is what is called a
"sufficient statistic." This means that the $\mu$ dependence of the probability of the data (*i.e.*,
the likelihood, equation [29]) depends on the data only through the value of the single number $\bar{x}$, and not on any further information in the sample; a single number summarizes all of
the information in the sample, regardless of the size of $N$. Second, the sampling probability
of $\bar{x}$, equation (26), depends on $\mu$ and $\bar{x}$ only through their difference. These properties
permitted the calculation of the coverage probability without requiring knowledge of the
true value of $\mu$. Unfortunately, not all distributions have sufficient statistics, and of those
that do, few depend on the the sufficient statistics and the parameters only through their
differences (Lindley 1958). In general, then, a frequentist confidence region can only be defined approximately. In contrast, a Bayesian can always calculate an HPD region exactly,
regardless of the existence of sufficient statistics and without special requirements on the
form of the sampling distribution.

As a final, fourth difference, we note that the Bayesian result that is identical to the
frequentist result used a *least informative prior*. As soon as there is any cogent prior information about unknown parameter values, the Bayesian result will differ from frequentist
results, since the latter have no natural means for incorporation of prior information.

In summary, Bayesian and frequentist results will only be mathematically identical if
(1) there is only least informative prior information, (2) there are sufficient statistics, and
(3) the sampling distribution depends on the sufficient statistic and the parameters only
through their differences. Bayesian/frequentist equivalence is thus seen to be something of
a coincidence (Jeffreys 1937). When these conditions are not met, Bayesian and frequentist
results will generally differ (if a frequentist result exists!), and the Bayesian result will be
demonstrably superior, incorporating prior information and evidence in the sample that is
ignored in frequentist theory (Jaynes 1976).

5.2.4. *Improper Priors.* The Bayesian posterior becomes precisely identical to the frequentist sampling distribution when $[\mu_{min}, \mu_{max}] \to [-\infty, \infty]$. Interestingly, in this limit both
the prior (28) and the marginal likelihood (30) vanish, but they do so in such a way that the
ratio $p(\mu \mid I)/p(D \mid I)$ is nonzero. In fact, in this infinite limit, we can set the prior equal to
any constant, say $p(\mu \mid I) = 1$, and we will get the same result. Such a prior is not normalized, and is therefore called *improper*. It is frequently true in estimation problems that use
of improper priors gives the result that would be found by using a proper (normalizable)

prior and taking the limit. Improper priors then become convenient expressions of prior ignorance of the range of a parameter. It is usually Bayesian results based on improper priors that are mathematically equivalent to frequentist results.

In some estimation problems, and more frequently in model comparison problems, allowing parameter ranges in least informative priors to become infinite leads to unnormalizable or vanishing posterior probabilities. This is a signal that prior information about the allowed ranges of parameters is important in the result. In principle, we will demand that all probabilities be proper. This is never a serious restriction, for we always know *something* about the allowed parameter range. For example, in measuring the length of an object in the laboratory with a caliper, we know it can't be larger than the earth, nor smaller than an atom. We can put these limits in our prior, and we will almost always find that the posterior is independent of them to many, many significant figures; the data "overwhelms" the information in the prior range. In these cases we might as well use an improper prior as a kind of shorthand. On the other hand, if the result depends sensitively on the prior range, BPT is telling us that the information in the data is not sufficient to "overwhelm" our prior information, and so we had better think carefully about just what we know about the prior range. Or alternatively, we could try to get better data!

5.2.5. *The Robustness of Estimation.* Not only does the information in the data usually overwhelm the prior range; it also often overwhelms the actual shape of the prior, even when it is informative. This is best illustrated by example.

Suppose in our gaussian problem that our prior information indicated that $\mu$ was likely to be within some scale $\delta$ about some value $\mu_0$. This state of knowledge could be represented by a gaussian prior with mean $\mu_0$ and standard deviation $\delta$,

$$p(\mu \mid I) = \frac{1}{\delta\sqrt{2\pi}} \exp\left[-\frac{(\mu - \mu_0)^2}{2\delta^2}\right]. \tag{32}$$

Repeating the posterior calculation above with this prior, we find that the posterior mean $\hat{\mu}$ and variance $\sigma_\mu^2$ are now

$$\hat{\mu} = \frac{\bar{x} + \mu_0 \frac{\alpha}{N}}{1 + \frac{\alpha}{N}}, \tag{33}$$

and

$$\sigma_\mu^2 = \frac{\sigma^2}{N + \alpha}, \tag{34}$$

where $\alpha = \sigma/\delta$. Therefore, unless $\delta \lesssim \sigma/N$ (so that $\alpha \gtrsim N$), the posterior will not be significantly different from that calculated with a least informative prior.

This is an interesting result of some practical importance. The gaussian prior is clearly much more informative than the uniform prior, but unless the prior probability is very concentrated, with $s \leq \sigma/N$, it will have little affect on the posterior. This is not a very deep result; it is just what we should expect. It merely tells us that unless our prior information is as informative as the data, it will have little effect on our inferences. Of course, it is seldom the case that we have such prior information when we analyze an experiment; our lack of such information is why we perform experiments in the first place!

The practical import of this result is that if it is not clear exactly what prior probability assignment expresses our prior information, we might as well go ahead and use some simple "diffuse" prior that qualitatively respects any prior information we have (it should vanish outside the allowed parameter range!) and see if the result depends much on the prior. Usually it will not. This phenomenon has been variously referred to as the "stability" (Edwards *et al.* 1963) or "robustness" (Berger 1984, 1985) of estimation. Berger (1984, 1985) has extensively studied the robustness of many Bayesian calculations.

This is a special case of a more general practical rule: if a problem is not well posed, in the sense of there not being obvious ways of converting information to probability assignments, just do a calculation using some approximation (a diffuse prior, a simple likelihood, a simple hypothesis space) that does not do too much violence to the information at hand. Such simplified problems are often of great use by themselves (see Section 8.3 for an example), and their solution may provide the insight one needs to put enough structure on the original problem to make it well posed.

5.2.6. *Reference Priors.* A number of investigators have developed procedures for constructing diffuse priors for estimation problems in which we are in a least informative state of knowledge about parameter values, but do not know how to find the corresponding prior distribution. The robustness of estimation implies that the detailed shape of the prior is unimportant as long as it is diffuse compared to the likelihood function, so these procedures use properties of the likelihood function to "automatically" create a diffuse prior. Such a prior is often generically referred to as a "reference prior" (Box and Tiao 1973; Zellner 1977; Bernardo 1979): it is an "off-the-shelf" diffuse prior that many consider to be a useful objective starting point for analysis.

All such priors are based on the idea that one can think of the least informative state of knowledge pragmatically as the state of having little information *relative to what the experiment is expected to provide* (Rosenkrantz 1977). Unfortunately, several different procedures can been created to express this qualitative notion. Fortunately, many of them lead to the same reference prior for many common statistical problems, and these priors are often identical to least informative priors, when the latter are known.

Though several of the proposed reference priors are often identical to least informative priors in specific problems, this will not be true in general. In particular, since the form of a reference prior depends on the likelihood function, if we are estimating the same parameter in two different experiments, the reference prior will in general be different for the two experiments. This emphasizes that a reference prior does not describe an absolute state of ignorance about a parameter, but rather specifies a state of ignorance with respect to the experiment. To the extent that we choose experiments based on our prior information about the quantity we wish to measure, we expect the prior to depend on *some* properties of the likelihood function. After all, the $I$ that appears in the prior is the same $I$ that appears in the likelihood; the role the parameter plays in the likelihood is an important part of our prior information about the parameter (Jaynes 1968, 1980a). But the form of the likelihood can be determined in part by information that is irrelevant to the parameter value, information that would have no influence on a least informative prior, but that could affect a reference prior.

Despite these problems, reference priors can play a useful role in Bayesian parameter estimation because they produce diffuse priors that qualitatively express ignorance about parameters, and estimation is often robust with respect to the detailed form of a diffuse prior. Some of the reference priors that have been advocated include the invariant priors of Jeffreys and Huzurbazar (Jeffreys 1939); the indifferent conjugate priors of Novick and Hall (1965); the maximal data informative priors of Zellner (1971, 1977); the data translated likelihood priors of Box and Tiao (1973); and the reference priors of Bernardo (1979, 1980).

The multiparameter marginalization priors of Jaynes (1980a), where the priors for each of the parameters in a multiparameter model are chosen to ensure that they are uninformative about the other parameters, may also be considered to be reference priors, in that they are diffuse priors determined by the form of the likelihood.

## 5.3 MODEL COMPARISON

We can use this signal measurement example to illustrate some key features of Bayesian model comparison. Suppose there is some model, $M_1$, that gives a precise prediction of the signal: $\mu_{\text{true}} = \mu_1$. Suppose further that an alternative model, $M_2$, specifies only that $\mu_{\text{true}}$ is in some interval, $[\mu_{\text{min}}, \mu_{\text{max}}]$. Model $M_2$ has a single parameter, and model $M_1$ is a simple hypothesis, with no parameters.

Now suppose that we obtain some data, $D$, with a sample mean of $\bar{x}$. Which model is more plausible in light of this data? We can answer this with Bayes' Theorem, in the form of equation (20), or in the form of posterior odds, equation (22). To use it, we need the marginal likelihoods for $M_1$ and $M_2$. Since $M_1$ has no parameters, the marginal likelihood is just the likelihood itself;

$$p(D \mid I_1) = \frac{1}{\sigma^N (2\pi)^{N/2}} \exp\left[-\frac{\sum_i (x_i - \mu_1)^2}{2\sigma^2}\right]. \tag{35}$$

Model $M_2$ is the model assumed for the estimation problem we solved above; its marginal likelihood is given by equation (30). Together, these give the Bayes factor in favor of model $M_1$,

$$\begin{aligned} B_{12} &= \frac{p(D \mid I_1)}{p(D \mid I_2)} \\ &\approx \frac{\mu_{\text{max}} - \mu_{\text{min}}}{\sigma/\sqrt{N}} \frac{1}{\sqrt{2\pi}} \exp\left[-\frac{N}{2\sigma^2}(\mu_1 - \bar{x})^2\right], \end{aligned} \tag{36}$$

where we have assumed that $\mu_{\text{max}}$ and $\mu_{\text{min}}$ are large compared to $\sigma/\sqrt{N}$, and are far enough away from $\bar{x}$ that the last factor in equation (30) is very nearly equal to $1/(\mu_{\text{max}} - \mu_{\text{min}})$. This assumption amounts to saying that the experiment has measured $\mu$ more accurately than $M_2$ predicted it.

This result is very interesting. If $\bar{x}$ happens to equal $\mu_1$, $B_{12}$ will be large, favoring model $M_1$ which predicts that the true mean is $\mu_1$. But $B_{12}$ will continue to favor $M_1$ even when $\bar{x}$ is somewhat different from $\mu_1$, despite the fact that model $M_2$ with best-fit $\mu = \bar{x}$ fits the data slightly better than $M_1$. In effect, $M_2$ is being penalized for having a parameter and therefore being more complicated than $M_1$.

We can see this better if we note that the ratio of the best-fit likelihoods of the models, from equations (35) and (29), is

$$R_{12} = \exp\left[-\frac{N}{2\sigma^2}(\mu_1 - \bar{x})^2\right]. \tag{37}$$

Thus the Bayes factor can be written

$$\begin{aligned} B_{12} &= \frac{1}{\sqrt{2\pi}} \frac{\mu_{\text{max}} - \mu_{\text{min}}}{\sigma/\sqrt{N}} R_{12} \\ &\equiv S_{12} R_{12}. \end{aligned} \tag{38}$$

The best-fit likelihood ratio, $R_{12}$, can never favor model $M_1$; the more complicated model almost always fits the data better than a simpler model. But the factor $S_{12}$ favors the simpler model; it is called the "simplicity factor" or the "Ockham factor", and is a quantification of the rule known as "Ockham's Razor": Prefer the simpler model unless the more complicated model gives a significantly better fit (Jeffreys 1939; Jaynes 1980b; Gull 1988; Bretthorst 1990).

We can understand how the penalty for complication arises by recalling that the Bayes' factor is the ratio of the prior predictive probabilities of the models. Thus BT compares models by comparing how well each predicted the observed data. Crudely speaking, a complicated model can explain anything; thus, its prior predictive probability for any particular outcome is small, because the predictive probability is spread out more or less evenly among the many possible outcomes. But a simpler model is more constrained and limited in its ability to explain or fit data. As a result, its predictive distribution is concentrated on a subset of the possible outcomes. If the observed outcome is among those expected by the simpler model, BT favors the simpler model because it has better predicted the data.

In this sense, BT is the proper quantitative expression of the notion behind $P$-values: Assess an hypothesis by how well it predicts the data. To do so, BT uses only the probability of the actually observed data; additionally, it takes into account all of the possible parameter values through marginalization. This is in stark contrast to frequentist GOF tests, which consider the probabilities of hypothetical data, and assume the truth of the best-fit parameter values.

Equation (36) has a sensitive dependence on the prior range of the additional parameter that at first seems disconcerting. But a little thought reveals it to be an asset of the theory, something we might have expected and wanted. For example, suppose the alternative to model $M_2$ was some model $M_3$ which was just like $M_2$, but had a smaller allowed range for $\mu$. If the sample mean, $\bar{x}$, fell in a region of overlap between the models, the likelihood ratio $R_{32}$ would be 1, but $S_{32}$ would lead BT to favor $M_3$. If the value of $\bar{x}$ fell outside of the range for $\mu$ specified by $M_3$, BT might still favor $M_3$, depending on how far $\bar{x}$ is from the prediction of $M_3$. In this way, BT "knows" that $M_3$ is simpler or more constrained than $M_2$, even though both models are very similar, and in particular have the same number of parameters. Such behavior could not result if the Bayes factor somehow ignored the prior ranges of model parameters. A consequence of this dependence on the prior range is that model comparison problems are not as robust as estimation problems with regard to the prior range.

Here and in other problems we can deal with sensitivity to the prior by "turning Bayes' Theorem around" and asking how different kinds of prior information would affect the conclusions. For example, if we report the likelihood ratio, $R_{12}$, and the posterior variance for $\mu$, $\sigma_\mu = \sigma/\sqrt{N}$, then we know that the prior range for $\mu$ in model $M_2$ would have to have been smaller than $\sigma_\mu(2\pi)^{1/2}/R_{12}$ for us to just favor the more complicated model.

This kind of analysis can give us some insight into the common practice of accepting a new parameter if its value is significant at the "$2\sigma$" level. Taking $\mid \bar{x} - \mu_1 \mid = 2\sigma_\mu$, then $R_{12} = e^{-2}$, and the prior range that would make the Bayes factor indifferent between the models (giving $B_{12} = 1$) has a size of $\sigma_\mu(2\pi)^{1/2}/R_{12} = 18.5\sigma_\mu$. Thus the common practice of accepting a parameter significant at about the $2\sigma$ level corresponds to an initial state of uncertainty regarding the parameter value that is about one to two orders of magnitude greater than the uncertainty after the experiment.

The simple example we have worked here is more sensitive to the prior range than most realistic model comparison problems. Good examples of realistic model comparison problems in the physical sciences are discussed by Bretthorst (1988b, 1989a,b,c,d). Many additional model comparison problems have been worked in the Bayesian literature under the name, "significance testing". Important references include Jeffreys (1939), Zellner (1980), and Bernardo (1980).

## 6. Case Study: Measuring a Weak Counting Signal

We need only generalize the gaussian measurement problem slightly to obtain a problem that is both astrophysically interesting and resistant to frequentist analysis. We will consider in this section the measurement of a signal in the presence of a background rate that has been independently measured. We will consider signals that are measured by counting particles (photons, neutrinos, cosmic rays), so that the Poisson distribution is the appropriate sampling distribution.

The usual approach to this problem is to obtain an estimate of the background rate, $\hat{b}$, and its and standard deviation, $\sigma_b$, by observing an empty part of the sky, and an estimate of the signal plus background rate, $\hat{r}$, and its standard deviation, $\sigma_r$, by observing the region where a signal is expected. The signal rate is then estimated by $\hat{s} = \hat{r} - \hat{b}$, with variance $\sigma_s^2 = \sigma_r^2 + \sigma_b^2$. This procedure is the correct one for analyzing data regarding a signal which can be either positive or negative, when the gaussian distribution is appropriate. Thus it works well when the background and signal rates are both large so that the Poisson distribution is well-approximated by a gaussian. But when the rates are small, the procedure fails. It can lead to negative estimates of the signal rate, and even when it produces a positive estimate, both the value of the estimate and the size of the confidence region are corrupted because the method can include negative values of the signal in a confidence region.

These problems are particularly acute in gamma-ray and ultra-high energy astrophysics, where it is the rule rather than the exception that few particles are counted, but where one would nevertheless like to know what these sparse data indicate about a possible source. Given the weaknesses of the usual method, it is hardly surprising that more sophisticated statistical analyses of reported detections conclude that "not all the sources which have been mentioned can be confidently considered to be present" (O'Mongain 1973) and that "extreme caution must be exercised in drawing astrophysical conclusions from reports of the detection of cosmic $\gamma$-ray lines" (Cherry et al. 1980).

Three frequentist alternatives to the above procedure have been proposed by gamma-ray astronomers (Hearn 1969; O'Mongain 1973; Cherry et al. 1980). They improve on the usual method by using the Poisson distribution rather than the gaussian distribution to describe the data. But they have further weaknesses. First, all three procedures interpret a likelihood ratio as the covering probability of a confidence region, and thus are not even accurate frequentist procedures. Second, none of the procedures correctly accounts for the uncertainty in the background rate. Hearn (1969) uses the best-fit estimate of the background in his calculation, correcting the result afterward by using the gaussian propagation of error rule. O'Mongain (1973) tries to find 'conservative' results by using as a background estimate the best-fit value plus one standard deviation. Cherry et al. (1980) try to more carefully account for the background uncertainty by a method similar to marginalization; but strangely they only include integral values of the product of the background rate and the observing time in their analysis.

There are several reasons for the difficulty in finding a unique, optimal frequentist solution to this problem. First, there is important prior information in this problem: neither the signal nor the background can be negative. Second, there is a nuisance parameter: we want to estimate the signal, but to do so we must also consider possible values of the background. Third, the appropriate distribution is not the gaussian distribution, and cannot be written as a function of the difference between sufficient statistics and the relevant parameters; thus frequentist methods for finding confidence regions and dealing with nuisance parameters in the gaussian case do not apply.

Bayesian probability theory can deal with all these complications straightforwardly. The Bayesian solution to this problem is as follows.

First, the background rate, $b$, is measured by counting $n_b$ events in a time $T$ from an "empty" part of the sky. If we were interested in the value of $b$, we could estimate it from these data by taking prior information $I_b$ specifying the connection between $b$, $n_b$, and $T$; $I_b$ will identify the Poisson distribution as the likelihood function (see Jaynes 1990a for an instructive Bayesian derivation of the Poisson distribution). The likelihood function is thus

$$p(n_b \mid bI_b) = \frac{(bT)^{n_b} e^{-bT}}{n_b!}. \tag{38}$$

The least informative prior for the rate of a Poisson distribution can be derived from a simple group invariance argument, noting that $1/b$ plays the role of a scale for measurement of time (Jaynes 1968). The result is

$$p(b \mid I_b) = \frac{1}{b}. \tag{39}$$

This is called the "Jeffreys prior", since it was first introduced in similar problems by Jeffreys (1939). It corresponds to a prior that is uniform in $\log b$, and expresses complete ignorance regarding the scale of the background rate. As written here, it is improper. We can bound $b$ to make the prior proper, and take limits after calculating the posterior for $b$, but as long as $n_b$ is not zero, the limit will exist and be the same as if we just used equation (39) throughout the calculation. Of course, the prior probability for negative values of $b$ will be taken to be zero.

Given these probability distributions, the marginal likelihood is

$$p(n_b \mid I_b) = \frac{T^{n_b}}{n_b!} \int_0^\infty db\, b^{n_b-1} e^{-bT}$$
$$= \frac{1}{n_b}. \tag{40}$$

The posterior density for $b$ is then,

$$p(b \mid n_b I_b) = \frac{T(bT)^{n_b-1} e^{-bT}}{(n_b - 1)!}. \tag{41}$$

If we are interested in the background, we might summarize this posterior by noting its mean, $\langle b \rangle = n_b/T$, and its standard deviation, $n_b^{1/2}/T$, the usual "root $N$" result expected from a Poisson signal. With a prior that is different from equation (39), these values would be different, but not substantially so if $n_b$ is reasonably large. For example, a uniform prior would give a mean value of $(n_b + 1)/T$ and a standard deviation of $\sqrt{n_b + 1}/T$.

Now we count $n$ events in a time $t$ from a part of the sky where there is a suspected source. This measurement provides us with information about both $b$ and the source rate $s$. From BT, the joint posterior density for $s$ and $b$ is,

$$p(sb \mid nI) = p(sb \mid I)\frac{p(n \mid sbI)}{p(n \mid I)}$$

$$= p(s \mid bI)p(b \mid I)\frac{p(n \mid sbI)}{p(n \mid I)}. \tag{42}$$

Of course, the information $I$ includes the information from the background measurement, as well as additional information $I_s$ specifying the possible presence of a signal. Symbolically, $I = n_b I_b I_s$.

The likelihood is the Poisson distribution for a source with strength $s + b$:

$$p(n \mid sbI) = \frac{t^n(s+b)^n e^{-(s+b)t}}{n!}. \tag{43}$$

The prior for $s$, $p(s \mid bI)$, is the least informative prior for a Poisson rate $(s + b)$, with the value of $b$ given,

$$p(s \mid bI) = \frac{1}{s+b}. \tag{44}$$

Again, we take the prior probability to be zero for negative values of $s$. The prior for $b$ in this problem is *informative*, since we have the background data available. In fact, since $I_s$ is irrelevant to $b$, the prior for $b$ in this problem is just the posterior for $b$ from the background estimation problem, and is given by equation (41). Ignoring the normalization for now, BT gives the dependence of the posterior on the parameters as

$$p(sb \mid nI) \propto (s+b)^{n-1}b^{n_b-1}e^{-st}e^{-b(t+T)}. \tag{45}$$

Usually, we are only interested in the source strength. To find the posterior density for the source strength, *independent of the background*, we just marginalize with respect to $b$, calculating $p(s \mid nI) = \int db\, p(sb \mid nI)$. After expanding the binomial, $(s+b)^{n-1}$, the integral can be easily calculated. The resulting normalized posterior is,

$$p(s \mid nI) = \sum_{i=1}^{n} C_i \frac{t(st)^{i-1}e^{-st}}{(i-1)!}, \tag{46}$$

with

$$C_i \equiv \frac{(1+\frac{T}{t})^i \frac{(n+n_b-i-1)!}{(n-i)!}}{\sum_{j=1}^{n}(1+\frac{T}{t})^j \frac{(n+n_b-j-1)!}{(n-j)!}}. \tag{47}$$

Note that $\sum_{i=1}^{n} C_i = 1$.

This result is very appealing. Comparing it with equation (41), we see that BT estimates $s$ by taking a weighted average of the posteriors one would obtain attributing 1, 2,..., $n$ events to the signal. The weights depend on $n$, $t$, $n_b$, and $T$ so that the emphasis is placed on a weak signal or a strong signal, depending on how $n/t$ compares with $n_b/T$. Further development of this result, including application to real data, will appear elsewhere.

## 7. Case Study: Neutrinos from SN 1987A

The simple example of the previous section shows how straightforwardly Bayes' Theorem provides a solution to a well-posed problem that, despite its simplicity, has so far evaded straightforward frequentist analysis. Now we will discuss another problem that at first appears to be much more complicated, but which we will see is no more complicated in principle than the gaussian estimation problem discussed in Section 5.

In February of 1987, a supernova was observed in the Large Magellanic Cloud. This supernova, dubbed SN 1987A, was the closest one observed in the history of modern astronomy. Setting it apart from all other supernovae ever observed—indeed, from all other astrophysical sources ever observed, except for the Sun—is the fact that it was detected, not only in electromagnetic radiation, but also in neutrinos. Roughly two dozen neutrinos were detected from the supernova by the Japanese Kamiokande II (KII), Irvine-Michigan-Brookhaven (IMB), and Soviet Baksan detectors.

Neutrinos are believed to carry away about ninety-nine percent of the energy released by a supernova; the KII, IMB, and Baksan detections thus represent the first direct measurement of the energy of a supernova. In addition, neutrinos interact with matter so weakly that once they leave the collapsing stellar core, they pass unimpeded through the massive envelope of the supernova. Thus the detected neutrinos provide astrophysicists with their first glimpse of a collapsing stellar core. The analysis of the observations is therefore of great significance for testing supernova theory.

In addition, important information about intrinsic properties of the neutrino, such as its rest mass and electric charge, is contained in the data. This is because the 50 kpc path length between the Large Magellanic Cloud and Earth is vastly larger than that accessible in terrestrial laboratories.

Unfortunately, the weakness of neutrino interactions responsible for their usefulness as probes of stellar core dynamics also makes them extremely difficult to detect once they reach Earth. Of the approximately $10^{16}$ supernova neutrinos that passed through the detectors, only about two dozen were actually detected. Even these few events were not detected directly, but only by detecting tertiary photons they produced in the detectors. The small size of the data set, and the complicated relationship between properties of the incident neutrino signal and properties of the detected tertiary particles, demand careful, rigorous analysis of the implications of these data.

### 7.1  A BEWILDERING VARIETY OF FREQUENTIST ANALYSES

Within days after the landmark detection, the first contributions to what would soon become a vast literature analyzing the detected neutrinos appeared. Today, the two dozen supernova neutrinos are probably the most analyzed data set in the history of astrophysics, the number of published analyses far outnumbering the number of data. Unfortunately, nearly all of these analyses have *ad hoc* methodological elements, due to their frequentist inspiration.

With the exception of several qualitative moment analyses, most investigators analyzed the data by comparing them with parametrized models for the neutrino signal. With so few data, only the simplest signal models can be justified. But despite the simplicity of the models, the complexity of the detection process greatly complicates any frequentist analysis of the data, because the sampling distribution is extremely complex even for simple models.

No obvious sufficient statistics exist, and it would be difficult, if not impossible, to analyze the frequency behavior of statistics to identify unbiased, efficient estimators. A consequence of the lack of sufficient statistics is that frequentist confidence regions for parameters can only be found approximately.

All the usual frequentist criteria therefore founder on this problem, and investigators have been forced to rely on their intuitions and their Monte Carlo codes to create and calibrate statistics for their analyses. It is no wonder, therefore, that a bewildering variety of statistics and statistical methodologies has been applied to these data, yielding a similarly bewildering variety of results (see Loredo and Lamb 1989 for a review). Though many investigators used the maximum likelihood method to find best-fit parameters—a method with a Bayesian justification— several employed Pearson's method of moments, or invented their own statistics. A wide variety of methods were invented to calculate "confidence regions" for parameters, most of them confusing GOF $P$-values with covering probabilities. The majority of these methods relied on one-dimensional or two-dimensional Kolmogorov-Smirnov (KS) statistics, or similar goodness-of-fit statistics based on the cumulative distribution for the events, rather than the likelihood, even when the likelihood was used to find best-fit parameter values. Finally, very few studies considered more than one model for the neutrino emission. Usually, the adequacy of a single model was assumed without question; in some cases, adequacy was justified with an "alternative-free" goodness of fit test. A few studies explored several models, attempting to compare them with maximum likelihood ratios, but more complicated models always had larger likelihoods.

Testimony to the robustness of this problem, the results of many of these studies agree, if not precisely, at least qualitatively. But there is still troubling variety in the conclusions reached. For example, some investigators conclude that the observations are in conflict with soft equations of state for neutron star matter, though most conclude that the data are consistent with all reasonable equations of state, soft or hard. Some investigators claim the data indicate a small, nonzero electron antineutrino mass of a few eV, while most claim that the data only indicate an upper limit on the mass in the 15 to 20 eV range. The wide variety of statistical methods used in these investigations, and the variety in the models assumed for the neutrino emission and detection processes, make the literature on the supernova neutrinos appear muddled and confused. In the context of frequentist theory, there is no compelling criterion for making a judgement about the relative soundness of one analysis compared to another. Some scientists, in an attempt to summarize the analyses, have been forced to do "statistical statistics", averaging the results of different studies.

The majority of these studies were not even good frequentist analyses. In particular, many investigators identified "95% confidence regions" with the range of parameter values that had goodness-of-fit $P$-values of greater than 5%, based on a flawed definition of a confidence region. These investigators did not notice that their best-fit $P$-values of $\approx 0.80$ implied that "confidence regions" with probability smaller than about 20% *could not even be defined* with their methods. But this is almost beside the point. The emphasis of frequentist statistics on averages over hypothetical random experiments, and the lack of a clear rationale for the choice of statistics, has led to a "Monte Carlo Mystique" in astronomical statistics whereby almost any calculation relying on a sufficient number of simulated data sets is deemed a "rigorous" statistical analysis.

Alone among these analyses is the work of Kolb, Stebbins, and Turner (KST, 1987). They focus on one interesting parameter—the mass of the electron antineutrino, $m_{\bar{\nu}_e}$,—and setting aside all of the fancy statistics and Monte Carlo codes, ask instead what careful intuitive reasoning about the data can reveal about $m_{\bar{\nu}_e}$. They conclude that at best, the data can put an upper limit on $m_{\bar{\nu}_e}$ of the order of 25 to 30 eV, not significantly better than current laboratory limits. Later detailed statistical studies found "95% confidence" limits ranging from 5 eV to 19 eV. Significantly, some recent reviews of the observations downplay these later studies and emphasize the qualitative KST limit, testimony to the lack of confidence scientists have in the statistical methods of astrophysicists.

## 7.2 THE BAYESIAN ANALYSIS

The Bayesian analysis of the neutrino data has been presented by Loredo and Lamb (1989; 1990a,b). They estimate parameters for simple neutrino emission models using Bayes' Theorem with uniform priors. This calculation is as straightforward in principle as the gaussian calculation of Section 5; the only complications are computational, arising from the complexity of the detector response and the dimensionality of the parameter spaces.

The data produced by the detectors are the detected energies, $\epsilon_i^{\text{det}}$, and arrival times, $t_i^{\text{det}}$, of the detected neutrinos. To analyze these data, Loredo and Lamb consider a variety of parametrized models for the neutrino emission, and use Bayes' Theorem to estimate the model parameters and to compare alternative models. Given a model for the neutrino emission rate, a predicted detection rate per unit time and unit energy, $d^2 N_{\text{det}}/d\epsilon^{\text{det}} dt^{\text{det}}$, can be calculated using the response function of the experiment. From this detection rate, the likelihood function needed in Bayes' Theorem can be constructed as follows.

The expected number of neutrinos detected in a small time interval, $\Delta t$, and a small energy interval, $\Delta \epsilon$, is just the detection rate times $\Delta t \Delta \epsilon$. From the Poisson distribution, the probability that no neutrinos will be detected within these intervals about a specified energy and time is

$$P_0(\epsilon^{\text{det}}, t^{\text{det}}) = \exp\left[-\frac{d^2 N_{\text{det}}(\epsilon^{\text{det}}, t^{\text{det}})}{d\epsilon^{\text{det}} dt^{\text{det}}} \Delta\epsilon\Delta t\right]. \tag{48}$$

Similarly, the probability that a single neutrino will be detected in the interval is

$$P_1(\epsilon^{\text{det}}, t^{\text{det}}) = \frac{d^2 N_{\text{det}}(\epsilon^{\text{det}}, t^{\text{det}})}{d\epsilon^{\text{det}} dt^{\text{det}}} \Delta\epsilon\Delta t \exp\left[-\frac{d^2 N_{\text{det}}(\epsilon^{\text{det}}, t^{\text{det}})}{d\epsilon^{\text{det}} dt^{\text{det}}} \Delta\epsilon\Delta t\right]. \tag{49}$$

The intervals are chosen small enough that the probability of detecting more than one neutrino is negligible compared to $P_0$ and $P_1$.

The likelihood of a particular observation is the product of the probabilities of detection of each of the $N_{\text{obs}}$ observed neutrinos, times the product over all intervals not containing a neutrino of the probability of no detection. That is,

$$\mathcal{L} = \left[\prod_{i=1}^{N_{\text{obs}}} P_1(\epsilon_i^{\text{det}}, t_i^{\text{det}})\right] \prod_j P_0(\epsilon_j^{\text{det}}, t_j^{\text{det}}), \tag{50}$$

where $j$ runs over all intervals not containing an event. It is more convenient to work with the log likelihood, $L = \ln(\mathcal{L})$. From the definitions of $P_0$ and $P_1$ it follows that

$$L = \sum_{i=1}^{N_{\text{obs}}} \ln\left[\frac{d^2 N_{\text{det}}(\epsilon_i^{\text{det}}, t_i^{\text{det}})}{d\epsilon^{\text{det}} dt^{\text{det}}} \Delta\epsilon\Delta t\right] - \sum_j \frac{d^2 N_{\text{det}}(\epsilon_j^{\text{det}}, t_j^{\text{det}})}{d\epsilon^{\text{det}} dt^{\text{det}}} \Delta\epsilon\Delta t, \tag{51}$$

where $j$ now runs over all intervals. In the limit of small $\Delta\epsilon$ and $\Delta t$, the second term becomes the integral of the rate function over all time and all energy. Thus the log likelihood is

$$L = \sum_{i=1}^{N_{obs}} \ln\left[\frac{d^2 N_{det}(\epsilon_i^{det}, t_i^{det})}{d\epsilon^{det} dt^{det}}\right] - \int_0^{t_{dur}} dt \int_0^\infty d\epsilon^{det} \frac{d^2 N_{det}(\epsilon^{det}, t^{det})}{d\epsilon^{det} dt^{det}}$$

$$= \sum_{i=1}^{N_{obs}} \ln\left[\frac{d^2 N_{det}(\epsilon_i^{det}, t_i^{det})}{d\epsilon^{det} dt^{det}}\right] - N_{det}, \tag{52}$$

where $t_{dur}$ is the duration of the time interval under study, and $N_{det}$ is the total number of events expected to be detected in that interval. In this equation, the intervals $\Delta\epsilon$ and $\Delta t$ have been omitted because they are constants that do not affect the functional dependence of $L$ on the detected rate function.

Equation (52) is the final form for the likelihood function. Combined with prior probability densities for the parameters (Loredo and Lamb [1989] assume uniform priors), it yields a posterior distribution for the model parameters. The calculation, though straightforward in principle, is complicated in practice because the response functions of the detectors are complicated. This is because the neutrinos are not detected directly; rather, tertiary photons produced in the detectors by the neutrinos are detected, leading to a complicated relationship between detected photon energy and the energy of the incident neutrino. As a result, calculation and study of the posterior distribution requires the resources of a supercomputer. Details are presented in Loredo and Lamb (1989, 1990a,b).

These calculations show that the observations are in spectacular agreement with the salient features of the theory of stellar collapse and neutron star formation which had developed over several decades in the absence of direct observational data. In particular, the inferred radius and binding energy of the neutron star formed by the supernova are in excellent agreement with model calculations based on a wide range of equations of state, despite earlier indications to the contrary.

These calculations also show that the upper limit on the mass of the electron antineutrino implied by the observations is 25 eV at the 95% confidence level, 1.5 to 5 times higher than found previously, and not significantly better than current laboratory limits.

This work demonstrates the value of using correct and rigorous Bayesian methods for the analysis of astrophysical data, and shows that such an analysis is not only possible, but straightforward, even when the data are related to the physical quantities of interest in a very complicated manner.

## 8. Where to Go from Here

Bayesian probability theory, as described here, is impressive in its simplicity and its scope. Desiderata of appealing simplicity lead to its rules for assignment and manipulation of probabilities, which are themselves extremely simple. Its identification of probability with plausibility makes it a theory of drastically broader scope than traditional frequentist statistics. This broad scope adds to the simplicity and unity of the theory, for whenever we wish to make a judgement of the truth or falsity of any proposition, $A$, the correct procedure is to calculate the probability, $p(A \mid E)$, that $A$ is true, conditional on all the evidence, $E$, available, regardless of whether $A$ refers to what would traditionally be called a random variable or a more general hypothesis (Jaynes 1990b). In most cases, this calculation will involve the use of Bayes' Theorem.

Because of its broad scope, BPT is more than merely a theory of statistics. It is a theory of *inference*, a generalization of deductive inference to cases where the truth of a proposition is uncertain because the available information is incomplete. As such, it deserves to be a familiar element of every scientist's collection of general methods and tools.

Of course, the theory is ideally suited for application to problems traditionally classified as "statistical". There, it promises to simplify and unify statistical practice. Indeed, it is already doing so in the fields of mathematical statistics, econometrics, and medicine. Astrophysicists have been slow to reap the benefits of the theory, but several applications relevant to astrophysics have been worked out. We will describe some here, as an entrance to the expanding literature on Bayesian methods.

## 8.1 ASTROPHYSICAL ESTIMATION AND MODEL COMPARISON PROBLEMS

Because of the prevalence of the gaussian distribution in statistical problems, many frequentist parameter estimation calculations will be equivalent to their Bayesian counterparts, provided that there are no nuisance parameters and that there is no important prior information about parameter values. But when there are nuisance parameters, or when there is important prior information, Bayesian methods should prove superior to frequentist methods, if the latter even exist for such problems. Also, if the relevant distributions are more complicated than gaussian, lacking obvious sufficient statistics, Bayesian methods will almost certainly prove superior to frequentist methods, and will be easier to derive.

Problems for which Bayesian methods will provide demonstrable advantages are only beginning to be be identified and studied. All such problems are approached in a unified manner using Bayes' Theorem, eliminating any nuisance parameters through marginalization. The signal measurement and supernova neutrino problems mentioned above are examples.

Another example is the analysis of "blurred" images of point sources in an attempt to resolve closely spaced objects (Jaynes 1988; Bretthorst and Smith 1989). In this problem, some of the parameters specifying the locations of objects are nuisance parameters, since it is their *relative* positions that are of interest. Further, the noise level is not always known; in the Bayesian calculation it, too, can be a nuisance parameter to be eliminated by marginalization, effectively letting Bayes' Theorem estimate the noise from the data. Finally, the brightnesses of the two or more possible objects can be marginalized away, leaving a probability density that is a function only of relative position between objects, and which answers the question, "Is there evidence in the data for an object at this position relative to another object?" In analyzing an image for the presence of two objects, the Bayesian procedure can thus reduce the dimensionality of the problem from seven (two two-dimensional positions, two brightnesses, and the noise level) to one (the relative separation of the objects). Of course, once the relative separation posterior is studied and found to reveal the presence of closely spaced objects, their intensities and positions can be calculated, using knowledge of their relative separation to simplify analysis of the full posterior.

Analytical work (Jaynes 1988) and numerical work analyzing simulated data (Bretthorst and Smith 1989) indicate that the Bayesian algorithm can easily resolve objects at separations of less than one pixel, depending on the signal-to-noise ratio of the data. Further, model comparison methods can be used to determine the number of point sources for which there is significant evidence in the data. Significantly, the calculation also reveals that the usual practice of apodizing an optical system to smooth out the sidelobes of the point spread function destroys significant information that the Bayesian calculation can use to resolve objects (Jaynes 1988). Apodizing leads to a smoother image that is less confusing to the eye, but it destroys much of the information in the sidelobes that probability theory can use to improve resolution. This work awaits application to real data, and extension to other similar problems, such as the analysis of data from optical interferometers.

## 8.2 BAYESIAN SPECTRUM ANALYSIS

One class of statistical problems is of such great importance in astrophysics that it deserves special consideration: the analysis of astrophysical time series data for evidence of periodic signals. This problem is usually referred to as *spectrum analysis*. In the past three years, new Bayesian spectrum analysis methods have been developed that offer order-of-magnitude greater frequency resolution than current methods based on the discrete Fourier transform (DFT). Additionally, they can be used to detect periodicity in amplitude modulated signals or more complicated signals with much greater sensitivity than DFT methods, without requiring the data to be evenly spaced in time.

Current frequentist methods seek information about the spectrum of the *signal* by calculating the spectrum of the *data* via the discrete Fourier transform (DFT). But the presence of noise and the finite length of the data sample make the data spectrum a poor estimate of the signal spectrum. As a result, *ad hoc* methods are used to "correct" the data spectrum, involving various degrees of smoothing (to eliminate spurious peaks). The statistical properties of the result are analyzed assuming the signal is just noise, to try to find the "false alarm" probability of an apparent spectral feature being due to noise. (Good reviews of these methods are in Press, *et al.* 1986, and van der Klis 1989.)

In contrast, Bayesian methods (Jaynes 1987; Bretthorst 1989, 1990) assess the significance of a possible signal by directly calculating the probabilities that the data are due to a periodic signal or to noise, and comparing them. To estimate the frequency of a signal, these methods simply calculate the probability of a signal as a function of its frequency, marginalizing away the phase and amplitude of the signal.

Using these methods, Jaynes (1987) derived the DFT as the appropriate statistic to use when analyzing a signal with a single sinusoid present. His work shows how to manipulate the DFT without smoothing to get an optimal frequency estimate that can have orders-of-magnitude greater resolution than current methods. Bretthorst (1989, 1990) has extended Jaynes' work, showing analytically and with simulated and actual data that the DFT is *not* appropriate for the analysis of signals with more complicated structure than a single sinusoid, and that Bayesian methods give much more reliable and informative results. In particular, Bayesian methods can easily resolve two frequencies that are so close together that there is only a single peak in the DFT of the data, simply by considering a model with more than one sinusoid present. As was the case in the analysis of blurred images just discussed, probability theory uses information in the sidelobes to improve resolution, information that is thrown away by the standard Blackman-Tukey smoothing methods. Model comparison calculations can be used to identify how many sinusoids there is evidence for in the data. Bretthorst (1988a,b) has applied these methods to Wolf's sunspot data, comparing the results of the Bayesian analysis with conventional DFT results.

For signals that are not stationary, such as chirped or damped signals, the DFT spreads the signal power over a range of frequencies. However, if the general form of the signal is known, Bayesian generalizations of the DFT can be constructed that take into account the possibility that the signal has some unknown chirp or decay rate, effectively concentrating all of the signal power into a single frequency, thereby greatly improving detection sensitivity for such signals. These methods should prove to be of immense value for the study of nonstationary astrophysical time series, such as those observed from the "quasi-periodic oscillator" x-ray sources, or those expected from sources of gravitational radiation. In particular, the gravitational radiation signal expected from coalescing binaries is chirped, so the "chirpogram" introduced by Jaynes (1987) and further studied by Bretthorst (1988a,b) should play an important role in the analysis of gravitational wave signals. An integrated circuit is currently being developed to facilitate rapid calculation of the chirpogram (Erickson, Neudorfer, and Smith 1989).

## 8.3 INVERSE PROBLEMS

Problems that are mathematically ill-posed in the sense of being underdetermined arise frequently in astrophysics; they are usually called *inverse problems*. Examples include calculating the interior structure of the sun from helioseismology data, calculating radio images from interferometric data, "deblurring" optical or x-ray images, or estimating a spectrum from proportional counter or scintillator data. Abstractly, all of these problems have the following form. Some unknown signal, $s$, produces data, $d$, according to

$$d = Rs + e, \tag{53}$$

where $R$ is a complicated operator we will call the response function of the experiment, and $e$ represents an error or noise term. Given $d$, $R$, and some incomplete information about $e$, we wish to estimate $s$. Such problems can be ill-posed in three senses.

First, the response operator is usually singular in the sense that a unique inverse operator, $R^{-1}$, does not exist. Put another way, there exists a class, $A$, of signals such that $Rs = 0$ for any $s$ in $A$. Thus $d$ contains no information about such signals, so that even the noiseless "pure inverse problem" of solving $d = Rs$ for $s$ does not have a unique solution: any element of $A$ can be added to any solution to give another solution. The set $A$ is called the *annihilator* of $R$. It exists because the "blurring" action of $R$ destroys information about finely structured signals.

Second, the presence of noise effectively enlarges the annihilator of $R$, since signals $s$ such that $Rs = \epsilon$, with $\epsilon$ small compared to the expected noise level, can be added to any possible solution to obtain another acceptable solution. In practice, this is revealed by instability in any attempt to directly invert equation (48), small changes in the data resulting in large changes in the estimated signal.

Finally, the data, $d$, are usually discrete and finite in number, and the signal, $s = s(x)$, is usually continuous. Thus, even if $R$ were not singular and there were no noise, estimating $s(x)$ from $d$ would still be severely underdetermined.

One approach to such ill-posed problems is to make them well-posed by studying simple parameterized models for the signal. The resulting estimation problem can be addressed straightforwardly with Bayes' Theorem. But often, one would like "model-independent" information about the signal, $s(x)$.

Frequentist approaches to this problem fall into two classes. *Regularization methods* estimate the signal by invoking criteria to select one member of the set of all possible signals that are consistent with the data as being "best" in some sense. *Resolution methods* try to determine what features all the feasible signals have in common by estimating resolvable averages of them. All such methods have obvious *ad hoc* elements—the choice of regularizer, or the choice of a measure of resolution—and there are usually many methods available for solving a particular problem. In recent years, the importance of using prior information to guide development of an inverse method has been greatly emphasized (Frieden 1975; Narayan and Nityanada 1986). Unfortunately, it is not clear how to optimally use even the simplest prior information, such as the positivity of the signal, to develop a frequentist inverse method.

The Bayesian approach to inverse problems is to *always* address them as estimation problems via Bayes' Theorem. They differ from other more common estimation problems only in the character of the model assumed. In particular, the model will usually have more parameters than there are data. Prior information, taken into account through prior probabilities, is what makes such problems well-posed despite the discrepancy between the number of data and the number of parameters.

Bayesian solutions to inverse problems are only beginning to be developed and understood. Only the simplest kinds of models and prior information have yet been explored. Surprisingly, the resulting methods are usually as good as any existing frequentist methods, and are sometimes significantly better. These methods are the *Maximum Entropy Methods* prominent in these Proceedings, though the "entropy" which plays such an important role in these methods is *not* the entropy described in Section 3, above.

Bayesian inversion methods, including the popular maximum entropy methods, can be developed as follows (Jaynes 1984a,b). Consider estimating a one-dimensional signal, $s(x)$. Begin by discretizing the problem, seeking to estimate the finite number of values $s_j \equiv s(x_j)$, $j = 1$ to $M$; $M$ may be much larger than the number of data. The "parameters" of our model are thus just the $M$ values of the discrete signal. Using Bayes' theorem, we can calculate the posterior probability of a signal, given the data, the response function, and information about the noise:

$$p(\{s_j\} \mid DI) = p(\{s_j\} \mid I)\frac{p(D \mid \{s_j\}I)}{p(D \mid I)}. \tag{54}$$

The likelihood function will be determined by our information about the noise; if the information leads to a gaussian noise distribution, the log likelihood will just be proportional to $\chi^2$. The critical element of the problem is the assignment of prior probabilities to the $s_j$. Uniform priors clearly will not do, for then all of the possible signals that fit the data will be equally likely, and the problem will remain underdetermined. Intuitively, we reject many of the possible signals—for example, wildly oscillating signals—because our prior information about the nature of the true signal makes it extremely unlikely that it could have been one of the many unappealing but possible signals. We must find a way to encode some of this information numerically in a prior probability assignment over the $s_j$.

The natural way to proceed is to specify precisely the available information, $I$, and use the principles discussed in Section 3.3 to assign the prior, $p(\{s_j\} \mid I)$. The information will probably be of the form of a specification of the nature of the alternatives, $I_0$, and some additional testable information, $E$. The information $I_0$ will lead to a least informative distribution, $p(\{s_j\} \mid I_0)$. For example, if the signal $s(x)$ must by nature be positive, the LIP distribution for $\{s_j\}$ might be a product of Jeffreys priors, $p(\{s_j\} \mid I_0) = \prod 1/s_j$. The testable information, $E$, could include, for example, information about the expected scale of detail in the signal, in the form of prior covariances among the $s_j$. This information would be used to identify the appropriate informative prior for the signal by MAXENT. The entropy of the distribution $p(\{s_j\})$ needed to use MAXENT is calculated by integrating over the values of the $s_j$ variables,

$$H[p(\{s_j\})] = -\int ds_1 \ldots \int ds_M \, p(\{s_j\}) \log \left[\frac{p(\{s_j\})}{m(\{s_j\})}\right], \qquad (55)$$

where $m(\{s_j\})$ is the LIP assignment for $\{s_j\}$. The informative distribution is the one with highest entropy, $H[p(\{s_j\})]$, among all those that satisfy the constraints imposed by $E$, and could be found (at least in principle) by the method of Lagrange multipliers.

For historical reasons, this is not the approach that has been taken in assigning a prior for the signal, though it is a promising direction for future research. Instead, a prior has been constructed by choosing an alternative space of hypotheses than the $s_j$, from which the $s_j$ values can be derived, but whose nature permits an unambiguous and appealing prior probability assignment.

The well-known maximum entropy inversion methods arise from a particularly simple alternative hypothesis space created as follows (Gull and Daniel 1978; Jaynes 1982, 1984a,b; Skilling 1986). First, discretize the $M$ signal values into some large number, $N$, of independent "signal elements" of size $\delta s$.* Then build a signal by taking the $N$ signal elements one at a time and putting them in one of the $M$ signal bins. A signal is built once each of the $N$ elements have been placed into a bin; we will call such a signal a "microsignal". The new hypothesis space is the set of the $M^N$ possible resulting microsignals, and as a least informative assignment, we will consider each of them to be equally probable, with probability $M^{-N}$. If we label each of the signal elements with an index, $\delta s_i$, then we can describe each microsignal by a set of $M$ lists of the indices corresponding to the elements in each of the $M$ bins. For example, for a two bin signal built from five signal elements, a particular microsignal could be described by the set $\{(2,3),(1,4,5)\}$.

Of course, the model leading to the microsignal hypothesis space is not the only model one could imagine for constructing a signal; further, it is not clear exactly what information about the signal is being assumed by this model. Nevertheless, the resulting prior for $\{s_j\}$ has some intuitively pleasing properties, and leads to inversion methods that have proved extremely useful for the analysis of complicated data.

The least informative distribution for microsignals implies a prior probability distribution for the "macrosignals" specified by the $M$ numbers, $s_j$, as follows. In terms of the basic signal element, we can write $s_1 = n_1\delta s$, $s_2 = n_2\delta s$, and so on, with $\sum_j s_j = N\delta s$. An element of the original hypothesis space can thus be specified by a set of integers, $n_j$. Now the key is to note that, in general, each of the possible macrosignals—each of the possible set of $n_j$ values—will correspond to *many* possible microsignals. For example, a macrosignal with $n_1 = 2$ signal elements in bin 1 is equally well described by microsignals with signal elements $(1,2)$ in bin 1, or $(1,3)$ in bin 1, or $(1437,3275)$ in bin 1.

Denote the number of microsignals that correspond to a given macrosignal by the *multiplicity* $W(\{n_j\})$ of the macrosignal. The prior probability we will assign to each macrosignal is just its multiplicity times the probability, $M^{-N}$, of each of its constituent microsignals;

---

\* These elements are not to be identified with any physical "quantum" in the problem; for example, they should not be identified with photons detected by an experiment. They should reflect our prior information about the interesting scale of variation in the *signal*, not the data.

$p(\{n_j\} \mid I) = W(\{n_j\})M^{-N}$. The multiplicity of a macrosignal is given by the multinomial coefficient,

$$W(\{n_j\}) = \frac{N!}{n_1! n_2! \ldots n_M!}. \tag{56}$$

Using Stirling's formula, the log of the multiplicity is well approximated by

$$\log W(\{n_j\}) \approx N \log N - \sum_{j=1}^{M} n_j \log n_j$$
$$= N\left[ -\sum_{j=1}^{M} \frac{n_j}{N} \log \frac{n_j}{N} \right]$$
$$= NH(\{n_j\}), \tag{57}$$

where we have defined the *combinatorial entropy of the signal*, $H(\{n_j\})$, as

$$H(\{n_j\}) \equiv - \sum_{j=1}^{M} \frac{n_j}{N} \log \frac{n_j}{N}. \tag{58}$$

In terms of the entropy, the prior probability of a macrosignal can now be written,

$$p(\{n_j\} \mid I) = M^{-N} e^{NH(\{n_j\})}. \tag{59}$$

This prior has some intuitively appealing properties. In particular, it favors smoothly varying signals in the following sense. A priori, the most probable signal using this particular signal model is the signal with maximum combinatorial entropy; a simple calculation shows that the completely uniform signal, with all $n_j$ equal, has maximum entropy. Similarly, a signal with all $N$ signal elements in one bin—the "least uniform" signal—is a priori the least probable; it has a multiplicity of one. When combined with a likelihood function, this prior assignment will thus tend to favor the most uniform of all those signals consistent with the data.

To use the entropic prior (59), the values of $M$ and $N$ must be specified. Their values should express prior information we have about the signal and the experiment's ability to measure it. $M$ will be related to the resolution we expect is achievable from our data. $N$ might be related to how well the data can resolve differences in the signal level; it therefore seems reasonable that the choice of $N$ should be tied to the noise level. Finding ways to convert prior information into choices for $M$ and $N$ is a current research problem (see, e.g., Jaynes 1985b, 1986b; Gull 1989). Fortunately, the results of inversion with entropic priors do not depend sensitively on these numbers.

Despite the simplicity of the information leading to entropic inversion, it has proved enormously successful for analyzing a wide variety of astrophysical data. Some impressive recent examples include the calculation of radio images from interferometric data (Skilling and Gull 1985); imaging accretion discs from emission line profiles (Marsh and Horne 1989); estimating distances to clusters of galaxies from angular positions and apparent diameters of galaxies (Lahav and Gull 1989); and deconvolution of x-ray images of the Galactic center region (Kawai *et al.* 1988). An extensive bibliography of earlier applications of entropic inversion in astronomy is available in Narayan and Nityanada (1986), and in the physical sciences in general in Smith, Inguva, and Morgan (1984).

Entropic inverses like that described here were first introduced in astrophysics by Gull and Daniel (1978), based on earlier work by Frieden (1972) and Ables (1974). In these works, entropic inverses are presented as regularization methods, that is, as methods for producing a single "best" estimate of the signal from the data. Most later work has emphasized this regularization interpretation of the combinatorial entropy of an image (see Narayan and Nityanada 1986 for a review). In this context, entropic inverses are referred to as "maximum entropy methods", since they focus attention on what we would here identify as the most probable (maximum entropy) signal. Only recently has the Bayesian interpretation of these methods been clarified (Jaynes 1984b, 1985b, 1986b; Gull 1989; Skilling 1986, 1989, 1990). As valuable as the regularization interpretation may be, the Bayesian interpretation should prove even more valuable, for the following reasons.

First, as a regularization method, it is not clear why maximum entropy methods should be preferred to other regularization methods. Many have argued that entropy should be preferred as a regularizer by making analogies between the combinatorial entropy of a signal and the entropy of a probability distribution. As we have shown above, a probability distribution with maximum entropy consistent with the available information is the uniquely correct distribution to choose to represent that information. The mathematical similarity of equations (12) and (58) has led some to claim the same status for a signal with maximum combinatorial entropy. But since a signal is not a probability distribution, the arguments identifying the entropy of a distribution as the uniquely correct measure of its information content do not apply to signals. (See Skilling 1989 for a different viewpoint.)

Second, when entropy is viewed as a regularizer and not a prior probability, the manner in which it should be used to address an inverse problem is not clear. It should be combined with some statistical measure of the goodness-of-fit of a signal to the data, but the choice of statistic and the relative weighting of the entropy factor and the goodness-of-fit is arbitrary in frequentist regularization theory. Thus entropy has been combined, not only with the likelihood of the signal, as dictated in the Bayesian approach, but also with other goodness-of-fit statistics, such as the Kolmogorov-Smirnov statistic, adding a new element of arbitrariness and subjectivity to the results. Further, the connection of the parameter $N$ with prior information is lost the regularization approach, where it plays the role of a relative weighting between entropy and goodness-of-fit. No compelling criteria for the specification of the value of such a "regularization parameter" have yet been introduced in regularization theory.

Third, as a regularization method, entropic inverses can provide only a single "best" signal. When viewed as Bayesian methods, however, they can not only produce a "best" (most probable) signal, but can also provide measures of the statistical significance of features in the inverted signal. This aspect of Bayesian entropic inverses is an important element of the "Quantified Maximum Entropy" approach described by Skilling (1990) and Sibisi (1990) in these proceedings.

Finally, the Bayesian interpretation of entropic inverses reveals their dependence on prior information and a specific model for the signal, indicating ways they may be improved for specific problems. For example, though maximum entropy methods impressively reconstruct signals with point sources against a weak background, it is well known that they often poorly reconstruct signals that have a strong smoothly varying component, producing spurious features (Narayan and Nityanada 1986). To deal with such situations, several *ad hoc* modifications have been advanced (see, *e.g.*, Frieden and Wells 1978; Narayan and Nityanada 1986; Burrows and Koornneef 1989). Yet from a Bayesian perspective, it is apparent that such poor behavior is simply the result of the minimal amount of information assumed in calculating entropic inverses. The microsignal model assumes little more than the positivity of a signal; in particular, it ignores possible correlations between values of the signal in adjacent bins. Incorporation of such information should improve restorations; initial studies by Gull (1989a) reveal the promise of such an approach.

Entropic inverses are only one particularly simple example of a Bayesian inverse method. Others can be created, either by incorporating additional information into the prior (59) through MAXENT, by considering some hypothesis space other than that of the microsignal model that leads to the entropic inverse (Jaynes 1984b, 1986b), or especially by using MAXENT to find the prior for the $s_j$ directly (using the entropy of the *distribution*, equation [55], not that of the signal). Further research into Bayesian inversion should yield methods superior to entropic inversion in particular problems, though the simplicity of the entropic inverse will no doubt recommend it as a useful "jackknife" method, useful in the preliminary analysis of a wide variety of problems.

## 8.4  JAYNESIAN PROBABILITY THEORY

Bayesian methods are playing an increasingly important role in many areas of science where statistical inference is important. They have had a particularly powerful impact in mathematical statistics and econometrics, and there is much a physical scientist can learn from the statistical and econometric Bayesian literature. Particularly rich sources of information are the books by Tribus (1969), Zellner (1971), Box and Tiao (1973), and Berger (1985), and the influential review article of Edwards *et al.* (1963). Many important references to the literature are available in the reviews of Lindley (1972), Zellner (1989), and Press (1989).

But with the exception of the much neglected work of Jeffreys (1939), Bayesian methods have had little impact in the physical sciences until very recently. This has been due in large part to the lack of compelling rationale for the assignment of prior probabilities. The majority of the Bayesian literature (including most of the references just mentioned) regards prior probabilities as purely subjective expressions of a person's opinions about hypotheses, allowing individuals in possession of the same information to assign different probabilities to propositions. With this subjective element, Bayesian probability theory was viewed as being of little value to physical science.

Virtually alone among statisticians, Jaynes has emphasized that an *objective* probability theory can be developed by requiring that probability assignments satisfy the desideratum that we have here called *Jaynes Consistency:* Equivalent states of knowledge should be represented by equivalent probability assignments. This principle is the key to finding objective solutions to the problem of assigning direct probabilities—both prior probabilities and sampling probabilities—which is fully half of probability theory. The resulting theory remains subjective in the sense that probabilities represent states of knowledge, and not properties of nature. But the theory is objective in the sense of being completely independent of personalities or opinions. It is this objective aspect that makes the *Jaynesian Probability Theory* outlined here the appropriate tool for dealing with uncertainty in astrophysics, and indeed in all sciences.

## 9. Acknowledgements

It is a great pleasure to thank Don Lamb, Larry Bretthorst, and Ed Jaynes for many very valuable discussions. This work was supported in part by NASA grants NGT-50189, NAGW-830, and NAGW-1284.

## 10. References

Ables, J.G. (1974) 'Maximum Entropy Spectral Analysis', *Astron. Astrophys. Supp.* **15**, 383.

Bayes, T. (1763) 'An Essay Towards Solving a Problem in the Doctrine of Chances', *Phil. Trans. Roy. Soc. London* **53**, 370. Reprinted in *Biometrika* **45**, 293, and in Press (1989).

Berger, J.O. (1984) 'The Robust Bayesian Viewpoint', in J.B. Kadane (ed.), *Robustness of Bayesian Analyses*, Elsevier Science Publishers, B.V., p. 63.

Berger, J.O. (1985) *Statistical Decision Theory and Bayesian Analysis*, Springer-Verlag, New York.

Berger, J.O., and D. A. Berry (1988) 'Statistical Analysis and the Illusion of Objectivity', *Amer. Scientist* **76**, 159.

Berger, J.O., and R. Wolpert (1984) *The Likelihood Principle*, Institute of Mathematical Statistics, Hayward, CA.

Bernardo, J.M. (1979) 'Reference Posterior Distributions for Bayesian Inference', *J. Roy. Stat. Soc.* **B41**, 113.

Bernardo, J.M. (1980) 'A Bayesian Analysis of Hypothesis Testing', in J.M. Bernardo, M.H. DeGroot, D.V. Lindley, and A.F.M. Smith (eds.), *Bayesian Statistics*, University Press, Valencia, Spain, p. 605.

Bevington, P.R. (1969) *Data Reduction and Error Analysis for the Physical Sciences*, McGraw-Hill Book Company, New York.

Birnbaum, A. (1962) *J. Amer. Statist. Assoc.* 'On the Foundations of Statistical Inference', **57**, 269; and following discussion.

Box, G.E.P., and G.C. Tiao (1973) *Bayesian Inference in Statistical Analysis*, Addison-Wesley Publishing Co., Reading, MA.

Bretthorst, G.L. (1988a) 'Excerpts from Bayesian Spectrum Analysis and Parameter Estimation', in G.J. Erickson and C.R. Smith (eds.), *Maximum-Entropy and Bayesian Methods in Science and Engineering, Vol. 1*, Kluwer Academic Publishers, Dordrecht, p. 75.

Bretthorst, G.L. (1988b) *Bayesian Spectrum Analysis and Parameter Estimation*, Springer-Verlag, New York.

Bretthorst, G.L. (1989a) 'Bayesian Model Selection: Examples Relevant to NMR', in J. Skilling (ed.), *Maximum-Entropy and Bayesian Methods*, Kluwer Academic Publishers, Dordrecht, p. 377.

Bretthorst, G.L. (1989b) 'Bayesian Analysis I: Parameter Estimation Using Quadrature NMR Models', *J. Magn. Reson.*, in press.

Bretthorst, G.L. (1989c) 'Bayesian Analysis II: Signal Detection and Model Selection', *J. Magn. Reson.*, in press.

Bretthorst, G.L. (1989d) 'Bayesian Analysis III: Applications to NMR Signal Detection, Model Selection and Parameter Estimation', *J. Magn. Reson.*, in press.

Bretthorst, G.L. (1990) 'An Introduction to Parameter Estimation Using Bayesian Probability Theory', these proceedings.

Bretthorst, G.L., and C.R. Smith (1989) 'Bayesian Analysis of Signals from Closely-Spaced Objects', in R.L. Caswell (ed.), *Infrared Systems and Components III*, Proc. SPIE 1050.

Burrows, C., and J. Koornneef (1989) 'The Application of Maximum Entropy Techniques to Chopped Astronomical Infrared Data', in J. Skilling (ed.), *Maximum-Entropy and Bayesian Methods*, Kluwer Academic Publishers, Dordrecht.

Cherry, M.L., E.L. Chupp, P.P. Dunphy, D.J. Forrest, and J.M. Ryan (1980) 'Statistical Evaluation of Gamma-Ray Line Observations', *Ap. J.* **242**, 1257.

Cox, R.T. (1946) 'Probability, Frequency, and Reasonable Expectation', *Am. J. Phys.* **14**, 1.

Cox, R.T. (1961) *The Algebra of Probable Inference*, Johns Hopkins Press, Baltimore.

Dawid, A.P. (1980) 'A Bayesian Look at Nuisance Parameters', in J.M. Bernardo, M.H. DeGroot, D.V. Lindley, and A.F.M. Smith (eds.), *Bayesian Statistics*, University Press, Valencia, Spain, p. 167.

Eadie, W.T., D. Drijard, F.E. James, M. Roos, and B. Sadoulet (1971) *Statistical Methods in Experimental Physics*, North-Holland Publishing Company, Amsterdam.

Edwards, W., H. Lindman, and L.J. Savage (1963) 'Bayesian Statistical Inference for Phychological Research', *Psych. Rev.* **70**, 193; reprinted in J.B. Kadane (ed.), *Robustness of Bayesian Analyses*, Elsevier Science Publishers, B.V., p. 1.

Efron, B. (1975) 'Biased Versus Unbiased Estimation', *Adv. Math.* **16**, 259.

Erickson, G.J., P.O. Neudorfer, and C.R. Smith (1989) 'From Chirp to Chip, A Beginning', in J. Skilling (ed.), *Maximum-Entropy and Bayesian Methods*, Kluwer Academic Publishers, Dordrect, p. 505.

Feigelson, E.D. (1989) 'Statistics in Astronomy', in S. Kotz and N.L. Johnson (eds.), *Encyclopedia of Statistical Science, Vol. 9*, in press.

Fougere, P.F. (1988) 'Maximum Entropy Calculations on a Discrete Probability Space', in G.J. Erickson and C.R. Smith (eds.), *Maximum-Entropy and Bayesian Methods in Science and Engineering, Vol. 1*, Kluwer Academic Publishers, Dordrecht, p. 205.

Fougere, P.F. (1989) 'Maximum Entropy Calculations on a Discrete Probability Space: Predictions Confirmed', in J. Skilling (ed.), *Maximum-Entropy and Bayesian Methods*, Kluwer Academic Publishers, Dordrect, p. 303.

Frieden, B.R. (1972) 'Restoring with Maximum Likelihood and Maximum Entropy', *J. Opt. Soc. Am.* **62**, 511.

Frieden, B.R. (1972) 'Image Enhancement and Restoration', in T.S. Huang (ed.), *Picture Processing and Digital Filtering*, Springer-Verlag, New York, p. 177.

Frieden, B.R., and D.C. Wells (1978) 'Restoring with Maximum Entropy. III. Poisson Sources and Backgrounds', *J. Opt. Soc. Am.* **68**, 93.

Good, I.J. (1980) 'The Contributions of Jeffreys to Bayesian Statistics', in A. Zellner (ed.), *Bayesian Analysis in Econometrics and Statistics*, North-Holland, Amsterdam, p. 21.

Grandy, W.T. (1987) *Foundations of Statistical Mechanics Vol. 1: Equillibrium Theory*, D. Reidel Publishing Company, Dordrecht.

Gull, S.F. (1988) 'Bayesian Inductive Inference and Maximum Entropy', in G.J. Erickson and C.R. Smith (eds.), *Maximum-Entropy and Bayesian Methods in Science and Engineering, Vol. 1*, Kluwer Academic Publishers, Dordrecht, p. 53.

Gull, S.F. (1989) 'Developments in Maximum Entropy Data Analysis', in J. Skilling (ed.), *Maximum-Entropy and Bayesian Methods*, Kluwer Academic Publishers, Dordrecht, p.

53.

Gull, S.F., and G.J. Daniell (1978) 'Image Reconstruction from Incomplete and Noisy Data', *Nature* **272**, 686.

Hearn, D. (1969) 'Consistent Analysis of Gamma-Ray Astronomy Experiments', *Nuc. Inst. and Meth.* **70**, 200.

Iverson, G.R. (1984) *Bayesian Statistical Inference*, Sage Publications, Beverly Hills, California.

Jaynes, E.T. (1957a) 'Information Theory and Statistical Mechanics', *Phys. Rev.* **106**, 620.*

Jaynes, E.T. (1957b) 'How Does the Brain Do Plausible Reasoning?', Stanford Univ. Microwave Laboratory Report No. 421, reprinted in G.J. Erickson and C.R. Smith (eds.), *Maximum-Entropy and Bayesian Methods in Science and Engineering, Vol. 1* (1988), Kluwer Academic Publishers, Dordrect, p. 1.

Jaynes, E.T. (1958) *Probability Theory in Science and Engineering*, Colloquium Lectures in Pure and Applied Science No. 4, Socony Mobil Oil Co. Field Research Laboratory, Dallas.

Jaynes, E.T. (1963) 'New Engineering Applications of Information Theory', in J.L. Bogdanoff and F. Kozin (eds.), *Proc. of the 1st Symp. on Engineering Applications of Random Function Theory and Probability*, John Wiley and Sons, Inc., New York, p. 163.

Jaynes, E.T. (1968) 'Prior Probabilities', *IEEE Trans.* **SSC-4**, 227.*

Jaynes, E.T. (1973) 'The Well-Posed Problem', *Found. of Phys.* **3**, 477.*

Jaynes, E.T. (1976) 'Confidence Intervals vs. Bayesian Intervals', in W.L. Harper and C.A. Hooker (eds.), *Foundations of Probability Theory, Statistical Inference, and Statistical Theories of Science*, D. Reidel Pub. Co., Dordrecht, p. 252.*

Jaynes, E.T. (1978) 'Where Do We Stand on Maximum Entropy', in R.D. Levine and M. Tribus (eds.), *The Maximum Entropy Formalism*, MIT Press, Cambridge, p. 15.*

Jaynes, E.T. (1980a) 'Marginalization and Prior Probabilities', in A. Zellner (ed.), *Bayesian Analysis in Econometrics and Statistics*, North-Holland, Amsterdam, p. 43.*

Jaynes, E.T. (1980b) 'Review of *Inference, Method, and Decision* (R.D. Rosenkrantz)', *J. Am. Stat. Assoc.* **74**, 740.

Jaynes, E.T. (1982) 'On the Rationale of Maximum Entropy Methods', *Proc. IEEE* **70**, 939.

Jaynes, E.T. (1983) *Papers on Probability, Statistics, and Statistical Physics* (ed. R.D. Rosenkrantz), D. Reidel Pub. Co., Dordrecht.

Jaynes, E.T. (1984a) 'The Intuitive Inadequacy of Classical Statistics', *Epistemologia* **VII**, 43.

Jaynes, E.T. (1984b) 'Prior Information and Ambiguity in Inverse Problems', *SIAM-AMS Proc.* **14**, 151.

Jaynes, E.T. (1985a) 'Some Random Observations', *Synthese* **63**, 115.

Jaynes, E.T. (1985b) 'Where Do We Go From Here?', in C.R. Smith and W.T. Grandy, Jr. (eds.), *Maximum-Entropy and Bayesian Methods in Inverse Problems*, D. Reidel Publishing Company, Dordrecht, p. 21.

Jaynes, E.T. (1985c) 'Highly Informative Priors', in J.M. Bernardo, M.H. DeGroot, D.V. Lindley, and A.F.M. Smith (eds.), *Bayesian Statistics 2*, Elsevier Science Publishers, Amsterdam, p. 329.

---

* Reprinted in Jaynes (1983).

Jaynes, E.T. (1986a) 'Bayesian Methods: General Background', in J.H. Justice (ed.), *Maximum-Entropy and Bayesian Methods in Applied Statistics*, Cambridge University Press, Cambridge, p. 1.

Jaynes, E.T. (1986b) 'Monkees, Kangaroos, and N', in J.H. Justice (ed.), *Maximum-Entropy and Bayesian Methods in Applied Statistics*, Cambridge University Press, Cambridge, p. 26.

Jaynes, E.T. (1987) 'Bayesian Spectrum and Chirp Analysis', in C.R. Smith and G.J. Erickson (eds.), *Maximum-Entropy and Bayesian Spectral Analysis and Estimation Problems*, D. Reidel Publishing Company, Dordrecht, p. 1.

Jaynes, E.T. (1988a) 'The Relation of Bayesian and Maximum Entropy Methods', in G.J. Erickson and C.R. Smith (eds.), *Maximum-Entropy and Bayesian Methods in Science and Engineering, Vol. 1*, Kluwer Academic Publishers, Dordrecht, p. 25.

Jaynes, E.T. (1988b) 'Detection of Extra-Solar System Planets', in G.J. Erickson and C.R. Smith (eds.), *Maximum-Entropy and Bayesian Methods in Science and Engineering, Vol. 1*, Kluwer Academic Publishers, Dordrecht, p. 147.

Jaynes, E.T. (1989a) 'Clearing Up Mysteries — The Original Goal', in J. Skilling (ed.), *Maximum-Entropy and Bayesian Methods*, Kluwer Academic Publishers, Dordrecht.

Jaynes, E.T. (1989b) 'Probability in Quantum Theory', in *Proceedings of the Workshop on Complexity, Entropy, and the Physics of Information*, in press.

Jaynes, E.T. (1990a) 'Probability Theory as Logic', these proceedings.

Jaynes, E.T. (1990b) *Probability Theory - The Logic of Science*, in preparation.

Jeffreys, H. (1937) 'On the Relation Between Direct and Inverse Methods in Statistics', *Proc. Roy. Soc.* **A160**, 325.

Jeffreys, H. (1939) *Theory of Probability*, Oxford University Press, Oxford (3d revised edition 1961).

Kawai, N., E.E. Fenimore, J. Middleditch, R.G. Cruddace, G.G. Fritz, and W.A. Snyder (1988) 'X-Ray Observations of the Galactic Center by Spartan 1', *Ap. J.* **330**, 130.

Kolb, E. W., A. J. Stebbins, and M. S. Turner (1987) 'How Reliable are Neutrino Mass Measurements from SN 1987A?', *Phys. Rev.* **D35**, 3598; **D36**, 3820.

Lahav, O., and S.F. Gull (1989) 'Distances to Clusters of Galaxies by Maximum Entropy Method', *M.N.R.A.S.* **240**, 753.

Lampton, M., B. Margon, and S. Bowyer (1976) 'Parameter Estimation in X-Ray Astronomy', *Ap. J.* **208**, 177.

Laplace, P.S. (1812) *Theorie Analytique des Probabilités*, Courcier, Paris.

Laplace, P.S. (1951) *Philosophical Essay on Probability*, Dover Publications, New York (originally published as the introduction to Laplace [1812]).

Lindley, D.V. (1958) 'Fiducial Distributions and Bayes' Theorem', *J. Roy. Stat. Soc.* **B20**, 102.

Lindley, D.V. (1965) *Introduction to Probability and Statistics from a Bayesian Viewpoint* (2 Vols.), Cambridge University Press, Cambridge.

Lindley, D.V. (1972) *Bayesian Statistics, A Review*, Society for Industrial and Applied Mathematics, Philadelphia.

Loredo, T.J. and D.Q. Lamb (1989) 'Neutrinos from SN 1987A: Implications for Cooling of the Nascent Neutron Star and the Mass of the Electron Antineutrino', in E. Fenyves (ed.), *Proceedings of the Fourteenth Texas Symposium on Relativistic Astrophysics, An. N. Y. Acad. Sci.* **571**, 601.

Loredo, T.J. and D.Q. Lamb (1990a) 'Neutrinos from SN 1987A: Implications for Cooling of the Nascent Neutron Star', submitted to *Phys. Rev. D*.

Loredo, T.J. and D.Q. Lamb (1990b) 'Neutrinos from SN 1987A: Implications for the Mass of the Electron Antineutrino', submitted to *Phys. Rev. D*.

Mardia, K.V. (1972) *Statistics of Directional Data*, Academic Press, London.

Marsh, T.R., and K. Horne (1989) 'Maximum Entropy Tomography of Accretion Discs from their Emission Lines', in J. Skilling (ed.), *Maximum-Entropy and Bayesian Methods*, Kluwer Academic Publishers, Dordrecht, p. 339.

Martin, B.R. (1971) *Statistics for Physicists*, Academic Press, London.

Mendenhall, W., R. L. Scheaffer, and D. D. Wackerly (1981) *Mathematical Statistics with Applications*, Duxbury Press, Boston.

Narayan, R., and R. Nityanada (1986) 'Maximum Entropy Image Restoration in Astronomy', *Ann. Rev. Astron. Astrophys.*, **24**, 127.

Novick, M., and W. Hall (1965) 'A Bayesian Indifference Procedure', *J. Am. Stat. Assoc.* **60**, 1104.

O'Mongain, E. (1973) 'Appplication of Statistics to Results in Gamma Ray Astronomy', *Nature* **241**, 376.

Press, S.J. (1989) *Bayesian Statistics: Principles, Models, and Applications*, John Wiley and Sons, New York.

Press, W.H., B.P. Flannery, S.A. Teukolsky, and W.T. Vetterling (1986) 'Numerical Recipes', Cambridge University Press, Cambridge.

Rényi, A. (1972) *Letters on Probability*, Wayne State University Press, Detroit.

Rosenkrantz, R.D. (1977) *Inference, Method and Decision: Towards a Bayesian Philosophy of Science*, D. Reidel Publishing Company, Dordrect.

Runcorn, K. (1989) 'Sir Harold Jeffreys (1891-1989)', *Nature* **339**, 102.

Shore, J.E., and R.W. Johnson (1980) 'Axiomatic Derivation of the Principle of Maximum Entropy and the Principle of Minimum Cross-Entropy', *IEEE Trans. Inf. Th.* **IT-26**, 26; erratum in **IT-29**, 942.

Sibisi, S. (1990) 'Quantified MAXENT: An NMR Application', these proceedings.

Skilling, J. (1986) 'Theory of Maximum Entropy Image Reconstruction', in J.H. Justice (ed.), *Maximum Entropy and Bayesian Methods in Applied Statistics*, Cambridge University Press, Cambridge, p. 156.

Skilling, J. (1989) 'Classic Maximum Entropy', in J. Skilling (ed.), *Maximum-Entropy and Bayesian Methods*, Kluwer Academic Publishers, Dordrecht, p. 45.

Skilling, J. (1990) 'Quantified Maximum Entropy', these proceedings.

Skilling, J. and S.F. Gull (1985) 'Algorithms and Applications', in C.R. Smith and W.T. Grandy, Jr. (eds.), *Maximum-Entropy and Bayesian Methods in Inverse Problems*, D. Reidel Publishing Company, Dordrecht, p. 83.

Smith, C.R., and G. Erickson (1989) 'From Rationality and Consistency to Bayesian Probability', in J. Skilling (ed.), *Maximum-Entropy and Bayesian Methods*, Kluwer Academic Publishers, Dordrecht, p. 29.

Smith, C.R., R. Inguva, and R.L. Morgan (1984) 'Maximum-Entropy Inverses in Physics', *SIAM-AMS Proc.* **14**, 151.

Tribus, M. (1962) 'The Use of the Maximum Entropy Estimate in the Estimation of Reliability', in R.E. Machol and P. Gray (eds.), *Recent Developments in Information and Decision Processes*, The Macmillan Company, New York, p. 102.

Tribus, M. (1969) *Rational Descriptions, Decisions and Designs*, Pergamon Press, New York.

Van Campenhout, J.M., and T.M. Cover (1981) 'Maximum Entropy and Conditional Probability', *IEEE Trans. on Info. Theory* **IT-27**, 483.

van der Klis, M. (1989) 'Fourier Techniques in X-Ray Timing', in H. Ögelman and E.P.J. van den Heuvel (eds.), *Timing Neutron Stars*, Kluwer Academic Publishers, Dordrect, p. 27.

Zellner, A. (1977) 'Maximal Data Informative Prior Distributions', in A. Aykac and C. Brumat (eds.), *New Developments in the Application of Bayesian Methods*, North-Holland Publishing Co., Amsterdam, p. 211; reprinted in A. Zellner (1984) *Basic Issues in Econometrics*, University of Chicago Press, Chicago, p. 201.

Zellner, A. (1971) *An Introduction to Bayesian Inference in Econometrics*, J. Wiley and Sons, New York.

Zellner, A. (1986) 'Biased Predictors, Rationality, and the Evaluation of Forecasts', *Econ. Let.* **21**, 45.

Zellner, A. (1988) 'A Bayesian Era', in J.M. Bernardo, M.H. DeGroot, D.V. Lindley, and A.F.M. Smith (eds.), *Bayesian Statistics 3*, Oxford University Press, Oxford, p. 509.

Zellner, A., and A. Siow (1980) 'Posterior Odds Ratios for Selected Regression Hypotheses', in J.M. Bernardo, M.H. DeGroot, D.V. Lindley, and A.F.M. Smith (eds.), *Bayesian Statistics*, University Press, Valencia, Spain, p. 585.

# ATMOSPHERIC $^{14}$C VARIATIONS: A BAYESIAN PROSPECT

Charles P. Sonett
Dept of Planetary Sciences and Lunar and Planetary Laboratory
University of Arizona, Tucson, Arizona 85711

ABSTRACT. The time spectrum of variations in the atmospheric radiocarbon inventory is reviewed. According to conventional analysis (discrete Fourier transform, periodogram, and maximum entropy or MEM), the spectrum is characterized by features at $\sim$ 2200, 900, 700, 207, 149, and 88 years as well possibly at other frequencies. Model fitting of sinusoids using the Bretthorst algorithm confirms most of these and places their periods on a more secure basis. The forcings of the spectral features are known to be seated in some combination of ocean-atmosphere interaction, the Sun, and the terrestrial magnetic field, but some specific line assignments are still regarded as tentative.

## 1. Introduction

The ultimate source of terrestrial radiocarbon is traceable to the cosmic ray (CR) flux upon the top of the atmosphere from which, by spallations, an atmospheric neutron sea is generated. Radioactive $^{14}$C is produced terrestrially primarily by the specific nuclear reaction

$$^{14}N + n \rightarrow ^{14}C + p \tag{1}$$

where $^{14}$N is atmospheric. $^{14}$C decays by

$$^{14}C \rightarrow ^{14}N + \nu^- + \beta^- \tag{2}$$

where $\nu^-$ is the antineutrino and $\beta^-$ the electron. The half life of radiocarbon is $\tau_{1/2} =$ 5730 years [Lederer, Hollander, and Perlman, 1967]. Thus it is neutrons which participate in the N(n,p) reaction yielding $^{14}$C (Lingenfelter and Ramaty, 1970; O'Brien et al in press). If the CR flux were constant, the atmospheric $^{14}$C inventory would be in secular equilibrium. Then radiocarbon would be an absolute archeological clock, but it would be of lesser interest geophysically.

The inventory of radiocarbon in the Earth's atmosphere is variable. The changes can, within constraints, be assigned to a number of periods some of which may be harmonically related. But in general neither the forcing source(s) nor spectral line relations are fully understood, though the spectrum is clearly of geophysical and geochemical and probably of solar significance. The state of affairs regarding $^{14}$C infers that spectral model estimation is a potentially important tool for establishing accurate line frequencies. Correspondingly, the spectrum is the basis for speculation regarding the putative forcing oscillators underlying the spectrum.

*P. F. Fougère (ed.), Maximum Entropy and Bayesian Methods*, 143–159.

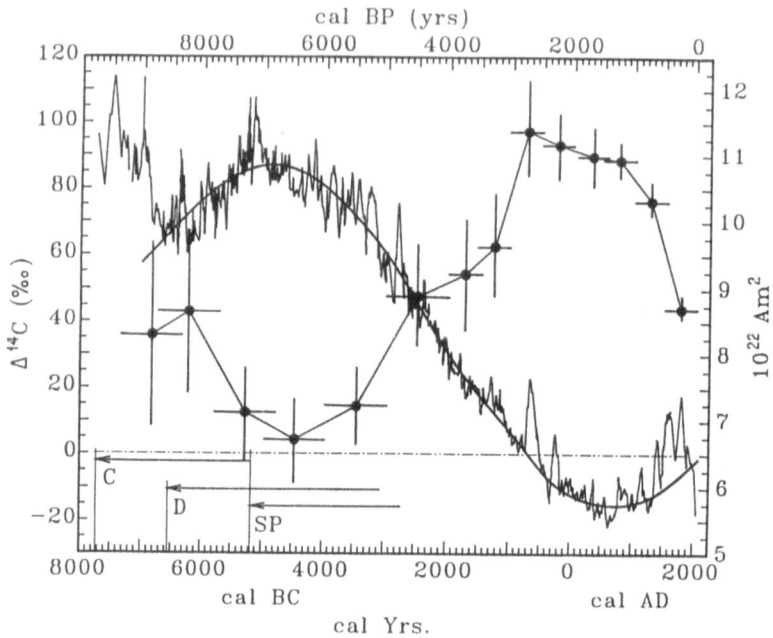

Fig. 1. Noisy trace with underlying sinusoid is the $\Delta^{14}$C time sequence for the past nine millennia from a compendium of Seattle (Stuiver and Pearson, 1986; Pearson and Stuiver, 1986; Pearson et al, 1986) to 7150 BP (before present) and Seattle (Stuiver et al, 1986), La Jolla (Linick et al, 1985), Tucson (Linick et al, 1986), and Heidelberg (Becker and Kromer 1986) radiocarbon data. The underlying absolute tree ring chronology exists to 6554 BC (Ferguson and Graybill, 1983). Solid line connecting dots is the estimated terrestrial dpole moment; values are from McElhinny and Senanayake, 1982). $^{14}$C dating of dipole ages to 6500 BP from Clark (1975). Earlier dipole dates are based on Stuiver et al, 1986) Arrow marked SP indicates extent of Seattle and Belfast data, D an independent chronology not 'wiggle matched', and C combined data of all laboratories. Parameters of underlying sine wave are offset=32 per mil, half amplitude=51 per mil, period=11,300 years, and phase lag=2.59 radians (131 deg) (from Damon and Sonett, 1989).

From the modern viewpoint a series of lines (either harmonically related or unrelated) may be taken as representing a quasi-periodic system and one in which the quasi-periodicity is based upon a non-linear dynamical entity perhaps even chaotic. This matter, which certainly needs to be resolved, is however far beyond the scope of this paper which addresses only the determination of periods together with some background on the geophysics thought to be associated with evolving periods in the record.

The primary source of information on the atmospheric inventory comes from measurements of the activity of radiocarbon in cellulose from trees together with an absolute chronology from counting of tree growth rings complementary to the wood whose activity is measured (Fig. 1). Through their respiratory and photosynthetic cycle, trees store carbon in the form of cellulose. In the U.S. Bristlecone pine, because of its extreme longevity and age, forms a cornerstone for much of the chronology upon which the radiocarbon record is based, e.g., Ferguson and Greybill (1983), though in Europe ancient oaks are commonly used. $\Delta^{14}C(\tau)$ is the age and fractionation corrected activity of wood at time $\tau$ in the past relative to mid-19th century wood with $\delta^{13}C = -25$ per mil [1]. The deviations from the mid-19th century value are per mil values that represent the fluctuation of the $^{14}C/^{12}C$ ratio in the atmosphere or the $\Delta^{14}C$ activity. It is the excess or deficiency in apparent age referred to the absolute chronology which yields the measure of variability.

## 2. The spectrum of $\Delta^{14}C$ variations

From a historical standpoint the first report of variability in the atmospheric inventory (from tree carbon) was by de Vries (1958, 1959) (See also de Jong and Mook, 1980). That the variations were periodic was initially reported by Houtermans (1971) who noted the $\sim 2200$ and 200 year periods while later Suess (1980) reported previously unpublished calculations of Kruse of the power spectrum of the La Jolla $\Delta^{14}C$ record showing spectral lines at 2200, 900, 498, 308, 202, and lesser years. (See also Suess, 1968 and 1986). Later Neftel et al (1981) studied the 208 year line exhaustively. The existence of this feature was confirmed by comparison of the La Jolla and Belfast records (Sonett, 1985).

The spectrum of $\Delta^{14}C$ is rich in detail though quite noisy. What makes this complex record potentially of special importance is that it is thought to be forced concurrently by several mechanisms (a) the oceanic transport of carbon dioxide through the ocean–atmosphere interface, (b) variations in the intensity of the geomagnetic field, and finally (c) variations of the flux of incoming cosmic rays incident upon the top of the atmosphere. The latter, in turn, arise from modulation by variations in the solar wind – ultimately traceable to the solar atmosphere, and less likely to changes in the interstellar flux upon the heliosphere (Sonett, 1984). Untangling and identification of the source of variation is a tool of substantial geophysical importance. Recent reviews covering the spectrum of

---

[1] $^{13}C$ is a stable carbon isotope of nuclear mass 13. Thus its susceptibility to kinetic fractionation is half that of radiocarbon; its stability makes it important in correcting for possible error–inducing fractionation in radiocarbon.

radiocarbon are found, e.g. in Damon, Lerman, and Long (1978), Sonett (1984), Sonett and Finney (in press), and Damon and Sonett (1989).

In the initial approach to the problem of the existence of periodic behavior of the time sequence of variations, we use the periodogram which discloses numerous periods ranging from $\sim$ 2200 years downwards to an approximate Nyquist period of 20 years, defined loosely by the average laboratory sampling (measurement) interval [2]. Masquerading of side lobes as primary lines in the radiocarbon spectrum is a distinct possibility; it is here especially that the importance of the Jaynes/Bretthorst algorithm is needed, e.g. Jaynes (1983), Bretthorst (1988, 1989).

Viewed from the admittedly somewhat flawed standpoint of periodograms, Fourier transforms, and sometimes MEM, radiocarbon has a well defined spectrum. Fig. 2 shows the discrete Fourier transform (DFT) of $\Delta^{14}C$ concatenated from data of several laboratories, La Jolla (Suess, 1970; 1978; 1980), Belfast (Pearson et al, 1986), Seattle (Stuiver and Pearson, 1986), and La Jolla/Hohenheim (Linick et al, 1985) which extend backwards in time to 7199 BC.

Since much of the data is unequally spaced in time, the combined sequence is interpolated to 10 or 20 year intervals using a cubic spline [3], followed by detrending via least squares using a third or fifth order reference polynomial (The average interval for much of the radiocarbon data is $\sim$ 10 years, thus the 'average' Nyquist period noted earlier. A more exact description of the Nyquist period is a substantially more complex subject than can be discussed within the confines of this article.) The insert of Fig. 2 is the extension by periodogram for frequencies in the range of $0.03 \leq f \leq 0.01$ year$^{-1}$ or in period $33.3 \geq \tau \geq 20$ years. Major line features are present at 2272, 909, 649, 207, 149, and 88 years in the large panel and 37.2, 26.3, 20.8, and 20.1 years in the inset.

The specific Bayesian model as a test of confidence for spectral features is due to Jaynes (1983) and to Bretthorst (1988, 1989); it involves the determination of a frequency dependent orthonormal set of basis vectors, $e_{1...k}$ (where k is the number of models to be fitted) upon which the data sequence, $x_{1...N}$, is projected, giving sets of frequency dependent orthonormal sequences (Bretthorst (1988, 1989). Line probabilities vs. frequency are Student t–distributed with maxima at the most probable frequency. This procedure has some similarity to the empirical orthogonal representation of time series (Wallace and Dickenson, 1972) involving the orthogonalization of the data covariance matrix and in a more general way corresponds to the dynamical transformation of a system's modes into normal coordinates. Significant differences can arise between the periodogram and the Bayesian estimate of line position for certain (periodogram) periods.

---

[2] In sampling at more than one rate, more than one Nyquist period is defined; dual sampling is discussed by Sonett (1968) and the more complex general case by Bracewell (1978) and by Linden (1959)

[3] The interpolation is only for computation of the Fourier transform; it is not needed for the periodogram (Fig. 3) or Bayesian estimation

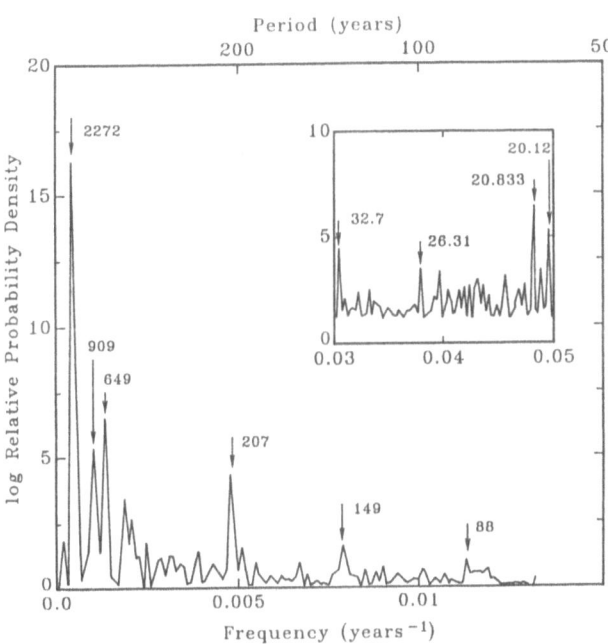

Fig. 2. DFT (discrete Fourier transform) of concatenation of La Jolla, Becker, Stuiver, and Pearson ¹⁴C sequences. Detrended using least squares fit to 5th order polynomial; data is interpo- lated using cubic spline. Major features are demarked by period in years with some periods of lesser amplitude unmarked. y axis is in $\log_e$ probability measure (Bretthorst, 1988, 1989). Insert is from high frequency segment periodogram of the BSP combined sequence concatenated further by combining with the La Jolla data (SBSP or Suess-Becker-Stuiver-Pearson) sequence extended to 20 years period to emphasize higher frequencies. Major high frequency lines are present at 20.12, 20.83, 26.31, and 32.7 years. The first two are tentatively identified with the Hale cycle; these lines especially the latter two are possibly aliased from higher frequencies.

## 3. The Secular trend

The dominant variation in the radiocarbon record is a secular(?) trend over the past 10,000 years; to avoid bias it must be removed prior to an accurate assessment of periodicities. This trend can be fitted (but non-uniquely) by a model consisting of a sinusoid with an offset = 32 per mil, half amplitude = 51 per mil, period = 11,300 years and phase lag = 2.29 radians (Damon and Linick, 1986). As the variation is equally well modelled by a polynomial (as is the case for the Bayesian algorithm used here), there is thus no a priori reason to suspect that this variation is actually periodic. Indeed the recent sequence of combined Becker, Stuiver, and Pearson data yields a local decrease during the first $\sim$ 1000 years (6000-7200 BP) [4].

The major radiocarbon variation is generally attributed to a change in the terrestrial dipole moment and is qualitatively supported by paleomagnetic data. Fig. 1 also shows the ca. 11,000 year sinusoid model curve upon which is superimposed data for the Earth's dipole moment assembled by McElhinny and Senanayake (1982). The dipole moment data from individual sites is inherently noisy, not only because of measurement uncertainties, but because of non—dipole components and, particularly, their westward drift, plus regional components of the non–dipole field and differences due to the magnetic properties of the Earth's crust at the sample locations. The additional dip in the $^{14}$C record from $\sim$ 8000 BC – 5000 BC which appears prominently in Fig. 1 is less well established, as the underlying tree ring record has not yet been fully certified as chronologically absolute for time earlier than 6556 BC.

4. Periodic features

Fig. 1 also displays a more or less continuous record of variations or fluctuations, whose analysis has been a central issue in radiocarbon analysis. Aside from the aforementioned long 'period', the dominant feature in the $\Delta^{14}$C record is a period of $\sim$ 2200 years. Its source is enigmatic for no periodic geomagnetic field change of the required amplitude has been detected. Various proposals have been made for a connection between the ca. 2200 year ($\delta^{18}$O and $\Delta^{14}$C) cycle and the ca. 2400 year cycle in the directional components and westward drift of the non-dipole part of the Earth's magnetic field (Creer, 1983).

There are only weak grounds for supposing that the 2200 year radiocarbon period might be associated with the Earth's field. But candidate sources for this variation still admit of both terrestrial and extraterrestrial forcing sources. The quasiperiodicity of ca. 2200 years is also found in the $\delta^{18}$O record in ice cores and foraminifera from ocean cores (Pestiaux et al, 1987; 1988). Glaciation discloses this period as does the Middle Europe oak dendroclimatic record, and Dansgaard et al (1984) report the most prominent period in their Camp Century $\delta^{18}$O core to be a line at 2550 years. A terrestrial origin can account for the radiocarbon line if it has the role of a tracer of globally varying $CO_2$ inventory and exchange rates. An additional factor to be considered in conjunction with this problem is the possibility of an association between the 2200 year period and a ubiquitous period at 208 years discussed below in conjunction with the overall spectrum of radiocarbon. All of these reports strengthen the supposition that the ca. 2200 year period is an ubiquitous climatic feature.

---

[4] BP means 'before present' used alternatively to AD and BC. Radiocarbon activity is actually measured relative to 1850 AD.

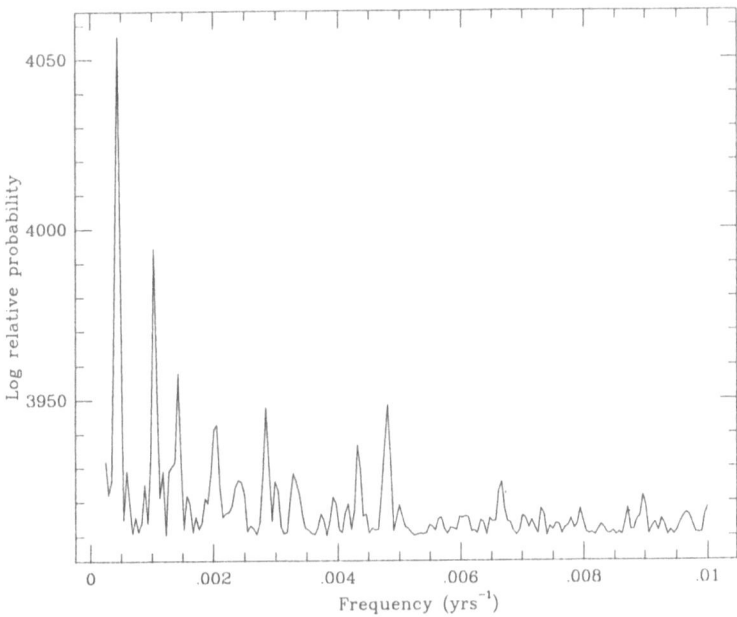

Fig. 3. Periodogram of the SBSP sequence to 100 year period. All the major features shown in Fig. 2 are present with some additional lines suggested. y axis is in $\log_e$ probability measure as for Fig. 2.

The four longest periods shown in Figs. 2 and 3 suggest relatively sharp line sources; this is supported by the very peaked probability distributions calculated using Bretthorst's algorithm. At the shorter period range of $\sim 80-90$ years initial tests with this algorithm identify the Gleissberg period at 88 years if the identification as the Gleissberg is correct. A similar period is obtained at 88.2 years by Sonett (1984) and by Feynman and Fougere (1984) using MEM (Fig. 4) and Bretthorst (1988, 1989) obtains a strong period at about 90 years in the spectrum of the yearly averaged sunspot index.

Feynman and Fougere (1985) obtain a period of $88.4 \pm 0.7$ years from the AD 450-1450 auroral record. Neighboring Bayesian spectral features are also identified at $\sim$ 75 and 77 years, but these line outside the range of periods estimated by various workers for the Gleissberg period. Siscoe (1980) and Attolini et al.(1988) observe periodicities of 88 and 131 years in the historical auroral record. The 131 year period also shows up in the $\Delta^{14}$C power spectra (Fig 2 and 3). In spite of intense atmospheric attenuation and damping of spectral lines, e.g. Houtermans (1973), the Bretthorst algorithm does identify the lines (Section 4) present in the neighborhood of 20 years (20.12 and 20.83 years) which we are tempted to identify with the Hale period! It is curious that two periods appear and that they lie close to 20 years rather than ca. 22 years; the Bayesian estimate from the sunspot index (Bretthorst, 1988) also shows periods near to 10 and 20 years. That two periods appear may be associated with the modulation proposed by Sonett (1984) where a number of spectral features are generated by a complicated form of AM. Indeed a large number of lines are noted in the conventional DFT of the sunspot index. (The additional lines appearing at 26.31 and 32.7 years in Fig. 3 are so close to a possible Nyquist folding axis as to give unreliable true frequency estimates). The irregular sampling period of most radiocarbon records increases the likelihood that alias 'noise' may corrupt the spectrum at high frequency. The average sampling period for the concatenated Becker (Hohenheim)–Stuiver (Seattle)–Pearson (Belfast) BSP is 5.8 years but the irregularity prevents definition of a specific Nyquist period.

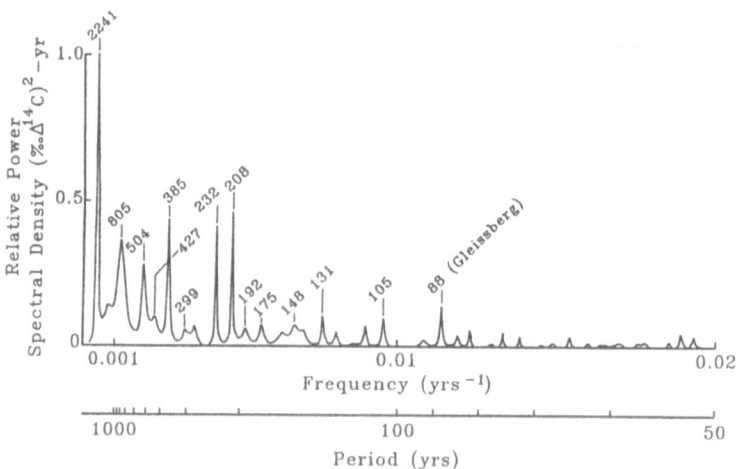

Fig. 4. Maximum entropy (MEM) power spectrum of combined Belfast and Seattle data. Detrended via least squares sine wave fit and cubic tension spline interpolated to 20 year intervals. Autoregressive order 120.

## 5.0 Possible solar associations

The Maunder minimum of the sunspot index (SI), an historically important minimum in the sunspot index ca. 1640-1720 AD, is also of special interest to the problem of the radiocarbon variations because it is usually associated with solar activity and therefore ought to be present in the radiocarbon record through variable modulation of the solar wind resulting in variations in the incident cosmic ray flux.

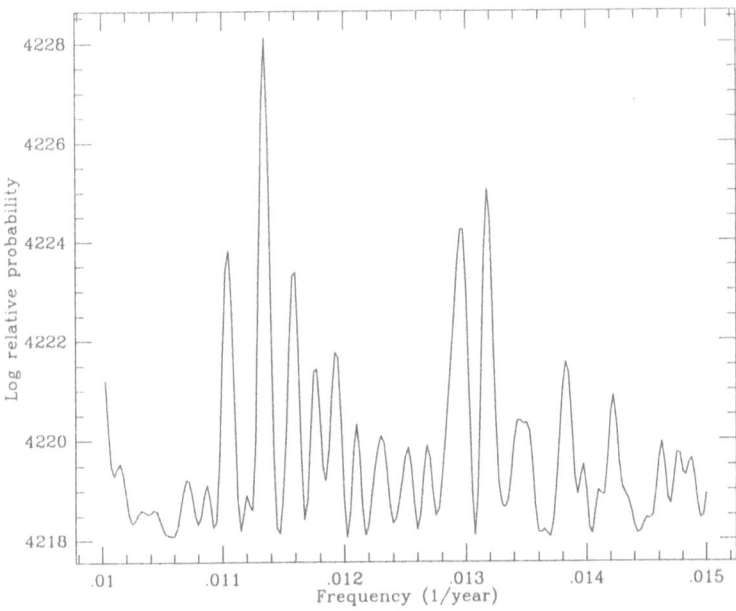

Fig. 5. Narrow band residue periodogram less Bayesian model estimate for one feature removed. This is the distribution function for the BSP spectrum showing major features at 75, 77, and about 87 years. Some idea of why the various estimates of the Gleissberg period differ may be found from the detailed model calculation restricted to the bandwidth range $0.01 < f < 0.014$ yr$^{-1}$ shown in Fig. 6 where a major lie appears at $(0.0114$ yr$^{-1})$ 87.7 years but with two lines at 75 and 77 years which are not present in the periodogram. Precise assignments of these lines is premature, but it is suggestive to note their approximate 2nd harmonic relation to the feature at about 150 years in the spectrum.

Measurements of radiocarbon by Stuiver (1961; 1965) and a reconstruction of the sunspot variability over the past millennium by Stuiver and Quay (1980) using Pacific Northwest wood provide a model reconstruction of sunspot activity from which it has been hypothesized that the Maunder minimum and its additional representations (Spörer and Wolf) at intervals of approximately 200 years are manifestations of the 206 year period in radiocarbon. However the 206 year line width seems overly narrow to match the variability seen in the model derived by Stuiver and Quay, and this issue is unresolved.

The persistence and putative solar-origin of the 208 year line in the spectrum of radiocarbon has been repeatedly stressed by Suess (1980). This 208 year line has been shown to extend over the entire 8500 year La Jolla $\Delta^{14}C$ record by Sonett (1984). Finally, frequency estimates updated from Sonett and Finney (in press) are given in Table 1. (The longest period listed in Table 1 is 2319 years ($4.3127 \times 10^{-4} \pm 5.64 \times 10^{-7}$ years$^{-1}$).

Table 1. $^{14}C$ Major Bayesian line position assignments[a]

| Frequency (yrs$^{-1}$) | $2\sigma$ | Period (years) |
| --- | --- | --- |
| $4.616 \times 10^{-4}$ | $\pm 0.47 \times 10^{-5}$ | 2166 |
| $1.106 \times 10^{-3}$ | $\pm 0.95 \times 10^{-5}$ | 904 |
| $1.400 \times 10^{-3}$ | $\pm 0.11 \times 10^{-4}$ | 714 |
| $4.833 \times 10^{-3}$ | $\pm 0.16 \times 10^{-4}$ | 207 |
| $1.905 \times 10^{-3}$ | $\pm 0.16 \times 10^{-4}$ | 525 |
| $2.804 \times 10^{-3}$ | $\pm 0.93 \times 10^{-5}$ | 357 |

[a] Line assignments differ somewhat from previous repors, i.e. Sonett and Finney (in press). Thus Table 1 should be regarded as tentative and subject to correction for reasons still under study but which include possible bias from variation in detrend polynomial order. Fortran progarm version by Elisabetta Pierazzo.

6. Non-linearity and 'mixing'

Non-linear forcing in geophysics is not a rare phenomenon. In the circumstances surrounding the generation of the atmospheric radiocarbon inventory, not only is the incident CR flux inconstant, but additional processes of terrestrial origin can influence the eventual activity. Simple non-linear (finite amplitude) forcing can lead to the generation of a harmonic spectrum, while a mixture of two or more periods in a non-linear device usually leads to amplitude modulation (AM). Frequency modulation (FM) is a more complex process and involves periodic or aperiodic changes in source forcing frequency, by and large a problem outside the scope of this chapter, though FM cannot easily be ruled out as an adjunct to some of the spectral features of radiocarbon. The ability to detect and understand modulation can lead to significantly enhanced likelihood of being able to identify forcing mechanisms and to separate those of terrestrial from solar origin.

Sonett (1984) showed how the non-linear dependence of the radiocarbon production rate upon changes in the Earth's dipole moment (Elsasser et al, 1956) could provide the essential basis for amplitude modulation. The corresponding statistical analysis, using a combination of the La Jolla and Workshop data provided marginal evidence for modulation. The 2200 year feature has also been argued (Damon and Linick, 1986; Damon, 1988; Damon and Sonett, 1989) to strongly modulates the ca. 210 year period. The evidence for modulation is strengthened by analysis of the more complete BSP time sequence. But direct detection of symmetric side bands about the 208 year period (viewed here as a 'carrier' wave) is negative, due possibly to a low signal/noise (S/N) ratio or interference between signal and window side lobes. But neither MEM nor the DFT discloses symmetric side bands, even the higher resolution of the Bretthorst algorithm fails this test.

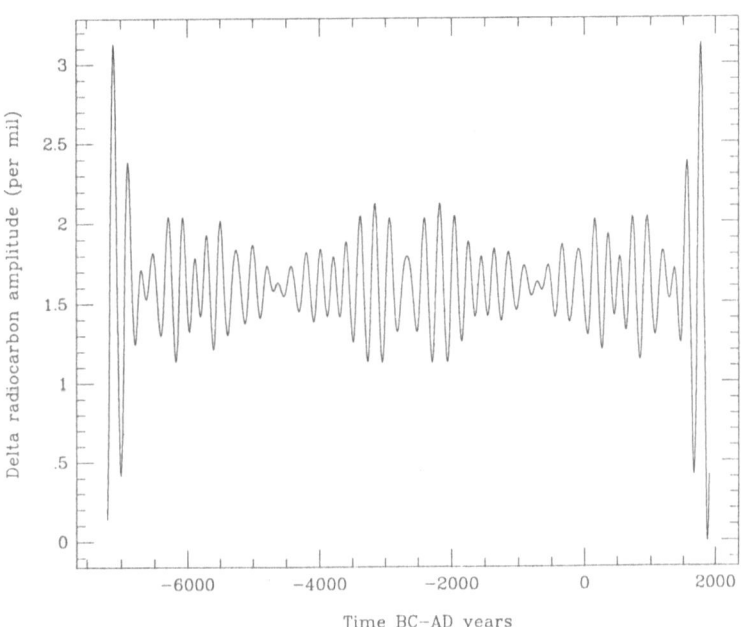

Fig. 6. $\Delta^{14}$C BSP time sequence narrow banded to $237 \leq \tau \leq 190$ where $\tau$ is period in years. Two major periods appear to be 208 years ('carrier') and $\sim$ 2200 year 'modulation' though the latter is noisy and the statistical certainty has not been resolved. The very large swings at the ends of the record are not understood, but might be a result of the very sharp phase shift associated with the sharp 'cornered' filter.

As an alternate approach to the problem of detection of modulation, a narrow banded version of the BSP sequence suggests the presence of modulation of the 208 year period by the 2200 year period. This is seen in Fig. 6 from the BSP sequence bandwidth limited to a range of $237 < \tau < 190$ years (where $\tau$ is the period) about a central period of 208 years showing intense modulation of approximately 100%.

The procedure for narrow banding is to select a narrow range of frequencies from the Fourier transform and then invert into the time domain. The bandwidth is just sufficient to accommodate the two side bands generated by amplitude modulation (AM) by a 2200 year period. We cannot with certainty rule out FM (frequency modulation) but as a working hypothesis AM modulation suggests two forcing functions, an extraterrestrial 208 year (carrier) period and a terrestrial 2200 year (modulation). Another restriction is that the modulation is very noisy and the confidence that it is due to the 2200 year period is still to be fully tested. Preliminary tests on an earlier data set (Sonett, 1984) were less than fully satisfactory with the difference in the distribution of modulation envelope values from a Rayleigh distribution of limited confidence.

7. Discussion

In identifying likely forcing mechanisms for the radiocarbon variations, certain other isotopes are useful. Oxygen–18 ($\delta^{18}O$), an uncommon isotope of oxygen is especially useful as a climatic indicator of evaporation and condensation of water because of its identifiable kinetic fractionation. Ice strata for the past 8500 years, a prime source of oxygen data, have been dated by stratigraphic methods to an accuracy that is considered to be better than 2% (Hammer et al., 1978), though the earlier ice core chronology is less accurate due to poorly known strain and accumulation rates. The MEM spectrum of the Camp Century ice core for the last 8500 years shows that, apart from the longer term trend, a 2550 year oscillation dominates the spectrum (containing 17% of the spectral power), noting that MEM amplitude estimates are notoriously uncertain.

Early publications of Dansgaard's group discuss periodicities in the $\delta^{18}O$ record for the time interval from 1200-2000 AD (Johnsen et al. 1970; Dansgaard et al., 1971). Their chronology was based on the assumed average accumulation rate of ice. Their power spectrum for this time interval shows prominent periods at 78 and 181 years suggesting an association with the Gleissberg sunspot period. (Note the confusion with the putative assignments of lines in the neighborhood of 80 years in Section 4.) In a longer section of the core, going back to about 10,000 years, they obtain a period of 350 years, again using the ice accumulation time scale (ibid., 1971), and they also find a persistent oscillation with period of $\sim 2000$ years.

Those geophysical parameters such as the aurorae and insofar as we can say, $\Delta^{14}C$ which appear at the ca. 88 year Gleissberg period infer an electrodynamic cycle which likely has its origin in the solar atmosphere. Other periods which appear in the radiocarbon record and elsewhere do face the problem of partitioning between climate and electrodynamics (cf. Sonett and Suess, (1984).

The gravest difficulty with the idea that the Sun is endowed with either/both long period bolometric and hydromagnetic variability rests upon the lack of a conceptual framework. All the known periods which we tend to associate with the Sun, save for the Hale cycle and conviction, e.g. 'seismic', supergranulation, rotation, etc. are less at most than the 27 day rotation period (Gough, 1988). Seismic transit times are certainly restricted to several hours though arguments can be made for longer hydromagnetic periods. The mean Alfven speed over the full solar radius for 83 years is $\sim 0.05$ cm/sec corresponding, for uniform density $\rho = 1$ gm/cm$^3$, to an internal magnetic field intensity $\sim 0.2$ gauss. Of course these values are not too useful since the solar interior is radially very inhomogeneous. Nevertheless it is possible to envision hydromagnetic modes within the radiative zone which very approximately satisfy transit times consistent with the radiocarbon observations.

But it is more difficult to reconcile such 'eigenmodes' with the sources necessary to excite them if one holds to the view that the convective zone is isothermal and hydrodynamically 'benign'. Among the few solar physicists who take seriously the possibility of long period modes, Dilke and Gough (1972) and Gough (1988) propose episodic breakdown in the solar core, but typical time scales are in the hundreds of millions of years.

8. Acknowledgements I have benefitted by discussions with P.E. Damon, H.E. Suess, G.L. Bretthorst, and Elisabetta Pierazzo whom I thank.

## 8. References

Attolini, M.R., M. Galli, and T. Nanni, Long and short cycles in solar activity during the last millennium, in "Secular, Solar, and Geomagnetic variations in the last 10,000 years", pp.49-68. eds. Stephenson and Wolfendale, Kluwer (1988).

Becker, B. and B. Kromer, Extension of the Holocene dendrochronology by the preboreal pine series, Radiocarbon, calibration issue, 28, 961–968 (1986).

Bracewell, R.N., The Fourier Transform and Its Applications, 2nd Ed., McGraw-Hill (1978).

Bretthorst, L.G., Bayesian Spectrum Analysis and Parameter Estimation, vol. 48, Lecture Notes in Statistics, eds. J. Berger, S. Fienberg, J. Gani, K. Krickeberg, and B. Singer, Springer-Verlag (1988).

Bretthorst, G.L., An introduction to parameter estimation using Bayesian probability theory, preprint (1989).

Creer, K., Geomagnetic field and radiocarbon activity through Holocene time, in "Secular, Solar, and Geomagnetic variations in the last 10,000 years", Stephenson and Wolfendale (1988).

Damon, P.E. and C.P. Sonett, Solar and Terrestrial Components of the Atmospheric Variation Spectrum, in "The Sun in Time", eds. C.P. Sonett, M.S. Giampapa, and M.S. Matthews, Univ. Arizona Press (in press).

Damon, P.E., J.C. Lerman, and A. Long, Temporal fluctuations of atmospheric $^{14}$C: causal factors and implications, Ann. Rev. Earth Planet. Sci., 6, 457-494 (1978).

Damon, P.E., Production and decay of radiocarbon and its modulation by geomagnetic field-solar activity changes with possible implications for global environment, pp. 267-285 in "Secular solar and geomagnetic variations in the last 10,000 years", eds. F.R. Stephenson and A.W. Wolfendale, Kluwer Acad. Publ. (1988).

Damon, P.E. and T. W. Linick, Geomagnetic-heliomagnetic modulation of atmospheric radiocarbon production, in Intl. radiocarbon conference, 12th Proc., eds. M. Stuiver and R. Kra, Radiocarbon, 28, 266-278 (1986).

Dansgaard, W., S.J. Johnsen, H.B. Clausen, D. Dahl-Jensen, N. Gunderstrup, C. Hammer, and H. Oeschger, North Atlantic climate oscillations revealed by deep Greenland ice cores, in "Climate Processes and Climate Sensitivity", pp. 288-298 AGU Maurice Ewing series. eds. J.E. Hansen, and . Takahashi (1984).

Dansgaard, W., S.S. Johnsen, H.B. Clausen. and C.C. Langway, Climatic record revealed by the Camp Century ice core, in "The Late Cenozoic Glacial Ages" pp. 37-46, ed. K. Turekian, Yale Univ. Press (1971).

de Vries, H., Variations in concentration of radiocarbon with time and location on Earth, Proc. K. Ned. Akad. Wet. Ser. B, 61, 94 (1958).

de Vries, H., Measurement and use of natural radiocarbon. In "Researches in Geochemistry", ed. P.H. Abelson, pp.169-189, (1959).

de Jong, A.F.M. and W.G. Mook, Medium-term atmospheric $^{14}$C variations, Radiocarbon, 22, 267-272 (1980).

Dilke, F. W. W. and D.O. Gough, The Solar Spoon, Nature, 240, 262- 294 (1972).

Elsasser, W. , E.P. Ney, and J.R. Winckler, Cosmic-ray intensity and geomagnetism, Nature, 178, 1226 (1956).

Ferguson, C.W. and D.A. Greybill, Dendrochronology of Bristlecone pine: A progress report, Radiocarbon, 25, pp.287-288 (1983).

Feynman, J. and P. Fougere, Eighty year cycle in solar terrestrial phenomena confirmed, J. Geophys. Res., 89, p. 2032 (1985).

Gough, D.O., Theory of solar variation, in "Solar-Terrestrial Relationships and the Earth Environment in the last Millennia", pp. 90-132, Proc. Enrico Fermi Intl. School of Physics, course XCV, ed. G. Cini Castagnoli, North-Holland (1988).

Hammer, C., H.B. Clausen, W. Dansgaard, N, Grunddestrup, S.S. Johnsen, and N. Reeh, Dating of Greenland ice cores by flow models, isotopes, volcanic debris, and continental dust, J. Glaciology, 20, pp. 3-26 (1978).

Houtermans, J.C., Geophysical interpretation of bristlecone pine radiocarbon measurements using a method of Fourier analysis of unequally spaced data, Ph.D. thesis, Univ. Berne (1971).

Houtermans, J.C., H.E. Suess, and H. Oeschger, Reservior models and production rate variations of natural radiocarbon, J. Geophys. Res., 78, 1897-1908 (1973).

Jaynes, E.T., "Papers on Probability, Statistics, and Statistical Physics", a reprint collection, D. Reidel (1983).

Johnsen, S.J., W. Dansgaard, and H.B. Clausen, Climatic oscillations 1200-2000 AD, Nature, 227, 482-483 (1970).

Lederer, C.M., J.M. Hollander, and I. Perlman, "Table of Isotopes", 6th ed., John Wiley, New York (1967).

Linden, D.A., A discussion of sampling theorems, Proc. IRE, 47, p.1219 (1959).

Lingenfelter, R. E. and R. Ramaty, Astrophysical and geophysical variations in C-14 production, Astrophysical and geophysical variations in [14]C production, in "Nobel symposium XII: Radiocarbon Variations and Absolute Chronology", pp. 513-537, Almquist and Wiksells, Uppsala (1970).

Linick, T.W., H.E. Suess, and B. Becker, La Jolla measurements of radiocarbon in south German oak tree–ring chronologies, Radiocarbon, 27, pp. 20-32 (1985).

Linick, T.W., A. Long, P.E. Damon, and C.W. Ferguson, High-precision radiocarbon dating of Bristlecone pine from 6554–5350 BC, Radiocarbon, calibration issue, 28, 943–953 (1986).

McElhinny, M.W. and W.E. Senanayake, Variations in the geomagnetic dipole: the past 50,000 years, Geomag. Geoelect, 34, p.39 (1982).

Neftel, A., H. Oeschger, and H.E. Suess, Secular non-random variations of cosmogenic carbon-14 in the terrestrial atmosphere, Earth, Planet. Sci. Lett., 56, 127 (1981).

O'Brien, K., M.A. Shea, D.F. Smart, and A. de la Zerda Lerner, The production in the Earth's atmosphere of cosmogenic isotopes and their inventories, in "The Sun in Time", eds. C.P. Sonett, M.S. Giampapa, and M.S. MAtthews, Univ. of Arizona Press (in press).

Pearson, G.W., J.R. Pilcher, M.G.L. Baillie, D.M. Corbett, and F. Qua, High-precision $^{14}C$ measurements of Irish oaks to show the natural $^{14}C$ variations from AD 1840-5120 BC, in Intl. Radiocarbon Conf., 12th Proc. pp. 911-934, eds. M. Stuiver and R. Kra, Radiocarbon 28, no.2B (1986).

Pearson, G.W. and M. Stuiver, High-precision calibration of the radiocarbon time scale, 500–2500 BC, Radiocarbon, calibration issue, 28, 839–862 (1986).

Pestiaux, P., I van der Mersch, and A. Berger, Paleoclimatic variability at frequencies ranging from 1 cycle per 0000 years to 1 cycle per 000 years: evidence for nonlinear behavior of the climate system, Climate Change, 12, ,9-37 (1988).

Siscoe, G. L., Evidence in the auroral record for secular solar variations, Rev. Geophys. Spa. Phys., 18, pp. 647-658 (1980).

Sonett, C.P., Multirate sampling and generalized alias as a source of errors in magnetometer experiments, IEEE Trans., GE-6, 3 (1968).

Sonett, C.P. and H.E. Suess, Correlation of Bristlecone pine ring widths with atmospheric carbon-14 variations: a climate–Sun relation, Nature, 307, pp. 141-143 (1984).

Sonett, C.P., Suess 'Wiggles' - A comparison between radiocarbon records, Meteoritics, 20, 383-394 (1985).

Sonett, C.P. and S.A. Finney, The spectrum of radiocarbon, Phil. Trans, Royal Soc., London (in press).

Sonett, C.P., Very long solar periods and the radiocarbon record, Rev. Geophys. Spa. Phys., 22, 239-254 (1984).

Stuiver, M. and P.D. Quay, Changes in atmospheric carbon-14 attributed to a variable Sun, Science, 207, 11, (1980).

Stuiver, M., Variations in radiocarbon concentration and sunspot activity, J. Geophys. Res., 66, 273-276 (1961).

Stuiver, M., Carbon-14 content of 18th and 19th century wood; variations correlated with sunspot activity, Science, 149, pp. 533-535 (1965).

Stuiver, M. and G.W. Pearson, High-preciosion calibration of the radiocarbon time scale, AD 1950–500BC, in Radiocarbon, calibration issue, 28, 805-838, (1986).

Suess, H.E., Bristlecone pine calibration of the radiocarbon time scale 5200 BC to the present, pp/ 303-311, in 12th Nobel symposium, ed. I.U. Ollson, Wiley (1970).

Suess, H.E., La Jolla measurements of radiocarbon in tree ring– dated wood, Radiocarbon, 20, pp. 1-18 (1978).

Suess, H.E., The radiocarbon record in tree rings of the last 8000 years, in Intl. ¹⁴C conf., 10th Proc., eds M. Stuiver and R. Kra, Radiocarbon 3, 1-4 (1980).

Suess, H.E., Secular variations of cosmogenic ¹⁴C on Earth: their discovery and interpretation, Radiocarbon, 28, 259-265 (1986).

Suess, H.E., Climate changes, solar activity, and cosmic-ray production rate of natural radiocarbon, Meteorol. Monographs 8, 146-150 (1968).

Wallace, J.M. and R.E. Dickenson, Empirical orthogonal representation of time series in the frequency domain. Part I: theoretical considerations, J. Applied Meteor., 11, 887-892 (1972).

# ON DECOUPLING PROBABILITY FROM KINEMATICS IN QUANTUM MECHANICS

David Hestenes
Department of Physics
Arizona State University
Tempe, AZ  85287-1504

ABSTRACT.  A means for separating subjective and objective aspects of
the electron wave function is suggested, based on a reformulation of
the Dirac Theory in terms of Spacetime Algebra.  The reformulation
admits a separation of the Dirac wave function into a two parameter
probability factor and a six parameter kinematical factor.  The
complex valuedness of the wave function as well as its bilinearity in
observables have perfect kinematical interpretations independent of
any probabilistic considerations.  Indeed, the explicit unit
imaginary in the Dirac equation is automatically identified with the
electron spin in the reformulation.  Moreover, the canonical momentum
is seen to be derived entirely from the rotational velocity of the
kinematical factor, and this provides a geometrical interpretation of
energy quantization.  Exact solutions of the Dirac equation exhibit
circular zitterbewegung in exact agreement with the classical
Wessenhoff model of a particle with spin.  Thus, the most peculiar
features of quantum mechanical wave functions have kinematical
explanations, so the use of probability theory in quantum mechanics
should not differ in any essential way from its use in classical
mechanics.

## INTRODUCTION

I believe that quantum mechanics, as generally understood and
practiced today, intermixes subjective and objective components of
human knowledge, and furthermore, that we will not understand the
subject fully until those components can be cleanly separated.  The
main purpose of this article is to propose a means by which that
separation might be effected.  As will be seen, my proposal has many
specific and surprising consequences as well as possibilities for
further development.

I regard the Dirac electron theory as the fundamental core of
current quantum mechanics.  It is from the Dirac theory that the most
precise and surprising consequences of quantum mechanics have been
derived.  Some would claim that quantum field theory is more
fundamental, but one can argue that field theory is merely a formal

*P. F. Fougère (ed.), Maximum Entropy and Bayesian Methods*, 161–183.

device for imposing boundary conditions of the single particle theory
to accommodate particle creation and annihilation along with the
Pauli principle [1]. For these reasons, it is to the Dirac theory
that I look to understand the role of probability in quantum
mechanics. We shall see that the Dirac theory supplies insights into
the significance of quantum mechanical wave functions that could not
possibly be derived from the Schrödinger theory.

To separate subjective and objective components of the Dirac
Theory I suggest that we need two powerful conceptual tools. The
first tool is the Universal Probability Calculus which has been
synthesized and expounded so clearly by Ed Jaynes and amply justified
by many applications discussed in these Workshops and elsewhere.
Fortunately, this calculus is so familiar to workshop participants
that I need not spell out any of the details, though the calculus is
still not appreciated by most physicists. However, I should
reiterate the major claims for the calculus which explains its
relevance to the interpretation of quantum mechanics. First of all,
the calculus is universal in the sense that it is applicable to any
problem involving interpretions and explanations of experimental data
with mathematical models. Thus, the calculus provides a universal
interface between theory and experiment. Secondly, the calculus
provides the basis for unambiguous distinctions between subjective
and objective knowledge. Probabilities are always subjective; they
describe limitations on objective knowledge about the real world
rather than properties of the real world itself. Jaynes has applied
the calculus brilliantly to cleanly separate subjective and objective
components of statistical mechanics. It should be possible to do the
same in quantum mechanics.

The very generality of the probability calculus is one of its
inherent limitations. Objective knowledge enters the calculus only
through the Boolean algebra of propositions, and this does not take
into account the general implications of spacetime structure for
probabilistic reasoning about the physical world. To remedy this
deficiency, I propose to employ another powerful conceptual tool
which I call the Universal Geometric Calculus. We shall see that it
suggests some extensions of the probability calculus.

Geometric Calculus is a universal mathematical language for
expressing geometrical relations and deducing their consequences. As
such, it is a natural language for most of mathematics and perhaps
all of physics ([2], [3], [4]). Of special interest here is a
portion of the general geometric calculus called Spacetime Algebra
(STA). It can be regarded as the minimally complete algebra of
geometrical relations in spacetime.

When the Dirac theory is expressed in terms of STA a hidden
geometric structure is revealed, and natural explanations appear for
some of the most peculiar features of quantum mechanics [5]. For
example, it can be seen that there are geometrical reasons for the
appearance of complex probability amplitudes and the bilinear
dependence of observables on them. The main ideas and insights of
this approach are reviewed below. Then they are applied to achieve
the proposed separation between subjective and objective features of
the Dirac theory, resulting in a new concept of "pure state". A new

The geometric product of vectors u and v can be decomposed into a symmetric part u·v and an antisymmetric part u∧v as defined by

$$u \cdot v = \frac{1}{2}(uv + vu) \tag{1.5}$$

$$u \wedge v = \frac{1}{2}(uv - vu) \tag{1.6}$$

$$uv = u \cdot v + u \wedge v \tag{1.7}$$

One can easily prove that the symmetric product u·v defined by (1.5) is scalar-valued. Thus, u·v is the usual <u>inner product</u> (or metric tensor) on spacetime. The quantity u∧v is neither scalar nor vector, but a new entity called a <u>bivector</u> (or 2-vector). It represents an oriented segment of the plane containing u and v in much the same way that a vector represents a directed line segment.

Let $\{\gamma_\mu, \mu = 0,1,2,3\}$ be a <u>righthanded orthonormal frame</u> of vectors; so

$$\gamma_0^2 = 1 \quad \text{and} \quad \gamma_1^2 = \gamma_2^2 = \gamma_3^2 = -1, \tag{1.8}$$

and it is understood that $\gamma_0$ points into the forward light cone. In accordance with (1.5), we can write

$$g_{\mu\nu} \equiv \gamma_\mu \cdot \gamma_\nu = \frac{1}{2}(\gamma_\mu \gamma_\nu + \gamma_\nu \gamma_\mu), \tag{1.9}$$

defining the components of the metric tensor $g_{\mu\nu}$ for the frame $\{\gamma_\mu\}$.

Representations of the vectors $\gamma_\mu$ by 4 × 4 matrices are called <u>Dirac matrices</u>. The <u>Dirac algebra</u> is the matrix algebra over the <u>field of the complex numbers</u> generated by the <u>Dirac matrices</u>. The conventional formulation of the Dirac equation in terms of the Dirac algebra can be replaced by an equivalent formulation in terms of STA. This has important implications. First, a representation of the $\gamma_\mu$ by matrices is completely irrelevant to the Dirac theory; the physical significance of the $\gamma_\mu$ is derived entirely from their representation of geometrical properties of spacetime. Second, imaginaries in the complex number field of the Dirac algebra are superfluous, and we can achieve a geometrical interpretation of the Dirac wave function by eliminating them. For these reasons we eschew the Dirac algebra and stick to STA.

A generic element of the STA is called a <u>multivector</u>. Any multivector M can be written in the <u>expanded form</u>

$$M = \alpha + a + F + ib + i\beta, \tag{1.10}$$

where $\alpha$ and $\beta$ are scalars, a and b are vectors, and F is a bivector. The special symbol i will be reserved for the <u>unit pseudoscalar</u>, which has the following three basic algebraic properties:

(a) it has negative square,

$$i^2 = -1, \tag{1.11a}$$

(b) it anticommutes with every vector a,

$$ia = -ai \tag{1.11b}$$

(c) it factors into the ordered product

$$i = \gamma_0 \gamma_1 \gamma_2 \gamma_3. \tag{1.11c}$$

Geometrically, the pseudoscalar i represents a unit oriented 4-volume for spacetime.

By multiplication the $\gamma_\mu$ generate a complete basis for the STA consisting of

$$1, \ \gamma_\mu, \ \gamma_\mu \wedge \gamma_\mu, \ i\gamma_\mu, \ i. \tag{1.12}$$

These elements comprise a basis for the 5 invariant components of M in (1.10), the scalar, vector, bivector, pseudovector and pseudoscalar parts respectively. Thus, they form a basis for the space of completely antisymmetric tensors on spacetime. It will not be necessary for us to employ a basis, however, because the geometric product enables us to carry out computations without it.

Computations are facilitated by the operation of reversion. For M in the expanded form (1.10), the reverse $\tilde{M}$ can be defined by

$$\tilde{M} = \alpha + a - F - ib + i\beta. \tag{1.13}$$

Note, in particular, the effect of reversion on scalars, vectors, bivectors and pseudoscalars:

$$\tilde{\alpha} = \alpha, \ \tilde{a} = a, \ \tilde{F} = -F, \ \tilde{i} = i.$$

Reversion has the general property

$$(MN)^\sim = \tilde{N}\tilde{M}, \tag{1.14}$$

which holds for arbitrary multivectors M and N.

Having completed the preliminaries, we are now equipped to state a powerful theorem of great utility: Every Lorentz transformation of an orthonormal frame $\{\gamma_\mu\}$ into a frame $\{e_\mu\}$ can be expressed in the canonical form

$$e_\mu = R\gamma_\mu \tilde{R}, \tag{1.15}$$

where R is a <u>unimodular spinor</u>, which means that R is an even multivector satisfying the unimodularity condition

$$R\tilde{R} = 1. \tag{1.16}$$

A multivector is said to be <u>even</u> if its vector and trivector parts are zero. The spinor R is commonly said to be a spin <u>representation</u> of the Lorentz transformation (1.15).

The set {R} of all unimodular spinors is a group under multiplication. In the theory of group representations it is called SL(2,C) or "the spin-1/2 representation of the Lorentz group". However, group theory alone does not specify its invariant imbedding in the STA. It is precisely this imbedding that makes it so useful in the applications to follow.

## Classical Electrodynamics and Particle Mechanics

The electromagnetic field $F = F(x)$ is a bivector-valued function on spacetime. The expansion of F in a bivector basis,

$$F = \frac{1}{2}F^{\mu\nu}\gamma_\mu\wedge\gamma_\nu \tag{2.1}$$

shows its relation to the usual representation of the electromagnetic field by tensor components $F^{\mu\nu}$. We have no need of this expansion, however, because STA enables to do everything in a completely coordinate-free manner.

The derivative with respect to a (vector) spacetime point x is a vector differential operator $\Box$ which is related to conventional coordinate derivatives by

$$\Box = \gamma^\mu\partial_\mu \tag{2.3}$$

where

$$\partial_\mu = \frac{\partial}{\partial x^\mu} = \gamma_\mu\cdot\Box. \tag{2.4}$$

It will recognized that the matrix representation of (2.3) where the $\gamma^\mu$ are replaced by Dirac matricies is the famous "Dirac operator". However, STA reveals that the significance of this operator lies in its role as the fundamental differential operator on spacetime, rather than in any special role in quantum mechanics or even spinor mechanics.

Since $\Box$ is a vector operator, we can use (1.7) to decompose the derivative of a vector field $A = A(x)$ into <u>divergence</u> $\Box\cdot A$ and <u>curl</u> $\Box\wedge A$; thus

$$\Box A = \Box\cdot A + \Box\wedge A. \tag{2.5}$$

class of solutions to the Dirac equations is identified, solutions
which clearly exhibit circular zitterbewegung and suggest that it is
an objective property of electron motion independent of probabilistic
aspects of the theory, in general agreement with earlier speculations
[6]. This is illustrated by a new kind of free particle solution to
the Dirac equation.

All of this together indicates a need to integrate probability
calculus with geometric calculus to form a single coherent conceptual
system, and it suggests how that might be accomplished. Geometric
calculus is needed to represent objective properties of real objects
in mathematical models and theories. Probability calculus is needed
to relate such models to real phenomena. Thus, the two must be
integrated to achieve an integrated world view.

## Spacetime Algebra

For the purposes of this paper, we adopt a flat space model of
physical spacetime, so each point event can be uniquely represented
by an element x in a 4-dimensional vector space. We call x the
location of the event, and we call the vector space of all locations
spacetime. Thus, we follow the usual practice of conflating our
mathematical model with the physical reality it supposedly
represents.

To complete the mathematical characterization of spacetime, we
define a geometric product among spacetime vectors u, v, w by the
following rules:

$$u(vw) = (uv)w, \tag{1.1}$$

$$u(v + w) = uv + uw, \tag{1.2a}$$

$$(v + w)u = vu + wu. \tag{1.2b}$$

For every vector u and scalar $\lambda$

$$u\lambda = \lambda u, \tag{1.3}$$

$$u^2 = \text{a scalar (real number).} \tag{1.4}$$

The metric of spacetime is specified by the allowed values for $u^2$.
As usual, a vector u is said to be timelike, lightlike or spacelike
if $u^2 > 0$, $u^2 = 0$, $u^2 < 0$ respectively.

Under the geometric product defined by these rules, the vectors
of spacetime generate a real associative algebra, which I call the
Spacetime Algebra (STA), because all its elements and operations have
definite geometric interpretations, and it suffices for the
description of geometric structures on spacetime. An account of STA
is given in [2], and extended in the other references. Only a few
features of STA with special relevance to the problem at hand can be
reviewed here.

Taking A to be the electromagnetic <u>vector</u> <u>potential</u> and imposing the "Lorentz condition"

$$\square \cdot A = 0, \tag{2.6}$$

we have

$$F = \square \wedge A = \square A \tag{2.7}$$

Maxwell's equation for the E.M. field can then be written

$$\square F = \square^2 A = J_e, \tag{2.8}$$

where $J_e = J_e(x)$ is the (electric) <u>charge</u> <u>current</u> (density).

Equation (2.8) describes the production and propagation of E.M. fields equally well in classical and quantum theories, but it must be complemented by an equation of motion for charged particles which describes the effects of the E.M. field. The ordinary classical equation of motion for a charged particle seems so different from the quantum mechanical equation that they are difficult even to compare. However, STA admits a new formulation of the classical equation which greatly clarifies its relation to quantum theory.

The equation

$$e_\mu = R\gamma_\mu \tilde{R} \tag{2.9}$$

can be used to describe the relativistic kinematics of a rigid body (with negligible dimensions) traversing a world line $x = x(\tau)$ with proper time $\tau$, if we identify $e_0$ with the proper velocity $v$ of the body (or particle), so that

$$\frac{dx}{d\tau} = v = e_0 = R\gamma_0 \tilde{R}. \tag{2.10}$$

Then $e_\mu = e_\mu(\tau)$ is a comoving frame traversing the world line along with the particle, and the spinor R must also be a function of proper time, so that, at each time $\tau$, equation (2.9) describes a Lorentz transformation of some fixed frame $\{\gamma_\mu\}$ into the comoving frame $\{e_\mu(\tau)\}$. Thus, we have a spinor-valued function of proper time $R = R(\tau)$ determining a 1-parameter family of Lorentz transformations.

The spacelike vectors $e_k = R\gamma_k\tilde{R}$ (for $k = 1,2,3$) can be identified with the principal axes of the body in some applications, but for a particle with an intrinsic angular momentum or <u>spin</u>, it is most convenient to identify $e_3$ with the spin direction $\hat{s}$; so we write

$$\hat{s} = e_3 = R\gamma_3\tilde{R}. \tag{2.11}$$

From the fact that R is an even multivector satisfying $R\tilde{R} = 1$, it can be proved that $R = R(\tau)$ must satisfy a <u>spinor equation of motion</u> of the form

$$\dot{R} = \frac{1}{2}\Omega R, \qquad (2.12)$$

where the dot represents the proper time derivative, and $\Omega = \Omega(\tau) = -\tilde{\Omega}$ is a bivector-valued function. Differentiating (2.9) and using (2.12), we see that the equations of motion for the comoving frame must be of the form

$$\dot{e}_\mu = \frac{1}{2}(\Omega e_\mu - e_\mu \Omega) \equiv \Omega \cdot e_\mu. \qquad (2.13)$$

Clearly $\Omega$ can be interpreted as a generalized <u>rotational velocity</u> of the comoving frame.

For a classical particle of mass m and charge e moving in an external E.M. field F, we take $\Omega = em^{-1}F$ and from (2.13) with $\mu = 0$ we get (in units with the speed of light e = 1)

$$m\dot{v} = eF \cdot v. \qquad (2.14)$$

This is the classical equation of motion where the right side is the <u>Lorentz force</u>. On the other hand, (2.12) gives us

$$\dot{R} = \frac{e}{2m}FR. \qquad (2.15)$$

This spinor equation of motion implies the conventional equation (2.14), so it determines the same world line, but it gives us much more. First, it is easier to solve; solutions for various external fields are given in [7]. Second, it gives immediately a classical model for a particle with spin; by (2.11) and (2.13) with $\mu = 3$, the equation of motion for the spin is

$$\dot{s} = \frac{e}{m}F \cdot s \qquad (2.16)$$

This is, in fact, the spin precession equation for a particle with gyromagnetic ratio g = 2, exactly the basic value of g implied by the Dirac theory. This is no accident, for the greatest advantage of (2.15) is that the spinor R can be related directly to the Dirac wave function, thus enabling a close comparison of classical and quantum equations of motion, as shown below.

## The Real Dirac Theory

This section summarizes (without proof) the formulation of the Dirac electron theory in terms of STA. Proofs that this is equivalent to the conventional matrix formulation are given in [8] and [5].

The Dirac wave function $\psi = \psi(x)$ is an even multivector-valued function on spacetime. It has the Lorentz invariant decomposition

$$\psi = (\rho e^{i\beta})^{1/2}R, \qquad (3.1)$$

where $R = R(x)$ is a unimodular spinor, i is the unit pseudoscalar (1.11c), and $\rho = \rho(x)$, $\beta = \beta(x)$ are scalar-valued functions.

The Dirac equation has the real form

$$\Box \psi i \hbar + eA\psi = m\psi\gamma_0, \tag{3.2}$$

where $A = A(x)$ is the vector potential, h is Planck's constant over $2\pi$, and the boldface

$$\underset{\sim}{i} = \gamma_2\gamma_1 = i\gamma_3\gamma_0 \tag{3.3}$$

is a constant bivector which corresponds to the unit imaginary in the matrix formulation of the Dirac equation. Equation (3.2) is called the real form of the Dirac equation, because it does not involve complex numbers; it involves only elements of STA, which all have definite geometrical meaning.

The physical meaning of $\psi$ is determined by assumptions which relate it to observables. From (3.1) we can construct the invariant

$$\psi\tilde{\psi} = \rho e^{i\beta}. \tag{3.4}$$

but this does not elucidate physical meaning. The Dirac probability current J is defined by

$$J = \psi\gamma_0\tilde{\psi} = \rho v, \tag{3.5}$$

where

$$v = R\gamma_0\tilde{R} \tag{3.6}$$

is the local velocity, so $\rho = \rho(x)$ can be interpreted as the proper probability density. The Dirac equation implies the probability conservation law

$$\Box \cdot J = \Box \cdot (\rho v) = 0. \tag{3.7}$$

The wave function determines everywhere a "comoving frame" $\{e_\mu = e_\mu(x)\}$ by

$$\psi\gamma_\mu\tilde{\psi} = \rho R\gamma_\mu\tilde{R} = \rho e_\mu, \tag{3.8}$$

which, of course, includes (3.5). The spin vector $s = s(x)$ is defined by

$$s = \frac{\hbar}{2}R\gamma_3\tilde{R} = \frac{\hbar}{2}e_3. \tag{3.9}$$

Actually, angular momentum is a bivector quantity and (as demonstrated in [8]) the appropriate spin bivector $S = S(x)$ is given by

$$S = \frac{1}{2}R i\hbar\tilde{R} = \frac{\hbar}{2}R\gamma_2\gamma_1\tilde{R} = \frac{\hbar}{2}e_2 e_1 = isv. \tag{3.10}$$

This shows exactly the sense in which the imaginary factor $i\hbar$ in the Dirac theory (and hence in the Schrödinger Theory) <u>can be interpreted as a representation of the electron spin</u>. The right side of (3.10) relates the spin vector to the spin bivector, so either of them can be used to describe spin.

The conservation law (3.7) implies that through each point x where $\psi(x) \neq 0$ there is a unique world line (or <u>bicharacteristic</u>) of $\psi$ with tangent v(x). Along each world line, by $\overline{(3.9)}$ and $\overline{(3.6)}$ the unimodular spinor R(x) uniquely determines the direction of electron spin as well as the velocity, exactly as in the classical case discussed previously. Thus, the spinor R can be given the same well-defined kinematic interpretation in both classical and quantum theory.

The assumptions made so far provide a kinematic interpretation for the factor R in the wave function. As detailed in [8], one more assumption is needed to complete the Dirac theory. That is the crucial assumption introducing the <u>energy-momentum</u> operator $\underline{p}_\mu$, defined in STA by

$$\underline{p}_\mu\psi = \partial_\mu\psi i\hbar - eA_\mu\psi. \tag{3.11}$$

(The underbar in $\underline{p}_\mu$ is meant to indicate a linear operator.) This operator can be introduced by the following definition of the electron <u>energy-momentum</u> <u>tensor</u> components

$$T_{\nu\mu} = \langle\gamma_0\tilde{\psi}\gamma_\nu \, \underline{p}_\mu\psi\rangle = \langle\gamma_\nu(\underline{p}_\mu\psi)\gamma_0\tilde{\psi}\rangle, \tag{3.12}$$

where <M> means "scalar part of M". Accordingly, the energy momentum flux through a hyper-surface with normal $\gamma_\nu$ is $T\gamma_\nu = T_{\nu\mu}\gamma^\mu$. So the proper <u>energy momentum density</u> $\rho p$, defined as the flux in the direction v, is given by

$$\rho p = \underline{T}v = v^\nu T_{\nu\mu}\gamma^\mu, \tag{3.13}$$

where $v^\nu = \gamma^\nu \cdot v$. The components of the energy-momentum density are therefore given by

$$\rho p_\mu = \rho p \cdot \gamma_\mu = v^\nu T_{\nu\mu} = \langle v(\underline{p}_\mu\psi)\gamma_0\tilde{\psi}\rangle. \tag{3.14}$$

Insight into the kinematic significance of this relation is obtained by introducing the canonical decomposition for $\psi$.

From the fact that R is a unimodular spinor, it can be proved that its partial derivatives have the form

$$\partial_\mu R = \gamma_\mu \cdot \Box R = \frac{1}{2}\Omega_\mu R, \tag{3.15}$$

where $\Omega_\mu = -\tilde{\Omega}_\mu$ is a bivector quantity representing the rate of rotation for a displacement in the direction $\gamma_\mu$. The derivatives of the frame $\{e_\mu\}$ in (3.9) can therefore be put in the form

$$\partial_\mu e_\gamma = \Omega_\mu \times e_\nu = \Omega_\mu \cdot e_\nu, \tag{3.16}$$

where $\times$ signifies the commutator product defined by $A \times B = \frac{1}{2}(AB - BA)$. Similarly, the derivatives of the spin (3.10) has the form

$$\partial_\mu S = \Omega_\mu \times S. \tag{3.17}$$

Now with the help of (3.10), (3.15) and (3.17) we can write

$$\partial_\mu R i \hbar \tilde{R} = \Omega_\mu S = P_\mu + i q_\mu + \partial_\mu S, \tag{3.18}$$

where $P_\mu$ and $q_\mu$ are defined by

$$P_\mu = \Omega_\mu \cdot S, \tag{3.19}$$

$$i q_\mu = \Omega_\mu \wedge S. \tag{3.20}$$

Finally, using (3.18) to evaluate $\partial_\mu \psi$ in (3.14) we get

$$p_\mu = P_\mu - e A_\mu = \Omega_\mu \cdot S - e A_\mu. \tag{3.21}$$

This result is striking, because it shows that $p_\mu$ has a purely kinematical dependence on the wave function independent of the parameters $\rho$ and $\beta$. It shows that $p_\mu$ depends on the rotation rate $\Omega_\mu$ only through its projection onto the spin plane. This, in turn, gives insight into the geometrical meaning of quantization in stationary states; for the requirement that the wave function be single-valued implies that the comoving frame (3.8) be single-valued, so along any closed curve it must rotate an integral number of times; according to (3.21), therefore, we should have

$$\oint (p + eA) \cdot dx = \oint P \cdot dx = \oint (dx^\mu \Omega_\mu) \cdot S = \frac{1}{2} n \hbar, \tag{3.22}$$

where n is an integer. Thus, we have a geometric interpretation of the electron energy-momentum vector and its relation to quantization. That is what we need to separate probabilistic and kinematical components of the Dirac theory in the next section.

## Separating Kinematics from Probabilities

The STA formulation of the Dirac theory in the preceding section suggests that the wave function decomposition $\psi = R(\rho e^{i\beta})^{1/2}$ has a fundamental physical significance besides being "relativistically invariant". For it reveals that the kinematics of electron motion (specifically the values of velocity, spin and energy-momentum) are completely determined by the unimodular spinor R and its derivatives $\partial_\mu R = \frac{1}{2} \Omega_\mu R$. On the other hand, $\rho$ has an obvious probabilistic interpretation and we shall see how $\beta$ might be given one as well.

Thus, it appears that ψ can be decomposed into a purely kinematic factor R and a probabilistic factor $(\rho e^{i\beta})^{1/2}$. These factors are coupled by the Dirac equation in a way which is difficult to interpret physically [8]. However, it appears possible to decouple them completely by using the superposition principle as explained below.

Let us suppose that unimodular solutions of the Dirac equation have a fundamental physical significance and sanctify this by calling them pure states. A pure state ψ = R satisfies $\tilde{\psi}\psi = 1$. The most common example of a pure state state is a plane wave, but there are more interesting examples as we shall see. For a plane wave it is ordinarily supposed that the probability density $\rho(x)$ is uniform, hence unnormalizable, so there is some question as to its physical significance. Anyway, it is argued that in a plane wave state the particle position is indeterminate so that, in accordance with the uncertainty principle, the momentum and velocity have definite values. We shall introduce a different interpretation of pure states.

Each pure state $\psi(x)$ determines a unique velocity $v(x)$, spin $S(x)$ and momentum $p(x)$ at each spacetime location x. Let us assume that the electron is a point particle and this is the state of motion to be attributed to it if it is at a given location x. Thus, the pure state assigns a definite state of motion to every possible electron location. The dynamics of electron motion are incorporated into the pure state by requiring that it be a solution of the Dirac equation. This much is a purely deterministic model of electron motion.

Probabilities enter the theory by assuming that we do not know precisely the electron's pure state or location so the state of our knowledge is best described as a weighted average of pure states indexed by some parameter λ, e.g.

$$\psi = \int d\lambda w_\lambda \psi_\lambda = R(\rho e^{i\beta})^{1/2} \qquad (4.1)$$

It is easy to prove that, in general, a superposition of pure states (with $\psi\tilde{\psi} = 1$) produces a state with $\psi\tilde{\psi} = \rho e^{i\beta}$, so the factor $e^{i\beta}$ arises naturally along with the probability density. Another way to see that the parameter ρ alone is not enough to describe the result of averaging process is to suppose that an expected velocity $\bar{v}$ (x) at x is to be obtained by averaging velocities $v_\lambda$ (x). To satisfy the relativistic constraint $\bar{v}^2 = v_\lambda^2 = 1$, the average must have the form

$$\int d\lambda \bar{v}_\lambda = \bar{v} \cos \alpha, \qquad (4.2)$$

where α could possibly be identified with β. This argument is meant to be suggestive only. The main idea is that the $e^{i\beta}$ in (3.1) is the statistical factor that arises from the unimodularity constraint on the kinematical factor of the wave function.

A similarity to Feynman's path integral formulation of quantum mechanics appears by interpreting the integral in (4.1) as a sum over paths with unit weight factor $w_\lambda = 1$.  Feynman assumed

$$\psi = e^{\underset{\sim}{i}\phi_\lambda/\hbar},$$

where $\phi_\lambda$ is the classical action along the $\lambda$-path.  Our theory
suggests that this phase factor should be generalized to a unimodular
spinor to account for spin.  In that case also the "statistical
factor" on the right side of (4.1) would arise from superposition.
This promising generalization of the Feynman approach will not be
pursued here, although the following sections contain hints on how to
carry it out.  The main point is that the unimodular spinors play a
fundamental role in the Feynman approach just as they do in the
present approach.  The basic statistical problem is how to assign
appropriate statistical weights to alternative world lines.

It remains to be seen whether the statistical notions suggested
here can be accommodated by the Universal Probability Calculus in a
natural way.  This issue should be set within the general question of
how best to reconcile probability calculus with relativity.  The
probability densities ordinarily used in probability calculus are not
relativistically invariant, so they must be replaced by probability
currents which are.  This generalization from scalar densities to
vectorial currents raises questions about how such quantities as
entropy should be defined in terms of currents.

On the other hand, since a probability current $J = J(x)$ is a
timelike vector field, it can always be written in the form

$$J = \psi\gamma_0\tilde{\psi}, \tag{4.3}$$

so the spinor field $\psi = \psi(x)$ can be taken as the fundamental
descriptor of probabilistic state.  This applies to classical as well
as quantum statistical mechanics.  The bilinearity of (4.3) is a
consequence of spacetime geometry alone.  However, it raises a
question about how to compute statistical averages.  In accordance
with the superposition principle, or at least with the linearity of
the Dirac equation, quantum mechanics constructs composite states by
averaging over spinor wave functions, whereas classical statistical
mechanics takes averages over vector and tensor quantities.  What
general principle of probability theory will tell us which kind of
average to consider?

Let me suggest that the answer is to be found in the way that
physical information about pure states is specified.  In The Dirac
Theory the relation among pure state observables is specified by the
Dirac equation.  Therefore, in constructing a composite state to
express our uncertainty as to which pure state describes a given
electron, we should require that the form of the Dirac equation be
preserved.  Considering the linearity of the Dirac equation, there is
evidently no alternative to the linear superposition (4.1).  This
makes the superposition principle of quantum mechanics seem much less
fundamental, and it directs us to a study of the pure state Dirac
equation to understand the underlying dynamics of quantum mechanics.

## Unimodular Solutions of the Dirac Equation

In this section we study general properties of unimodular (pure state) solutions of the Dirac equation. In particular, we seek to determine the equations of motion for a particle with spin moving along a bicharacteristic of the wave function.

As noted before, the unimodular condition $\psi\tilde{\psi} = 1$ implies that

$$\partial_\mu\psi = \frac{1}{2}\Omega_\mu\psi, \tag{5.1}$$

where the $\Omega_\mu$ are bivectors. Consequently the rate of change of $\psi$ along a bicharacteristic is given by

$$\dot{\psi} = v \cdot \Box\psi = \frac{1}{2}\Omega\psi, \tag{5.2}$$

where $\Omega = v^\mu\Omega_\mu$. The equations of motion (2.13) for a comoving frame $\{e_\mu = \psi\gamma_\mu\tilde{\psi}\}$ on a bicharacteristic $x = x(\tau)$ are determined by expressing $\Omega$ as a function of $x(\tau)$. To that end, we note that $\Omega$ can be obtained from the identity (derived in [8])

$$\Omega = 2\dot{\psi}\tilde{\psi} = (\Box\psi)\tilde{\psi}v + v(\Box\psi)\tilde{\psi} - \Box v. \tag{5.3}$$

Now, using (3.6) and (3.10), the Dirac equation (3.2) can be put in the form

$$\Box\psi i\hbar\psi = 2(\Box\psi)\tilde{\psi}S = mv - eA. \tag{5.4}$$

Using this in (5.3) we obtain

$$\Omega = - \Box v + (m - ev \cdot A)S^{-1} \tag{5.5}$$

Since $\Box v = \Box\wedge v + \Box \cdot v$, the scalar part of (5.5) gives the Dirac current conservation equation

$$\Box \cdot v = 0. \tag{5.6}$$

Thus (5.5) reduces the problem of evaluating $\Omega$ to evaluating $\Box\wedge v$.

We can express the Dirac equation as a constitutive relation among observables by using (3.18) and

$$p = P + eA \tag{5.7}$$

to put the Dirac equation (5.4) in the form

$$p - iq + \Box S = mv. \tag{5.8}$$

Separately equating vector and trivector parts, we obtain

$$p = mv - \Box \cdot S, \tag{5.9}$$

and

$$iq = \Box \wedge S. \tag{5.10}$$

We can regard (5.9) as a constitutive equation for the momentum p in terms of the velocity and spin.

Further properties of observables can be derived by diffentiating the Dirac relation (5.8) to get

$$\Box p + i\Box q + \Box^2 S = m\Box v. \tag{5.11}$$

Separating this into scalar, pseudoscalar and bivector parts and using (5.6), (5.7) we obtain

$$\Box \cdot p = \Box \cdot P + e\Box \cdot A = m\Box \cdot v = 0, \tag{5.12}$$

$$\Box \cdot q = 0, \tag{5.13}$$

$$m\Box \wedge v = \Box^2 S - eF + (\Box \wedge P + i\Box \wedge q). \tag{5.14}$$

This last equation shows explicity how $\Box \wedge v$ depends on the E.M. field $F = \Box \wedge A$, but we still need to evaluate the terms in parenthesis. That can be done as follows.

By (5.1), the "integrability condition"

$$\partial_\nu \partial_\mu \psi = \partial_\mu \partial_\nu \psi$$

can be cast in the form

$$\partial_\nu \Omega_\mu - \partial_\mu \Omega_\nu = \Omega_\gamma \times \Omega_\mu. \tag{5.15}$$

And by (3.18) it can be cast in the alternative form

$$\partial_\nu P_\mu - \partial_\mu P_\gamma + i(\partial_\nu q_\mu - \partial_\mu q_\nu) = (\partial_\mu S \times \partial_\nu S) \, S^{-1}. \tag{5.16}$$

Whence

$$\Box \wedge P = (\partial_\mu S \times \partial_\nu S) \cdot S^{-1}(\gamma^\mu \wedge \gamma^\nu), \tag{5.17}$$

$$i\Box \wedge q = (\partial_\mu S \times \partial_\nu S) \wedge S^{-1} (\gamma^\mu \wedge \gamma^\nu). \tag{5.18}$$

This enables the far right part of (5.14) to be expressed solely in terms of the spin S and its derivatives. To make further simplifications we need to know something specific about the spin derivatives $\partial_\mu S$. That's next.

## Classical Solutions of the Dirac Equation

Now we look for a special class of pure state solutions to the Dirac equation by requiring that variations in the spin $S = S(x)$ be due solely to motion along the bicharacteristics. This can be expressed mathematically by the requirement

$$\partial_\mu S = v_\mu v \cdot \Box S = v_\mu \dot{S} \tag{6.1}$$

Let us call this the <u>decoupling condition</u>, because it implies that neighboring bicharacteristics are not coupled by an exchange of spin angular momentum. It follows that

$$\Box S = v\dot{S}, \tag{6.2}$$

$$\Box^2 S = \ddot{S}. \tag{6.3}$$

Furthermore, from (5.17) and (5.18)

$$\Box \wedge P = 0, \tag{6.4}$$

$$\Box \wedge q = 0. \tag{6.5}$$

It must be understood that these are local conditions which may be violated at singular points; otherwise (6.4) is inconsistent with the quantization condition (3.23) because of Stokes' Law

$$\int d^2x \cdot (\Box \wedge P) = \oint dx \cdot P, \tag{6.6}$$

where $d^2x$ is a bivector-valued directed area element.
    Substituting (6.3) to (6.5) into (5.14) we get

$$m\Box \wedge v = \ddot{S} - eF, \tag{6.7}$$

so (5.5) yields the desired result

$$m\Omega = eF - \ddot{S} + m(m - eA \cdot v)S^{-1}. \tag{6.8}$$

This gives us immediately the bicharacteristic equations of motion for velocity and spin:

$$m\dot{v} = (eF - \ddot{S}) \cdot v, \tag{6.9}$$

$$m\dot{S} = (eF - \ddot{S}) \times S \tag{6.10}$$

    Since (6.9) and (6.10) are well-defined, deterministic equations of motion along any specific bicharacteristic, they may be regarded as classical equations of motion for a point particle. Accordingly, let us refer to the corresponding pure state spinor $\psi$ as a <u>classical solution</u> of the Dirac equation.

To find out more about these classical solutions, we look to the Dirac equation itself.  Applying (6.2) to (5.9) we get the "Wessenhoff relation"

$$p = mv + \dot{S} \cdot v \qquad (6.11)$$

This shows that the momentum of the particle is not generally collinear with the velocity, because it includes a contribution from the spin.  Nevertheless, (6.11) implies

$$p \cdot v = m. \qquad (6.12)$$

Using this along with (5.7) we can now put (6.8) in the form

$$m\Omega = eF - \ddot{S} + m(v \cdot P)S^{-1}. \qquad (6.13)$$

Now $\Box \wedge p = 0$ implies $p = \Box \phi$, where $\phi = \phi(x)$ may be recognized as the phase of the wave function $\psi$.  Consequently $v \cdot P = v \cdot \Box \phi = \dot{\phi}$ is the rate of phase change along a bicharacteristic.  Since

$$\dot{\phi} = v \cdot P = m - eA \cdot v, \qquad (6.14)$$

we can find the phase change by direct integration after the bicharacteristics have been found.

Equations (5.13) and (6.5) combine to give

$$\Box q = 0, \qquad (6.15)$$

while (6.2) and (5.10) give

$$iq = v \wedge \dot{S}. \qquad (6.16)$$

It can be shown that equation (6.15) implies that q is constant if and only if (6.15) holds everywhere without any singularity.  It seems likely, therefore, that we can impose the simplifying condition

$$q = 0 \qquad (6.17a)$$

without unduly restricting the class of physically significant solutions.  So let us consider its implications.  According to (6.16), then, we have

$$v \wedge \dot{S} = 0. \qquad (6.17b)$$

Adding this to (6.11), we obtain

$$p = mv + \dot{S}v. \qquad (6.18)$$

Multiplying by v and taking the bivector part, we obtain

$$\dot{S} = p \wedge v. \qquad (6.19)$$

Equations of motion for the spin and velocity of the form (6.19) and
(6.9) have been derived by Wessenhoff [9] and Corben [10] as a
classical model of a particle with spin analogous to the Dirac
electron.  Here we have ascertained for the first time the conditions
under which these equations may be <u>exact</u> consequences of the Dirac
theory.  However, the present theory differs from their model in some
important respects.  First, the mass m, which is rigorously constant
in (6.9) and (6.18), is allowed to be variable in the Wessenhoff-
Corben equations.  Second, we have here the additional feature of a
wave function with variable phase determined by the last term on the
right side of (6.8).

The momentum $p = P - eA$ is most easily related to the rotational
velocity by (3.18), which gives us

$$\partial_\mu \psi i \hbar \tilde{\psi} = \Omega_\mu S = P_\mu + v_\mu \dot{S}, \tag{6.20}$$

whence

$$\Omega S = P \cdot v + \dot{S}. \tag{6.21}$$

For later use, it should be noted that the geometric product of two
bivectors $\Omega$ and S can be decomposed into scalar, bivector and
pseudoscalar parts by means of the identity $\Omega S = \Omega \cdot S + \Omega \times S + \Omega \wedge S$.  On
using (6.18) to eliminate S (6.21) yields

$$\Omega S = pv + eA \cdot v, \tag{6.22}$$

an algebraic relation among all the basic observables which is
readily solved for any one of them in terms of the others.
Alternatively, elimination of $\Omega$ from (6.21) by (6.13) yields

$$m\dot{S} = (eF - \ddot{S})S. \tag{6.23}$$

The bivector part of this has already been found in (6.10).  However,
the scalar and pseudoscalar parts yield the relations

$$\ddot{S} \cdot S = eF \cdot S = -\dot{S}^2 \tag{6.24}$$

$$\ddot{S} \wedge S = eF \wedge S. \tag{6.25}$$

The last equality in (6.24) follows from the fact that $S^2$ is constant
so $S \cdot \dot{S} = 0$.  The pseudoscalar part of (6.21) gives

$$\Omega \wedge S = 0. \tag{6.26}$$

This is a consequence of the condition $q = 0$, though it also follows
from the weaker condition $q \cdot v = 0$, which is already entailed by
(6.16).  To understand its significance, consider the identity

$$v \cdot (\Omega \wedge S) = (v \cdot \Omega) \wedge S + \Omega \wedge (v \cdot S).$$

Since $v \cdot S = 0$, this shows that (6.26) implies

$$\dot{v} \wedge S = 0. \tag{6.27}$$

This means that $\dot{v}$ lies in the S-plane.
Combining (6.27) with (6.17b) we find that

$$v \cdot \Box (v \wedge S) = 0. \tag{6.28}$$

Thus, the condition $q = 0$ implies that the trivector $S \wedge v = Sv =$ is is a constant of motion along the bicharacteristics. Evidently it is too strong a condition for motion in arbitrary E.M. fields, but it may hold in special cases. It can be shown that (6.25) is a consequence of this condition.

   To find an equation of motion for the momentum we differentiate (6.11) and use (6.9) to get

$$\dot{p} = eF \cdot v + \dot{S} \cdot \dot{v}. \tag{6.29}$$

The last term here is an unusual one in dynamics, but note that

$$v \cdot \dot{p} = v \cdot \dot{S} \cdot \dot{v} = \dot{S} \cdot (\dot{v} \wedge v). \tag{6.30}$$

Moreover, from the condition (6.17b) we get

$$v \cdot (v \wedge \dot{S}) = \dot{S} - v \wedge (v \cdot \dot{S}) = 0, \tag{6.31}$$

whence

$$v \cdot \dot{S} = (v \cdot S \cdot \dot{v})v. \tag{6.32}$$

So (6.29) becomes

$$\dot{p} = eF \cdot v + (\dot{p} \cdot v)v. \tag{6.33}$$

The absence of the last term in (6.33) is responsible for the variable mass in the Wessenhoff-Corben model.

   Finally, to justify the interpretation of S as intrinsic angular momentum, we define the total angular momentum

$$M = x \wedge p + S, \tag{6.34}$$

where $x \wedge p$ is the orbital angular momentum. Differentiating with $v \cdot \Box$ and using (6.19) we obtain the equation for <u>angular</u> <u>momentum</u> <u>conservation</u>

$$\dot{M} = x \wedge \dot{p}, \tag{6.35}$$

the right side being the generalized torque.

The bicharacteristics can be found by integrating (7.8), but a
better way is to combine (7.3) with the total angular momentum as
follows

$$x \wedge p + x \cdot p = xp = M - S + m\tau.$$

Multiplication by $p^{-1} = p/p^2$ gives

$$x = (M - S)p^{-1} + m\tau p^{-1}, \tag{7.10}$$

which becomes an explicit function $x = x(\tau)$ when (7.9) is inserted.
To show the character of the solution more explicitly, note that
(7.1) yields $p^{-1}\Omega = vS^{-1}$, so $v \cdot S = 0$ implies

$$p \cdot \Omega = 0, \text{ or } p\Omega = \Omega p. \tag{7.11}$$

Consequently,

$$p \cdot S = e^{\frac{1}{2}\Omega\tau} p \cdot S_0 e^{-\frac{1}{2}\Omega\tau}$$

$$= [(p \cdot S_0) \wedge \Omega + e^{\Omega\tau}(p \cdot S_0) \cdot \Omega]\Omega^{-1}. \tag{7.12}$$

Noting that $M = x_0 p + S_0$ on the hyperplane $x_0 \cdot p = 0$ and defining

$$r_0 = [(p^{-1} \cdot S_0) \cdot \Omega]\Omega^{-1}, \tag{7.13}$$

we can put (7.10) in the form

$$x = x_0 - r_0 + e^{\Omega\tau}r_0 + m\tau p^{-1}. \tag{7.14}$$

This is the parametric equation $x = x(\tau)$ for a timelike helix with
axis $x_0 - r_0 + m\tau p^{-1}$ and squared radius $|r_0|^2 = -r_0^2$. This is the same
type of helical orbit found by Wessenhoff and Corben.

This peculiar helical motion, unsupported by an external E.M.
field, can be identified with the zitterbewegung originally
attributed to the electron by Schrödinger[6]. Its physical
significance is problematic. The radius of the zitterbewegung
vanishes when $\Omega S = \Omega \cdot S = m$, so

$$\Omega = mS^{-1} = \frac{-2m}{\hbar} R_0 i \tilde{R}_0. \tag{7.15}$$

In that case (7.7) becomes

$$\psi = R_0 e^{-i p \cdot x/\hbar}, \tag{7.16}$$

which will be recognized as the usual free particle solution to the
Dirac equation. It has been suggested [6] that the phase factor in
(7.16) also describes a helical zitterbewegung and that the Dirac
theory should be modified to show it. If that is correct, then the
zitterbewegung must be the source of the electron's magnetic moment

and other feature of quantum theory [6]. The zitterbewegung must then be truly fundamental. That remains to be seen!

CONCLUSIONS

Reformulation of the Dirac Theory in terms of Spacetime Algebra reveals a hidden geometric structure and opens up a possibility for separating objective and subjective components of quantum mechanics. More specifically, we have noted the following:

(1) The Dirac wave function $\psi$ has a Lorentz invariant decomposition

$$\psi = R(\rho e^{i\beta})^{\frac{1}{2}},$$

where R is a unimodular spinor which completely characterizes the kinematics of electron motion.

(2) The Dirac current J is given by

$$J = \psi\gamma_0\tilde{\psi} = \rho R\gamma_0\tilde{R}.$$

In fact, any timelike vector field $J = J(x)$ can be expressed in terms of a spinor field $\psi = \psi(x)$ in exactly this way. Therefore, the bilinear dependance of the probability density on the wave function is not a special feature of quantum mechanics. Rather, it is a consequence of spacetime geometry (as represented by the spacetime algebra). This decomposition into spinors could be applied as well to probability currents in classical relativistic statistical mechanics.

(3) The electron spin angular momentum S is given by

$$S = \frac{1}{2}R i\hbar\tilde{R}$$

This reveals that the unit imaginary $i$ in the quantum mechanics of electrons (at least!) is a bivector quantity. The ubiquitous factor $\frac{1}{2}i\hbar$ represents the spin in a standard orientation, and the spinor field $R = R(x)$ rotates it into the local spin direction at each spacetime location. That is why $i$ and $\hbar$ always appear together in the fundamental equations of the Dirac theory (and, perhaps, of quantum mechanics in general). This supports and expands Dirac's insight that the most fundamental aspect of quantum mechanics is the role of $i = \sqrt{-1}$ [11].

(4) The electron energy-momentum vector $p = p(x)$ is given by:

$$(p - eA)\cdot\gamma_\mu = \Omega_\mu\cdot S = 2[(\partial_\mu R)\tilde{R}]\cdot S,$$

where $A = A(x)$ is the external E.M. potential. Thus, it has a purely kinematic interpretation.

(5) The bicharacteristics of the Dirac wave function (tangent to the Dirac current) are interpreted as predicted electron world lines. This is not a new idea. Bohm and Hiley, among others, have argued forcefully that the identification of bicharacteristics of the Schrödinger wave function with possible electron paths leads to sensible particle interpretations of electron interference and tunneling as well as other aspects of Schrödinger electron theory [12].

(6) It is suggested that unimodular (pure state) solutions of the Dirac equation have a purely objective physical interpretation, while the factor $(\rho e^{i\beta})^{1/2}$ has a subjective probabilistic interpretation, and it arises from pure states by superposition.

(7) It is suggested that the superposition principle is simply a consequence of requiring that the form of the Dirac equation be preserved in the construction of statistical composites of pure states.

(8) Pure states exhibit circular zitterbewegung which may be the origin of the electron spin and magnetic moment, but the Dirac theory must be modified if that interpretation is to be upheld [6].

## REFERENCES

[1] J. Dorling, Schrödinger's original interpretation of the Schrödinger equation: a rescue attempt. In C.W. Kilmister (ed.), SCHRÖDINGER, Cambridge U. Press, Cambridge (1987), p. 16-40.

[2] D. Hestenes, Space Time Algebra, Gordon and Breach, NY (1966).

[3] D. Hestenes and G. Sobczyk, Clifford Algebra to Geometric Calculus, a unified language for mathematics and physics, D. Reidel, Dordrecht/Boston (1984).

[4] D. Hestenes, New Foundations for Classical Mechanics, D. Reidel, Boston/Dordrecht (1986).

[5] D. Hestenes, Clifford Algebra and the Interpretation of Quantum Mechanics. in J.S.R. Chisholm and A.K. Common (eds.), Clifford Algebras and their Applications in Mathematical Physics, D. Reidel, Boston/Dordrecht (1986), p. 321-346.

[6] D. Hestenes, Quantum Mechanics from Self Interaction, Found. Phys. 15, 63-87 (1985).

[7] D. Hestenes, Proper Dynamics of a Rigid Point Particle, J. Math. Phys. 15, 1778-1786 (1974).

[8] D. Hestenes, Local Observables in the Dirac Theory, J. Math. Phys. 14, 893-905 (1973).

[9] J. Wessenhoff and A. Raabe, Acta Phys. Pol. 9, 7 (1947).

[10] H. Corben, Classical and Quantum Theories of Spinning Particles, Holden-Day, San Francisco (1948). See especially p. 72.

[11] C.N. Yang, Square root of minus one, complex phases and Erwin Schrödinger. In C.W. Kilmister (ed.), SCHRÖDINGER, Cambridge U. Press, Cambridge (1987), p. 53-64.

[12] D. Bohm and B.J. Hiley, Unbroken Realism, from Microscopic to Macroscopic Levels, Phys. Rev. Let. 55, 2511-2514 (1985).

## The Zitterbewegung

Now we can throw new light on the most basic "classical solution" of the Dirac equation, the free particle. In the absence of an external E.M. field, we have immediately from (6.33) and (6.35) that the momentum p and the total angular momentum $M = x \wedge p + S$ are constants of motion. From (6.22) combined with (6.18) we have

$$\Omega S = pv = m + \dot{S}. \tag{7.1}$$

Its scalar part

$$\Omega \cdot S = p \cdot v = m \tag{7.2}$$

integrates immediately to

$$p \cdot x = m\tau. \tag{7.3}$$

This defines a proper time $\tau = \tau(x)$ on the bicharacteristic passing through any give location x.

Multiplying (7.1) by its reverse, we obtain

$$p^2 = m^2 - \dot{S}^2 = \Omega^2 S^2 = -\frac{\hbar^2}{4}\Omega^2. \tag{7.4}$$

This determines $|\Omega| = 2|p|\hbar^{-1}$ and incidentially shows that $\dot{S}^2$ is constant. Solving (7.1) for $\Omega$ we obtain

$$\Omega = pvS^{-1} = p(v \wedge S)S^{-2}. \tag{7.5}$$

According to (6.28), $v \wedge S$ is a constant of the motion, so $\Omega$ is constant as well. Consequently the spinor equation

$$\dot{\psi} = \frac{1}{2}\Omega\psi \tag{7.6}$$

integrates immediately to

$$\psi = e^{\frac{1}{2}\Omega\tau}R_0 = e^{\frac{1}{2}m^{-1}\Omega p \cdot x}R_0, \tag{7.7}$$

where $R_0$ is a constant unimodular spinor. Specifically, $\psi = R_0$ on the spacelike hyperplane $m\tau = p \cdot x = 0$. The velocity and spin are thus given by

$$v = e^{\frac{1}{2}\Omega\tau}v_0 e^{-\frac{1}{2}\Omega\tau}, \tag{7.8}$$

$$s = e^{\frac{1}{2}\Omega\tau}s_0 e^{-\frac{1}{2}\Omega\tau}, \tag{7.9}$$

where $v_0 = R_0\gamma_0\tilde{R}_0$ and $S_0 = \frac{1}{2}R_0 i\hbar\tilde{R}_0$. It should be mentioned that (7.7) is not necessarily a full solution of the Dirac equation, but only the result of integrating (7.6) along bicharacteristics.

# BAYESIAN MODEL SELECTION AND PARAMETER ESTIMATION APPLIED TO SEA FLOOR PRESSURE DATA

D. BURTON, G. J. MOORE, W. J. FITZGERALD
*Marconi Maritime Applied Research Laboratory*
*Unit 33 - 35, Cambridge Science Park*
*Cambridge CB4 4FX*
*The United Kingdom*

ABSTRACT. A Bayesian model and parameter estimation algorithm is described which determines both the model order and the model parameters. The Bayesian approach also allows the removal of parameters, which are of no interest, from the models, thus concentrating attention on only those parameters which are of interest. Artificial time series are analyzed with the algorithm, which is shown to be robust in the presence of noise, and the results are compared with those obtained using a Fourier transform technique. An original time series consisting of 383 hourly measurements of the sea floor pressure beneath the Bellinghausen Sea, Antarctica is also examined. The frequency components of the pressure variations are known accurately from related calculations, but the amplitudes and phases are not well known. The Bayesian algorithm is used to determine both the number of sinusoidal components and their individual frequencies, amplitudes and phases. It is concluded that the Bayesian approach is superior, to the commonly used Fourier transform approach, in the identification of the frequency components of both real and artificial time series.

## 1. Introduction

This paper is concerned with the problem of choosing, from a set of competing models, the model which best accounts for the data and accurately estimating the parameters of the model. Parameter estimation is an optimisation problem involving the search of an $m$ - dimensional parameter space and is accomplished by a number of available techniques. However, such an approach does not allow a distinction to be made between the relative merits of different models. It is the use of Bayes' theorem which formulates the problem in terms of probabilities, which allows this choice between models to be made. The probabilistic formulation depends upon the use of pri-

*P. F. Fougère (ed.), Maximum Entropy and Bayesian Methods, 185–194.*

ors for correct normalisation. Further use of priors is made when integrating nuisance parameters from the models, where they are used to supply additional information. This paper proceeds by first describing the role of Bayes' theorem in the removal and estimation of model parameters, this is then followed by a description of the model selection procedure. The techniques described are then applied to the analysis of real and artificial time series.

## 2. The Bayesian Methodology

The Bayesian approach adopted in this paper follows that outlined by Bretthorst [1], and begins with Bayes' theorem, which for our purposes is written as

$$P(H \mid DI) = \frac{P(H \mid I)\ P(D \mid HI)}{P(D \mid I)} \tag{1}$$

where $H$ is any hypothesis to be tested, $D$ is the data and $I$ is the prior information. The term $P(H \mid DI)$ is the posterior probability of the hypothesis given the data and prior information, $P(H \mid I)$ is the prior probability of the hypothesis given only the prior information, $P(D \mid HI)$ is the direct probability of the data given the hypothesis and the prior information and $P(D \mid I)$ is the probability of the data given only the prior information. In order to evaluate the probability of our hypothesis $H$ we must calculate the three terms on the right hand side of equation (1). If we consider only one hypothesis then the term $P(H \mid I)$ is simply a normalisation term, as is the term $P(D \mid I)$, which leaves the calculation of the remaining term $P(D \mid HI)$.

This last term is calculated by considering the noise, which is assumed to contaminate the data. We wish to obtain the most general probability distribution for the noise assuming only that it has a finite second moment. To do this we maximize the entropy of the distribution subject to the constraints that the second moment is finite and that the probability distribution is normalised to unity [2]. This results in a Gaussian form for the probability distribution given by

$$P(e_t \mid I) = \frac{1}{(2\pi\sigma^2)^{1/2}}\ \exp\left\{\frac{-e_t^2}{2\sigma^2}\right\} \tag{2}$$

where $e_t$ is a single noise value at time $t$, and $\sigma^2$ is the second moment of the noise. It should be stressed that the noise process was not assumed to be Gaussian, only that it had a finite second moment. Applying the product rule of probability theory leads to the probability of a sequence of noise values $\{e_1, \ldots, e_N\}$,

$$P(e_1, \ldots, e_N \mid I) = \prod_{t=1}^{N} \left( \frac{1}{(2\pi\sigma^2)^{1/2}} \exp\left\{ \frac{-e_t^2}{2\sigma^2} \right\} \right) \tag{3}$$

Associating the sequence of noise values with the difference between the data and the model hypothesis, we obtain the probability of the data given the model hypothesis $H$ and prior information $I$. Setting the normalisation term $P(D \mid I)$ to unity for convenience, we obtain the probability of the model hypothesis given the data and prior information. We are initially interested in keeping the model constant and varying the parameters and write the probability of obtaining the set of parameters $\{\omega\}$ as

$$P(\{\omega\} \mid DI) \propto \sigma^{-N} \exp\left\{ \frac{-1}{2\sigma^2} \sum_{i=1}^{N} [d(i) - f(i)]^2 \right\} \tag{4}$$

where $\sigma^2$ is the assumed variance of the noise, $\{d(i)\}$ is the data set to be analyzed and $\{f(i)\}$ is the model to be tested. The parameters of the model function which are of no interest, referred to as nuisance parameters and denoted $\{\omega^*\}$, are removed from the calculation by integrating the posterior probability (3) to yield a marginal posterior probability distribution

$$P(\{\omega : \omega \neq \omega^*\} \mid D I) = \int d\omega^* \, P(\{\omega\} \mid DI) \tag{5}$$

Typical nuisance parameters are the amplitude of each model function, the phases of any sinusoidal terms and the noise variance $\sigma^2$, for which the integration may be performed analytically, the result of which is a probability distribution function (p.d.f.) for the remaining parameters which has the form of a student-t

$$P(\{\omega : \omega \neq \omega^*\} \mid D I) \propto \left[ 1 - \frac{m\,\overline{h^2}}{N\,\overline{d^2}} \right]^{(m-N)/2} \tag{6}$$

where $h^2$ is the sum of the squared projections of the data onto the model functions which have been previously orthogonalised to assist in the removal of the nuisance parameters. This p.d.f. (6) is maximized with respect to the remaining set of parameters $\{\omega : \omega \neq \omega^*\}$, to yield a set of the most probable parameter values which best account for the data.    The optimisation is accomplished by a continuous search procedure,  for example the method of Hooke and Jeeves,   or may just involve the evaluation of the probability distribution at a set of discrete points in a region of the parameter space where the parameter values are thought to lie.  In practice,  one may wish to consider competing models,  composed of several functions,  each with their own set of parameters.  To go beyond the estimation of model parameters and actually determine the model which best accounts for the data,  requires the use of proper priors in the derivation of the probability distribution (4).  This means taking into account all the prefactors and using normalised probability distributions.

The problem of selecting one model from a set,  consisting of $s$ models begins by assuming that although we are only looking within a finite set of models we can assume

$$\sum_{k=1}^{s} P(f_k|DI) = 1 \tag{7}$$

where $P(f_k|DI)$ is the posterior probability of the model $f_k$ and calculated using Bayes' theorem

$$P(f_k|DI) = \frac{P(f_k|I)P(D|f_kI)}{P(D|I)} \tag{8}$$

where $P(D|f_k I)$ is the (normalised) likelihood of the data.  Usually,  before the data is available we have little knowledge about the model function and it would appear *a priori* that each of the models in the set were just as likely,  in this case we can take $P(f_k|I)$ as locally uniform.  We thus obtain the probability of the model $f_k$ as

$$P(f_k|DI) = \frac{P(D|f_kI)}{\sum_{k=1}^{s} P(D|f_kI)} \tag{9}$$

The relative probabilities depend solely on the likelihoods,  their calculation is essentially the same as in (2),  but we must integrate out the $\{\omega^*\}$ parameters,  keeping track of all normalisation constants (which means using *proper* priors).  In the remainder of this paper results are presented for the estimation of parameters and the selection of models for both real and artificial time series.  In particular,  the resolution of sinusoids using the Bayesian approach is compared with the Fourier transform.

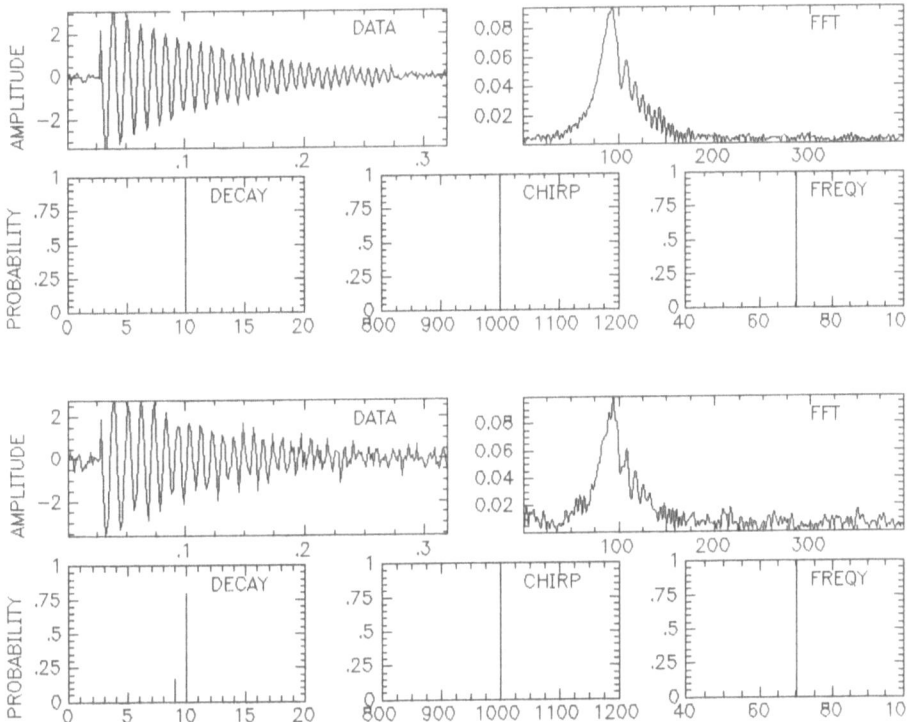

Figure 1. Two sets of results for SNR values of 20 dB (upper set) and 10 dB (lower set), each set of figures composes the data shown contaminated with noise (upper left), the Fourier transform padded to 2048 points (upper right).

## 3. Analysis Of Artificial Data

In general, the type of data one is interested in analyzing are time series of short duration, usually composed of sinusoidal signals with additional chirp and/or decay. To complicate matters further, the arrival time of the signal is often unknown, and with the signal contaminated with noise is often ill-defined. To investigate the benefits of the Bayesian approach in the analysis of such signals, a delayed, chirped, decaying sinusoid was generated of short duration and contaminated with varying degrees of noise, the model function taking the form

$$f(t) = A(t)\, e^{-\alpha t} \sin(\phi + 2\pi f t + \beta t^2) \tag{10}$$

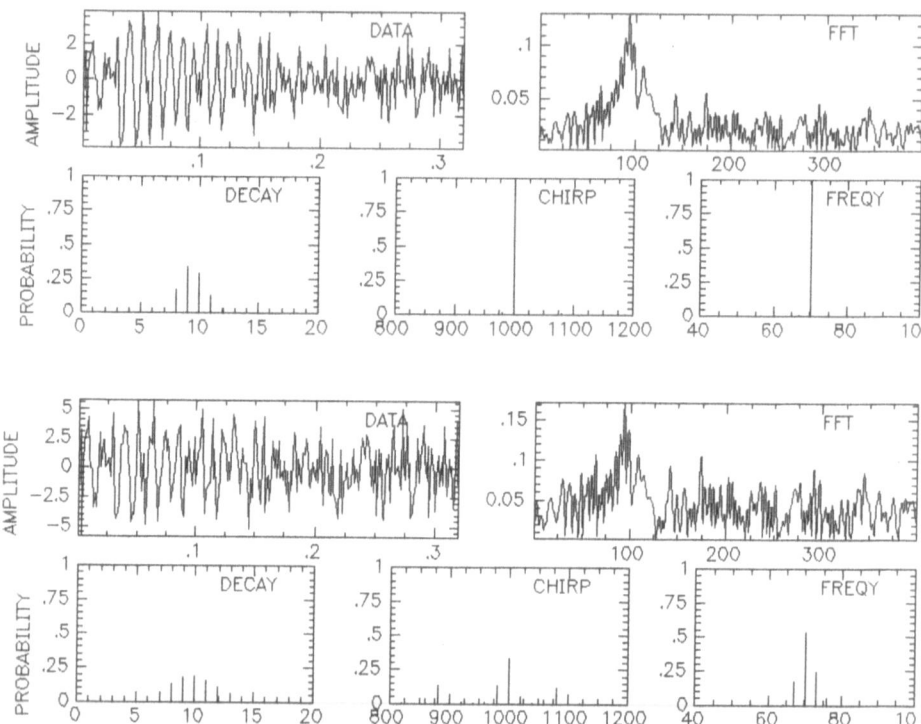

Figure 1 cont.   Further results for SNR values of 0 dB (upper set) and -6 dB (lower set).   The time series consisted of 256 samples, sampled at 400 Hz,   with $f = 70$ Hz, $\alpha = 10$ s$^{-1}$, $\beta = 1$ KHz$^2$.

where $A(t) = 1$ for $t_1 \leq t \leq t_2$, and zero elsewhere.   The amplitude,  phase and temporal localization of the signal were regarded as nuisance parameters and integrated from the model function to yield a probability function dependent only upon the decay, chirp and frequency.   A discrete search was performed in this three dimensional parameter space, for the signal contaminated with varying amounts of Gaussian noise. The results for each parameter,  with the other two integrated out,  are displayed in figure 1.

For high SNR values,  the Bayesian parameter estimates coincide with the actual parameter values used to generate the signal,  while the Fourier transform is already showing a bias in its frequency estimate,  returning the mean value of the chirped frequency.   At lower SNR values the Bayesian parameter estimates show a broadening of the probability distribution, but still peak around the actual parameter values.

## 4. Analysis Of The Northern Ice Front Data

In a remote and isolated area like the Antarctic, tidal records are often of short length and their analysis has previously required assumptions regarding the relationship between unresolved components. This approach can give rise to appreciable error. The response to this problem has been to record longer and longer data records, which is an expensive solution. The Bayesian parametric approach adopted in this paper is presented as an alternative solution.

The Northern Ice Front (NIF) data consists of 383 hourly measurements of the sea floor pressure, and shows periodic variations due to gravitational tidal forcing. Many harmonics are known to be present in the time series from the analysis of much longer records. In particular, the tidal spectrum in this region has two main components, a doublet $(O_1, K_1)$ in the diurnal band and a doublet $(M_2, S_2)$ in the semi-diurnal band [3]. Each doublet is separated by a period of approximately half a lunar month and with the time series itself being only half a lunar month long, the resolution of the doublets sets a challenging problem.

A visual inspection of the NIF data, figure 2.1, reveals several harmonic components to be present, and this is borne out by calculating the Fourier transform of the time series, figure 2.2, which is zero padded to 2048 points. Taking the sampling frequency to be 1 Hz for convenience, the Fourier transform reveals three components at frequencies of 0.03873 Hz, 0.0418 Hz and 0.08310 Hz. The first two frequencies compose the diurnal doublet $(O_1, K_1)$, with known frequencies of 0.0387306 Hz and 0.0417807 Hz. The third component at 0.08310 Hz falls between the two frequencies of the $(M_2, S_2)$ doublet with known frequencies of 0.0805114 Hz and 0.0833333 Hz. Figure 2.3 shows the Bayesian estimates of the frequency content on a probabilistic scale using a single sinusoid model. The similarity between the Bayesian and Fourier transform results reflects the fact that the Fourier transform is, in the Bayesian sense, the sufficient statistic for determining the presence of a single sinusoid, or widely spaced sinusoid, but is the wrong approach for closely spaced sinusoids.

Using a model composed of two sinusoids, figure 3, shows that the choice of model function affects the frequency estimates such that improved frequency resolution is obtained. The upper plots show the Fourier transform estimates over a frequency range around each of the two doublets. The lower plots show the Bayesian frequency estimates over the corresponding frequency ranges and yields frequencies for the two doublets of 0.03875 and 0.0418 Hz for the lower diurnal doublet $(O_1, K_1)$ and 0.0802 and 0.0832 Hz for the upper, semi-diurnal doublet $(M_2, S_2)$ to be compared with known values of 0.03873, 0.04178, 0.08051 and 0.08333 Hz [3].

Following on from the two sinusoid model, models composed of up to twenty sinusoids were evaluated, and the Bayesian approach identified the fourteenth order model as the one best accounting for the data. The frequencies identified, along with

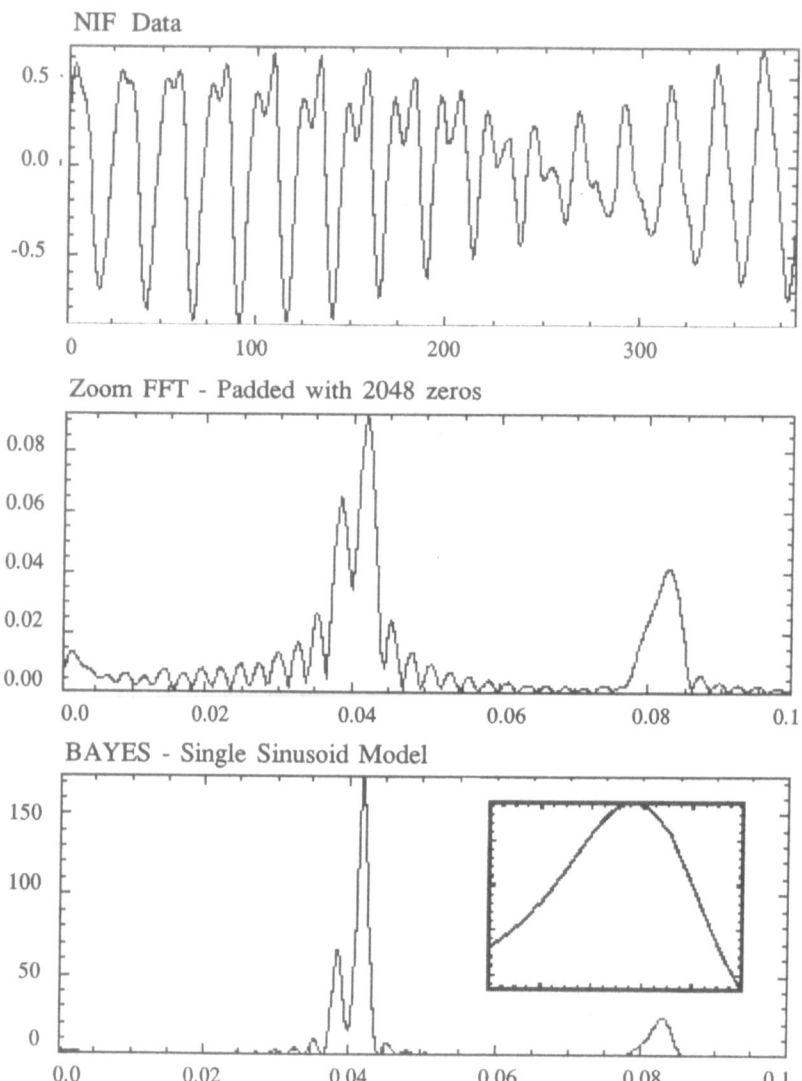

Figure 2.   The Northern Ice Front data consisting of 383 hourly measurements of the sea floor pressure beneath the Bellinghausen Sea, Antarctica is shown in figure 2.1. The Fourier transform, zero padded to 2048 points, is shown in figure 2.2 over the frequency interval (0.0,0.1) Hz, calculated taking a 1 Hz sampling frequency. Bayesian results for the single sinusoid model are shown in figure 2.3 over the same frequency range as the Fourier transform. The inset figure covers the frequency interval (0.080,0.085) Hz and shows that a finer grid search still fails to split the doublet.

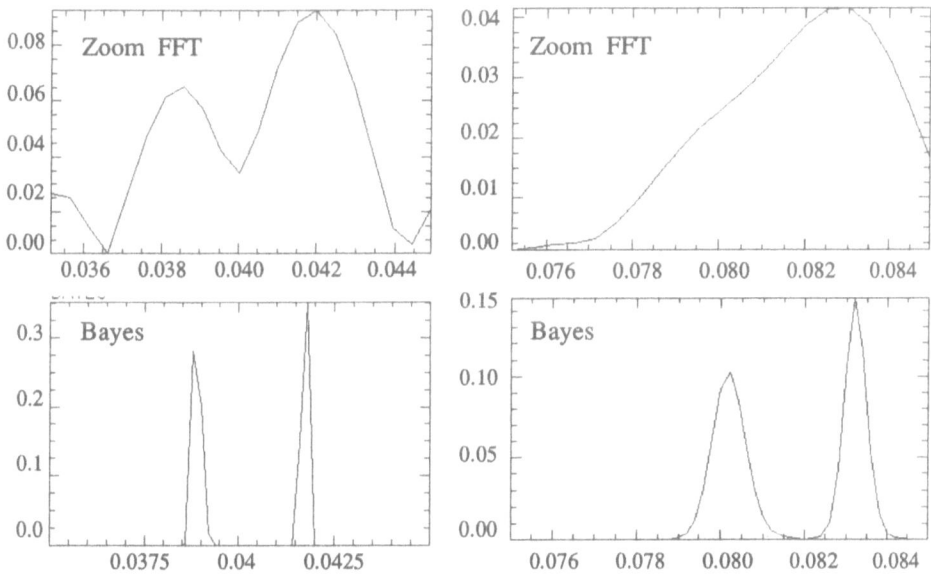

Figure 3. Bayesian probability distributions, lower plots, for the frequencies of the two sinusoid model (with one sinusoid integrated out) shown in the frequency range around each doublet. The upper plots show the Fourier transform over the same frequency range.

the relative amplitudes and phases are shown in Table 1, where the known frequency components are also listed. From the fourteen sinusoid model, identified by Bayes as best accounting for the data, eleven components were recognized from the excellent agreement between the Bayesian frequency estimates and the known frequency content of the time series. In addition, the Bayesian method also provided the relative amplitudes and phases of the components, which are important because of the way in which they vary over the face of the Earth.

Table 1. Bayesian estimates of the frequencies, relative amplitudes and phases for the 14 sinusoid model, along with the known frequencies of 11 identified components.

| | Estimated Freqy. (Hz) | Known Freqy. (Hz) | Relative Amplitude | | | Phase (radians) | | | Tidal Component |
|---|---|---|---|---|---|---|---|---|---|
| 1 | 0.00262 | 0.0789992 | 0.0508 | ± | 0.0014 | -0.827 | ± | 0.028 | $M_f$ |
| 2 | 0.00575 | | 0.0304 | ± | 0.0015 | 0.703 | ± | 0.048 | |
| 3 | 0.00994 | | 0.0208 | ± | 0.0014 | -0.440 | ± | 0.069 | |
| 4 | 0.01371 | | 0.0112 | ± | 0.0011 | 1.521 | ± | 3.503 | |
| 5 | 0.03712 | 0.0372185 | 0.0530 | ± | 0.0016 | -1.029 | ± | 0.030 | $Q_1$ |
| 6 | 0.03871 | 0.0387306 | 0.2818 | ± | 0.0016 | 0.615 | ± | 0.006 | $O_1$ |
| 7 | 0.04174 | 0.0417807 | 0.4052 | ± | 0.0014 | -1.079 | ± | 0.003 | $K_1$ |
| 8 | 0.04429 | 0.0448308 | 0.0114 | ± | 0.0014 | 0.918 | ± | 0.126 | $OO_1$ |
| 9 | 0.07872 | 0.0789999 | 0.0511 | ± | 0.0015 | 0.672 | ± | 0.029 | $N_2$ |
| 10 | 0.08076 | 0.0805114 | 0.1301 | ± | 0.0014 | -1.142 | ± | 0.011 | $M_2$ |
| 11 | 0.08339 | 0.0833333 | 0.1807 | ± | 0.0011 | 1.470 | ± | 0.006 | $S_2$ |
| 12 | 0.12000 | 0.1207670 | 0.0200 | ± | 0.0017 | -1.089 | ± | 0.085 | $M_3$ |
| 13 | 0.12146 | 0.1222921 | 0.0222 | ± | 0.0018 | -0.954 | ± | 0.080 | $SO_3 / MK_3$ |
| 14 | 0.12527 | 0.1251140 | 0.0183 | ± | 0.0014 | 0.592 | ± | 0.080 | $SK_3$ |

## 5. Conclusions

Three main points arise from the use of the Bayesian approach to model selection and parameter estimation which immediately suggest its superiority over such diverse methods as the Fourier transform approach and least-squares algorithms. The first is the ability to include correctly formulated priors which reflect ones prior knowledge, which results in enhanced estimates. The second point is the removal of nuisance parameters, which reduces the dimension of the parameter space to be searched and thus improves the efficiency of the method. Finally, the ability to choose between different models according to which model best accounts for the data.

## References

1. Bretthorst, G. L. (1989) 'Bayesian Spectrum Analysis and Parameter Estimation', Springer-Verlag.
2. Jaynes, E. T. (1989) 'Information Theory and Statistical Mechanics', in R. D. Rosenkrantz (ed.). E. T. Jaynes: Papers on Probability, Statistics and Statistical Physics, Kluwer Academic Publishers, Dordrecht, pp 6-16.
3. Melchior, P. (1983) 'The Tides of the Planet Earth', Pergammon Press, Oxford.

# Applications of Maximum Entropy and Bayesian Methods in Neutron Scattering

*Devinderjit Singh Sivia*
Theoretical Division & Manuel Lujan Jr. Neutron Scattering Center
Los Alamos National Laboratory
Los Alamos, New Mexico 87545, U.S.A.

*Abstract:* We report on the use of Maximum Entropy (MaxEnt) and Bayesian methods applied to problems in neutron scattering at Los Alamos over the past year. Although the first applications were straight-forward deconvolutions, the work has been extended to make routine use of multi-channel entropy to additionally determine (broad) unknown backgrounds. A more exotic example of the use of MaxEnt involves the study of aggregation in a biological sample using Fourier-like data from small angle neutron scattering. We have also been considering the question of how to optimise instrumental hardware, leading to the derivation of better "figures-of-merit" for spectrometers and moderators, which may result in a far-reaching revision of ideas on the design of neutron scattering facilities.

## 1. Introduction

Neutron scattering is a tool for the study of condensed matter, from superconductors to biological samples. In Section 2, we give a brief review of the neutron properties which make it a versatile probe of matter and outline how the experimentally measured data are related to the physical quantities of interest.

In Section 3, we show three examples of the use of MaxEnt for the analysis of neutron scattering data. The first is a straight-forward convolution problem (although the response function is not invariant), the second involves the study of aggregation of macromolecules in a biological sample using Fourier-like data and the third illustrates a more advanced use of MaxEnt — multi-channel entropy — to simultaneously perform a deconvolution and determine a (broad) unknown background signal.

The previously mentioned examples of the use of MaxEnt are a case of trying to do the "best" with data we have, but *how should we design instrumentation to obtain "better" data*? In Section 4, a simple Bayesian analysis is shown to lead to the derivation of a much more robust "figure-of-merit" (for spectrometers and moderators) than is commonly used, with potentially serious implications for the design of neutron scattering facilities.

## 2. Why Scatter Neutrons ?

Fig.1 shows a schematic picture of scattering: a beam of particles (be they neutrons, x-rays, or whatever), with wave-vector $k_i$, impinges upon a sample, with which it interacts in some fashion, and is scattered. If $k_f$ is the wave-vector of the out-going particle, then the relationships for the energy transfer E and momentum transfer $Q$ are given by:

*P. F. Fougère (ed.), Maximum Entropy and Bayesian Methods*, 195–209.
© 1990 *Kluwer Academic Publishers.*

$$Q = \hbar.(k_f\text{-}k_i) \quad \text{and} \quad E = \frac{\hbar^2}{2m}.(|k_f|^2\text{-}|k_i|^2) \ ,$$

where m is the mass of the scattered particle. The number of particles scattered with energy transfer E—>E+dE and momentum transfer Q—>Q+dQ is proportional to S(Q,E)dQdE, where S(Q,E) is the *scattering law* (assuming the first Born approximation is valid). The structural properties of the sample (e.g. where are the atoms?) are given by elastic scattering (E=0) and are related to the scattering law through a Fourier transform. The dynamical properties of the sample (e.g. how strong are the "springs" holding the atoms together?) are given by inelastic scattering (E≠0) and are derived directly from the scattering law (e.g. dispersion curves). The experimental data are usually a blurred version of the scattering law with Poisson noise.Thus, data analysis for dynamical studies constitutes a convolution problem where as structural studies entail a Fourier (& convolution) problem.

But why should we go to the great expense of scattering neutrons in addition to the more readily available x-rays, for example? The answer is that neutrons are a very versatile probe of matter, providing information not always accessible by other means. The main neutron properties are listed below:

**(a) Zero charge:** results in deep penetration of the sample, allowing an investigation of bulk properties rather than just those of the surface layers.

**(b) Wavelength:** the wavelengths of neutrons with thermal energies is of the order of the atomic scale and longer (like x-rays). This makes them ideal for structural analysis, the longer wavelengths being particularly suited for studying the gross features of biological macromolecules as opposed to the details of their crystal structure.

**(c) Scattering lengths:** the neutron scattering lengths (or strengths) of atoms, which represent the complex interaction between a nucleus and a neutron by a single number, bear no monotonic relationship to the atomic number (unlike x-rays). This allows the possibility of locating light atoms (like hydrogen) and having good contrast between atoms of similar atomic weight (even isotopes).

**(d) Energy:** the energies of thermal neutrons are comparable with lattice and molecular excitations, making them ideal for dynamical studies.

**(e) Dipole moment:** rather than the neutron interacting with the nucleus through the strong force, it can also scatter as a result of the electromagnetic interaction of its magnetic dipole moment with the valence electrons responsible for the magnetic properties of the sample. Thus, neutrons can be used to study both structural and dynamical properties associated with magnetic phenomena.

You might wonder where we get the neutrons from in the first place. Well, there are

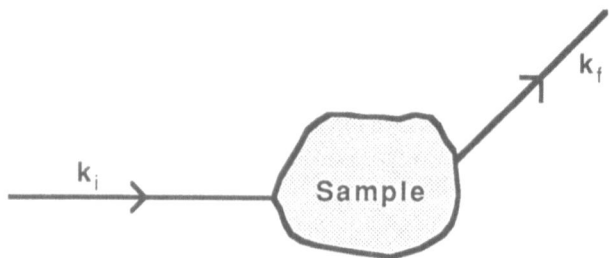

**Figure 1:** A schematic picture of scattering: a beam of particles is incident on the sample with wave-vector $k_i$ and is scattered to a final wave-vector $k_f$.

two types of neutron facilities: reactors (e.g. the ILL at Grenoble) and accelerators (e.g. LANSCE at Los Alamos). Reactors are steady-state sources, whereas accelerator-based facilities have neutrons which come in pulses. The nature of pulsed sources leads to a natural choice of time-of-flight and scattering angle as experimental variables; the scattering law measured in this way has to be transformed, therefore, into the physically meaningful units of Q & E for interpretation. For further details about neutron scattering theory and hardware the reader is referred to Bacon (1955), Windsor (1981), and references therein.

# 3. Applications of MaxEnt at LANSCE

## (a) The Filter Difference Spectrometer

The first example of the use of MaxEnt at the *Manual Lujan Jr. Neutron Scattering Center* (LANSCE) was for a standard convolution problem on the Filter Difference Spectrometer (FDS) (Sivia, Vorderwisch & Silver, 1989). The FDS (Taylor *et al.*, 1984) is an inelastic scattering instrument used for molecular vibrational spectroscopy (the neutron analogue of infra-red, or Raman, spectroscopy). Fig. 2(a) shows data taken with a Be filter on the FDS at LANSCE, in which we can see the sharpish edge and long decaying tail of the Be filter response function (varies along the spectrum). The earliest method used to remove this response function, in order to see the detail of the underlying spectrum, was a hardware "filter difference" solution which lends its name to the instrument. A BeO filter gives a response function with almost the same shape as the Be filter, but with a slightly different off-set for the sharp edge. A BeO filter, therefore, gives data similar to the Be filter but shifted by a small amount. When data taken using these two filters are subtracted, the long decaying tails (and background) tend to cancel out leaving only the significant features near the sharp rising edges. An example of such a "filter difference" spectrum, corresponding to the Be filter data in Fig. 2(a), is given in Fig. 2(b). It should be noted that the reconstructed spectra (Figs. 2(b)-(d)) are inverted relative to the data; this is because the data channels represent increasing time-of-flight which is equivalent to decreasing energy transfer (for energy loss). Given the data from the Be filter alone, and a knowledge of the response function and background, it should be possible to carry out the deconvolution mathematically (in software). The MaxEnt reconstruction thus obtained is shown in Fig. 2(c) and is overlaid on a conventional reconstruction (which can be interpreted as a filtered inverse) due to Mezei (Mezei & Vorderwisch, 1989) in Fig. 2(d). As expected, the MaxEnt reconstruction is an improvement over both the filter difference and the Mezei solutions in that we can see more fine detail and fewer noise artefacts. The improvement is obvious but not dramatic in this case, however, because we had data with good statistical accuracy.

## (b) The Low-Q Diffractometer

The next example involves the analysis of Fourier-like data, from the Low-Q Diffractometer (LQD), to study the aggregation of biological macromolecules. The LQD (Seeger, Hjelm & Nutter, 1989) is a small-angle neutron scattering (SANS) instrument, useful for studying large-scale structures in the range 10 - 1000Å. The spatial distribution of particles (what is their size & shape, and where are they?) is related to the scattering data through a Fourier transform —in general, a complex quantity. The neutron counts we measure are, of course, given by the Fourier intensities (or a blurred and noisy version thereof, because of instrumental contributions), bringing us face-to-face with the dreaded *Fourier phase problem*! Luckily, for our problem, we are not interested in *where* the particles are but *how many* there are, of a given size & shape. This is similar to wanting the auto-correlation (or Patterson) function, instead of the electron density map, in x-ray crystallography, which is derived from the Fourier intensities only.

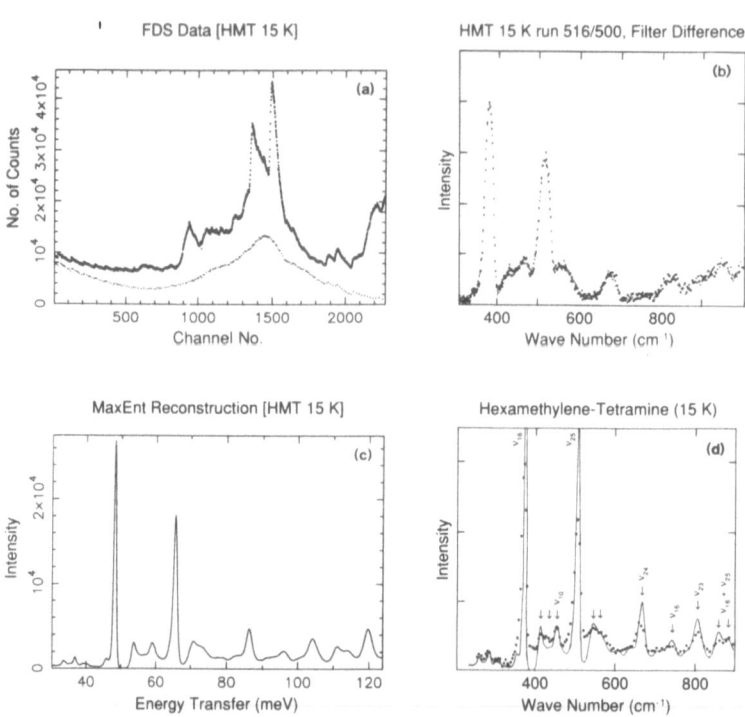

**Figure 2**: **(a)** FDS data taken with a Be filter. The channels are in increasing time-of-flight or decreasing energy transfer. **(b)** The "filter difference" spectrum. **(c)** The MaxEnt Reconstruction. **(d)** The filtered inverse-based "Mezei" solution (dots) overlaid on the MaxEnt Reconstruction.

My biologist colleague at LANSCE, Rex Hjelm, told me:"Your body is mostly water. If you visit your favourite ice cream parlour then the fat in the ice cream will sit as a greasy blob at the bottom of your stomach and you will soon die!" Bile salts, produced in the liver, I was told, had hydrophylic heads and cholesterol-like tails (Fig. 3(a)). "So the body dumps in some bile salts to act as detergents", I remarked (somewhat relieved). "No, that's what an engineer would do!", came the reply. For reasons that we do not fully understand, nature uses a conglomerate of bile salts and a fat lecithin (Fig. 3(b)).

As a step towards understanding the action of bile in lipid digestion and in the transport of liver products such as cholesterol (elucidation of which has potential applications in industrial processes and in the development of drug delivery systems & model membranes), Hjelm *et al.* (1989) have been investigating the nature of particle growth in aqueous solutions of lecithin-bile salt mixtures. Figs. 4(a)-(c) show three SANS data-sets for increasingly dilute aqueous solutions of the sample. Figs. 4(d)-(f) are the corresponding size distributions for the conglomerate particles, derived by MaxEnt, on the basis that they can be approximated as cylinders of uniform density (this defines the hypothesis space for the analysis). In Fig. 4(d) there is evidence only for a single type of particle, with radius≈25Å & height≈50Å. As the sample is diluted, Fig.4(e) shows evidence for a second type of particle with height≈100Å: a rod-like structure twice the length of the original particles. Fig. 4(f) suggests that a further dilution leads to the formation of even more elongated structures of height≈170Å, or three to four times the original length. These results lead us to believe that growth occurs through the aggregation of preformed sub-units of size ~50Å, corresponding very nicely to the thickness of lecithin bilayers, rather than the addition of individual bile salt or lecithin molecules.

### (c) The Constant-Q Spectrometer

Our last example comes from the Constant-Q Spectrometer (CQS), an instrument designed to measure inelastic processes (such as magnons and phonons) in single crystal samples (Robinson, Pynn & Eckert, 1985). It illustrates a more advanced use of MaxEnt — *multi-channel entropy* — needed for the convolution problem in which one has, in addition to the (sharp) scattering law of interest, a (broad) unknown background signal. We will begin with a simple simulation to show the need for multi-channel entropy, explain the basic idea, outline an algorithm for its implementation and, finally, demonstrate its use on real data from the CQS.

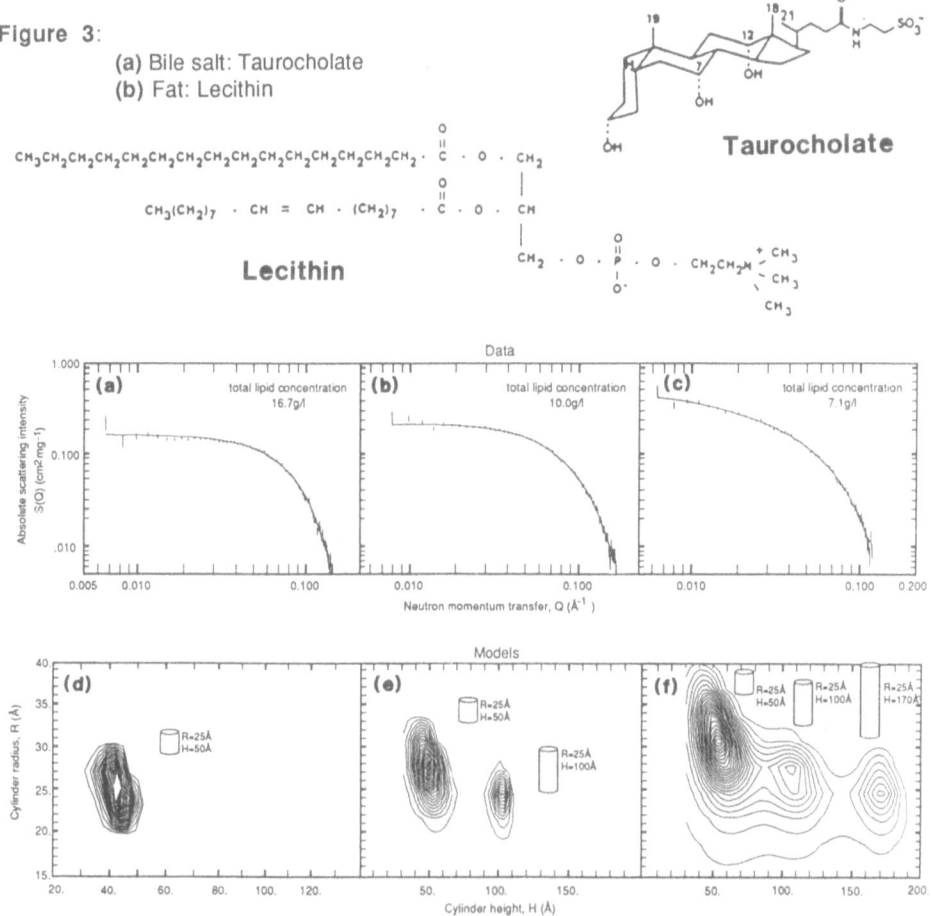

Figure 3:
    (a) Bile salt: Taurocholate
    (b) Fat: Lecithin

**Figure 4**: (a)-(c) Small angle neutron scattering data from increasingly dilute aqueous solutions of lecithin and bile salts (Hjelm et al.,1989). (d)-(f) The corresponding size distributions of the conglomerate particles (modelled as cylinders of uniform density) derived using MaxEnt.

Let us start with a simple simulation of a straight-forward convolution, computed on a grid of 128 pixels. The "true" object, shown in Fig. 5(a), consists of two spikes separated by a small plateau on the left and a broader peak on the right. This was convolved with the FDS-like response function shown in Fig. 5(b), and a small background and (√N) noise added to generate the data-set in Fig. 5(c). Given these data, and a knowledge of the background & response function, Fig. 5(d) shows the MaxEnt reconstruction of the object — a very satisfactory one at that. Next, we repeat the same procedure except that our "true" object is slightly different: it now includes a large underlying bump in addition to the previous features, as in Fig 6(a). The corresponding data and MaxEnt reconstruction are shown in Figs. 6(b) & (c) respectively. The MaxEnt reconstruction is now a very poor representation of the truth: the general level of the large, broad, bump is correct but all the details of the sharp features has been lost; although there is some evidence for the sharp structure in the MaxEnt image, it is swamped by the ringing artefacts.

So, what went wrong? Well, perhaps nothing went wrong exactly: we tried to make our best inference of the object given the experimental information (data & response function) but only the prior knowledge that the object was a positive and additive entity (hence the use of MaxEnt, Skilling & Gull 1989). The fact that the object was positive (and additive) was true for both Figs. 5 & 6, but this information was not as overwhelming for Fig. (6) as it was for Fig. (5). If we have further prior information then this should be used, and it will improve the reconstruction.

Let us now try to use the additional knowledge that the object is the sum of two entities, each positive & additive, one of which is broad while the other consists only of (relatively) sharp structure. What we are doing is an example of multi-channel entropy, where we try to make our best inference about several different features of the object simultaneously (Newton, 1985). In our case, we have only two channels because we want

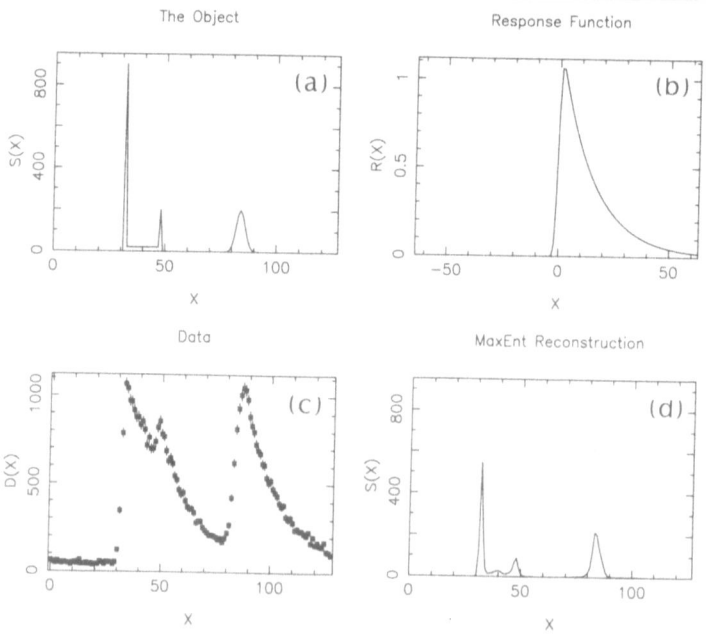

**Figure 5**: A straight-forward convolution problem. **(a)** "True" object. **(b)** FDS-like response function. **(c)** Noisy data. **(d)** MaxEnt reconstruction.

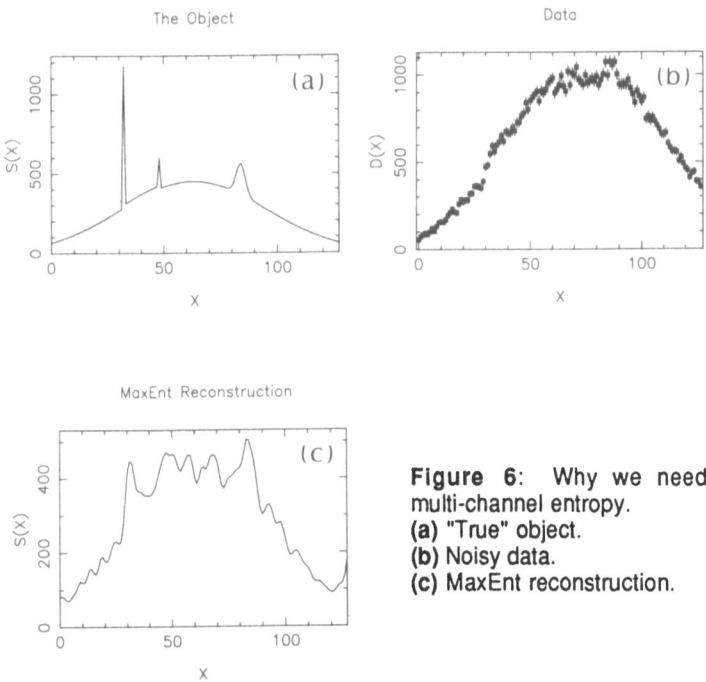

**Figure 6**: Why we need multi-channel entropy.
**(a)** "True" object.
**(b)** Noisy data.
**(c)** MaxEnt reconstruction.

to separate broad features from sharp ones. What we would like to do is to create two image channels with the following properties: (a) one channel is only allowed to have broad structure, while the other is permitted high resolution, and (b) it should be (entropically) cheaper to put structure in the broad channel, rather than in the high-resolution channel. If we can arrange this, then it will have the following (desirable) effects: if broad structure is required to account for the data, it will appear in the broad-feature channel; if sharp structure is required, it can only appear in the high-resolution channel. An algorithm designed to implement this idea, inspired by a conversation with John Skilling, is sketched in Fig. 7: the high-resolution channel is the same as before — a fine grid with many pixels. The background channel, however, consists of only a few "fuzzy" pixels. The joint entropy, which is to be maximised, is defined in the usual fashion as $\sum f_i - m_i - f_i \log(f_i/m_i)$, summed over all the pixels (in both channels). This setup has the desired properties since only broad features can appear in one channel and because it is (entropically) cheaper to put structure in this channel (requiring only a few pixels to change from their default value). In Figs. 8(a) and (b) we show the MaxEnt reconstructions for the broad and sharp channels, using the data of Fig. 6 and the algorithm described above. This is a considerable improvement over Fig. 6(c), and almost miraculous given the data of Fig. 6(b)!

Finally, we show the use of this two-channel entropy algorithm on real data taken on the CQS by Yethiraj *et al.* (1989). The data are shown in Fig. 9(a), as a function of the experimental variables: time-of-flight and detector angle. The data suffer from a combination of broadening and an unknown background signal (in addition to $\sqrt{N}$ noise), which obscures the scattering law of interest. It may be noted that the problem here is very similar, but not identical, to our simulation because the broad signal (which we now identify with the background) is merely added rather than added & convolved; this only requires a slight modification in the formalism, but not the idea or the algorithm. Fig. 9(b) is the MaxEnt reconstruction of the high-resolution channel, showing a dramatic

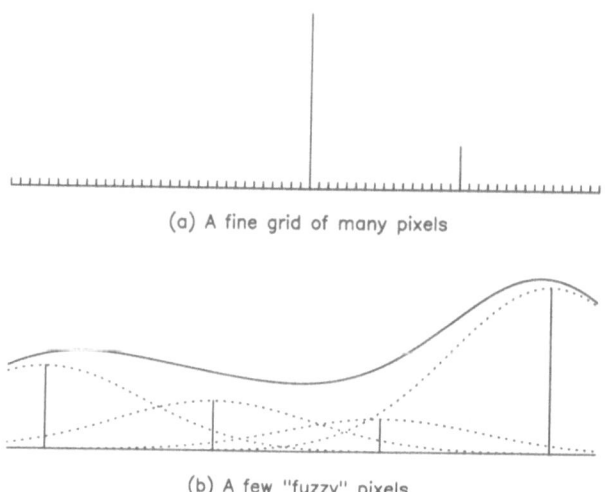

(a) A fine grid of many pixels

(b) A few "fuzzy" pixels

**Figure 7**: An algorithm for multi-channel entropy . **(a)** "Signal", or high-resolution, channel. **(b)** "Background", or broad-feature, channel.

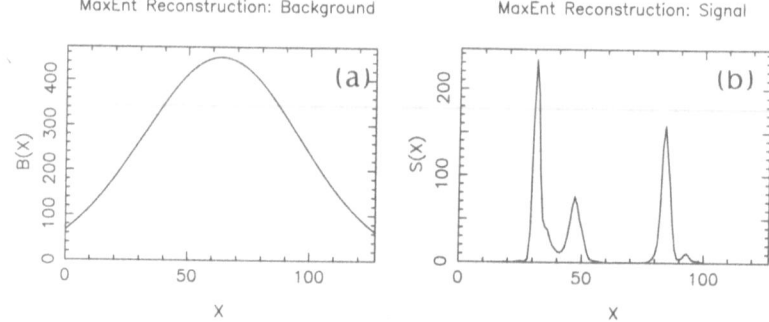

**Figure 8**: Multi-channel entropy reconstructions. **(a)** "Background". **(b)** "Signal".

improvement in both the detail seen in the scattering law and in the reduction of background artefacts. When this scattering law is transformed to the physically meaningful coordinates of energy and momentum transfer, as in Fig. 9(c), we can then identify the resulting dispersion curves with magnon and phonon excitations characteristic of the sample.

## 4. Instrument Design

The examples of the use of MaxEnt given in the last section are all a case of doing the "best" with the data we have. Usually, this is all we can do: a user goes to do an experiment at a facility like LANSCE; since the instrumentation & hardware already exists, often the only freedom he or she has in terms of the data is the time allocated to do the experiment (which governs the statistical accuracy of the data). Let us suppose, however, that we are going to build a new facility, or just a new spectrometer. How should we design it to get the "best" data? This is an important question since a new facility can cost a hundred million dollars or more (even a single spectrometer can cost a million or two)!

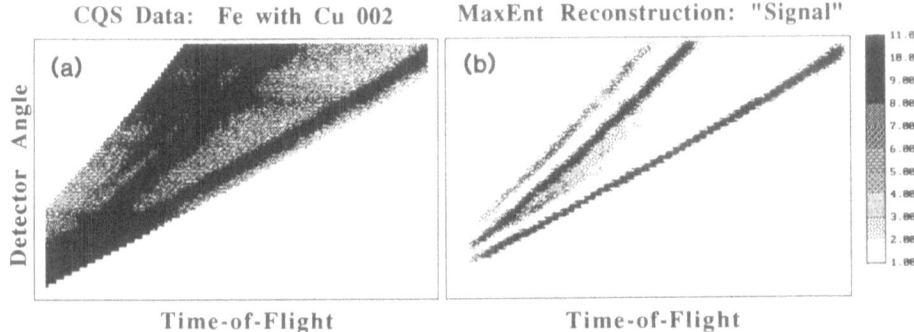

CQS Data:  Fe with Cu 002          MaxEnt Reconstruction: "Signal"

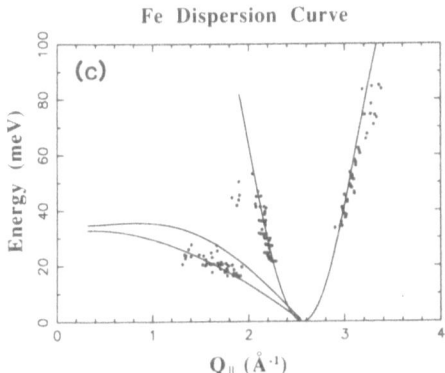

Fe Dispersion Curve

**Figure 9**:  Multi-channel entropy with real data. **(a)** CQS data (Yethiraj *et al.*, 1989), as a function of detector angle and time-of-flight, corrupted by instrumental broadening, an unknown background and (√N) noise. **(b)** MaxEnt reconstruction of the "signal" channel. **(c)** Dispersion curves (dots), showing both branches of a magnon and some phonons, obtained by transforming the lines in (b) to the physically meaningful coordinates of energy  and momentum transfer (the continuous lines are from Brockhouse, Abou-Helal & Hallman (1967) and Lynn (1975)).

Silver, Sivia & Pynn (1989)  have addressed this question from a heuristic viewpoint, and have also suggested a quantitative answer (Sivia, Silver & Pynn, 1989) based on elementary signal-to-noise ratio arguments from a power spectrum error analysis. We now provide a formal Bayesian rationale for their results.

Let us begin by setting up a simple framework for the neutron scattering data analysis problem. Most problems in neutron scattering involve a convolution with an instrumental response function:

$$D(x_k) = \int_{-\infty}^{\infty} S(t).R(x_k-t)\, dt + B(x_k) \pm \sigma(x_k) \ ,$$

where $D(x_k)$ is the $k^{th}$ datum, S is the scattering law of interest, R is the response function, B is the background signal and $\sigma$ is noise in the measurement. Given the data and a knowledge of the response function, background and the noise, we wish to infer the scattering law — i.e. we need to compute the posterior probability distribution function:

$$\text{prob}[S(x)|\{D(x_k)\},\{\sigma(x_k)\},R(x),B(x)] \ .$$

**Question:** Given that the response function, background & noise (and  hence the data) are functions dependent on the instrumental parameters, how should we design the spectrometer (and moderator) to give the most reliable estimate of the scattering law?

*This is what we mean by designing our instrument to get "better" data!*

**Answer:** Well, this is given by Bayes' theorem:

$$\text{prob}(S|D) \propto \text{prob}(D|S).\text{prob}(S) .$$

Our *prior* state of knowledge (or the lack thereof) about S, prob(S), is modified by the experimental data through the *likelihood function*, prob(D|S), to yield our *posterior* state of knowledge, prob(S|D). Since the data enter the posterior probability (only) through the likelihood function, we need to look at its sharpness, or spread. The sharper the likelihood function, the greater the "information content" of the experiment in the sense that the data impose a severe restriction on what S could be.

The Bayesian answer for optimising instrumental design is (we believe) to adjust the instrumental parameters to give the sharpest likelihood function possible. Although this principle always applies, there are (at least) two fundamental difficulties in deriving a universal "figure-of-merit" (FOM) for instrument design:

**(a)** The likelihood function exists in the space of parameters $\{b_i\}$ which define $S(x)$. For example, if we know (or assume) that $S(x)$ consists of a single Lorentzian then the likelihood function exists in a 3-dimensional space defined by the height, width and position of a Lorentzian. If we have no functional form for $S(x)$, then we might digitise it into a large number of M pixels, where upon the likelihood function exists in a large M-dimensional space defined by the flux in each pixel. The problem is that the shape of the likelihood function varies not only with the instrumental parameters but also with the hypothesis space in which we work — i.e. "what is the question?".

**(b)** If the expected noise $\langle\sigma(x_k)^2\rangle$ depends on the value of the datum $D(x_k)$, as it does for a Poisson process $\langle\sigma(x_k)^2\rangle = D(x_k)$, then the shape (rather than just the position) of the likelihood function will depend, in part, on $S(x)$ itself — i.e. the reliability of the answer is determined not only by the instrumental parameters (& the question) but also by the answer!

Given these difficulties, let us investigate how the reliability of the inferred scattering law depends on the shape of the response function using three simple examples and making suitable approximations to obtain analytic solutions. First of all we will assume that the data are independent and Gaussian, enabling us to write the likelihood function as:

$$\log_e[\text{prob}(D|S)] = -\frac{\chi^2}{2} = -\frac{1}{2}\sum_{k=1}^{N}\frac{[D(x_k)-F(x_k)]^2}{\sigma(x_k)^2} ,$$

where $\{F(x_k)\}$ are the data which a given (trial) $S(x)$ would produce in the absence of noise. Although neutron counts are Poisson they can be approximated as Gaussian because the numbers involved are large (due to background signals etc.). We will also make the (commonly used) assumption that the likelihood can be adequately represented by the quadratic term in the Taylor series expansion of $\chi^2$ about its minimum; then, using matrix/vector notation we have:

$$\text{prob}(D|S) \approx c.e^{-\frac{1}{4}\underline{\delta b}^T \nabla\nabla\chi^2 \underline{\delta b}} ,$$

where c is the normalisation constant, $\underline{\delta b}$ is the deviation of the parameters describing $S(x)$ from their optimal value (given by $\nabla\chi^2 = \partial\chi^2/\partial b_j = 0$) and $\nabla\nabla\chi^2$ is the Hessian matrix. The likelihood function can then be described either by the eigenvalues & eigenvectors of the Hessian matrix or by the covariance matrix:

$$<\delta b_i \delta b_j> = 2.(\underline{\nabla\nabla\chi^2})^{-1}_{ij} \; .$$

Now let us consider some specific situations. First the simplest case: estimating the position of a δ-function of known magnitude from data resulting from a convolution with a Gaussian response function of width q.

**Case 1:** Given suitable data and that $R(x) = I.\exp(-x^2/2q^2)$ and $S(x)=\delta(x-x_0)$, what is the uncertainty in our estimate of $x_0$?

After some algebra we find:

$$\left\langle \delta x_0^2 \right\rangle \propto \frac{(I+B)}{I^2}.q \; ,$$

where B is a measure of the background signal. A related figure-of-merit (FOM) is widely quoted in neutron scattering (Michaudon, 1963; Day & Sinclair, 1969; Windsor, 1981):

$$\text{FOM}_{\text{conventional}} = \frac{\text{Total number of neutrons}}{(\text{FWHM})^2} \sim \frac{I}{q} \sim \left\langle \delta x_0^2 \right\rangle^{-1} \; ,$$

where FWHM is the full-width-half-maximum of the response function (and $N_{\text{total}} \propto I.q$). We now give a couple of simulated examples to make the point that, although this FOM is the correct answer to the question posed above, it is quite unsuitable for general use. The procedure is much the same as that used for the examples in Section 3(c): the test object of Fig. 10(a) is convolved, independently, with the two response functions shown in Fig. 10(b); the MaxEnt reconstructions derived from the resulting data-sets are given in Fig. 10(c). Even though the response functions had identical FOMs according to conventional

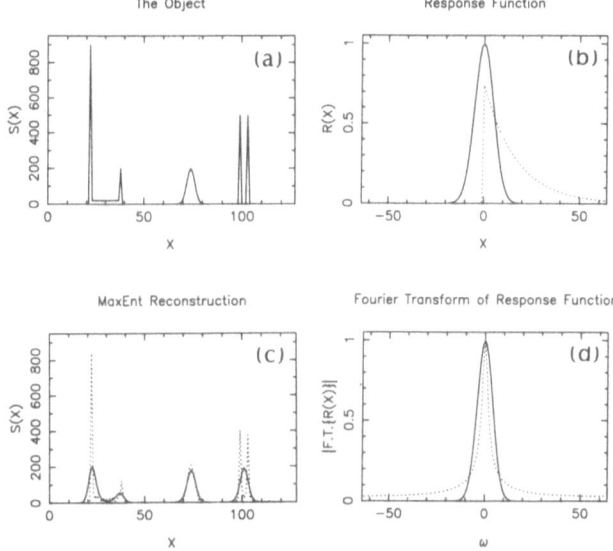

**Figure 10:** (a) "True" object. (b) Two response functions having identical conventional FOMs. (c) MaxEnt reconstructions from noisy data generated by a convolution of (a) with the response functions in (b). (d) Fourier transforms of the response functions.

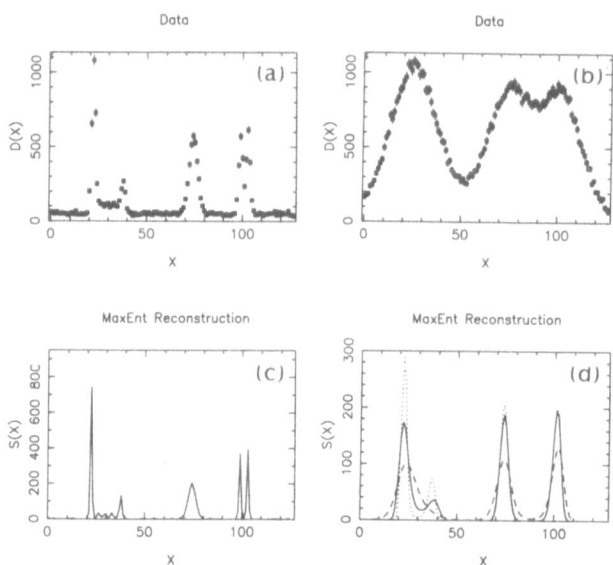

**Figure 11**: **(a)** Noisy data resulting from the convolution of the object in Fig.10(a) with a narrow Gaussian. **(b)** Corresponding data for a wide Gaussian. **(c)** MaxEnt reconstruction from data in (a). **(d)** Dashed line is the MaxEnt reconstruction from data in (b); solid line was from data with 100 times , and dotted line 10000 times, the number of counts as in (b).

wisdom, the reconstruction from the sharp-edged response function is clearly far superior to the other. But the FOM above was based on a Gaussian response function, you might complain, and so is not valid here. In Fig. 11 we show the result of applying the same test to data from two response functions which are both Gaussian but with a FWHM ratio of 10:1. Again, according to conventional thinking, the FOMs can be equalised by increasing the total number of counts for the wide Gaussian by a factor of 100. Fig. 11 illustrates the point that this is not the case in general — to recover the sharpest structure with the wide Gaussian one would need to increase the number of neutrons counts by many orders of magnitude!

Let us move on to consider the second simplest case: estimating both the position and magnitude of a $\delta$-function.

**Case 2:**  Given suitable data and that $R(x) = I.\exp(-x^2/2q^2)$ and $S(x)=A.\delta(x-x_0)$, what is the uncertainty in our estimate of A and $x_0$?

For this problem, we find that the covariance matrix is given by:

$$\langle \delta x_0, \delta A \rangle \propto \frac{(I+B)}{I^2} \begin{bmatrix} q & 0 \\ 0 & 1/q \end{bmatrix} .$$

Now the question arises: *what do we mean by a FOM?* To improve our estimate of  the position of the $\delta$-function, we should make the Gaussian response function as narrow as possible; to improve our estimate of its magnitude, we need to make the response function as wide as possible!

Rather than going through specific problems, and coming up with the conclusion that "different questions have different answers", let us try to ask a *generalised question*. We accept that it will not give the accurate answer for every specific case, but hope that it will yield a sensible FOM for a wide range of situations.

**Case 3:** Given suitable data and some R(x), how reliably can we infer the scattering law assuming no particular functional form for S(x)?

We will outline the algebra for this problem. The Hessian operator H is given by:

$$H(x,x') = \frac{\partial^2 \chi^2}{\partial S(x)\partial S(x')} = \sum_{k=1}^{N} \frac{2}{<\sigma(x_k)^2>}.R(x_k-x).R(x_k-x') \quad .$$

Assuming that $<\sigma(x_k)^2> \approx$ constant $= \sigma$ (because of the background, the variation is not too big), and taking the integral limit (i.e. assuming that we have a lot of finely-sampled data), the Hessian operator is given by the auto-correlation function of R(x):

$$H(x,x') = \frac{2}{\sigma^2} \int_{-\infty}^{\infty} R(t-x).R(t-x')\, dt = H(|x-x'|) \quad .$$

The shape of the likelihood function is described by the eigenvalues $\lambda$ and eigenfunctions $\eta$ of the Hessian operator:

$$\int_{-\infty}^{\infty} H(x,x').\eta(x')\, dx' = \lambda.\eta(x) \quad ,$$

which is easiest to solve in Fourier space. Denoting the Fourier transform of G(x) by $\tilde{G}(\omega)$, and substituting for the Hessian, we have:

$$\frac{2}{\sigma^2}.|\tilde{R}(\omega)|^2.\tilde{\eta}(\omega) = \lambda.\tilde{\eta}(\omega) \quad .$$

Since R(x) is real, $|\tilde{R}(\omega)|^2$ is symmetric; this leads to the solution:

Eigenfunctions:     $\eta_\omega(x) = \text{Cos}(\omega x) \,\&\, \text{Sin}(\omega x)$ , with

Eigenvalues:     $\lambda_\omega = \frac{2}{\sigma^2}.|\tilde{R}(\omega)|^2$ .

As the uncertainty in S(x) along the eigen-direction $\eta_\omega(x)$ is $\sqrt{[2/\lambda_\omega]}$, $\lambda_\omega$ can be used as a FOM for inferring structure in the scattering law with resolution $\omega \approx 1/\Delta x$.

The implications of this analysis for a figure-of-merit for instrument design are as follows:

(a) A versatile FOM depends largely on the Fourier transform of the response function rather than on its full-width-half-maximum — this is illustrated by our example of Fig. 10 (the Fourier transforms of the response functions being given in Fig. 10(d)).

(b) The FOM is not constant for a given response function, but depends on the amount of detail required in the inferred scattering law.

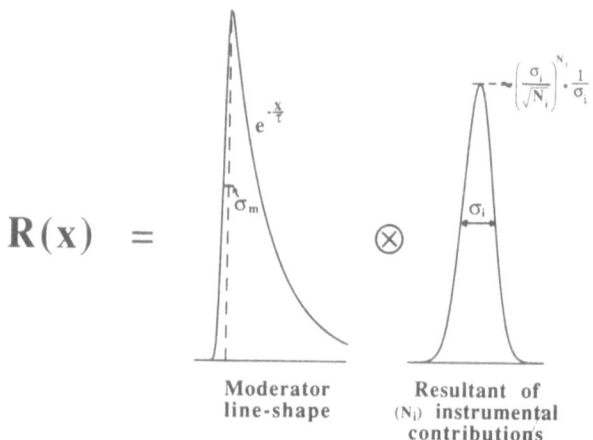

$$R(x) \quad = \quad$$

Moderator
line-shape

Resultant of
$(N_i)$ instrumental
contributions

**Figure 12**: A schematic picture of response *matching* at a pulsed source. Instrument design questions include: what is the optimal value of $\sigma_i$ (i.e. flight-path, collimation etc.) given $\sigma_m$ & $\tau$ (the moderator) ? What is the best choice of $\sigma_m$ (e.g. moderator material) and $\tau$ (moderator poison) ?

(c) The background signal (which is ignored in the conventional FOM) enters the FOM through its $\sigma^2$ dependence: $<\sigma^2> \approx <D(x_k)> \approx <R\otimes S+B> \sim N_{total}$.

Since the response function in neutron scattering depends on the details of the spectrometer & moderator, there is potential for a revision of ideas on the design of neutron scattering facilities. Take, for example, *matching* at an acclerator-based source, which is illustrated in Fig. 12. The response function for an experiment is the resultant of a convolution between the moderator pulse-shape, which has a sharp rising edge and a long decaying tail, and a roughly a Gaussian component from the instrumental contributions (e.g. flight-path, collimation, etc.). The question is how to "match" these two components. Conventional wisdom recommends that we should make the width of the instrumental, Gaussian-like, contribution comparable to the width of the moderator pulse-shape. The analysis above, however, suggests that following this advice could seriously impair our ability to infer (reliably) the scattering law at high resolution and that what we should probably do is to match to the sharp leading edge. How this translates into the optimal choice for collimation and flight-path length, or the moderator material chosen to control the sharpness of the leading edge, or whether we should "poison" the moderator to reduce the decaying tail, are the subject of on-going research.

## Acknowledgements

The work reported here involved close collaboration with several colleagues, including: Drs. Richard N. Silver, Roger Pynn, Peter Vorderwisch, Rex P. Hjelm Jr. and Mohana Yethiraj. This research was supported by the Office of Basic Energy Sciences of the U.S. Department of Energy.

## References

Bacon, G.E. (1955)*Neutron Diffraction*, Oxford University Press (3rd edition, 1975).
Brockhouse, B.N., Abou-Helal, H.E., Hallman, E.D. (1967). *Lattice Vibrations in Iron at 296 K*, Solid State Communications, Vol. 5, 211-216, Pergamon Press Ltd., GB.

Day, D.H. & Sinclair, R.N. (1969). *Neutron Moderator Assemblies for Pulsed Thermal Neutron Time-of-Flight Experiments*, Nucl. Instr. Meth., 72, 237-253.

Hjelm , R.P., Thiyagarajan, P., Sivia, D.S., Lindner, P., Alken, H., Schwahn, D. (1989). *Small-Angle Neutron Scattering From Aqueous Mixed Colloids of Lecithin and Bile Salts,* Progress in Colloid and Surface Science 82 (in press).

Lynn, J.W. (1975). *Temperature Dependence of Magnetic Excitations in Iron,* Phys. Rev. B, 11, 2624-2637.

Mezei, F. & Vorderwisch, P. (1989). *Spectroscopy With Asymmetric Resolution Functions: Resolution improvement by an on-line algorithm,* Physica B, 156 and 157, 678.

Michaudon, A (1963). *Reactor Science and Technology,* Journal of Nuclear Energy A/B, 17, 165-186.

Newton, T.J. (1985). *Blind Deconvolution and Related Topics,* Ph.D. Thesis, Cambridge University.

Robinson, R.A., Pynn, R., Eckert, J. (1985). *An improved Constant-Q Spectrometer for Pulsed Neutron Sources,* Nucl. Instr. Meth., A241, 312-324.

Skilling, J. & Gull, S.F. (1989). *Bayesian Maximum Entropy Image Reconstruction,* Proceedings of AMS-SIAM Summer School on Spatial Statistics and Imaging, ed A. Possolo (in press). Also in Maximum Entropy and Bayesian Methods: Cambridge 1988, ed. J. Skilling, Kluwer Academic Publishers.

Seeger, P.A., Hjelm, R.P., Nutter, M.J. (1989). *The Low-Q Diffractometer at the Los Alamos Neutron Scattering Center,* Mol. Cryst. Liq. Cryst. (in press).

Silver, R.N., Sivia, D.S., Pynn, R. (1989). *Information Content of Lineshapes,* International Collaboration on Advanced Neutron Sources X, Los Alamos, Institute of Physics, Conf. 97 (in press).

Sivia, D.S., Vorderwisch, P., Silver, R.N. (1989). *Deconvolution of Data from the Filter Difference Spectrometer: From Hardware to Maximum Entropy,* submitted to Nucl. Instr. Meth.

Sivia, D.S., Silver, R.N., Pynn, R. (1989). *Optimization of Resolution Functions for Neutron Scattering,* Nucl. Instr. Meth. (in press).

Taylor, A.D., Wood, E.J., Goldstone, J.A. & Eckert, J. (1984). *Lineshape Analysis and Filter Difference Method for a High Intensity Time-of-Flight Inelastic, Neutron Scattering Spectrometer,* Nucl. Instr. Meth., 221, 408-418.

Windsor, C. (1981). *Pulsed Neutron Scattering,* Taylor & Francis Ltd., London.

Yethiraj, M., Robinson, R.A., Sivia, D.S., Lynn, J.W., Mook, H.A. (1989). In preparation.

# THE CONDITIONAL ENTROPY OF A CANONICAL CONSTRAINT

Randall Barron
Laboratory Technologies Corp.
400 Research Drive
Wilmington, MA 01887
USA

**ABSTRACT.** The conditional entropy of a family of canonical constraints,

$$S(x,y,...) = \sup \{ H(\rho) : <\rho,X> = x , <\sigma,Y> = y , ... \}$$

enjoys concavity properties analogous to those of the phenomenological entropy in equilibrium thermodynamics. These concavity properties are established by a simple argument not involving the usual machinery of partition functions. A variational principle based on $S$ itself yields interval-type estimates for statistical quantities in cases where the constraints, although consistent, may not determine a representative distribution $\rho^*$ of maximum entropy. Mathematical examples exhibiting such behavior are easy to construct; typically, it happens beyond a "barrier" where the associated partition function becomes singular. The physical interpretation remains uncertain, but the author's speculations favor disordered systems like amorphous materials, turbulent flows, etc.

## NOTATION

We are concerned with probabilities of the form $P(\rho, X, \Delta) \in [0,1]$, where $\rho$ is a complete probability model, or state; $X$ is a Real random variable, or **observable**; and $\Delta$ is an interval (or, more generally, a Borel subset) of the set $R$ of Real numbers. $P(\rho, X, \Delta)$ is the probability, in the state $\rho$, that the observable $X$ manifests a value in the interval $\Delta$. We will not need to discuss joint probabilities except, perhaps, those for a **compatible** set of observables expressible as Borel functions of some common observable $Z$. In quantum mechanics, at least, the existence of joint probabilities for incompatible sets of observables is problematical.

The **expectation-value** of an observable $X$ in the state $\rho$ is defined as the average over the probability distribution,

$$<\rho, X> = \int_R x \, P(\rho, X, dx)$$

The bra-ket notation emphasizes the dependence of the expectation-value on both arguments, the observable $X$ and the state $\rho$. In general, $<\rho,X>$ takes values in the extended set of scalars, $\widetilde{R} = R \cup \{-\infty, \sim, +\infty\}$, where $\sim = \infty - \infty$ is the Real Indefinite magnitude that I spoke of last year in Cambridge [1]. This is the value we assign to the expectation-value whenever the

P. F. Fougère (ed.), Maximum Entropy and Bayesian Methods, 211–220.

defining integral "does not exist" in the usual Lebesgue sense.

The information-theoretic entropy is a function on the states, given by well-known formulas in the three standard models.

**classical discrete:**

$$H(\rho) = - \sum_n p_n \log p_n$$

**classical continuous:**

$$H(\rho) = \begin{cases} - \int \left[\dfrac{d\rho}{d\mu}\right] \log\left[\dfrac{d\rho}{d\mu}\right] d\mu & \text{if } \rho << \mu \text{ and the integral exists in } \overline{R} \\ \\ -\infty & \text{otherwise} \end{cases}$$

**quantum separable:**

$$H(\rho) = - \operatorname{Tr} \rho \log \rho$$

The entropy functional is strictly $\overline{R}$-valued, where $\overline{R} = R \cup \{-\infty, +\infty\}$ is the usual set of Extended Real numbers. In the classical continuous case, $\mu$ is some sigma-finite reference measure, for instance, the Liouville measure or "phase volume" of classical mechanics.

## CONVEXITY AND CONCAVITY

The probability models or states constitute a **convex set**, with the operation of convex combination

$$\rho, \sigma \rightarrow t\rho + (1-t)\sigma$$

for $t \in (0,1)$. In the ensemble interpretation, the operation of convex combination represents the mixing of two ensembles with relative weights $t{:}1{-}t$.

The probability and expectation-value functionals are **affine** in their dependence on the states,

$$P(t\rho + (1-t)\sigma, X, \Delta) = tP(\rho, X, \Delta) + (1-t)P(\sigma, X, \Delta)$$

$$<t\rho + (1-t)\sigma, X> = t <\rho, X> + (1-t) <\sigma, X> .$$

(But for the positivity of $t$, we might say "linear" instead of "affine"). The set $\widetilde{R}$ carries a generalized "convex structure" allowing for the notion of points at infinity [7]. Any finite admixture of a point at infinity in a convex combination yields another point at infinity.

An $\overline{R}$-valued function $H$ defined on a convex set is **concave** if it satisfies the inequality

$$H(t\rho + (1-t)\sigma) \geq tH(\rho) + (1-t)H(\sigma) \qquad\qquad \text{for all } t \in (0,1) ,$$

whenever $H(\rho) > -\infty$ and $H(\sigma) > -\infty$. Similarly, another $\overline{R}$-valued function $G$ is **convex** if $-G$ is concave. These are the accepted mathematical definitions for the terms convex and concave as applied to functions; I would be just as happy as you to have them the other way around.

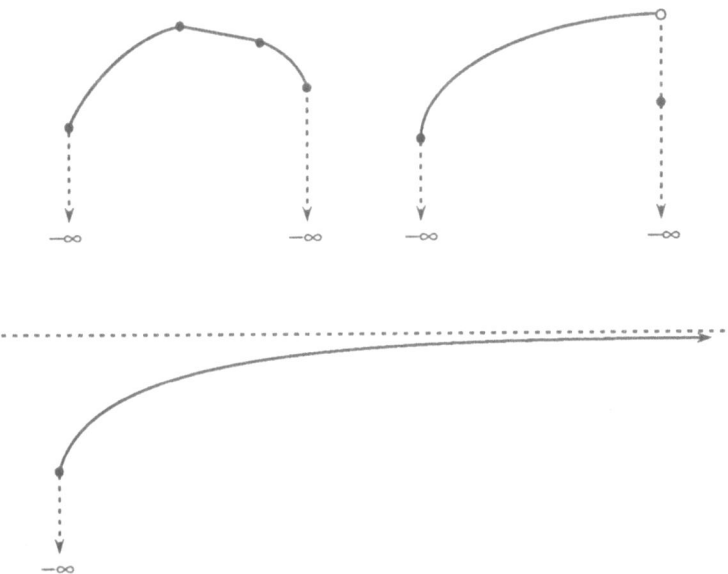

FIG. 1: REPRESENTATIVE CONCAVE FUNCTIONS

The figure shows three concave functions of a single Real variable. The set on which each differs from $-\infty$ is an interval, called its **effective domain**. The first example illustrates the point that the concavity may be degenerate, i.e., not strict, in some subinterval — the curve has a "flat" part. The other two examples illustrate the point that a concave function, though bounded above, need not attain its maximum. (A "drop-down" discontinuity at the end of the effective domain is allowed by the definition of concavity.) There is something insubstantial about these one-dimensional examples, but they make the point. In infinitely many dimensions, there is much more room for Monkey Business.

## CONCAVITY INEQUALITIES FOR THE INFORMATION-THEORETIC ENTROPY

It is natural that the entropy functional should be concave on the states. In the ensemble interpretation, this expresses a loss of information that occurs on the mixing of ensembles — information that may distinguish the members of the ensemble on the basis of their origin. We will state the inequalities for any finite convex combination,

$$\rho_1, \rho_2 \ldots, \rho_N \rightarrow \Sigma_n \, t_n \, \rho_n \; ,$$

where $t_1, t_2, \ldots t_N \in (0,1)$ satisfy $\Sigma_n \, t_n = 1$. If $H(\rho_n) > -\infty$ for $n = 1, 2, \ldots, N$ then

$$H(\Sigma_n \, t_n \rho_n) \geq \Sigma_n \, t_n \, H(\rho_n) \; ,$$

with finite equality only if $\rho_n = \rho$ for all $n$; and

$$H(\Sigma_n t_n \rho_n) \le \Sigma_n t_n H(\rho_n) - \Sigma_n t_n \log t_n \, ,$$

with finite equality only if $\rho_m \perp \rho_n \ (m \ne n)$.

In the ensemble interpretation, the first inequality says that the entropy of mixing is non-negative; and the second, that it is bounded above by the Shannon entropy of the weights. The orthogonality relation $\perp$ indicated is the one appropriate to the model, e.g., in the classical continuous case, the $N$ probability measures $\rho_1, \rho_2, ..., \rho_N$ have disjoint support. For proofs of these relations see the paper [2] by Ochs and the review article [3] by Wehrl.


## STATISTICAL CONSTRAINTS

A **statistical constraint** is a condition that determines a convex set $C$ of states,

$$\rho, \sigma \in C \Rightarrow t\rho + (1-t)\sigma \in C \, ,$$

for all $t \in (0,1)$. Thus, if two states satisfy a statistical constraint, then so does any convex combination of the two. The ensemble interpretation is obvious. Two kinds of statistical constraint are familiar from statistical mechanics. A **canonical constraint**,

$$<\rho, X> = x \, ,$$

fixes the expectation-value of an observable $X$; a **microcanonical constraint**,

$$P(\rho, X, \Delta) = 1 \, ,$$

asserts that (with probability 1) the value of an observable $X$ lies in a particular interval $\Delta$. Since the intersection of any family of convex sets is again convex, the conjunction of any family of statistical constraints is again a statistical constraint.

In his writings on the Principle of Maximum Entropy, a theme central to these meetings, E. T. Jaynes has introduced a notion of **testable information**, meaning a condition which characterizes some definite set $C$ of probability models or states. Although he has not mentioned the convexity property, so far as I know, in all examples Jaynes has given the set of states so characterized is convex. The examples tabulated below are taken from ref. [4]; convexity of each set of states follows in elementary fashion from the fact that the probability and expectation-value functionals are affine. I propose that the two notions, of testable information and statistical constraint, should be identified.

|      | TESTABLE INFORMATION | STATISTICAL CONSTRAINT |
|------|----------------------|------------------------|
| I1:  | $X < 6$ (with probability 1) | $P(\rho, X, (-\infty, 6)) = 1$ |
| I2:  | The mean value of $\tanh^{-1}(1-X^2)$ in previous measurements was $1.37$. | $<\rho, \tanh^{-1}(1-X^2)> = 1.37$ |
| I3:  | ... Laplace summarized his analysis of the mass of Saturn [X] ..."It is a bet of 11000:1 that the error of this estimate is not 1/100 of its value." He estimated this mass as 1/3512 of the Sun's mass. | $P(\rho, \|X-M\|/M, [0,1/100)) = 11000/11001$  $M = M_\odot/3512$ |
| I4:  | There is at least a 90 percent probability that $X > 10$. | $P(\rho, X, (10, +\infty)) \ge 0.9$ |

## LEAST UPPER BOUNDS

If an $\overline{R}$-valued function bounded above does not attain its maximum, there is, nevertheless, an upper limit to which it approaches arbitrarily closely. We digress briefly to introduce the notion of the least upper bound.

A bounded subset of $R$ need not contain a greatest element. The set of all rational numbers r for which $r^2 \leq 2$ provides the classical example. Given a set $B \subseteq R$, say that a is a **least upper bound** of $B$, provided:

(1) a is an **upper bound** for $B$, $a \geq x$ for all $x \in B$;
(2) if b is any other upper bound for $B$, then $b \geq a$.

If it exists, the least upper bound is obviously unique. By the completeness of the Real Numbers (a theorem or a definition, depending how you look at it), any non-empty set $B \subseteq R$ that is bounded above has a least upper bound. The term **supremum** is synonymous with least upper bound, leading to the notation

$$a = \sup B .$$

By convention, we extend the notion to arbitrary subsets $B \subseteq \overline{R}$. If B is not bounded above, then $\sup B = +\infty$; if B is empty, then $\sup B = -\infty$.

## CONDITIONAL ENTROPY

Consider a statistical constraint which characterizes some convex set $C$ of states. Its **conditional entropy** $S(C)$ is defined by

$$S(C) = \sup \{ H(\rho) : \rho \in C \} .$$

This is the greatest value approached by $H(\rho)$ as $\rho$ ranges over the set $C$ of states compatible with the constraint; it is not necessarily attained. If $H$ attains its maximum on $C$, say at $\rho^* \in C$, then $S(C) = H(\rho^*)$; but the existence of $\rho^*$ is not guaranteed!

Just like the physical entropy in thermodynamics, the conditional entropy increases on relaxation of a constraint,

$$C \subseteq D \Rightarrow S(C) \leq S(D) .$$

Associated with a constraint of the canonical form, introduced earlier, is a **conditional entropy function**,

$$S(x) = \sup \{ H(\rho) : <\rho, X> = x \}, \qquad\qquad x \in \widetilde{R} .$$

Just like its counterpart in thermodynamics, the conditional entropy function associated with a canonical constraint is concave in the constraint variables. I will present a simple and elegant proof, which follows directly from the concavity of the information-theoretic entropy $H$.

Concavity is a regularity condition almost as strong as differentiability. A function that is finite and concave on an open interval is continuous throughout that interval, and differentiable at all but a countable number of points in it; furthermore, even at those singular points, both left-hand and right-hand derivatives exist. I refer you to Rockafellar's book [5] for a proof of this assertion, and its generalization to the finite-dimensional case.

## CONCAVITY OF THE CONDITIONAL ENTROPY

THEOREM. $S(tx + (1-t)y) \geq tS(x) + (1-t)S(y)$ for all $t \in (0,1)$, whenever $S(x), S(y) > -\infty$.

Proof: By hypothesis $S(x) > -\infty$, so there is a state $\rho$ such that $<\rho, X> = x$ and $H(\rho) > -\infty$. Similarly, there is a state $\sigma$ such that $<\sigma, X> = y$ and $H(\sigma) > -\infty$. We have already seen that $<\bullet, X>$ is affine,

$$<t\rho + (1-t)\sigma, X> \;=\; t<\rho, X> + (1-t)<\sigma, X> \,,$$

and that $H(\bullet)$ is concave,

$$H(t\rho + (1-t)\sigma) \geq tH(\rho) + (1-t)H(\sigma) \,.$$

From the definition of $S$, it follows that

$$S(tx + (1-t)y) \geq H(t\rho + (1-t)\sigma) \geq tH(\rho) + (1-t)H(\sigma) \,.$$

This is true for all $\rho$ such that $<\rho, X> = x$ and $H(\rho) > -\infty$, but these are the only states that matter in the evaluation of

$$S(x) \;=\; \sup \{ H(\rho) : <\rho, X> = x \} \,,$$

whence it follows that

$$S(tx + (1-t)y) \geq tS(x) + (1-t)H(\sigma) \,.$$

We can make the same argument with $\sigma$, leading to the desired result,

$$S(tx + (1-t)y) \geq tS(x) + (1-t)S(y) \,.$$

Two corollaries follow from the strictness of the concavity inequalities for $H(\bullet)$. The first is well-known, the second, perhaps not so well-known.

Corr. 1: If $S(x)$ is finite and a MAXENT state $\rho^*$ exists, satisfying $<\rho^*, X> = x$ as well as $H(\rho^*) = S(x)$, then $\rho^*$ is unique.

Corr. 2: If $S$ is finite on some convex set $D \subseteq \widetilde{R}$, and a MAXENT state $\rho^* = \rho^*(x)$ exists for every $x \in D$, then the concavity of $S(x)$ is strict on $D$, i.e.,

$$S(tx + (1-t)y) > tS(x) + (1-t)S(y) \,, \qquad\qquad \text{for all } t \in (0,1),$$

for all distinct $x, y \in D$.

The proof given above requires none of the "niceness" properties of the partition function—existence, continuity and differentiability—which feature so prominently in the usual textbook demonstration. These niceness properties, which are assumed more often than they are proved, are rather difficult to establish even in the one dimensional case. The extension of the theorem to finitely many, countably many, or even a continuum of canonical constraints is trivial. Under the same conditions, the partition function would have to be replaced by a partition functional, whose niceness properties—existence, continuity and differentiability—would be even more difficult to establish.

## PROOF UTILIZING THE PARTITION FUNCTION

I will review the conventional proof of the concavity of $S$ just as you might find it in a textbook on statistical mechanics. Let us consider the classical continuous model, with a finite number of canonical constraints,

$$<\rho, \mathbf{X}> = x ,$$

where the boldface font indicates an N-component vector,

$$\mathbf{X} = (X_1, X_2, ..., X_N), \qquad x = (x_1, x_2, ..., x_N).$$

In the usual way we write down the Gibbsian ansatz for a MAXENT state,

$$\rho^* = \exp(-\beta \bullet \mathbf{X}) / Z(\beta), \qquad Z(\beta) = \int \exp(-\beta \bullet x) \, d^N x ,$$

and find that

$$S(x) = H(\rho^*) = \beta \bullet \mathbf{X} + \log Z(\beta) ,$$

where $\beta = (\beta_1, \beta_2, ..., \beta_N)$ is a function of $x$ to be determined by solving the equations

$$-(\partial/\partial\beta_i) \log Z(\beta) = x_i , \qquad i = 1, 2, .., N .$$

The general criterion for solvability is that the Jacobian matrix,

$$J_{ij} = \partial x_i / \partial \beta_j = -(\partial^2/\partial\beta_i\partial\beta_j) \log Z(\beta) ,$$

be non-singular. From the definition of the partition function $Z(\beta)$, it follows on differentiating twice that

$$J_{ij} = -<\rho^*, X_i X_j> + <\rho^*, X_i> <\rho^*, X_j> ,$$

which shows that $J_{ij}$ is negative semi-definite. Assuming $J_{ij}$ is non-singular, we deduce that it must be negative definite. From the formula $S(x) = \beta \bullet \mathbf{X} + \log Z(\beta)$, we can derive the relation

$$(\partial/\partial x_j) S(x) = \beta_j .$$

The concavity of $S$ is then established by the argument that the matrix of second derivatives,

$$K_{ij} = (\partial^2/\partial x_i \partial x_j) S(x) = \partial\beta_j/\partial x_i = (J^{-1})_{ij} ,$$

is likewise negative definite. In the quantum mechanical model, the argument goes through pretty much unchanged, except that a more complicated formula is obtained for $J_{ij}$ in case the observables $X_1, X_2, ..., X_N$ do not commute.

This proof shows us a lot of interesting relations between the partition function and its derivatives, the correlation matrix, and so on. But, when all is said and done, I am not quite as convinced as I was by the first proof that $S(x)$ really is concave. Moreover, this proof applies only to the region of the parameter space $\widetilde{R}^N$ accessible by means of Gibbsian MAXENT states associated with the partition function. As we will see in the example to follow, there may be non-trivial regions of the parameter space where the constraints, although consistent (in the sense that they admit some states), do not admit any MAXENT state.

I have shown you what I believe is a superior way of proving certain general results, without the machinery of partition functions. However, when it comes to performing detailed computations in a concrete model, I will cheerfully use that machinery to get the job done. In particular, the following example is analyzed with the aid of a partition function.

### EXAMPLE

An example due to Dowson and Wragg [6], illustrates the point that the Gibbsian MAXENT states may not cover the parameter space of consistent constraints. Consider $X \geq 0$ with a probability density $f(x)$ that we would like to estimate by maximizing the entropy

$$H = -\int_0^\infty f(x) \log f(x) \, dx$$

subject to the constraints

$$<X> = \int_0^\infty x \, f(x) \, dx = \mu, \qquad <X^2> = \int_0^\infty x^2 \, f(x) \, dx = \lambda.$$

The constraints are consistent for $\mu \geq 0$ and $\lambda \geq \mu^2$, but admit a MAXENT distribution only if $\lambda \leq 2\mu^2$. Let us look at the region of the parameter space—coordinatized by $(\mu,\lambda)$ —accessible using MAXENT distributions of the familiar truncated Gaussian form,

$$f(x) = \exp(-\alpha x - \beta x^2) / Z(\alpha,\beta), \qquad x \geq 0.$$

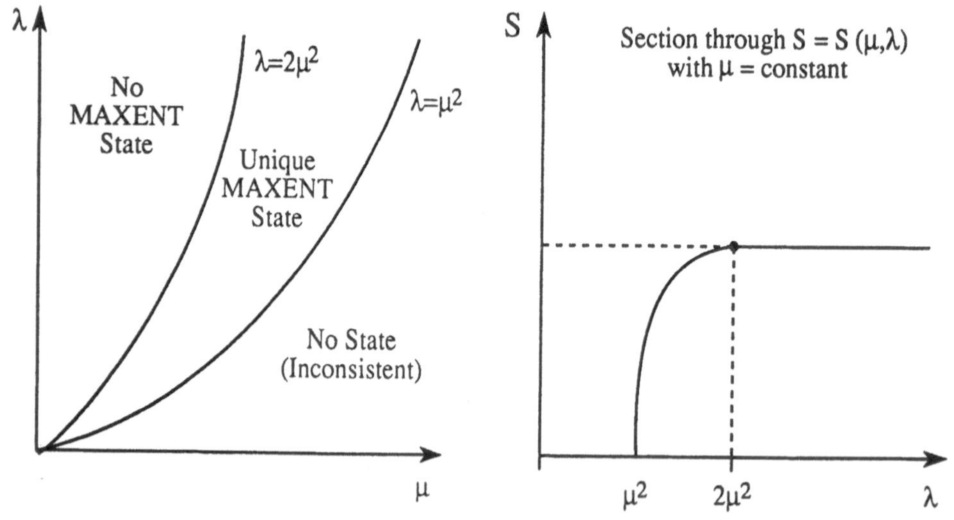

FIG. 2: PARAMETER SPACE FOR THE EXAMPLE OF DOWSON & WRAGG

## MAXIMUM ENTROPY ESTIMATION WITHOUT MAXENT STATES

The foregoing example illustrates the point that a rather innocuous-looking family of statistical constraints may fail to determine a MAXENT state. By a minor extension of the usual methodology we can, nevertheless, obtain maximum entropy estimates for statistical quantities of interest. In his writings on the subject, Jaynes has alluded to the transitory nature of our need for the MAXENT state. It is an intermediate construction, useful for generating the estimates in which we are really interested.

Let us pose the following problem: Given a convex set $C$ of states, estimate $<X>$ subject to the statistical constraint $\rho \in C$. We make use of the conditional entropy function,

$$S(x, C) = \sup \{ H(\rho) : \rho \in C, <\rho, X> = x \},$$

which is well-defined whether or not the constraints admit a MAXENT state. Note that

$$\sup \{ S(x, C) : x \in \widetilde{R} \} = \sup \{ H(\rho) : \rho \in C \} = S(C).$$

In what follows, we suppress the dependence on $C$ and write $S(x, C) = S(x)$, $S(C) = s^*$.

Suppose there is no MAXENT state $\rho^*$ in $C$. In a practical optimization problem, we might be content with a single approximating state $\rho \in C$ whose entropy $H(\rho)$ is sufficiently close to the maximum, $s^*$. In a theoretical discussion, we might consider an approximating sequence, $\rho_1, \rho_2,... \in C$ such that $H(\rho_n) \to s^*$ as $n \to \infty$. The limit points in $R$ of the sequence $\{ <\rho_n, X> \}$ belong in any estimate that we might offer for $<X>$. The same can be said for the set of limit points so obtained if we start from any other approximating sequence of states. Therefore, we offer the union of all such sets of limit points generated by such approximating sequences as a generalized, interval-type estimate, $J \subseteq \widetilde{R}$, for the expectation-value $<X>$. This set can be characterized in terms of a variational principle involving the conditional entropy function $S(x)$ or, more precisely, the closure of $S$ as a concave function.

THEOREM: $J = \{ x : \overline{S}(x) = s^* \}$, where $\overline{S}$ is the upper envelope, or concave closure of $S$,

$$\overline{S}(x) = \begin{cases} \max \{ S(x), \limsup_{y \to x} S(y) \} & x \in \overline{R} \\ \max \{ S(\sim), \min \{ \overline{S}(+\infty), \overline{S}(-\infty) \} \} & x = \sim \end{cases}$$

I will not attempt to give the proof here [7], but I will discuss the consequences. As the closure of a concave function, $\overline{S}$ coincides with $S$ except possibly at the endpoints of its effective domain, the interval of $R$ where $S(x) > -\infty$. $\overline{S}$ is concave on $R$. $\overline{S}$ is upper semicontinuous on $\overline{R}$, and therefore attains its maximum on $\overline{R}$ and, consequently, on $\widetilde{R}$. But this maximum value is

$$\sup \{ \overline{S}(x) : x \in \widetilde{R} \} = \sup \{ S(x) : x \in \widetilde{R} \} = s^*,$$

and the estimation set $J$ is non-empty. From the concavity of $\overline{S}$, $J \cap R$ is an interval.


## CONCLUSION

The example presented here does not pretend to any physical interpretation. I have others which do, but they are weak, and susceptible to attack on physical grounds—which would be beside the point. The example presented suffices to demonstrate a mathematical possibility. The door is opened to the prospect that such examples may arise in the statistical description of real physical systems, and may represent some interesting physical behavior. Instead of a unique maximum entropy state which dominates the statistics in autocratic fashion, we would find an egalitarian assembly of submaximal entropy states competing for partial ascendency. This sounds like a prescription for turbulence, disorder and chaos (or, perhaps, a democracy). The idea seems to have been anticipated by John von Neumann, in an old ONR report on the nature of turbulence [8].

## PHYSICAL INTERPRETATION

We should look for such behavior in physical systems not ordinarily considered to be in thermodynamic equilibrium but which may, nevertheless, be describable by means of a constrained maximum entropy principle. As Jaynes argued, in his earliest publication on the subject [9], it is the *experimental reproducibility* of a macroscopic phenomenon that justifies a statistical description based on the maximum entropy principle. There is no requirement of temporal stationarity, and the method, so successful in the description of thermodynamic equilibrium, should be equally applicable to reproducible irreversible phenomena.

I suggest the following "syndrome" of characteristics: (1) A large degree of disorder; fluctuations in space and/or time; interval-type predictions for extensive physical quantities. We are already familiar with interval-type predictions in the vicinity of a phase transition. (2) Marked susceptibility to certain perturbations of an adiabatic nature. If all of the states have about the same entropy, it doesn't cost much to move quasi-reversibly from one to another. (3) Hysteresis or "memory effect", in which present behavior depends on past history as well as present conditions.

## REFERENCES

[1] Barron, R., 'The Paradox of the Money Pump: A Resolution', *Maximum Entropy and Bayesian Methods, Cambridge, England, 1988*, ed. J. Skilling, pub. Kluwer (1989).

[2] Ochs, W., 'Basic Properties of the Generalized Boltzmann-Gibbs-Shannon entropy', *Repts. Math. Phys.* **9**, 135-155 (1976), Th. 9.

[3] Wehrl, A., 'General Properties of Entropy', *Rev. Mod. Phys.* **50**, 221-260 (1978).

[4] Jaynes, E.T., 'Prior Probabilities', *IEEE Transactions* **SSC-4**, 227-241 (1968).

[5] Rockafellar, R.T., **Convex Analysis**, Princeton University Press (1970), cf. Th. 25.3.

[6] Dowson, D.C., and A. Wragg, 'Maximum Entropy Distributions having Prescribed First and Second Moments', *IEEE Transactions* **IT-19**, 689-693 (1973).

[7] Barron, A.R., *Integrals, Expectation-Values and Entropy*, PhD Thesis (Physics), Brandeis University, 1981. *Dissertation Abstracts* **42B**, 4828 (1982).

[8] von Neumann, J., 'Recent Theories of Turbulence', report to the Office of Naval Research (1949). Reprinted in *John von Neumann: Collected Works*, Vol. 6, ed. A.H. Taub, pub. Pergamon Press (1963), cf. §1.2, §11.2.

[9] Jaynes, E.T., 'Information Theory and Statistical Mechanics', *Phys. Rev.* **106**, 620-630, (1957), Sec. 4.

# SOLVING OVERSAMPLED DATA PROBLEMS BY MAXIMUM ENTROPY.

R. K. BRYAN
*European Molecular Biology Laboratory,*
*Meyerhofstrasse 1,*
*6900 Heidelberg,*
*West Germany.*

ABSTRACT. A numerical algorithm for the solution of the Classic Maximum Entropy problem is presented, for use when the data are considerably oversampled, so that the amount of independent information they contain is very much less than the actual number of data points. Examples of problems for which this algorithm is particularly appropriate are dynamic light scattering, solution scattering and fibre diffraction. The application of a general purpose entropy maximisation program is then comparatively inefficient. In the new algorithm the independent variables are in the singular space of the transform between map (or image or spectrum) and data, and much fewer in number than either the data or the reconstruction. This reduction in the dimension allows a direct evaluation of the posterior probability of the solution, and thus enables the 'Classic Maxent' problem to be solved completely.

## 1. Introduction.

The motivation for developing a new numerical algorithm to solve the maximum entropy problem was twofold. For some high-accuracy datasets, such as those resulting from dynamic light scattering (DLS), a previous algorithm (Bryan, 1980, Skilling & Bryan, 1984), designed for general image processing problems, such as deconvolution, converged disappointingly slowly. Such datasets are often oversampled, and an algorithm working in the singular space of an appropriate linear transform is then more efficient than one which reconstructs the spectrum directly. Secondly, the work of Skilling (1989a) and Gull (1989), presented at last year's workshop, has embedded Maximum Entropy in a more general Bayesian framework. The advantages of this approach include a firm axiomatic foundation for the entropy as the prior of the distribution of intensity over a positive, additive image (or map or spectrum); an estimate of the covariance of the solution; and, most important for practical computations, a Bayesian estimate of the value of the Lagrange multiplier $\alpha$, which balances the relative weights of the entropy and data constraint, replacing the '$\chi^2 = N$' criterion (Gull & Daniell, 1978).

We start by quoting the results of the Skilling-Gull theory which we will need. We wish to reconstruct a positive, additive, density, represented by an $N$-dimensional vector $\mathbf{f}$, from data represented by an $N_{data}$-dimensional vector $\mathbf{D}$. The physics of the experiment allow us to define the likelihood $p_r(\mathbf{D} \mid \mathbf{f})$, which we write as $p_r(\mathbf{D} \mid \mathbf{f}) = \exp{-L}$. To use Bayes' theorem to obtain $p_r(\mathbf{f} \mid \mathbf{D})$, the prior on $\mathbf{f}$ is needed, which Skilling (1989a) has shown to be

$$p_r(\mathbf{f} \mid \alpha, \mathbf{m}) = \exp{\alpha S}/Z_S(\alpha),$$

where $S$ is the entropy of $\mathbf{f}$ relative to an initial map $\mathbf{m}$,

$$S = \sum_i f_i - m_i - f_i \log f_i/m_i.$$

*P. F. Fougère (ed.), Maximum Entropy and Bayesian Methods, 221–232.*
© 1990 *Kluwer Academic Publishers.*

and $Z_S(\alpha)$ is a normalising factor. A further consequence of his derivation is that a measure $\prod f^{-1/2}$ should be applied in f-space. $\alpha$ is initially undetermined, but Gull (1989), and also Sibisi (1989) in the context of quadratic regularisation, show how Bayes' theorem may be used to find the joint posterior probability of $\mathbf{f}$ and $\alpha$, which may be written as

$$p_r(\mathbf{f}, \alpha \mid \mathbf{D}, \mathbf{m}) = p_r(\mathbf{f} \mid \alpha, \mathbf{D}, \mathbf{m}) p_r(\alpha \mid \mathbf{D}, \mathbf{m}).$$

Gaussian approximations to the exponentials are made, so that, with unimportant normalisation constants omitted, and $p_r(\alpha)$ as the prior on $\alpha$,

$$p_r(\mathbf{f} \mid \alpha, \mathbf{D}, \mathbf{m}) = \prod_i (\alpha + \lambda_i)^{1/2} \exp \frac{1}{2} \delta \mathbf{f}^T \nabla \nabla Q \delta \mathbf{f}, \qquad \text{with} \quad \delta \mathbf{f} = \mathbf{f} - \hat{\mathbf{f}}, \qquad (1)$$

$$p_r(\alpha \mid \mathbf{D}, \mathbf{m}) = \prod_i \left( \frac{\alpha}{\alpha + \lambda_i} \right)^{1/2} \exp \hat{Q} \; p_r(\alpha), \qquad (2)$$

where $\hat{\mathbf{f}}(\alpha)$ maximises $Q = \alpha S - L$, $\hat{Q}(\alpha) = Q(\hat{\mathbf{f}}(\alpha))$, and the $\{\lambda_i\}$ are the eigenvalues of $\mathrm{diag}\{f^{\frac{1}{2}}\} \nabla \nabla L \, \mathrm{diag}\{f^{\frac{1}{2}}\}$ evaluated at $\hat{\mathbf{f}}(\alpha)$. Gull then shows that when the number of observations is large, (2) gives a sharp optimum for $\alpha$, at $\hat{\alpha}$, say. Assuming that any reasonable prior for $\alpha$ will be overwhelmed by the data, this leads, after rearrangement of the derivative $dp_r(\alpha \mid \mathbf{D})/d\alpha = 0$, to the condition

$$-2\hat{\alpha}\hat{S} = \sum_i \frac{\lambda_i}{\hat{\alpha} + \lambda_i}.$$

Each eigenvalue $\lambda$ which is significantly larger than $\hat{\alpha}$ contributes one to the right hand side, which is therefore a count of the number of good observations, $N_g$. The posterior distribution of $\mathbf{f}$ is then approximately the Gaussian (1) about $\hat{\mathbf{f}}(\hat{\alpha})$. As will be shown in a later section, the problems of interest here may have a low $N_g$, a wide distribution of reasonably probable $\alpha$, and possibly multiple maxima. The full form of $p_r(\mathbf{f}, \alpha \mid \mathbf{D}, \mathbf{m})$ is needed, and expectation values may be obtained by integrating over $d^N f$ using the Gaussian approximation (1), and over $\alpha$ numerically.

## 2. Numerical Algorithm.

The theory described in §1 results in two numerical problems; the maximisation of $Q = \alpha S - L$, and the evaluation of the eigenvalues of $\mathrm{diag}\{f^{\frac{1}{2}}\} \nabla \nabla L \, \mathrm{diag}\{f^{\frac{1}{2}}\}$.

The algorithm presented here is intended for application to problems where the likelihood $p_r(\mathbf{D} \mid \mathbf{f})$ can be written as $\exp -L(\mathbf{F}, \mathbf{D})$, where $\mathbf{F}$ is linearly related to $\mathbf{f}$, $\mathbf{F} = T\mathbf{f}$. Many practical problems allow the likelihood to be cast in this form. In particular, a purely linear problem with uncorrelated Gaussian noise has $L(\mathbf{F}, \mathbf{D}) = \frac{1}{2} \sum_k (F_k - D_k)^2 / \sigma_k^2$, which may be recognised as '$\frac{1}{2}\chi^2$'. It is assumed that the rank of $T$ is sufficiently small that full matrices of this dimension may be handled. If $\mathrm{rank}(T) < N_{\mathrm{data}}$ we call the data *oversampled*. That is not to say some are redundant; they all will contribute to the statistical accuracy of the solution. We are particularly interested in the case $\mathrm{rank}(T) \ll N_{\mathrm{data}}$.

The key to the method is the singular value decomposition (SVD) of $T$,

$$T = V \Sigma U^T,$$

where $V$ is an $N_{\text{data}} \times N_{\text{data}}$ orthogonal matrix, $U$ an $N \times N$ orthogonal matrix, and the $N_{\text{data}} \times N$ matrix $\Sigma$ is zero except for the elements $\Sigma_{ii} = \sigma_i$, $i = 1, \ldots, s$. The $\sigma_i$, conventionally ordered $\sigma_1 \geq \sigma_2 \geq \ldots \geq \sigma_s > 0$, are the singular values, and $s = \text{rank}(T)$. $T$ is defined by the geometry of the problem: the coordinates of the data-points; the domain of f-space selected for reconstruction; and the functional form of $T$, which is assumed known. All this information is implicitly included in the 'other information', which all the probabilities are conditional on, and is fixed for the problem.

## 2.1. EIGENVALUES.

The non-zero eigenvalues of $\text{diag}\{f^{\frac{1}{2}}\}\nabla\nabla L\,\text{diag}\{f^{\frac{1}{2}}\}$ are required for the computation of the probabilities (1, 2). This is an eigenproblem in the full space, but we show now that by a careful change of basis the non-zero eigenvalues may be computed from matrices of only singular-space dimension. From the definition of $L$,

$$\nabla\nabla L \;=\; T^T\frac{\partial^2 L(\mathbf{F},\mathbf{D})}{\partial \mathbf{F}^2}T \;=\; U\Sigma V^T\frac{\partial^2 L(\mathbf{F},\mathbf{D})}{\partial \mathbf{F}^2}V\Sigma U^T.$$

We define $M = \Sigma V^T \partial^2 L(\mathbf{F},\mathbf{D})/\partial\mathbf{F}^2 V\Sigma$, so $\nabla\nabla L = UMU^T$. Hence we must solve

$$\text{diag}\{f^{\frac{1}{2}}\}UMU^T\,\text{diag}\{f^{\frac{1}{2}}\}X = X\Lambda, \qquad \Lambda = \text{diag}\{\lambda\}.$$

which can be manipulated to

$$KMKU^T\,\text{diag}\{f^{-\frac{1}{2}}\}X = K\,U^T\,\text{diag}\{f^{-\frac{1}{2}}\}X\Lambda, \qquad \text{with} \quad K = U^T\,\text{diag}\{f\}U. \quad (3)$$

$K$ is positive definite, so its Cholesky decomposition $K = CC^T$, where $C$ is lower triangular and non-singular, exists. Premultiplying (3) by $C^{-1}$ gives the eigenproblem

$$C^TMCZ = Z\Lambda, \qquad \text{where} \quad Z = C^TU^T\,\text{diag}\{f^{-\frac{1}{2}}\}X.$$

Since the only non-zero elements of $\Sigma$ are the first $s$ elements of the diagonal, it is clear that only the leading $s \times s$ submatrix of $M$ is non-zero, which we denote by $M_s$. Together with the pattern of zeros in $C$, this means that

$$C^TMC = \begin{pmatrix} C_s^T M_s C_s & 0 \\ 0 & 0 \end{pmatrix}$$

where $C_s$ is the leading $s \times s$ submatrix of $C$, and the eigenproblem may be written as

$$\begin{pmatrix} C_s^T M_s C_s & 0 \\ 0 & 0 \end{pmatrix}\begin{pmatrix} Z_s & 0 \\ 0 & \Phi \end{pmatrix} = \begin{pmatrix} Z_s & 0 \\ 0 & \Phi \end{pmatrix}\begin{pmatrix} \Lambda_s & 0 \\ 0 & 0 \end{pmatrix}.$$

where new $s$-subscripted quantities have obvious meanings, and $\Phi$, which forms a basis for the null-space, is any $(N - s)$-dimensional orthogonal matrix. Clearly

$$C_sC_s^T = K_s = U_s^T\,\text{diag}\{f\}U_s,$$

and the non-zero eigenvalues are given by the eigenproblem

$$K_sM_sK_sY_s = K_sY_s\Lambda_s, \qquad Y_s = C_s^{-T}Z_s. \quad (4)$$

$K_s$ and $M_s$ are calculated from $f$ and $\partial^2 L(\mathbf{F}, \mathbf{D})/\partial \mathbf{F}^2$, by operations involving only the first $s$ columns of $U$ and $V$ respectively.

In practical calculations, if $f$ has a high dynamic range, $K_s$ may be close to singularity and its Cholesky decomposition may fail numerically. It is better to work with its eigendecomposition,

$$K_s = P \Xi P^T, \qquad \Xi = \text{diag}\{\xi\}, \qquad P^T P = I, \tag{5}$$

correct slightly negative $\xi$'s to zero, and solve the problem

$$\text{diag}\{\xi^{\frac{1}{2}}\} P^T M_s P \, \text{diag}\{\xi^{\frac{1}{2}}\} R = R \Lambda_s, \qquad R^T R = I.$$

However, the Cholesky decomposition is important for the proof of the method, as it performs a change of basis which preserves the singular space whilst diagonalising the $K$ matrix. The eigenvectors $Y$ of the generalised eigenproblem (4) do not necessarily exist, but

$$Y_s^{-1} = R^T \, \text{diag}\{\xi^{\frac{1}{2}}\} P^T, \tag{6}$$

always does, and for future reference

$$K_s = Y_s^{-T} Y_s^{-1}, \tag{7}$$

$$\Lambda_s = Y_s^{-1} M_s Y_s^{-T}. \tag{8}$$

## 2.2. MAXIMISATION OF Q.

As the eigenproblem can be solved in an $s$-dimensional space, which incidentally shows that $N_g \leq s$, it seems reasonable to expect that the solution for $f$ can be defined in terms of $s$ parameters.

Writing the condition for the maximum of $Q$ at given $\alpha$ as $\nabla Q = 0$, or

$$-\alpha \log f/m \quad = \quad T^T \frac{\partial L(\mathbf{F}, \mathbf{D})}{\partial \mathbf{F}} \quad = \quad U \Sigma V^T \frac{\partial L(\mathbf{F}, \mathbf{D})}{\partial \mathbf{F}},$$

we see that $\log f/m$ must lie in the column space of $U_s$, and the solution may be represented by an $s$-dimensional vector $\mathbf{u}$, with

$$f_i = m_i \exp \sum_{t=1}^{s} U_{it} u_t. \tag{9}$$

Then

$$-\alpha U_s \mathbf{u} \quad = \quad U_s \Sigma_s V_s^T \frac{\partial L(\mathbf{F}, \mathbf{D})}{\partial \mathbf{F}}.$$

Hence, since the columns of $U$ are orthogonal, and dropping the $s$ subscripts,

$$-\alpha \mathbf{u} \quad = \quad \Sigma V^T \frac{\partial L(\mathbf{F}, \mathbf{D})}{\partial \mathbf{F}} \quad = \quad \mathbf{g}, \quad \text{say}, \tag{10}$$

thus defining the entropy maximum for given $\alpha$ in terms of $\mathbf{u}$, where the singular space gradient $\mathbf{g}$ is a function of $\mathbf{u}$ through (9) and $\mathbf{F} = T\mathbf{f}$. $\mathbf{F}$ may be computed efficiently by successive application of $U^T$, $\Sigma$ and $V$ to $\mathbf{f}$, which takes $s(N + 1 + N_{\text{data}})$ operations, as opposed to the $N N_{\text{data}}$ needed for the direct application of $T$, and the exponential representation of $\mathbf{f}$ avoids the need to protect the algorithm against points going negative during the calculation.

Although it is tempting to try to solve equation (10) by fixed-point iteration, to do so is doomed to failure, as such a procedure will converge only if the modulus of the largest singular value of

$\partial g/\partial u$ is less than $\alpha$. Straightforward damping helps only if these non-zero singular values are all of the same magnitude, which is certainly not the case here. However, a Newton method can be applied to equations (10), the increment at each iteration being given by $J\delta u = -\alpha u - g$, where $J = \alpha I + \partial g/\partial u$ is the Jacobian of the system. Now

$$\frac{\partial g}{\partial u} = \Sigma V^T \frac{\partial^2 L(\mathbf{F}, \mathbf{D})}{\partial \mathbf{F}^2} V \Sigma U^T \operatorname{diag}\{f\} U = M K$$

hence

$$(\alpha I + M K)\delta u = -\alpha u - g. \tag{11}$$

At each iteration the size of the increment $\delta u$ must be restricted so that the second-order approximation used in (11) remains accurate. Previously, (Skilling & Bryan, 1984), $-\nabla\nabla S = \operatorname{diag}\{1/f\}$ was used as a metric in $f$-space, a better justification being given by Skilling (1989a). Now

$$\delta f^T \operatorname{diag}\{1/f\}\delta f = \delta u^T \frac{\partial f}{\partial u} \operatorname{diag}\{1/f\} \frac{\partial f}{\partial u} \delta u = \delta u^T K \delta u,$$

so $K$ is the equivalent metric in $u$-space. The step-length restriction is achieved by augmenting $J$ with a multiple of the unit matrix (Levenberg, 1944, Marquardt, 1963), giving

$$((\alpha + \mu)I + M K)\delta u = -\alpha u - g, \tag{12}$$

where $\mu$ is chosen so that $\delta u^T K \delta u \le O(\sum m)$.

As in Skilling & Bryan (1984), the values of $\alpha$ and $\mu$ are adjusted so that the iteration proceeds, depending on requirements, either towards the maximum probability (2) or towards the maximum of $Q$ for a specific value of $\alpha$, whilst imposing the step-length constraint. This search may be made efficiently if (12) is diagonalised, so that only $O(s)$ operations are required for each trial $\alpha$-$\mu$ pair, rather than $O(s^3)$ if (12) is solved directly. Substituting the expressions (7, 8) into (12) gives

$$((\alpha + \mu)I + \Lambda)Y^{-1}\delta u = -\alpha Y^{-1}u - Y^{-1}g, \tag{13}$$

giving $s$ independent equations for the components of $Y^{-1}\delta u$, and the step-length

$$\delta u^T K \delta u = (Y^{-1}\delta u)^T (Y^{-1}\delta u).$$

As mentioned above, $K$ may become near-singular (some $\xi$ approximately zero) and $\delta u$ cannot be found by applying $Y$ to the $Y^{-1}\delta u$ calculated by (13). Nevertheless, $Y^{-1}\delta u$ itself is correctly determined. To investigate how $\delta u$ can be found, equations (12) may be partitioned into the null (subscript 0) and non-null (subscript 1) space of $K$, by writing $P = (P_0 \quad P_1)$, and $p = P^T u$. Hence, multiplying (12) on the left by $P^T$,

$$(\alpha + \mu)\begin{pmatrix} \delta p_0 \\ \delta p_1 \end{pmatrix} + \begin{pmatrix} P_0^T \\ P_1^T \end{pmatrix} M ( P_0 \quad P_1 ) \begin{pmatrix} 0 & 0 \\ 0 & \Xi_1 \end{pmatrix} \begin{pmatrix} \delta p_0 \\ \delta p_1 \end{pmatrix} = -\alpha p - P^T g,$$

so that

$$(\alpha + \mu)\delta p_0 + P_0^T M P_1 \Xi_1 \delta p_1 = -\alpha p_0 - P_0^T g, \tag{14}$$

$$(\alpha + \mu)\delta p_1 + P_1^T M P_1 \Xi_1 \delta p_1 = -\alpha p_1 - P_1^T g. \tag{15}$$

and

$$\delta \mathbf{u}^T K \delta \mathbf{u} = \delta \mathbf{p}^T \Xi \delta \mathbf{p} = \delta \mathbf{p}_1^T \Xi_1 \delta \mathbf{p}_1. \tag{16}$$

The $P_1$ space equations (15) are of the same form as (12), and may be solved in the same way. The null-space increment, $\delta \mathbf{p}_0$, is now directly given by (14) in terms of $\delta \mathbf{p}_1$, and does not contribute to the step-length estimate (16). However, this still leaves the numerical problem of deciding when an eigenvalue of $K$ is sufficiently small for it to be treated as zero. Fortunately, (12) can be rewritten as

$$(\alpha + \mu)\delta \mathbf{u} = -\alpha \mathbf{u} - \mathbf{g} - MK\delta \mathbf{u},$$
$$= -\alpha \mathbf{u} - \mathbf{g} - MY^{-T}Y^{-1}\delta \mathbf{u}, \tag{17}$$

so finally, (12) is solved for $Y^{-1}\delta \mathbf{u}$, the result inserted in (17), and the true $\delta \mathbf{u}$ recovered. For the $P_1$ components, (17) is an identity; for the $P_0$, it is the same as the partitioned equation (14); the $Y^{-T}$ on the right eliminates any contribution from its null space. The two spaces merge smoothly. At all stages, the calculations are stable, requiring no division by small quantities.

When $T$ is very ill-conditioned, as in the exponential-decay problem, this algorithm performs much better than one which increments $\mathbf{f}$ directly (Skilling & Bryan, 1984). The latter algorithm computes the increment $\delta \mathbf{f}$ as a linear combination of $\text{diag}\{f\}\nabla S$, $\text{diag}\{f\}\nabla L$, and powers of $\text{diag}\{f\}\nabla\nabla L$ acting on these two initial vectors. Except for the single direction $\text{diag}\{f\}\nabla S$, they are all dominated by $\text{diag}\{f\}U_1$, the direction associated with the largest singular value of $T$. Thus, almost irrespective of the number of directions generated, this algorithm searches in only a two-dimensional subspace, whereas the current algorithm works in all $s$ significant directions.

## 2.3. CONVERGENCE TEST.

To test for convergence, the magnitude of the vector difference of $\alpha \partial S/\partial \mathbf{u}$ and $\partial L/\partial \mathbf{u}$ is compared with the sum of their magnitudes, using

$$t = 2\left|\alpha \frac{\partial S}{\partial \mathbf{u}} - \frac{\partial L}{\partial \mathbf{u}}\right|^2 \Big/ \left(\left|\alpha \frac{\partial S}{\partial \mathbf{u}}\right| + \left|\frac{\partial L}{\partial \mathbf{u}}\right|\right)^2,$$

again evaluated using the $K$ metric (in fact, $K^{-1}$, as the gradients are covariant). Since $\partial S/\partial \mathbf{u} = -K\mathbf{u}$ and $\partial L/\partial \mathbf{u} = K\mathbf{g}$, these quantities are easily evaluated. This test checks for equal lengths of the gradients, and not only for parallelism, so is stricter than one used previously (Skilling & Bryan, 1984), and, when zero, confirms that the maximum of $Q(\mathbf{f} \mid \alpha)$ is attained. All the results in §3 have $t < 10^{-4}$.

## 2.4. COVARIANCE MATRIX.

A further result of the analysis of Gull (1989) and Skilling (1989b, 1990) is the covariance matrix of $\mathbf{f}(\alpha)$, $-(\nabla\nabla Q)^{-1}$. Algebra similar to that of Skilling (1989b) is used to obtain an expression in terms of singular space quantities. Some manipulation gives

$$-\nabla\nabla Q = -\alpha\nabla\nabla S + \nabla\nabla L,$$
$$= \text{diag}\{1/f\}UY^{-T}(\alpha I + \Lambda)Y^{-1}U^T \text{diag}\{1/f\},$$

so

$$-(\nabla\nabla Q)^{-1} = \text{diag}\{f\}UY\,\text{diag}\Big\{\frac{1}{\alpha+\lambda}\Big\}Y^T U^T\,\text{diag}\{f\},$$

$$= \frac{1}{\alpha}\,\text{diag}\{f\} - \text{diag}\{f\}UY\,\text{diag}\Big\{\frac{\lambda}{\alpha(\alpha+\lambda)}\Big\}Y^T U^T\,\text{diag}\{f\}, \qquad (18)$$

and, using a similar triangularisation argument as in §2.1, the matrices in the second term can again be restricted to the singular space.

### 2.5. ALPHA.

The expression (2) for the posterior probability of $\alpha$ derived by Gull (1989), has its maximum where the derivative with respect to $\log\alpha$

$$\frac{d\log p_r(\alpha\mid D,m)}{d\log\alpha} = \frac{1}{2}N_g - \frac{\alpha}{2}\sum_i\Big(\frac{d\lambda_i/d\alpha}{\alpha+\lambda_i}\Big) + \alpha\hat{S} + \frac{d\log p_r(\alpha)}{d\log\alpha} \qquad (19)$$

is zero. Gull's '$-2\alpha S = N_g$' criterion is obtained if $d\lambda_i/d\alpha$ and the prior may be ignored in comparison with $N_g$ and $\alpha\hat{S}$. For notational convenience, the posterior of $\alpha$ without the prior is defined as $p_q(\alpha) = \prod_i(\alpha/(\alpha+\lambda_i))^{1/2}\exp\hat{Q}$. If $N_g$ is low, $p_q(\alpha)$ can have a broad peak, and the position of the optimum can be affected both by the prior on $\alpha$ and by fluctuations in the values of the $\lambda$'s. Since $\alpha$ is a scale factor, the Jeffreys prior $p_r(\alpha) = 1/\alpha$ is appropriate (Jaynes, 1968, Gull, 1989), which reduces $N_g$ by two, and thus causes the data to be fitted rather more closely. The examples in §3 demonstrate these effects.

The width of the peak in $p_q(\alpha)$ means that expectation values must be taken over $\alpha$ as well as over $f$, whose distribution at fixed $\alpha$ is the Gaussian (1). The strategy we have adopted is to perform integrals over $\alpha$ numerically, using equal steps in $\log\alpha$ over the range $\alpha_{\min}\le\alpha\le\alpha_{\max}$, where $\alpha_{\min}$ and $\alpha_{\max}$ are chosen such that $p_q(\alpha) < cp_q(\hat{\alpha})$, $\alpha\notin[\alpha_{\min},\alpha_{\max}]$. Typically $c = \exp(-8)$. The integrals we wish to evaluate are those required for normalisation of the probability, estimation of $\bar{\alpha}$, $\bar{f}$, etc. For example, using (1), (2) and the product rule,

$$\bar{f} = \int f\,p_r(f,\alpha\mid D,m)\,d^N f\prod f^{-1/2}\,d\alpha,$$

$$\propto\int\hat{f}(\alpha)p_q(\alpha)\,d\log\alpha,$$

so $\bar{f}$ is found by averaging the maximum entropy maps $\hat{f}$ found for each $\alpha$. The prior on $\alpha$ now comes in via the measure in the integral. $\bar{f}$ itself is not a maximum entropy map for any $\alpha$. In general, the derivative is dominated by the $\frac{1}{2}N_g$ term when $\alpha\ll\hat{\alpha}$, and by the $\alpha\hat{S}$ term when $\alpha\gg\hat{\alpha}$. Hence $p_q(\alpha)$ decreases with a power law as $\alpha\to 0$, and the truncation of the integral can be estimated as

$$\int_0^{\alpha_{\min}} p_r(\alpha\mid D,m)\,d\alpha \approx \int_{-\infty}^{\log\alpha_{\min}} p_q(\alpha_{\min})(\alpha/\alpha_{\min})^{N_g/2}\,d\log\alpha,$$

$$= \frac{2}{N_g}p_q(\alpha_{\min}),$$

which is negligible. The integral for $\alpha\to\infty$ is improper (Gull, 1989), but for any reasonable upper-limit of $\alpha$ the contribution to the integral is almost entirely from the range of $\alpha$ around $\hat{\alpha}$.

TABLE 1. Singular values of the example cosine transform.

| No. | Value. | No. | Value. | No. | Value. |
|-----|--------|-----|--------|-----|--------|
| 1 | 17.3682 | 7 | 15.8114 | 13 | 0.1993 |
| 2 | 16.4758 | 8 | 15.8110 | 14 | 0.0196 |
| 3 | 15.8114 | 9 | 15.7796 | 15 | 0.0016 |
| 4 | 15.8114 | 10 | 14.7091 | 16 | 0.0001 |
| 5 | 15.8114 | 11 | 7.6513 | >17 | Negligible |
| 6 | 15.8114 | 12 | 1.5621 | | |

## 3. Applications.

The SVD of the matrix $T$, of dimension $N \times N_{\text{data}}$ is performed once and for all at the beginning of the calculation, and thereafter only the components of $U$ and $V$ associated with non-zero singular values are required. The cutoff on singular values is not critical, provided sufficient vectors are included to span the row-space of $T$; including additional components which correspond to zero singular values simply results in the extra components of u being calculated as zero. Otherwise, the main computational cost is in the evaluation of the $M$ and $K$ matrices each iteration, the other scalar and singular space calculations being small in comparison. Indeed, if the relation between f and D is completely linear, $\nabla\nabla L$ is constant, and so is $M$.

### 3.1. OVERSAMPLED FOURIER PROBLEM.

Many scattering experiments produce data as a function of scattering angle, and are thus related to the object of interest by a Fourier transform. Special geometries, such as cylindrical or spherical symmetry, give rise to analogous problems, as in fibre diffraction and solution scattering. A continuous distribution of scattered intensity may be measured, as distinct from the crystallographic problem where discrete reflections are recorded (and $V = \Sigma = I$), for which this algorithm has no particular advantages. The SVD of the Fourier transform relating an image of restricted range to the data may be analysed in terms of prolate spheroidal wavefunctions (PSWs) (Slepian & Pollak, 1961, Slepian, 1964), but it is usually just as convenient to perform the SVD by linear algebra techniques (Golub & van Loan, 1983) as to evaluate the PSWs themselves. The results obtained in numerical tests agree closely, as do the singular value spectra.

As a simple illustration, a reconstruction from cosine transform data is presented, with $T_{jk} = \cos \pi x_j \omega_k$, $x_j = j\delta$, $j = 0, \ldots, N-1$, $\omega_k = k\Delta$, $k = 0, \ldots, N_{\text{data}} - 1$. The reconstruction is thus made on the interval $0 \leq x \leq X = (N-1)\delta$. The real-space grid spacing, $\delta$, must be selected to give adequate sampling for the highest frequency signal in the data, $N_{\text{data}}\Delta\delta \ll 1$. The Shannon sampling limit is $\Delta = 1/X$. For the example, we take $X = 1$, oversample five times, so $\Delta = 1/5$, and provide data for $N_{\text{data}} = 50$ samples. Hence we require $\delta \ll 1/10$, so we take $N = 100$, $\delta = 1/100$. The SVD analysis gives the singular values in Table 1; the first 10 are of the same magnitude, and then there is a rapid drop off to zero, as expected.

The results are shown in Fig. 1. Data are constructed from a simulated object consisting of two $\delta$-functions, with noise added, standard deviation 0.5% of the maximum data value. The fluctuation of the eigenvalues (particularly those with values near $\alpha$) cause the true derivative of $\log p_q(\alpha)$ to depart considerably from the $\frac{1}{2}N_g + \alpha S$ approximation, shifting the position of the maximum, and creating a subsidiary maximum removed by an order of magnitude in $\alpha$, whose probability is somewhat lower. The probability of the $2L = N_{\text{data}}$ solution is very low, and shows less sharp structure. $\hat{\alpha}$ is close to $\bar{\alpha}$, although $\hat{f}$ and $\bar{f}$ are not identical, $\bar{f}$ showing sharper peaks but not much difference in the wings. The maximum pointwise probability when

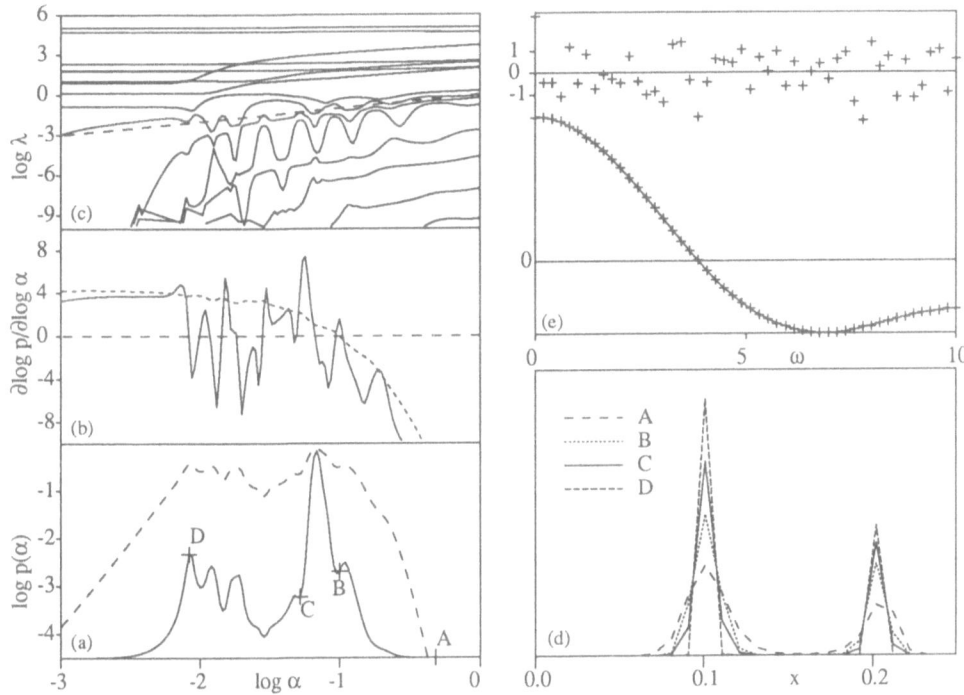

Figure 1. Plots of quantities calculated for the cosine problem. (a). Posterior probability of $\alpha$ (continuous) and its log (base 10) (dashed), as functions of $\log_{10}\alpha$, evaluated pointwise with a flat prior (the integrated area is thus correct as if a Jeffreys prior were used). The crosses are at the positions given by A: $2L=N_{data}$, B: $-2\alpha S=N_g$, C: $\bar{\alpha}$, D: maximum of probability evaluated pointwise with a Jeffreys prior. (b). The numerical derivative of $\log p_q(\alpha)$ (solid), $0.5 N_g + \alpha S$ (dashed). (c). Eigenvalues $\lambda$. Dashed line is $\alpha = \lambda$. (d). Part of restorations $\mathbf{f}$, labelling same as (a), except that C is $\bar{\mathbf{f}}$. For $0.25 \leq x < 1.00$, $\mathbf{f}$ is flat. (e). Data (crosses) and transform of restoration (continuous), with residuals (in standard deviations) at top.

the Jeffreys prior on $\alpha$ is used leads to an unexpected result. The solution using the approximate derivative moves to a nearby point, as would be expected. However, the true maximum is an order of magnitude in $\alpha$ away and gives a considerably sharper solution. The integrals of $\mathbf{f}$ over the peaks are essentially the same for all the maps with significant probabilities. For $\mathbf{f}(\hat{\alpha})$ they are 10.0987 and 4.9134 with the integral over the rest of $\mathbf{f}$ being 0.0396. Equation (18) yields the covariance matrix of these quantities (analogous to the method described by Sibisi, 1990),

$$\text{cov}(\text{peak}_1, \text{peak}_2, \text{background}) = \begin{pmatrix} .0638 & -.0346 & -.0286 \\ -.0346 & .0653 & -.0313 \\ -.0286 & -.0313 & .0606 \end{pmatrix},$$

and not surprisingly they are strongly negatively correlated, as the total integrated intensity, $\sum f$, is extremely well determined.

## 3.2. DYNAMIC LIGHT SCATTERING.

This example illustrates a likelihood expression $L(\mathbf{F}, \mathbf{D})$ which is non-quadratic and contains an unknown parameter which is also estimated. Livesey et. al. (1986) describe an earlier application of maximum entropy to this problem. The counts obtained in a DLS experiment are proportional to $F(t)^2 + A$, where $F(t)$ is a multiexponential decay curve, related to the spectrum of decay rates $f(\tau)$ by $F(t) = \int f(\tau) \exp -t/\tau \, d\tau$, and $A$ an unknown baseline count. The counts may be very large ($10^6$ or more), so the Poisson distribution of noise is approximated by Gaussian, giving the discretised form

$$L = \frac{1}{2} \sum_k \frac{1}{\sigma_k^2} (F_k^2 + A - D_k)^2,$$

with $F_k = F(t_k)$, $\sigma_k = \sqrt{D_k}$. Following other authors (Provencher, 1979, Livesey et. al., 1986), $\mathbf{f}$ is represented on a grid of points distributed uniformly in $\log \tau$, with a uniform prior on this grid, whose value is calculated as the best-fit constant. The baseline level may be estimated from measurements at large $t$, when all components have decayed, but like other 'nuisance parameters', it may be integrated out of the likelihood (Jaynes, 1987) to give the 'quasi-likelihood' defined by

$$\exp - \tilde{L}(\mathbf{F}, \mathbf{D}) = \int \exp -L(\mathbf{F}, A, \mathbf{D}) \, dA.$$

$L$ in the Gaussian noise approximation is quadratic in $A$, so the integral is easily performed. $\tilde{L}$ and $\partial \tilde{L}/\partial \mathbf{F}$ are the same functional form as $L$ and $\partial L/\partial \mathbf{F}$, except that $A$ is replaced by $\tilde{A}$, its 'best estimate' at the current $\mathbf{F}$,

$$\tilde{A} = -\sum_k \frac{1}{\sigma_k^2} (F_k^2 - D_k) \bigg/ \sum_k \frac{1}{\sigma_k^2}.$$

$\partial^2 L/\partial F_j \partial F_k$ undergoes a rank-1 modification to give

$$\frac{\partial^2 \tilde{L}}{\partial F_j \partial F_k} = \text{diag}\{\mathbf{W}\} - \mathbf{w}\mathbf{w}^T / \sum_k \sigma_k^{-2}$$

where

$$W_k = \frac{2}{\sigma_k^2}(3F_k^2 + \tilde{A} - D_k) \quad \text{and} \quad w_k = \frac{2}{\sigma_k^2} F_k.$$

Computationally, this modification is better performed in the singular space, rather than data space. $\partial^2 \tilde{L}/\partial F_j \partial F_k$ has at most one extra negative eigenvalue; otherwise the calculation is unchanged.

The posterior distribution of $A$ may also be calculated, by a further application of Bayes' theorem. Assuming a flat prior for $A$, the result is

$$\text{pr}(A \mid \mathbf{D}) = \int \text{pr}(A, \mathbf{f}, \alpha \mid \mathbf{D}) \, d\alpha \, d^N f \prod f^{-1/2},$$

$$= \int \text{pr}(A \mid \mathbf{f}, \alpha, \mathbf{D}) \text{pr}(\mathbf{f} \mid \alpha, \mathbf{D}) \text{pr}(\alpha \mid \mathbf{D}) \, d\alpha \, d^N f \prod f^{-1/2}.$$

TABLE 2. Singular values of the exponential decay transform.

| No. | Value. | No. | Value. | No. | Value. |
|---|---|---|---|---|---|
| 1 | 65.8017 | 7 | $3.875 \times 10^{-2}$ | 13 | $1.490 \times 10^{-5}$ |
| 2 | 17.3744 | 8 | $9.231 \times 10^{-3}$ | 14 | $2.735 \times 10^{-6}$ |
| 3 | 5.9095 | 9 | $2.994 \times 10^{-3}$ | 15 | $1.375 \times 10^{-6}$ |
| 4 | 1.9106 | 10 | $1.815 \times 10^{-3}$ | 16 | $4.009 \times 10^{-7}$ |
| 5 | 0.5665 | 11 | $4.050 \times 10^{-4}$ | 17 | $6.276 \times 10^{-8}$ |
| 6 | 0.1480 | 12 | $7.971 \times 10^{-5}$ | 18 | $9.360 \times 10^{-9}$ |

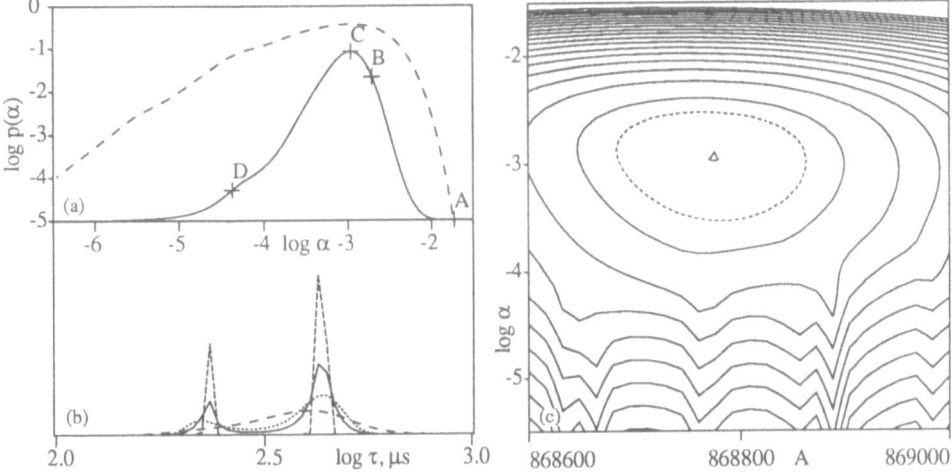

Figure 2. Plots of quantities calculated for the exponential decay problem. (a). Posterior probability of $\alpha$ (continuous) and its log (dashed). Other details as for fig. 1a. (b). Part of restorations $\mathbf{f}$, labelling as before. $\mathbf{f}$ is flat over the rest of the reconstructed spectrum. (c) Contour map of $\log_e p_r(A, \alpha)$. Contour interval of continuous lines, 1, dashed, 0.5.

Now $p_r(A \mid \mathbf{f}, \alpha, \mathbf{D}) \propto \exp -\frac{1}{2} \sum_k \sigma_k^{-2} (A - \tilde{A})^2$, so using the Gaussian approximation for $p_r(\mathbf{f} \mid \alpha, \mathbf{D})$ (1), the covariance of $\mathbf{f}$ (18), and linearising $\tilde{A}(\mathbf{f})$, the variance of $A$ at fixed $\alpha$ is

$$\left( \sum_k \sigma_k^{-2} + \mathbf{w}^T V \Sigma Y^{-T} \text{diag} \left\{ \frac{1}{\alpha + \lambda} \right\} Y^{-1} \Sigma V^T \mathbf{w} \right) \bigg/ \left( \sum_k \sigma_k^{-2} \right)^2. \tag{20}$$

An example is shown here, using data (kindly supplied by J. Langowski, EMBL, Grenoble) from a mixture of two sizes of polystyrene beads in solution. The data are collected at correlation times which are switched in interval, so that $\log t$ space (3–800 $\mu$s) is roughly evenly covered, plus six counts at large $t$ (13 ms), to help establish the baseline. The correlator scales the counts by a factor of 16, so the signal to noise ratio (SNR) is actually four times that expected from Poisson statistics. The spectrum is reconstructed on $\log_{10} \tau \in [1.3, 3.3]$. The singular values (Table 2) drop off rapidly, compared with the level spectrum for the Fourier problem.

$N_g$ is found to be approximately 4, and again a wide range of probable $\alpha$ is found (fig. 2a). The $2L = N_{\text{data}}$ solution again has low probability, and fails to resolve the two peaks. The expectation baseline count, $<A> = 868763$, with variance (20) 7700. $1/\sum_k \sigma_k^{-2} = 400$, so the variance is dominated by the uncertainty in $\mathbf{f}$, not by the distribution of $A$ about $\tilde{A}$.

As an alternative to integrating $A$ out of the likelihood, the joint posterior $p_r(A, \alpha \mid D, m)$ may be obtained (fig. 2c) showing good agreement with the previous results, although, away from the maximum, the distribution, even at fixed $\alpha$, is distinctly non-Gaussian. This method is more general than forming the quasi-likelihood, as it does not rely on analytic integration, which may be harder for other functions of parameters (*e.g.*, if the likelihood for Poisson noise were used), nor on the assumption that the posterior of $A$ is also Gaussian.

## References.

Bryan, R. K. (1980). Maximum Entropy Image Processing. PhD Thesis, University of Cambridge.

Golub, G. H. & van Loan, C. F. (1983). Matrix Computations. Johns Hopkins, Baltimore.

Gull, S. F. (1989). Developments in maximum entropy data analysis. *In:* J. Skilling (ed.), Maximum Entropy and Bayesian Methods. Kluwer Academic, Dordrecht, pp. 53–71.

Gull, S. F. & Daniell, G. J. (1978). Image reconstruction from incomplete and noisy data. *Nature,* 272, 686–690.

Jaynes, E. T. (1968). Prior probabilities. *IEEE Trans.,* SCC-4, 227–241.

Jaynes, E. T. (1987). Bayesian spectrum and chirp analysis. *In:* C. Ray Smith & Gary J. Erickson (ed.), Maximum Entropy and Bayesian Spectral Analysis and Estimation Problems. Kluwer Academic, Dordrecht, pp. 1–37.

Levenberg, K. (1944). A method for the solution of certain non-linear problems in least squares. *Quart. Appl. Math.,* 2, 164–168.

Livesey, A. K., Licinio, P. & Delaye, M. (1986). Maximum entropy analysis of quasielastic light scattering from colloidal dispersions. *J. Chem. Phys.,* 84, 5102–5107.

Marquardt, D. W. (1963). An algorithm for least squares estimation of non-linear parameters. *SIAM J. Appl. Math.,* 11, 431–441.

Provencher, S. W. (1979). Inverse problems in polymer characterization: Direct analysis of polydispersity with photon correlation spectroscopy. *Makromol. Chem.,* 180, 201–209.

Sibisi, S. (1989). Regularization and inverse problems. *In:* J. Skilling (ed.), Maximum Entropy and Bayesian Methods. Kluwer Academic, Dordrecht, pp. 389–396.

Sibisi, S. (1990). Quantified Maxent: an NMR application. These Proceedings.

Skilling, J. (1989a). Classic maximum entropy. *In:* J. Skilling (ed.), Maximum Entropy and Bayesian Methods. Kluwer Academic, Dordrecht, pp. 45–52.

Skilling, J. (1989b). The eigenvalues of mega-dimensional matrices. *In:* J. Skilling (ed.), Maximum Entropy and Bayesian Methods. Kluwer Academic, Dordrecht, pp. 455–466.

Skilling, J. (1990). Quantified maximum entropy. These Proceedings.

Skilling, J. & Bryan, R. K. (1984). Maximum entropy image reconstruction: general algorithm. *Mon. Not. R. astr. Soc.,* 211, 111–124.

Slepian, D. (1964). Prolate spheroidal wave functions, Fourier analysis and uncertainty, IV: Extensions to many dimensions; Generalized prolate spheroidal functions. *Bell Syst. Tech. J.,* 43, 3009–3058.

Slepian, D., Pollak, H. O. (1961). Prolate spheroidal wave functions, Fourier analysis and uncertainty, I. *Bell Syst. Tech. J.,* 40, 43–64.

# CONSTRUCTING PRIORS IN MAXIMUM ENTROPY METHODS[*]

N. RIVIER*, R. ENGLMAN[+] and R. D. LEVINE[#]
*Argonne National Laboratory
MSD-223
Argonne, IL 60439, USA
+Soreq Nuclear Research Center
Yavne 70600, Israel
#Fritz Haber Research Center for Molecular Dynamics
The Hebrew University
Jerusalem 91904, Israel

ABSTRACT. We show how to construct the best prior for a Maximum Entropy procedure when two or more priors are conceivable or are proposed. The prior is a weighed sum of the conceivable priors with weights that depend exponentially on the overlap of the prior with the exponential part of the maximum entropy probability. With additional information, one can iteratively improve the prior and sharpen the choice between alternative priors. Our construction can be used to predict in some physical cases the probability distribution functions, and to make quantitative decisions in the presence of conflicting expert opinions.

## 1. Introduction

This paper is concerned with the construction of prior probability, the measure in the space of events, by Maximum Entropy methods. Maximum Entropy methods can be used to validate hypotheses and predict probabilities on the basis of some given, but limited information. This information takes the form of constraints which may be physical or geometrical. The probability distributions predicted by Maximum Entropy are maximally non-committal, and it is this variational requirement which grants unicity to an otherwise under-constrained problem. (See e.g., Levine and Tribus (1978), Rosenkrantz (1983)).

The situation is different in practice because the MaxEnt probability distribution depends on a distribution called "the prior" which is neither specified by the constraints, nor given by the theory (Jaynes 1968). For example, in predicting the distribution of fragment sizes in explosive disintegration (Grady and Kipp (1985), Englman et al. (1988 a,b), Holian and Grady (1988)), are the events uniform a priori on a linear scale of radius, area or volumes, or any other for that matter? The use of sieves to sort out the fragments might suggest area, but why should the physical distribution depend on the methods of measurement? Here the prior is a measure of the space of geometrical events, an essentially ill-defined problem illustrated by the Bertrand paradox, and solved by expressing the ill-definition as an invariance group which the solution must passes if it is to be free of prejudice (Jaynes (1973), Rosenkrantz (1983)). Only statistical mechanics poses no problem because events are uniformally distributed a priori in phase space, in cells of volume $\Delta p \Delta q = h$, Planck's constant.

---

[*] Work supported by the U.S. Department of Energy, BES-Materials Sciences, under contract W-31-109-ENG-38.

P. F. Fougère (ed.), Maximum Entropy and Bayesian Methods, 233–242.
© 1990 Kluwer Academic Publishers.

In this paper, we shall use MaxEnt methodology to determine the prior in a specific, but frequently encountered class of problems: Given several possible priors conflicting expert opinions, which one, or what prior should be used to construct the MaxEnt distribution? For this class of priors, there is no obvious invariance group, but a unique, minimally prejudiced solution, is obtainable directly by MaxEnt methods.

We are interested in predicting a probability distribution, rather than in hypothesis testing. Our formalism can be extended to the latter, where the hypotheses to be tested are the experts' opinions, but in that case, the constructed "prior", which depends on the experimental information encoded in the constraints, should not be called a prior.

An extended version of this paper, available as a preprint for some time, will hopefully appear in print soon (Englman et al.(1989)).

## 2. Formalism

Let us label the events on a discrete scale, by using abstract bins, labelled by $n=1,2,...N$. The information entropy, or uncertainty corresponding to the probability set $\{p_n\}$, where $p_n$ is the probability of observing an event collected by bin $n$, is

$$S = - \sum_n p_n \log p_n , \qquad (1)$$

with $\sum_n p_n = 1$ as normalization. Equation (1) is the unique functional form satisfying Shannon's (1949) axioms or desiderata of consistency. (cf. also Khinchin 1957).

The physical scale (weight, diameter, area...) is labelled by $i=1,2,...$, and for a sufficiently large number of bins, the distribution of bins on the physical scale is given by the distribution (histogram) $\Pi(i)$, called here the prior. The entropy (1) is rewritten by changing labels from abstract ordering to physical scale,

$$S = - \sum_i \Pi(i) \, p_i \log p_i = - \sum_i P(i) \log [P(i)/\Pi(i)] \qquad (2)$$

with $P(i) = p_i \Pi(i)$, and $\sum_i \Pi(i) = 1$ following from normalization.

Given a set of physical constraints,

$$<\phi_l> = \sum_i P(i) \, \phi_l(i) , \qquad\qquad \phi_0 = 1 = <\phi_0> \qquad (3)$$

and the prior $P(i)$, MaxEnt infers the probability distribution

$$P(i) = \Pi(i) \exp [- \sum_l \lambda_l \, \phi_l(i)] . \qquad (4)$$

$\lambda_l$ are Lagrange multipliers, determined by eq. (3).

Suppose now that several $(k=1,2,...N_\pi)$ specified bin arrangements

$$\Pi(i,k) , \qquad\qquad \sum_i \Pi(i,k) = 1 \qquad (5)$$

are conceivable, or suggested by "experts". Weighted superpositions

$$\Pi(i) = \sum_k p_k \, \Pi(i,k) , \qquad\qquad \sum_k p_k = 1 \qquad (6)$$

are also bins arrangements giving rise to the MaxEnt distribution (4), but with $\lambda_l$

depending on the weights $\{p_k\}$. Note that compromise ($p_k \neq 0,1$) between different expert advices is permissible. It also turns out to be computationally useful (Hopfield and Tank (1985)), and gives a higher entropy (see § 4), that is a less biased distribution than exclusivism. The various $\Pi(i,k)$ are given a priori. The problem is to determine $\{p_k\}$.

## 3. MaxEnt Weighting of Expert Opinion

> *"The law of averages, if I have got this right, means that if six monkeys were thrown in the air for long enough they would land on their tails about as often as they would land on their---".* (Stoppard (1967))

The MaxEnt solution consists in maximizing an information entropy (2) plus the uncertainty in the expert selection $-\sum_k p_k \log p_k$, subject to the physical (3) and normalization (6) constraints, that is by extremizing the "lagrangian" $L$,

$$L[\{P(i)\},\{p_k\}] = - \sum_i P(i) \log [P(i)/\Pi(i)] - \sum_k p_k \log p_k$$
$$- \sum_l \lambda_l \{\sum_i P(i) \phi_l(i) - <\phi_l>\} - (\mu-1) \{\sum_k p_k - 1\} , \qquad (7)$$

with respect to the independent variables $\{P(i)\}$ and $\{p_k\}$. When the constraints are satisfied, $\{\ \}=0$ and $L$ is the total information entropy. It is a convex function of its variables. $\Pi(i)$ (see (6)) depends explicitly on the weights $\{p_k\}$, which also affect the Lagrange multipliers $\lambda_l$ and $\mu$, implicitly through the constraint equations.

[3.1 *Proof of Eq. (7)*. To prove that (7) is indeed the relevant measure of uncertainty, imagine two "thought experiments" involving a large number n of independent trials (we are not turning into "frequentists"). In the first experiment, the outcome of each trial is the observation of the physical variable i (i=1,2,...M) with probability $P(i)$ and, <u>independently</u>, the random selection of expert k with probability $p_k$. The outcomes of the n independent trials are $v_i$ observations of i and $\mu_k$ selections of expert k, with

$$\sum_i v_i = \sum_k \mu_k = n . \qquad (8)$$

The probability of these events is $P_n = P(1)^{v_1} P(2)^{v_2}...P(M)^{v_M} p_1^{\mu_1} p_2^{\mu_2}...p_{N\pi}^{\mu_{N\pi}}$, as if the experts had no influence on the physical observations. The expectation values of $v_i$ and $\mu_k$ are $E(v_i) = n P(i)$ and $E(\mu_k) = n p_k$, respectively, and the uncertainty, or amount of information (Renyi 1970) of the first thought experiment is given by

$$- E [(1/n) \log P_n] = - \sum_i P(i) \log P(i) - \sum_k p_k \log p_k \qquad (9)$$

The second thought experiment involves again n independent trials, whose outcome is now the observation of the physical variable i only, with probability $\Pi(i)$ given by (6). The compromise matches perfectly the outcome of the trials as if the (fictitious) data had been "cooked", so the outcome is still $E(v_i) = n P(i)$. The probability of these n events is $P_n = \Pi(1)^{v_1} \Pi(2)^{v_2}...\Pi(M)^{v_M}$, and the uncertainty is then,

$$- E [(1/n) \log P_n] = - (1/n) \sum_i E(v_i) \log \Pi(i) = - \sum_i P(i) \log \Pi(i) . \qquad (10)$$

The difference in uncertainty in the two thought experiments gives the total information entropy ((10)-(9)) equal to the first two terms in (7)), which is to be maximized with respect to {P(i)} and {$p_k$}, in order to minimize bias associated with expert opinion.]

Note that the first term in (7) is always non-positive, and reaches its maximum (zero) when $P(i) = \Pi(i)$. The second is always non-negative and reaches its maximum (log $N_\pi$) when all experts are given equal weight. But the weights occur explicitly in $\Pi(i)$ in the first term, and must also be adjusted in order for the probability distribution $P(i)$ to satisfy the constraints. The highest possible lagrangian (7) is therefore $L_M = \log N_\pi$. The MaxEnt distributions {P(i)} and {$p_k$} are given by solving

$$\delta L/\delta P(i) = 0, \qquad\qquad\qquad \delta L/\delta p_k = 0 \qquad\qquad (11)$$

irrespective of the order, together with the constraint eq. (3) ($\delta L/\delta \lambda_l = 0$) and (6) ($\delta L/\delta \mu = 0$). The distribution $P(i)$ is given by (4), as indicated already in § 2, while the weights are

$$p_k = \exp X_k /[\Sigma_k \exp X_k], \qquad\qquad (12)$$

depending exponentially on

$$X_k = \Sigma_i \Pi(i,k) \exp [-\Sigma_l \lambda_l \phi_l(i)] \geq 0, \qquad\qquad (13)$$

which is the overlap between the proposed prior $\Pi(i,k)$ and what would be the MaxEnt distribution in a prior-free situation (with the proviso that the $\lambda_l$ depend implicitly, though the constraints, on {$p_k$}). The larger the overlap, the greater (exponentially) the weight of this particular prior. Normalization yields also

$$\Sigma_k p_k X_k = 1, \qquad\qquad (14)$$

so that {$X_k$-1} is the barycenter of $N_\pi$ masses $p_k$ located on the vertices {$X_k$} of a ($N_\pi$-1)-simplex. The masses $p_k$ depend on $X_k$ through (12).

The system of equations (4), (6), (12), (13), (3) is highly nonlinear, but not too difficult to solve in practice. Because a small change of $X_k$-$X_0$ implies a large change of $p_k/p_0$ in (12), one can vary $X_k[\lambda_l(\{p_k\})]$ in (13) as a function of $p_k$ until it matches (12). Alternatively, perturbation theory (quadratic approximation) about the highest maximum entropy $L_M$ can be used.

## 4.  Perturbation Theory

For $N_\pi$ conceivable priors, there are values of the constraints, $<\phi_l>_M$ for which the lagrangian is the highest possible $L_M = \log N_\pi$. These values are obtained automatically (without having to enforce the constraints by Lagrange multiplier $\lambda_l \neq 0$) by setting the weights in $\Pi(i)$ all equal, $p_k = 1/N_\pi$,

$$<\phi_l>_M \equiv \Sigma_i \Pi(i) \phi_l(i) = (1/N_\pi) \Sigma_{i,k} \Pi(i,k) \phi_l(i), \qquad\qquad (15)$$

(democratic mean). For these constraints values, $P(i) = \Pi(i)$ in (3), and $L = L_M = \log N_\pi$.

Convexity of the entropy enables us to set up a perturbation theory. Take, for simplicity, only one constraint $<\phi_1> = <i>$, besides normalization. The small parameter

$$z = [\exp(-\lambda_1)-1]\{1-[\exp(-\lambda_1)-1](1+4<i>_M)/2\} = \delta<i>/\sigma^2 + O(z^2) , \qquad (16)$$

is proportional to the difference $\delta<i> = <i>_{obs}-<i>_M$ between observed, $<i>_{obs}$, and demo-cratic mean $<i>_M$ (15). Then,

$$L \approx \log N_\pi - (1/2) [\delta<i>]^2/\sigma^2 + O(\delta<i>^3) , \qquad (17)$$

where

$$\sigma^2 = <<i^2>_k>+<<i>_k^2>-2<i>_M^2 = <<(i-<i>_M)^2>_k>+<(<i>_k-<i>_M)^2> \geq 0 , \quad (18)$$

with $<f>_k \equiv \Sigma_i \Pi(i,k) f(i)$ and $<<f>_k> \equiv (1/N_\pi) \Sigma_k <f>_k$, so that $<i>_M = <<i>_k>$.

Equation (17) has several uses. In practice, it yields the MaxEnt probability distribution and the weights rapidly, in closed form, in the non-trivial cases where there is competition between expert advices. Theoretically, the highest maximum entropy confirms the relevance of our lagrangian (7). In particular, all the alternatives discussed in § 5, [one of which, (25), requiring that each expert advice is a physical solution on its own (even if a compromise is eventually decided)], have lower maximum entropy, and have therefore introduced additional assumptions. Finally, it has amusing consequences: if the experts are stupid, and all predict $<i>_k = 0$ but the observed value $<i>_{obs} \neq 0$, then

$$L \approx \log N_\pi - (1/2) <i>_{obs}^2/<<i^2>_k> , \qquad (19)$$

so that among stupid experts, the noisiest are the least biased ($<<i^2>_k>$ maximum).

The MaxEnt solution (4) and (12) can be given in closed form, to second order in z,

$$p_k = (1/N_\pi)\{1 + z[<i>_k-<i>_M] + (z^2/2)[<i^2>_k-<<i^2>_k>+<i>_k^2-<<i>_k^2>] + O(z^3)\}$$
$$\lambda_1 = - z - 2 z^2 <i>_M + O(z^3) \qquad (20)$$
$$\exp(-\lambda_0) = 1 - z<i>_M - (z^2/2)[2<<i^2>_k>+<<i>_k^2>] + O(z^3) .$$

By contrast, the constraint equations are transcendental in the Lagrange multipliers, and a non-perturbative solution can only be obtained numerically.


## 5. Alternative Constructions

If each suggested prior $\Pi(i,k)$ is physically realizable, and gives rise to a distribution $P(i|k)$, one could argue for maximizing simply the entropy,

$$S = - \Sigma_{i,k} P(i,k) \log [P(i,k)/\Pi(i,k)] \qquad (21)$$

instead of the lagrangian $L$ given in (7), with $P(i) = \Sigma_k P(i,k) = \Sigma_k p_k P(i|k)$. $P(i|k)$ is indeed a conditional probability, with

$$\Sigma_i P(i|k) = 1 , \qquad (22)$$

and this normalization condition must be imposed on each P(i|k) separately with $N_\pi$ Lagrange multipliers $\lambda_0^{(k)}$. (Levine and Kosloff, 1979, Levine 1986). Then the weights

$$p_k = X'_k/(\Sigma_k X'_k) = \exp \lambda_0^{(k)} \tag{23}$$

depend, now linearly, on an overlap $X'_k$ similar to (13), but with $l = 0$ excluded from the sum in the exponent. This leads to a smaller maximal entropy (there are now $N_\pi - 1$ additional constraints) and assigns weights which are too democratic and insufficiently discriminating.

One could go further and impose that the probability distribution satisfies all constraints regardless of the assigned weights, so that the constraints

$$<\phi_l>^{(k)} \equiv \Sigma_i P(i|k) \phi_l(i) , \tag{24}$$

should be satisfied by each prior $\Pi(i,k)$ separately (Gull 1987). The number of constraints and that of Lagrange multipliers $\lambda_l^{(k)}$ are multiplied by $N_\pi$. The weights are then

$$p_k = \exp S_k / \Sigma_k \exp S_k , \qquad\qquad S_k = \Sigma_l \lambda_l^{(k)} <\phi_l>^{(k)} . \tag{25}$$

Here the weights are discriminating, but they do not have a simple and physically satisfying expression in terms of overlap as in (12)-(13). Furthermore, because constraints have been added, the maximal entropy is smaller than $L_M$

$$S \leq L_M , \tag{26}$$

with equality if and only if all $\Pi(i,k)$ give the same mean $<\phi_l>_k = <\phi_l>$ to every $\phi_l$.

## 6.  Iterations

The compromise prior (6) can be used as a new bin-distribution to make further measurements, which can be used in turn to improve the prior and refine and extend the range of the exponential part of the probability distribution. Label the iteration stage by the integer $r=1,2,....$ The physical information (the constraints) may or may not remain the same at each stage. There are two ways of using the distribution obtained at stage (r-1) as prior for stage r of the iteration:

i) Use the (r-1)-weights $\{p_k^{(r-1)}\}$ as prior weights for stage r. That is, replace the second term in $L$ (7) by $-\Sigma_k p_k^{(r)} \log [p_k^{(r)}/p_k^{(r-1)}]$. Now, the prior distribution $\Pi(i)^{(r)}$ (6) involves $p_k^{(r)}$, which must be calculated at stage r as in § 3, rather than the prior weight $p_k^{(r-1)}$. Let $p_k^{(0)} = 1/N_\pi$. Then, simply,

$$p_k^{(r)} = p_k^{(r-1)} \exp X_k^{(r)} / \Sigma_k p_k^{(r-1)} \exp X_k^{(r)} , \tag{27}$$

in obvious generalization of the notation. If the constraints remain the same at every stage r

$$p_k^{(r)} \propto [p_k^{(1)}]^r , \tag{28}$$

so that the iteration converges to give weight 1 to the prior $\Pi(i,k)$ with the largest overlap, and zero to the others. What happens when the constraints change at every stage (for example when British universities are awarded less money after each report of a commission of experts) will of course depend on the particular constraints and on the type of models $\Pi(i,k)$.

ii) One could use also the whole MaxEnt probability distribution obtained at stage (r-1) as prior for stage (r), thus

$$\Pi(i)^{(r)} = \Sigma_k \, p_k^{(r)} \, \{\Pi(i,k) \, \exp \, [-\Sigma_l \, \lambda_l^{(r-1)} \, \phi_l(i) - \lambda_0^{(k)}]\} \,, \qquad (29)$$

where $\lambda_0^{(k)}$ normalizes { } to unity, $\Sigma_i\{ \ \}=1$ as in (5). This simply updates $\Pi(i,k)$ to { }. If the constraints remain the same at every stage, the weights are only confirmed

$$p_k^{(r)} = p_k^{(r-1)} \,, \qquad \lambda_k^{(r)} = \lambda_k^{(r-1)} \,, \qquad X_k^{(r)} = 1 \,. \qquad (30)$$

On the other hand, new empirical data $<\phi_l>^{(r)}$ at every stage will change $p_k^{(r)}$ as befits acquisition of new information.

It is interesting that the two modes of iteration yields different weights. If, as in most physics situations, the prediction problem can be separated into a theory-based prior and an empirical component (the constraint and the posterior probability distribution), mode i) of iteration is appropriate. If, on the other hand, prior and posterior probability distributions are of the same nature, as, for example, measures of rational belief, mode ii) should be used.


## 7. Compromise and Iteration

Note that all the solutions ($p_k\neq0,1$) are compromises between expert advices, except in the final stage of iteration in mode i). The compromise entertains several solutions, which may be mutually exclusive in reality (as in the traveling salesman problem) or even physically inaccessible. Iteration (i) leads to rapid convergence to the best solution (28).

An example where both individual priors and their selected superposition are actual states of the system occurs in the process of computation of an optimal strategy for the traveling salesman by using an analogue electrical network (Hopfield and Tank 1985). Any valid tour of the salesman corresponds to one particular prior ($p_k=\delta_{kj}$) but in the physical process of computation the electrical network entertains a superposition of many conceivable tours. "It is as though the logical operations of a calculation could be given continuous values between "True" and "False", and evolve toward certainty only near the end of the calculation" (Hopfield and Tank 1985, p.149, see also their Fig. 2).


## 8. Example: Fragmentation of a (two-dimensional) Shell

The problem, reviewed by Grady and Kipp (1985) is to obtain the distribution of particle sizes in explosive fragmentation. There are two issues to be addressed, one dynamical or physical, the other geometrical. In the MaxEnt formalism, the dynamical issue consists in identifying the constraints, which are here the energy of a fragment and, the total area of the shell. The fragment energy consists of several terms with various dependences on the linear dimension i of the fragment, reminiscent of the droplet model of nuclear matter (Englman et al. 1988 a,b).

The second issue is geometrical. It focuses on the essence of a prior, that of providing a geometrical measure of the space of events. What measure should be ascribed to the random position of, here, a given area?

8.1 *Geometrical Measure.* A priori homogeneity, isotropy and scale invariance (Jaynes 1973) of the fragmented mosaic impose a measure which is an arbitrary power of the linear dimension i of the (two-dimensional) fragments (Rivier 1986). Consequently, we shall consider the priors

$$\Pi(i,k) \, di = [(D_k)^k/\Gamma(k)] \, i^{k-1} \, \exp(-D_k i) \, di \,, \qquad k = 0,1,2 \,. \tag{31}$$

i is now a continuous variable, and the exponential is included here solely as a simple normalization device. It is not essential. Besides the natural linear (k=1) and areal (k=2) measures, for which $D_k = k/<i>_k$ and $<i^2>_k = [(k+1)/k] <i>_k^2$, we have included Jeffreys' (1961) uninformative prior

$$\Pi(i,0) \, di \sim di \, / \, i \,. \tag{32}$$

Its moments are all negligible because the moment generating function $\int di \, \Pi(i,0) \exp(-xi)$ = 1 is independent of x (as well as $D_0$), as befits a non-informative prior. The overlap $X_0$ = $\exp(-\lambda_0)$ is independent of the Lagrange multiplier $\lambda_1$.

The alternative priors (31) are now combined into an improved prior $\Pi(i) = \Sigma_k \, p_k \, \Pi(i,k)$, on the basis of the experimental observation of $<i>_{obs}$. Perturbation theory of § 4 gives the weights explicitly:

$$p_0 = (1/N_\pi) \, \{1 - z <i>_M - z^2 \, (\kappa/2) + 0(z^3)\} \tag{33}$$

$$p_{1/2} = (1/N_\pi) \, \{1 + (z/2) \, [(N_\pi-2) <i>_M \pm \alpha] + (z^2/4) \, [N_\pi-2] \, \kappa \pm \xi] + O(z^3)\}$$

with the upper/lower sign corresponding to prior 1 or 2, respectively. There are $N_\pi = 3$ or 2 experts, depending on whether Jeffreys' uninformative prior k=0 is included.

The expert's advice is summarized in $<i>_k$, with $<i>_0 = 0$ for the non-informative prior. The small parameter z is given by (16), $z = [<i>_{obs}-<i>_M]/\sigma^2 + O(z^2)$ where $<i>_{obs}$ is the observed, and $<i>_M = (<i>_1+<i>_2)/N_\pi$, the democratic mean (15). $\sigma^2 = \kappa - 2<i>_M^2$ is given by (18), with

$$\kappa = <<i^2>_k> + <<i>_k^2> = [3<i>_1^2 + (5/2) <i>_2^2] / N_\pi \,. \tag{34}$$

The other parameters are the differences in mean, $\alpha$, and fluctuation, $\xi$,

$$\alpha = <i>_1 - <i>_2 \,, \qquad \xi = (<i^2>_1 + <i>_1^2) - (<i^2>_2 + <i>_2^2) = 3<i>_1^2 - (5/2)<i>_2^2 \,. \tag{35}$$

The compromise prior $\Sigma_k \, p_k \, \Pi(i,k)$, is a pre-exponential polynomial in the full MaxEnt distribution.

8.2 *Poisson Measure.* The geometrical issue has been addressed differently in the literature (Grady and Kipp 1985): The alternative priors are Poisson distributions of events which can either be point seeds (e.g. centers of Voronoi-like fragments) on the shell's surface, or long cracks across (orthogonal) directions (Mott and Linfoot 1943, of which we only have second-hand knowledge). The essential difference between alternative priors is now in the exponent of the Poison distribution, which is proportional to the area $i^2$ or to its square root $i$, respectively. This exponent is the vestige of a constraint, that of the finite area, or the finite length, on which the fragments or cracks seeds are distributed.

Application of the method of § 3 to construct a compromise between the two alternative Poisson priors, yields weights which remain close to 1/2 (Englman et al., 1989), a further indication that Poisson distributions are insufficiently discriminating as geometrical priors at least if the only experimental observations are the average fragment size, $<i>_{obs}$ and the total number of fragments.

## 9. Conclusions

Maximum entropy methodology does yield, in principle, a unique solution to an under-constrained or ill-defined problem, by minimizing any bias, apart from the physical information which can be encoded in the constraints. In practice, in order to quantify the bias, one must introduce priors or hypotheses which constitute advance information not expressible as constraints, and are therefore bias. We have shown that this vicious circle can be broken by MaxEnt methods.

In this paper, we have dealt with the particular, but very common situation where there is a finite number of conceivable hypotheses, and no, or not enough symmetry in the problem to impose their selection. Choice of a particular prior does introduce some bias, which must be minimized. This is done by adding to the entropy to be maximized a term measuring the arbitrariness of this choice of prior (the second term of eq. (7)).

We obtain by that procedure a prior which is a compromise between the various hypotheses. The weight given to a particular hypothesis is exponential in the overlap between the hypothesis and the prior-free MaxEnt distribution (which would have been obtained if all events had equal measure a priori). It is therefore discriminatory, but non-exclusive. It is the compromise of minimum bias, because it gives the highest maximum entropy (eq. (16) and (25)). This results together with the convexity of entropy, produces explicit expressions for the weights and the posterior probability distribution, as expansions in powers of a sensible small parameter. Finally, the construction procedure for the compromise can be iterated in a controllable way, which can accommodate new information input between iterations. When there is no new information, iteration either confirms the original compromise, or converges towards the single best hypothesis, the choice between the two modes of iteration depending on the nature of the problem.

The constructed "prior" is a compromise between the various hypotheses advanced, or a combination between two or more possible factors. This is not uncommon in natural sciences, where it is often the only way of making sense of experimental data, as in fragmentation which is usually not homogeneous (the average fragment size is a function of its position because of inhomogeneity in the material or in the stressing conditions (Grady and Kipp 1985)), or in the structure of coal, where various, non-covalent forces are known to link clusters of covalent network.

The weighing between the hypotheses depends on the experimental information encoded in the constraints, so that the constructed "prior" is not strictly a prior. On the other hand, the various hypotheses are true priors. But, and this is the essential step in our constru-ction, they are never taken as mutually exclusive. Accordingly, an expert opinion which turns out to accommodate most closely the experimental information (largest overlap $X_k$ in eq. (12)) dominates but never excludes the others in the final compromise. Further prior knowledge, in the form of additional expert opinions, improves the compromise by increasing the entropy, but only very little if the opinions fit poorly the experimental observation. In spite of this apparent lack of leadership, our constructed prior is (exponentially) discriminating between poor and adequate experts. In so far as statistical mechanics has been the physical prototype of the Maximum Entropy formalism in ergodic (events equally probable a priori) situations, one should look at simulated annealing through Monte-Carlo methods, physical systems with broken ergodicity like glasses, as practical models on which the somewhat formal results of this paper can be tested.

The authors would like to acknowledge the hospitality of Imperial College, London, where this work originated for N.R. and R.E.

## 10.  References

Englman, R., Levine, R.D. and Rivier, N. (1989) 'On the construction of priors in maximum entropy methods', submitted to IEEE Trans. Info. Theory; preprint, Imperial College (1987).

Englman, R., Rivier, N. and Jaeger, Z. (1988 a) 'Fragment-size distribution in disintegration by maximum-entropy formalism', Phil. Mag.B 56, 751-769.

Englman, R., Rivier, N. and Jaeger, Z. (1988b) 'Size-distribution in sudden breakage by the use of entropy maximization', J. Appl. Phys. 63, 4766-4768.

Grady, D.E. and Kipp, M.E. (1985) 'Geometric statistics and dynamic fragmentation', J. Appl. Phys. 58, 1210-1222.

Gull, S. (1987), seminar, Imperial College (private communication).

Holian, B.L. and Grady, D.E. (1988) 'Fragmentation by molecular dynamics: The microscopic "big bang"', Phys. Rev. Letters 60, 1355-1358.

Hopfield, J.J. and Tank, D.W. (1985) '"Neural" computation of decisions in optimization problems', Biol. Cybern. 52, 141-152.

Jaynes, E.T. (1968) 'Prior probabilities', IEEE Trans. Syst. Sci. Cybernetics SSC-4, 227-241.

Jaynes, E.T. (1973) 'The well-posed problem', Found. Physics 3, 477-492.

Jeffreys, H. (1961) Theory of Probability, 3rd. ed., Oxford Univ. Press.

Khinchin, A.I. (1957) Mathematical Foundations of Information Theory, Dover, NY.

Levine, R.D. (1986) 'The theory and practice of the maximum entropy formalism', in J.H. Justice (ed.), Maximum Entropy and Bayesian Methods in Applied Statistics, Cambridge Univ. Press, 59-84.

Levine, R.D. and Kosloff, R. (1979) 'The well-reasoned choice: An information-theoretic approach to branching ratios in molecular rate processes', Chem. Phys. Letters 28, 300-304.

Levine, R.D. and Tribus, M. (eds.) (1971) The Maximum Entropy Formalism, M.I.T. Press, Cambridge, MA.

Mott, N.F. and Linfoot E.H. (1943), Ministry of Supply, AC 3348

Renyi, A. (1970) Probability Theory, North-Holland, Amsterdam, ch. IX 5.

Rivier, N. (1986) 'Distribution of shapes and sizes by maximum entropy methods', Ann. Isr. Phys. Soc. 8, 560-567.

Rosenkrantz, R.D. (ed.) (1983) E.T. Jaynes, Papers on Probability Statistics and Statistical Physics, Reidel, Dordrecht.

Shannon, C.E and Weaver, W. (1949) The Mathematical Theory of Communication, Univ. of Illinois Press, Urbana.

Stoppard, T. (1967) Rosenkrantz and Guildenstern are Dead, Faber, London.

# MAXIMUM ENTROPY WITH NONLINEAR CONSTRAINTS: PHYSICAL EXAMPLES

A.J.M. GARRETT
Department of Physics and Astronomy
University of Glasgow
Glasgow G12 8QQ
U.K.

ABSTRACT. The Maximum Entropy literature has hitherto recognised only problems with constraints linear in the required probability distribution. However, nonlinear-constraint problems go back unrecognised for over half of the 120-year lifetime of Maximum Entropy. The classic example is calculation of the charge density $\rho$ in plasma in terms of the potential $\phi$, where the energy constraint is quadratic in $\phi$. In fact the Boltzmann distribution $\rho \propto \exp(-\beta_q \phi)$, traditionally derived under linear constraints, still holds; the condition for this is merely that a conserved quantity *exists*. The distribution can be combined with Poisson's equation to give a single equation for $\phi$ or $\rho$. Uniqueness of the entropy maximum must be examined case-by-case in nonlinear problems.

## 1. ANALYSIS OF NONLINEAR CONSTRAINTS

The Boltzmann distribution for the probability $p_i$ of occupancy of a fixed energy level $E_i$ is

$$p_i = Z(\beta)^{-1} e^{-\beta E_i}, \tag{1}$$

where the *partition function* $Z(\beta)$ (from the German, Zustand) is given by insisting that (1) be normalised:

$$Z(\beta) = \sum_i e^{-\beta E_i}, \tag{2}$$

and $\beta$ is in principle expressible in terms of the expectation of the total energy, $U$, by substituting (1) into the equation of energy conservation

$$U = U[\mathbf{p}] \equiv \sum_i p_i E_i. \tag{3}$$

Derivation of the Boltzmann distribution is a commonplace application of the Principle of Maximum Entropy, by which probability distributions are assigned by maximising the information

243

P. F. Fougère (ed.), Maximum Entropy and Bayesian Methods, 243–249.

entropy

$$S \equiv -\sum_i p_i \ln p_i, \tag{4}$$

subject to the constraints of normalisation $\sum_i p_i = 1$ with Lagrange multiplier $(\ln Z - 1)$, and energy conservation with Lagrange multiplier $\beta$. In the light of (1), the constraint (3) is equivalent to

$$U = -\frac{\partial}{\partial \beta} \ln Z(\beta), \tag{5}$$

whence $d(\ln Z) = -U d\beta$. Substitution of (1) into (4) gives for the maximised entropy

$$S_M = \ln Z + \beta U, \tag{6}$$

whence $dS_M = \beta dU$. The present derivation of (1), together with a combinatorial justification for Maximum Entropy, was first given by Boltzmann some 120 years ago [1]; in modern notation see [2]. For statistical-mechanical applications to systems of order $10^{23}$ particles, and provided all relevant constraints are included, the occupancy is proportional to the probability of occupancy to outstanding accuracy.

This analysis exemplifies the general ("canonical") formalism for assigning probabilities under constraints linear in the probability distribution [2]. The maximum of $S$ found by the calculus of variations is unique and global [2].

The Maximum Entropy principle was next taken up by Gibbs [3], who clearly understood its substance as a tool of inference. In the work of Jaynes [4] its power and generality have reached full fruition. Modern literature includes comments on generalization to constraints nonlinear in the desired probability distribution [5], but in the seeming absence of physical examples the matter has not been seriously pursued.

The purpose of this paper is to demonstrate that Maximum Entropy with nonlinear constraints has been with us unrecognised for some decades; and to give physical examples of it. In the context of energy levels, these are considered now to depend on their occupancies $p_i$. This happens, for example, in the random placement of a collection of masses onto shelves which are supported by springs in a uniform gravitational field. The energy of the shelf is dependent on its height, which is in turn dependent on the number of masses already resting on it. (Strictly, the system should be treated in a full $10^{23}$-dimensional phase space using energies and probabilities of entire configurations. This is how the virial expansion of the gas equation of state is generated [6]; the analysis is linear in the distribution. However, it is more involved, in considering probabilities of occupancy of levels rather than taking occupancy of a level proportional to its probability.)

To find the appropriate energy constraint, assemble the system from component charges, at infinity. The increment in energy on bringing element $dp_i$ in, with $p_i$ already in place, is

$$dU = \sum_i E_i[\mathbf{p}] dp_i. \tag{7}$$

A necessary condition for energy conservation is that this should be a perfect differential, so that

$$E_i = \frac{\partial}{\partial p_i} U[\mathbf{p}] \tag{8}$$

and consequently

$$\frac{\partial E_i}{\partial p_j} = \frac{\partial E_j}{\partial p_i}. \tag{9}$$

Otherwise, the system could be assembled in one way and dismantled in another with net energy change. In fact (9) is a *sufficient* condition. In all problems of this sort the energy function $U[\mathbf{p}]$ and its derivative $E_i[\mathbf{p}]$ are specified by the physics of the problem. Extension to the functional case, with continuous suffixes, is immediate.

Constrained maximisation of $S$ leads to

$$d\{S - (\ln Z - 1)\sum_i p_i - \beta U[\mathbf{p}]\} = 0, \tag{10}$$

with $dU$ given by (5), whence

$$\sum_i \{(-1 - \ln p_i) - (\ln Z - 1) - \beta E_i[\mathbf{p}]\}dp_i = 0 \quad \forall \, dp_i, \tag{11}$$

so that

$$p_i = Z^{-1} e^{-\beta E_i[\mathbf{p}]}. \tag{12}$$

Therefore the canonical form of the relation between $p_i$ and $E_i$ is unchanged, but now it is a transcendental equation for either set. It is no longer trivial to express $Z$ in terms of $\beta$, and solutions for $p_i$ may be non-unique or complex. Complex solutions imply the absence of a physically meaningful maximum: the concept of an ordering is restricted to the reals, and we do not know how to interpret complex entropy. There is no general theorem guaranteeing uniqueness, globality or reality of the maximum when constraints are nonlinear, and every problem must be examined individually.

## 2. EXAMPLES

The first example is the determination of the equilibrium disposition of a confined thermal plasma consisting of a single species of $N$ particles of charge $q$. It is supposed that there are no other charges in the vicinity; generalization to allow for a background of fixed charges and to multi-species plasmas is quoted from the companion paper [7]. Clearly, this problem is nonlinear: the plasma (charge) density $\rho(\mathbf{r})$, suitably normalised, plays the role of the probability distribution, with spatial coordinate $\mathbf{r}$ as index $i$; and the electrostatic energy as the container fills with charge depends on the charge already present. The electrostatic potential $\phi(\mathbf{r})$ takes the part of $E_i$. Energy density is proportional to the square of the electric field, $-\nabla\phi$, and hence the conserved energy is

$$U = \tfrac{1}{2}\epsilon_0 \int_V d\mathbf{r} |\nabla\phi|^2. \tag{13}$$

Here the integral is throughout all space and not merely the container. It is rewritten using Green's first identity as

$$U = \tfrac{1}{2}\epsilon_0 \int_V d\mathbf{r}[\nabla \cdot (\phi\nabla\phi) - \phi\nabla^2\phi]. \tag{14}$$

The first term vanishes on transforming it to a surface integral at infinity (or on the container if a conductor) and choosing $\phi = 0$ there, while the second term is handled using the physical relation between potential and charge density, Poisson's equation

$$\epsilon_0\nabla^2\phi = -\rho. \tag{15}$$

Therefore

$$U = \tfrac{1}{2} \int_{V_p} d\mathbf{r}\phi(\mathbf{r})\rho(\mathbf{r}), \tag{16}$$

where the integral is now only taken inside the container, since the charge density vanishes outside it. Denote the Green's function for Poisson's equation $G(\mathbf{r},\mathbf{r}')$:

$$\nabla_r^2 G(\mathbf{r},\mathbf{r}') = -\delta(\mathbf{r} - \mathbf{r}'), \quad G = 0 \quad \text{at} \quad r = \infty, \tag{17}$$

which is symmetric in $\mathbf{r}$ and $\mathbf{r}'$ since the Laplacian operator $\nabla^2$ is self-adjoint. This leads to the continuous version of the reprocity requirement (9). (Of course, in three dimensions and with a non-conducting container, $G(\mathbf{r},\mathbf{r}') = (1/4\pi)|\mathbf{r} - \mathbf{r}'|^{-1}$.) It now follows that

$$\phi(\mathbf{r}) = \epsilon_0^{-1} \int d\mathbf{r}' G(\mathbf{r},\mathbf{r}')\rho(\mathbf{r}'), \tag{18}$$

whence

$$U = \tfrac{1}{2}\epsilon_0^{-1} \iint d\mathbf{r}, d\mathbf{r}' G(\mathbf{r},\mathbf{r}')\rho(\mathbf{r})\rho(\mathbf{r}'). \tag{19}$$

Therefore we have found a conserved quantity which is quadratic in the probability distribution. If there is also a fixed array of charges, a linear term comes in. The functional version of the nonlinear-constraint analysis gives

$$\rho(\mathbf{r}) \propto e^{-\beta q \phi(\mathbf{r})}, \tag{20}$$

and on considering the charged particle distribution in velocity space, which has separated out cleanly in this problem, the Lagrange multiplier $\beta$ is identified as $(kT)^{-1}$ where $T$ is the temperature:

$$\rho(\mathbf{r}, \mathbf{v}) \propto e^{-(\tfrac{1}{2}mv^2 + q\phi)/kT}. \tag{21}$$

The plasma itself decides the kinetic:electrostatic energy ratio.

The Boltzmann distribution (20) can be combined with Poisson's equation to give the *Poisson-Boltzmann equation*:

$$\epsilon_0 \nabla^2 \phi = -\nu(\mathbf{r}) N q n e^{-q\phi/kT}, \tag{22}$$

where $\nu(\mathbf{r})$ is one within the container ($\mathbf{r} \in V_p$) and zero outside it, and $n$ is determined self-consistently so as to preserve normalisation:

$$n^{-1} = \int_{V_p} d\mathbf{r}\, e^{-q\phi/kT}. \tag{23}$$

In a multi-species plasma with species label $s$, and with fixed background ("test") charge density $\rho_t(\mathbf{r})$, the equation generalizes to

$$\epsilon_0 \nabla^2 \phi = -\rho_t - \nu(\mathbf{r}) \sum_s N_s q_s n_s e^{-q_s\phi/kT}, \tag{24}$$

where

$$n_s^{-1} = \int_{V_p} d\mathbf{r}\, e^{-q_s\phi/kT}. \tag{25}$$

If there is more than one disjoint volume of plasma, a separate $\nu$-function and temperature is defined for each. Equation (24) became well-known when Debye and Hückel wrote it down and linearised it [8].

The Poisson-Boltzmann equation (22) is a nonlinear partial differential equation. For solutions to conform to the integral version (18)/(20) it must be supplemented with the boundary condition $\phi(r=\infty) = 0$. This is equivalent to solving for the $E_i$'s; eliminating $\phi$ in favour of $\rho$ would give us the $p_i$'s, with the integral form of the resulting equation having logarithmic nonlinearity. However, the differential form — and we know more about solving differential than integral equations — is simpler. To find it, define

$$\zeta = (Nq)^{-1}\rho/n. \tag{26}$$

Then

$$\nabla^2 \ln \zeta = (Nq^2/\epsilon_0 kT)n\zeta, \quad (\mathbf{r} \in V_p), \tag{27}$$

which can be re-expressed with only polynomial nonlinearity:

$$\zeta\nabla^2\zeta - |\nabla\zeta|^2 = (Nq^2/\epsilon_0 kT)n\zeta^3. \tag{28}$$

The solution for $\phi$ and $\rho$, once found, completely characterises the equilibrium plasma when there is no other significant constraint and the "reduced" description in the one-particle phase space is accurate.

Solutions of the Poisson-Boltzmann equation are studied in detail in [7, 9]. In [9], Garrett and Poladian prove uniqueness of the solution at constant, positive temperature; the distribution takes the same form at constant temperature as at constant energy. (Grandy [10] proves this under linear constraints.) This is not of course a proof of uniqueness or globality of the entropy maximum.

The second example is the motion of guiding centres of charged particles in the two dimensions perpendicular to a magnetic field [11, 12]. The field should be strong enough for the Larmor radius of gyration to be far less than other relevant parameters. It is also assumed that particles have large velocities parallel to the field and are reflected back and forth between two planes perpendicular to the field. This is most easily arranged by concentrating the field lines into a magnetic mirror [13]. Although the component of velocity parallel to the field is small near the mirror point, this region can be made arbitrarily small with judicious field geometry.

The point about this system is that the particles can be viewed as smeared out along the field lines, and that the gyrorotational magnetic moments of these "particle lines" interact with energy quadratic in the density of particles perpendicular to the field. The situation is similar to the previous example, and gives rise to a two-dimensional Poisson-Boltzmann equation. Here, though, there is no associated velocity-space distribution and no automatic apportioning of the (fixed) total energy into kinetic and electro-magneto-static components. Only the latter is present, and if the energy so dictates it the temperature parameter becomes negative.

## 3. CONCLUSIONS

Maximum Entropy with nonlinear constraints is a proven tool of classical plasma analysis, and should also play a role in the quantum version — solid state theory. Only recognition of nonlinear constraints is novel. With the failure of the canonical formalism, uniqueness and existence of any entropy maximum must be examined on an individual basis.

A preliminary version of this material has been given in [14].

## REFERENCES

[1] Boltzmann, L. (1868, 1871), in: Sitzungsberichte der kaiserliche Akademie der Wissenschaften in Wien, Mathematisch-Naturwissenschaftliche Klasse IIa: 'Studien über das Gleichgewicht der lebendige Kraft zwischen bewegten materiellen Punkten', **58**, 517-560 (1868); 'Über das Wärmegleichgewicht zwischen mehratomingen Gasmolekülen', **63**, 397-418 (1871); 'Einige allgemeine Sätze über Wärmegleichgewicht', **63**, 679-711 (1871); 'Analytischer Beweis des zweiten Hauptsatzes der mechanischen Wärmetheorie aus den Sätzen über das Gleichgewicht der lebendigen Kraft', **63**, 712-732 (1871). Reprinted in: L. Boltzmann (1909), *Wissenschaftliche Abhandlungen 1*, F. Hasenöhrl (ed.), Leipzig, 49-96; 237-258; 259-287; 288-308.

[2] Jaynes, E.T. (1963) 'Information Theory and Statistical Mechanics', in: K.W. Ford (ed.), *Statistical Physics* (1962 Brandeis Lectures), Benjamin, New York, pp. 181-218. Reprinted as Chapter 4 of reference [4].

[3] Gibbs, J.W. (1902) *Elementary Principles in Statistical Mechanics*, Scribner, New York. Reprinted in *Collected Works* (1936), Yale University Press, New Haven.

[4] Jaynes, E.T. (1983) *E.T. Jaynes: Papers on Probability, Statistics and Statistical Physics*, R.D. Rosencrantz (ed.), Synthese Library **158**, Reidel, Dordrecht, Netherlands.

[5] Evans, R.B. (1979) 'A New Approach for Deciding Upon Constraints in the Maximum Entropy Formalism', in: R.D. Levine & M. Tribus (eds.), *The Maximum Entropy Formalism*, MIT Press, Cambridge, Massachusetts, pp. 169-206. We deny Evans' assertion that the Maximum Entropy procedure is only consistent when used with linear constraints.

[6] See almost any advanced statistical mechanics text, for example Grandy, W.T. Jr. (1987) *Foundations of Statistical Mechanics, Volume 1: Equilibrium Theory*, Reidel, Dordrecht, Netherlands, Chapter 7.

[7] Garrett, A.J.M. 'Maximum Entropy Description of Plasma Equilibrium'; these Proceedings.

[8] Debye, P. & Hückel, E. (1923) 'Zur Theorie der Electrolyte, I: Gefrierpunktserniedrigung und verwandte Erscheinungen', *Physik. Zeits.* **24**, 185-206.

[9] Garrett, A.J.M. & Poladian, L. (1988) 'Refined Derivation, Exact Solutions and Singular Limits of the Poisson-Boltzmann Equation', *Ann. Phys.* **188**, 386-435.

[10] reference [6], pp. 67-68.

[11] Joyce, G. & Montgomery, D. (1973) 'Negative Temperature States for the Two-Dimensional Guiding-Centre Plasma', *J. Plasma Phys.* **10**, 107-121.

[12] Ting, A.C., Chen, H.H. & Lee, Y.C. (1987) 'Exact Solutions of a Nonlinear Boundary Value Problem: the Vortices of the Two-Dimensional Sinh-Poisson Equation', *Physica D.* **26**, 37-66.

[13] see almost any plasma physics discussion of adiabatic invariants, for example Nicholson, D.R. (1983) *Introduction to Plasma Theory*, Wiley, New York, pp. 25-30.

[14] Garrett, A.J.M. (1988) 'Validity of the Boltzmann Distribution Under Non-Linear Constraints', *J. Phys.* **A 21**, 1467-1469.

# MAXIMUM ENTROPY DESCRIPTION OF PLASMA EQUILIBRIUM

A.J.M. GARRETT
Department of Physics and Astronomy
University of Glasgow
Glasgow G12 8QQ
U.K.

ABSTRACT. The self-consistent equation for the potential $\phi$ in an equilibrium plasma is found by combining Poisson's equation $\epsilon_0 \nabla^2 \phi = -\rho$ with the Maximum Entropy formula relating the charge density $\rho$ to the potential. On ignoring interparticle correlations this takes the form of the Boltzmann distribution, $\rho \propto \exp(-\beta_q \phi)$. The resulting 'Poisson-Boltzmann' equation for the potential is studied in various geometries with differing combinations of charge species. It is nonlinear, inducing a surprising result: a fixed line charge immersed in plasma may attract a non-zero amount of charge of opposite polarity arbitrarily close to it; neutralisation of fixed point charges is exact; and a fixed point dipole causes the entire plasma to condense onto it! Failure of these results in practice is due to particle correlations and the non-zero size of real multipoles.

## 1. INTRODUCTION

An equation for the electrostatic potential $\phi(r)$ in an ideal plasma in thermal equilibrium, in the presence of fixed charges, is derived by combining Poisson's equation $\epsilon_0 \nabla^2 \phi = -\rho$, where $\rho$ is the total charge density, with the statistical formula relating the charge density to the potential. Clearly charge is more likely to place itself in a trough than at a peak of potential; yet that same charge partly determines the potential, through Poisson's equation. A self-consistent scheme is needed.

The relevant statistical formula is the Boltzmann distribution. For the $s$-th particle species,

$$\rho_s \propto e^{-q_s \phi / kT} \tag{1}$$

where $T$ is the plasma temperature. This is derived by assuming that the density is proportional to the probability of finding a particle at a given potential, and then assigning this probability by maximising the information entropy $- \int dr \rho_s \ln \rho_s$ under the constraints of normalisation and energy conservation, using the calculus of variations. The electrostatic energy

$$\tfrac{1}{2} \epsilon_0^{-1} \iint dr, dr' G(\mathbf{r}, \mathbf{r}') \rho(\mathbf{r}) \rho(\mathbf{r}'). \tag{2}$$

P. F. Fougère (ed.), Maximum Entropy and Bayesian Methods, 251–271.
© 1990 Kluwer Academic Publishers.

is quadratic in $\rho$. Here $G$ is the Green's function for Poisson's equation, such that $\phi = \epsilon_0^{-1} \int G\rho'$:

$$\nabla_r^2 G(\mathbf{r}, \mathbf{r}') = -\delta(\mathbf{r} - \mathbf{r}') \tag{3}$$

with appropriate homogeneous boundary conditions; for example in three dimensions with $G = 0$ at infinity, $G = (1/4\pi)|\mathbf{r} - \mathbf{r}'|^{-1}$. The quantity $kT$ in (1) is the reciprocal of the Lagrange multiplier conjugate to electrostatic energy, and is interpreted as the plasma temperature via a fuller analysis including particle kinetic energy, in which the distribution in the phase space of one particle is $\exp -(\frac{1}{2}m_s v^2 + q_s \phi)/kT$. The plasma itself apportions energy between kinetic and electrostatic components.

In the presence of a fixed, background or "test" charge density $\rho_t(\mathbf{r})$, the full charge density is $\rho_t + \rho_p$ where $\rho_p$ is the plasma charge density. The electrostatic energy now contains terms quadratic and linear in plasma density. The Boltzmann distribution (1) is unchanged, however. It is derived in detail in the companion paper [1] and in the review of the problem, [2].

The Boltzmann distribution (1) is combined with Poisson's equation to give a self-consistent equation for the electrostatic potential — the Poisson-Boltzmann (PB) equation. To set it up, define a function $\nu(\mathbf{r})$ to be unity inside the volume $V_p$ accessible to the plasma and zero outside it, and let $N_s$ be the total number (not concentration) of particles of the $s$-th species. Then the PB equation, valid inside and outside the plasma, is

$$\epsilon_0 \nabla^2 \phi = -\rho_t(\mathbf{r}) - \nu(\mathbf{r}) \sum_s N_s q_s n_s e^{-q_s \phi / kT}, \tag{4}$$

where the factor $n_s$ is determined a posteriori so as to preserve normalisation:

$$n_s^{-1} = \int_{V_p} d\mathbf{r} e^{-q_s \phi(\mathbf{r})/kT}. \tag{5}$$

This factor ensures gauge invariance of (4). In a disconnected plasma, separate $\nu$-functions and temperature are defined for each volume.

Once (4)/(5) is solved with appropriate boundary conditions for the potential, the plasma density can be reconstructed from (1) to give a complete description of the equilibrium configuration. The purpose of the present paper is to study the PB equation as a classic consequence of maximum entropy. No general analytical method of solving equations this nonlinear is known yet in two or more dimensions and numerical and analytical approximation is required. In Section 2 uniqueness of any solution known is proven. Section 3 examines the high and low temperature limits and presents a series expansion of potential in inverse powers of temperature. The exact solution in planar geometry for single-species plasma, and discussion of generalization to several particle species, is given in Section 4, while Section 5 presents the single species solution in cylindrically symmetrical geometry and discusses its multi-species generalization. It is shown that a line charge immersed in the plasma may cause a non-zero amount of plasma charge to condense on to and partly neutralize it. In Section 6 the general single species solution in two dimensions is presented, while Section 7 examines the three dimensional and spherically symmetrical cases. The equation is not now Painlevé and only a series solution is known for the single species. It is proven that a fixed point charge immersed in plasma, according to the PB equation, attracts an equal and opposite charge arbitrarily close to itself, thereby becoming locally electrically invisible. This charge condensation phenomenon is studied more fully in Section 8; conclusions are presented and the validity of Poisson-Boltzmann analysis examined in Section 9. The present paper comprises a summary of the detailed work of Garrett and Poladian [2] on the Poisson-Boltzmann equation.

## 2. THE POISSON-BOLTZMANN EQUATION: UNIQUENESS

Here we prove the uniqueness of any given solution of (4)/(5). Existence is not studied, though it has been established in special cases [3, 4, 5] and studied for the more general equation $\nabla^2\phi = -F[\phi]$ [6, 7]. Define the difference between two solutions $\phi_1, \phi_2$ as $\psi \equiv \phi_1 - \phi_2$, so that the aim is to prove $\psi = 0$. We begin with the vector identity

$$\int_{\partial V} ds \cdot (\psi\nabla\psi) = \int_V dr\, \nabla \cdot (\psi\nabla\psi) = \int_V dr(|\nabla\psi|^2 + \psi\nabla^2\psi). \tag{6}$$

Choose the gauge of the solutions equal at infinity (or on the plasma vessel if conducting). The surface integral then vanishes, and on substituting for $\nabla^2\psi$ from the PB equation we have

$$0 = \int_V dr|\nabla(\phi_1 - \phi_2)|^2 - \epsilon_0^{-1}\sum_s N_s q_s$$
$$\int_{V_p} dr(\phi_1 - \phi_2)(n_s^{(1)}e^{-q_s\phi_1/kT} - n_s^{(2)}e^{-q_s\phi_2/kT}). \tag{7}$$

Define $\xi^s \equiv \exp(-q_s\phi/kT)$ and denote integration over the plasma volume by angled brackets $\langle\,\rangle$, so that (7) can be written

$$0 = \int_V dr|\nabla(\phi_1 - \phi_2)|^2 + (kT/\epsilon_0)\sum_s N_s\Big\langle [\ln(\xi_1^s/\langle\xi_1^s\rangle)$$
$$- \ln(\xi_2^s/\langle\xi_2^s\rangle) + \ln(\langle\xi_1^s\rangle/\langle\xi_2^s\rangle)] \, [\xi_1^s/\langle\xi_1^s\rangle - \xi_2^s/\langle\xi_2^s\rangle] \Big\rangle. \tag{8}$$

The $\ln(\langle\xi_1^s\rangle/\langle\xi_2^s\rangle)$ term is a constant and does not contribute to the overall expression after the outer brackets are taken. Upon discarding it the square brackets take the same sign whatever the values of $\xi_1^s/\langle\xi_1^s\rangle$ and $\xi_2^s/\langle\xi_2^s\rangle$, so that the outermost bracket is non-negative. Since temperature is non-negative the LHS of (8) is the sum of two non-negative contributions, which must separately be zero. The first term is zero only if $\phi_1 = \phi_2$ almost everywhere, and this condition also renders the angled brackets zero. Uniqueness is proved.

## 3. HIGH AND LOW TEMPERATURE SOLUTIONS

For an overall neutral plasma ($\sum_s N_s q_s = 0$) in the absence of fixed charges ($\rho_t = 0$), the solution of the PB equation (4)/(5) is trivially that the potential $\phi$ is everywhere constant and the electric field $-\nabla\phi$ zero. The plasma is uniform. At infinite temperature the plasma will also be uniform, even if charged, for then the kinetic energy overwhelms the electrostatic energy and the particles dash about without regard to the potential. This suggests an expansion of the potential in inverse powers of temperature about the uniform state:

$$\phi(r) = \sum_{j=0}^{\infty}(kT)^{-j}\phi_j(r) \tag{9}$$

and

$$n_s = \sum_{j=0}^{\infty} (kT)^{-j} n_{sj}. \tag{10}$$

We work with the integral form of the PB equation, using the Green's function from (3) so as to incorporate the boundary conditions:

$$\phi(\mathbf{r}) = \phi_t(\mathbf{r}) + \epsilon_0^{-1} \sum_s N_s q_s n_s \int_{V_p} d\mathbf{r}' G(\mathbf{r}, \mathbf{r}') e^{-q_s \phi(\mathbf{r}')/kT}, \tag{11}$$

where $\phi_t(\mathbf{r})$ is the potential set up by the fixed charge distribution $\rho_t$; schematically $\phi_t = \epsilon_0^{-1} \int G \rho_t'$. On putting in the expansions (9) and (10) and collecting terms one finds that

$$\phi_0(\mathbf{r}) = \phi_t(\mathbf{r}) + \epsilon_0^{-1} \sum_s N_s q_s n_{s0} \int_{V_p} d\mathbf{r}' G(\mathbf{r}, \mathbf{r}'); \tag{12}$$

$$\phi_1(\mathbf{r}) = \epsilon_0^{-1} \sum_s N_s q_s \int_{V_p} d\mathbf{r}' G(\mathbf{r}, \mathbf{r}') [n_{s1} - q_s n_{s0} \phi_0(\mathbf{r}')]; \tag{13}$$

$$\phi_2(\mathbf{r}) = \epsilon_0^{-1} \sum_s N_s q_s \int_{V_p} d\mathbf{r}' G(\mathbf{r}, \mathbf{r}')$$
$$[n_{s2} - q_s \{ n_{s1} \phi_0(\mathbf{r}') + n_{s0} \phi_1(\mathbf{r}') \} + \tfrac{1}{2} q_s^2 n_{s0} \phi_0(\mathbf{r}')^2] \tag{14}$$

and so on to higher order. Collecting terms in (5) gives

$$n_{s0} = \left\{ \int_{V_p} d\mathbf{r} \right\}^{-1} = V_p^{-1}; \tag{15}$$

$$n_{s1} = q_s n_{s0}^2 \int_{V_p} d\mathbf{r} \phi_0(\mathbf{r}); \tag{16}$$

$$n_{s2} = (n_{s1}^2/n_{s0}) + q_s n_{s0}^2 \int_{V_p} d\mathbf{r} [\phi_1(\mathbf{r}) - \tfrac{1}{2} q_s \phi_0(r)^2]; \tag{17}$$

and so on. The advantage of using the Green's-integral form (11) is that $\phi_{j+1}$ is an integral involving only the functions $\phi_j, \phi_{j-1}, \ldots, \phi_0$, so that recursion immediately gives $\phi_{j+1}$ by quadrature. The coefficient $n_{s(j+1)}$ which appears in the quadrature can be determined in advanve of $\phi_{j+1}$. Graphical techniques might provide the solution for $\phi_{j+1}$ in terms of integrals over products of $G(\mathbf{r}, \mathbf{r}')$ and $\phi_k(\mathbf{r})$.

A related procedure is to expand not the potential in inverse powers of temperature but the Poisson-Boltzmann equation itself. The zero-th order equation has solution $\phi_0(\mathbf{r})$, while the next approximation, which is linear in potential, is the well-known Debye-Hückel equation [8]. Its homogeneous solutions asymptotically behave as exponential functions of $|\mathbf{r}|$. Expansion of the PB equation in $(kT)^{-1}$, followed by solution by any available means, gives faster convergence than the original expansion scheme for the potential. This is because the domain of the accuracy of the Debye-Hückel solution extends well into the region in which the Debye-Hückel equation aproximates the PB equation poorly. To understand this, write the equation $\nabla^2 \phi = -F[\phi]$ where $F[\phi]$ has linear approximation $\mathcal{L}[\phi]$. The solution is $\phi = \int GF[\phi]$, and the criterion for accuracy is not $F \approx \mathcal{L}$ but $\int GF \approx \int G\mathcal{L}$. The major contribution to the integral should come from the linear regime, a less stringent requirement. More intuitively, $\nabla^2$ represents curvature, and two functions of differing curvature can approxiamte each other closely in some region before diverging outside it.

Traditionally the Debye-Hückel equation has been viewed as a weak-field approximation in neutral plasmas, and $\sum_s N_s q_s n_s$ taken as zero. But neutrality is $\sum_s N_s q_s = 0$ and is only the same if all the $n_s$ are equal, a condition valid only for zero electric field (and gauge $\phi = 0$). The equation is easily patched up, but because one cannot know if the field is weak until after solving, Debye-Hückel theory is more logically seen in terms of the inverse temperature expansion. A common related confusion is to take $N_s n_s$ as a single variable and interpret it as the "concentration at infinity", or the mean concentration of the $s$-th species; but these are not the same and $N_s n_s$ is only the mean concentration $N_s/V_p$ for zero field (and gauge $\phi = 0$).

Turn now to the low temperature regime. Here the kinetic energy is small and electrostatic energy dominates. Minimising it, according to a variational principle equivalent to Poisson's equation, implies reduction of the electric field within the bulk of the plasma to as small a value as possible. Provided only that there is sufficient plasma charge to do it, all impermeable fixed charge distributions within the plasma are neutralized by an appropriate plasma charge frozen onto their surfaces. (Any volume charge distribution is indistinguishable, outside itself, from an appropriate charge on its boundary.) Permeable volume charges are directly neutralized. The remainder of the plasma charge then distributes itself over the outer boundary of the plasma volume — or at infinity — as it would over the surface of a conductor of the same shape. The potential is constant on this boundary, and consequently within it: there is no electric field inside the plasma.

Expansion about this singular limit is not pursued here.

## 4. SOLUTIONS IN PLANAR GEOMETRY

For variation in a single spatial dimension $z$, corresponding in 3D to planar realisation, the PB equation within the plasma in the absence of fixed charge is

$$\epsilon_0 \frac{d^2 \phi}{dz^2} = -\sum_s N_s q_s n_s e^{-q_s \phi/kT} \tag{18}$$

where $N_s$ is now the number of particles per unit area perpendicular to the $z$-axis and $n_s^{-1}$ is the integral of the exponential along it. The first integral of (18) is

$$\tfrac{1}{2}\epsilon_0 \left(\frac{d\phi}{dz}\right)^2 = -kT \sum_s N_s n_s e^{-q_s \phi/kT} = -\tfrac{1}{2}\epsilon_0 A^2, \tag{19}$$

where the RHS is a constant. Direct integration and reversion then gives $\phi(z)$. Since particles charge is quantised such that $q_s = l_s Q$ where $l_s$ is an integer, and with $Q$ made as large as possible so that the $|l_s|$ have no common factor, (19) is reduced to polynomial nonlinearity by taking $\exp(-Q\phi/kT)$ as dependent variable. The new variable is a fractional power of the density of any particle species. For small $l_s$ the quadrature is often an elliptic integral [2, 9].

For a single species plasma this strategy is readily implemented, giving as general solution

$$\phi(z) = -\frac{kT}{q} \ln \left[ \frac{\epsilon_0 A^2}{2NnkT} \sec^2 \frac{qA}{2kT}(z - z_0) \right].$$ (20)

The integration constants $A$ and $z_0$ are determined together with $n$ by the conditions at each bounding plane of the plasma, and the self-consistency requirement for $n$. Should $A^2$ be negative, trigonometric functions are replaced by hyperbolics: (20) remains real if $z_0$ is given an imaginary part $\pi kT/qA$ and $\sec^2$ is replaced by $\operatorname{cosech}^2$. The solution for $A = 0$ can be found by a limiting process, or directly.

We apply this result to find the distribution of a single charged species confined between $z = \pm a$ in the absence of fixed charges. Determination of the constants is routine [2] and the result is

$$\phi(z) = \phi(0) - \frac{kT}{q} \ln \left[ \sec^2 \left( \frac{\pi}{2} \gamma \frac{z}{a} \right) \right]$$ (21)

where defining the (Bjerrum or Landau) length $L$ as

$$L = q^2/2\epsilon_0 kT,$$ (22)

$\gamma$ is the solution of

$$\frac{2}{NLa} \cdot \frac{\pi}{2} \cdot \gamma = \cot \frac{\pi}{2} \gamma, \quad 0 \le \gamma \le 1.$$ (23)

The electric field $-d\phi/dz$ is

$$E(z) = \frac{Nq}{2\epsilon_0} \tan \left( \frac{\pi}{2} \gamma \frac{z}{a} \right) \Big/ \tan \left( \frac{\pi}{2} \gamma \right)$$ (24)

and the particle density $Nn \exp(-q\phi/kT)$ is

$$\frac{N}{2a} \cdot \frac{\frac{\pi}{2}\gamma}{\tan \frac{\pi}{2}\gamma} \cdot \sec^2 \left( \frac{\pi}{2} \gamma \frac{z}{a} \right).$$ (25)

At high temperature the plasma becomes uniform, so that by Gauss' law the electric field increases linearly with distance from the origin. At low temperature the plasma piles up equally on the two boundaries and the field between is zero. The field outside the plasma is constant and independent of the plasma disposition, a feature unique to the one-dimensional situation. Even if fixed charges are placed outside $|z| < a$ or if the boundaries are conductors, the solution changes only through differences in the integration constants.

Solutions in planar geometry have been studied also for plasmas composed of two particle species of equal and opposite charge: $q_1 = -q_2 = q$ (and typically equal numbers of each). The introduction of further species beyond this does not lead to anything qualitatively new. The first integral (19) is then expressible as

$$\tfrac{1}{2}\epsilon_0 \left( \frac{d\phi}{dz} \right)^2 = -4kT(N_1 N_2 n_1 n_2)^{1/2} \sinh^2 \left( \tfrac{1}{2} \frac{q\phi}{kT} = \tfrac{1}{4} \ln \frac{N_1 n_1}{N_2 n_2} \right) = \text{CONST}.$$ (26)

The general solution involves elliptic functions, but if the RHS is zero the result is simply

$$\phi(z) = \frac{kT}{q} \ln\left[\left(\frac{N_1 n_1}{N_2 n_2}\right)^{1/2} \coth^2\left\{(N_1 N_2 n_1 n_2)^{1/4} L^{1/2}(z - z_0)\right\}\right], \tag{27}$$

where $z_0$ is the translational constant of integration.

This solution is traditionally associated with semi-infinite plasma, for which it is held to represent a "sheath" solution shielding a boundary held at a specified potential relative to infinity. However, this is obtained by assuming $N_1 n_1 = N_2 n_2 = $ the density of either species at infinity in a net neutral plasma, and there is an immediate problem: shielding implies a local excess of one charged species over the other, and in planar geometry the effect of the "leftover" charge does not diminish with distance. It cannot be ignored even if banished to infinity. Solution (27) therefore corresponds to a plasma with an excess of one species of exactly the right amount to screen the boundary: a highly contrived situation. A proper treatment of shielding must start from the full elliptic solution. The problem is studied more fully in [2].

## 5. CYLINDRICAL SYMMETRY:
## THE SINGLE SPECIES SOLUTION AND CHARGE CONDENSATION

In this section we examine the PB equation for plasmas confined between concentric cylinders of radii $a$ and $b$ ($> a$). Fixed charges are distributed within $r < a$ with cylindrical symmetry about the common axis; there is a total fixed charge $Q'_i$ per unit axial length. By Gauss' law the radial electric field at $r = a$, $E_a$, is $Q'_i/2\pi\epsilon_0 a$. Since $\phi = 0(\ln r)$ at large radius we cannot choose the gauge $\phi(\infty) = 0$ and instead specify the potential at $r = a$ to be $\phi_a$. Shrinking $a$ to zero permits us to study the shielding of line charges.

Much of this section is devoted to the single species problem, for which exact solutions are available. Within the plasma, the PB equation then becomes

$$\frac{1}{r}\frac{d}{dr}\left(r\frac{d\phi}{dr}\right) = -(Nqn/\epsilon_0)e^{-q\phi/kT}, \tag{28}$$

where $N$ is the number of particles per unit axial length and

$$n^{-1} = \int_a^b dr\, 2\pi r\, e^{-q\phi/kT}. \tag{29}$$

Equation (28) is invariant under a stretch of $r$ and a translation of $\phi$ which together leave the combination $r^2 \exp(-q\phi/kT)$ invariant. On calling this $\exp(-qX/kT)$ the equation becomes

$$\frac{d^2 X}{d(\ln r)^2} = -(Nqn/\epsilon_0)e^{-qX/kT}, \tag{30}$$

which is identical to the planar single-species equation just solved. We therefore apply its integral (19) and solution (20) to find that

$$\tfrac{1}{2}\epsilon_0\left(r\frac{d\phi}{dr} - \frac{2kT}{q}\right)^2 - NnkTr^2 e^{-q\phi/kT} = -\tfrac{1}{2}\epsilon_0 A^2, \tag{31}$$

and

$$\phi(r) = -\frac{kT}{q} \ln\left[\frac{\epsilon_0 A^2}{NnkTr^2} \sec^2\left(\frac{qA}{2kT} \ln\frac{r}{r_0}\right)\right].$$

(32)

This solution was first given in 1951 [10, 11].
   From (31) at $r = a$,

$$A^2 = 2N(kT/\epsilon_0)a^2 ne^{-q\phi_a/kT} - (aE_a + 2kT/q)^2.$$

(33)

The field $E_a$ can be chosen to make the second term either vanish or dominate; consequently both $A > 0$ and $A < 0$ must be examined. The gauge boundary condition is that $\phi = \phi_a$ at $r = a$, leaving only $n$ to be determined self-consistently from (29). In order to set up the result, define first the dimensionless parameters

$$\eta = 1 + \frac{qaE_a}{2kT}, \quad \mu = \eta \ln\frac{b}{a}, \quad \nu = 1 + \frac{NL}{2\pi\eta}$$

(34)

and the transcendental equation

$$\tan(\theta + \mu \cot\theta) = \nu \tan\theta,$$

(35)

where it is required that the interval $[\theta, \theta + \mu \cot\theta]$ not include a half-integer multiple of $\pi$, ensuring the normalisation integral (29) is convergent. The potential (32) within the plasma is [2]

$$\phi(r) = \phi_a - \frac{kT}{q} \ln\left[\frac{a^2}{r^2} \frac{\sec^2\{\theta + \eta \cot\theta \ln(r/a)\}}{\sec^2\theta}\right]$$

(36)

and the radial electric field is

$$E(r) = \frac{2kT}{qr}[\eta \cot\theta \tan\{\theta = \eta \cot\theta \ln(r/a)\} - 1]$$

(37)

which reduces at $r = b$, using (34), (35), to $(Nq + 2\pi\epsilon_0 aE_a)/2\pi\epsilon_0 b$ as demanded by Gauss' law. The particle density is

$$\frac{1}{Lr^2}\eta^2 \cot^2\theta \sec^2\{\theta + \eta \cot\theta \ln(r/a)\}.$$

(38)

Appearance of the field of the fixed charges in the combination $\eta = 1 + qaE_a/2kT$ indicates that this field competes against or enhances (depending on the sign) a geometric repulsion effect away from the axis.
   Results for negative integration constant $A^2$ in (31) are now quoted: let $\theta'$ satisfy the analogue of (35),

$$\coth(\theta' - \mu \tanh\theta') = \nu \coth\theta'$$

(39)

where the interval $[\theta' - \mu \tanh \theta', \theta']$ excludes the origin. Then

$$\phi(r) = \phi_a - \frac{kT}{q} \ln\left[\frac{a^2}{r^2} \frac{\cosech^2\{\theta' - \eta \tanh \theta' \ln(r/a)\}}{\cosech^2 \theta'}\right], \tag{40}$$

$$E(r) = \frac{2kT}{qr}[\eta \tan \theta' \coth\{\theta' - \eta \tanh \theta' \ln(r/a)\} - 1], \tag{41}$$

with density

$$\frac{1}{Lr^2}\eta^2 \tanh^2 \theta' \cosech^2\{\theta' - \eta \tanh \theta' \ln(r/a)\}. \tag{42}$$

It is this combination which takes over at sufficiently high temperature, corresponding in the limit to uniform plasma, with the field given via Gauss' law and the potential by integrating it [2].

Let us now consider a line charge immersed in the plasma. This corresponds to shrinking of the inner radius $a$ to zero while the product $aE_a = Q_i'/2\pi\epsilon_0$, and hence $\eta = 1 + qQ_i'/4\pi\epsilon_0 kT$ and $\nu$, remain constant. From (34) we take $\mu \to \pm\infty$ according to the sign of $\eta$. Consider first $\eta > 0$ and (hence) $\nu > 1$, which includes the case of line, plasma charges of like polarity $\eta > 1$. The hyperbolic form of solution is appropriate and as $\mu$ becomes large and positive (39) solves as $\theta' \approx \mu + \coth^{-1} \nu$, which on substitution into (41) gives the limit [2]:

$$E(r) = \frac{2kT}{qr}[\eta \coth\{\coth^{-1} \nu + \eta \ln(b/r)\} - 1]. \tag{43}$$

This expression has the property

$$\lim_{r \to 0} 2\pi \epsilon_0 r E(r) = Q_i', \tag{44}$$

as expected from Gauss' law. The plasma density near the line charge is proportional to the $(2\eta - 2)$-th power of radius, predictable by: assuming it is small on the basis of like charge repulsion, taking the potential as due entirely to the line charge, and the density then proportional to $\exp(-q\phi/kT)$. All of this remains valid for plasma polarized oppositely to the line charge provided that $\eta > 0$ so that $-qQ_i' \le 4\epsilon_0 kT$, a criterion independent of the amount of plasma! Here, the attraction of opposite charges competes with the geometric repulsion effect from $r = 0$.

We quote the result for zero line charge ($\eta = 1$), equivalent to a plasma introduced into a cylinder of radius $b$:

$$E(r) = \frac{4kT}{qb} \frac{r/b}{(4\pi/NL) + 1 - (r/b^2}, \tag{45}$$

with plasma density

$$\frac{4}{Lb^2}\left(\frac{4\pi}{NL} + 1\right)\left[\frac{4\pi}{NL} + 1 - \left(\frac{r}{b}\right)^2\right]^{-2}. \tag{46}$$

To see what happens when the line and plasma charge polarities are opposite and the plasma temperature is sufficiently low that $\eta < 0$, examine the limit $\mu \to -\infty$ of (39) – (42). This case is further subdivided into $0 < \nu < 1$ and $\nu < 0$, where from (34) $\nu = 1 + NL/2\pi\eta$. All values of fixed:plasma charge ratio are covered by these and by (43). For $0 < \nu < 1$ there are two solutions for $\theta'$ in the limit $a \to 0, \mu \to -\infty : \theta' = \pm \tanh^{-1} \nu$, both of which yield the same physics. Substitution of the upper branch into (41) gives

$$E(r) = \frac{2kT}{qr}[\eta\nu \coth\{\tanh^{-1}\nu - \eta\nu \ln(r/a)\} - 1] \tag{47}$$

which on again taking $a \to 0$ reduces finally to

$$E(r) = 2(\eta\nu - 1)kT/qr \tag{48}$$

$$= (Q'_t + Nq)/2\pi\epsilon_0 r. \tag{49}$$

The line charge has attracted the entire, oppositely charged plasma on to itself and the concentration is a delta function, even though the temperature is non-zero. The condition $0 < \nu < 1$ implies that

$$-\frac{Q'_k}{Nq} > 1 + \frac{4\pi\epsilon_0 kT}{Nq^2}, \tag{50}$$

so that the magnitude of the plasma charge is smaller by a non-zero amount than that of the line charge which it tries to neutralize. Moreover the criterion is temperature-dependent. This can best be understood by observing that the density is proportional to $\exp(-q\phi/kT)/\langle\exp(-q\phi/kT)\rangle$, and whenever $\exp(-q\phi/kT)$ has a non-integrable singularity the denominator pulls this expression to zero elsewhere while ensuring that it is normalised: the defining properties of the delta function. By integrating (49) is is found that $\exp(-q\phi/kT)$ is equal to the $p$-th power of $r$ where $p = q(Q'_t + Nq)/2\pi\epsilon_0 kT < 0$; non-integrability at the origin on multiplying by the element $2\pi r dr$ corresponds to $p \le -2$, giving immediate rise to condition (50). Charge condensation is revealed as a consequence of the nonlinearity of the PB equation. Rubinstein [12] treats it as a peculiar type of second order phase transition and provides historical references.

It remains to study $\nu < 0$, which includes the case of sufficient oppositely charged plasma to be capable of neutralizing the line charge. Equation (39) has now no real solution for $\theta'$ in the limit $a \to 0, \mu \to -\infty$ and the trigonometric solution (36)–(38) is appropriate. The limiting solution of (35) is

$$\theta = (j + \frac{1}{2})\pi \pm \left[\frac{\pi}{\mu} + \left(1 - \frac{1}{\nu}\right)\frac{\pi}{\mu^2}\right] + O(\gamma^{-3}) \tag{51}$$

where $j$ is an integer. On writing $\eta \ln(r/a)$ in (37) as $\mu - \eta \ln(b/r)$, substituting (51) for $\theta$ and taking $\mu \to -\infty$, the field is

$$E(r) = \frac{2kT}{qr}\left[\frac{\eta\nu}{1 + \eta\nu \ln(b/r)} - 1\right] \tag{52}$$

with a central line charge per unit length

$$\lim_{r \to 0} 2\pi\epsilon_0 r E(r) = -\frac{4\pi\epsilon_0 kT}{q}, \tag{53}$$

independent of (though including) the line charge! The density (38) other than at the origin becomes

$$\frac{1}{Lr^2}\left[\frac{\eta\nu}{1 + \eta\nu \ln(b/r)}\right]^2, \tag{54}$$

which integrates over the plasma volume to give a charge $(Nq + Q'_i + 4\pi\epsilon_0 kT/q)$; upon multiplication by $2\pi r dr$ and integration, (54) "just" converges: there is no safety margin in the power index. Added to the central charge (53), the volume charge gives just $Nq + Q'_i$, the total (plasma + fixed) charge. A variable amount of plasma charge condenses onto the fixed line charge, such that the net central charge is independent of the magnitude of the line charge (provided that $\nu < 0$). But the condensate always falls short of neutralization.

In summary, as the fixed line charge varies from zero to $-4\pi\epsilon_0 kT/q$ there is no condensation; when it exceeds $-(4\pi\epsilon_0 kT/q + Nq)$ the entire plasma condenses; and in between the amount of condensate lags behind the fixed charge by this amount $-4\pi\epsilon_0 kT/q$.

All of these results for $a = 0$ could have been derived *ab initio*. The PB equation and the form (32) of its solutions are unchanged. The constants of integration are now disposed in terms of $E(b)$, given trivially by Gauss' law, and $\phi(b)$. The consistency condition for $n$ yields a density which integrates to $Nq$ less any condensate; combining this with the requirement that $\lim_{r\to 0}(-2\pi\epsilon_0 r d\phi/dr)$ equals $Q'_i$ plus any condensate allows the amount of condensate to be found.

This discussion of the PB equation in cylindrical geometry concludes with a brief discussion of the two species plasma with equal and opposite particle charge: $q_1 = -q_2 = q$. In planar geometry this case was far more involved than the single species and the same is true here. No first integral is known; but the power-law transformation of the equation is related to a special case of the third Painlevé transcendent $P$ III [13]. It therefore has the Painlevé property that any branch points or essential singularities are fixed in being independent of the integration constants. (Detailed classification of Painlevé differential equations is given in [14, 15].) The equation in Painlevé form may be regarded as the compatability condition for existence of a solution to a system of linear homogeneous first order differential equations, just as the existence condition for a solution to a system of linear homogeneous algebraic equations is that the determinant of coefficients must vanish. Solutions to the linear differential system are characterised completely by their behaviour at singular points, called the *monodromy data*, and analyticity elsewhere. A great deal of progress can thereby be made without need to solve the PB equation [13].

Numerical solutions of the equations are given in [13, 16] and in [17], wherein condensation is exhibited numerically. Condensation onto a fixed line charge can clearly be no greater, on inclusion of a new species of like polarity to the line charge, than before. In fact, condensation of oppositely charged species *must* occur on to line charges of magnitude greater than $4\pi\epsilon_0 kT/|q|$. For suppose that condensation does not occur. Then, sufficiently close to the line charge, $\phi \sim -(Q'_i/2\pi\epsilon_0)\ln r$, and the density factor $\exp(-q\phi/kT)$ has a non-integrable singularity at the origin if $|Q'_i|$ exceeds the asserted amount. (Density of the like charges vanishes as a positive power of and can be asymptotically neglected.) Once again the condensate is expected to be the minimum bringing $\exp(-q\phi/kT)$ to integrability, provided that there is enough plasma charge to do it; otherwise it all condenses. This reasoning generalises obviously to multi-species plasmas.

For small $r$ the solution approaches that of the single species situation. If the plasma has no outer boundary the potential at large $r$ is asymptotic to solutions of the linear, Debye-Hückel equation. Connection of solutions at small and large $r$ is established in [13] using techniques of monodromy, and otherwise in [16, 18]. Connection must take heed of condensation; the Debye-Hückel equation fails dramatically at small $r$ and cannot account for the phenomenon. Approximate solutions are given in [4, 5, 12], and bounds on exact solutions in [3, 19, 20, 21].

## 6. THE TWO-DIMENSIONAL PROBLEM

Here we examine the PB equation for a single and a two (oppositely) charge species plasma, allowing for arbitrary two-dimensional spatial variation. First the single species problem, which in the absence of fixed charges is

$$\epsilon_0 \left( \frac{\partial^2}{\partial x^2} + \frac{\partial^2}{\partial y^2} \right) \phi = -Nqne^{-q\phi/kT}. \tag{55}$$

The general solution of this was found by Liouville [22]. Here is is solved using Bäcklund transformations, following the text of Drazin [23]. Recast it in characteristic form, with independent variables

$$u = \tfrac{1}{2}(x + iy), \quad v = \tfrac{1}{2}(x - iy) \tag{56}$$

and work for convenience with the dimensionless version

$$\frac{\partial^2 \phi}{\partial u \partial v} = e^\phi. \tag{57}$$

Now introduce the coupled first order equations

$$\frac{\partial \phi}{\partial u} + \frac{\partial \psi}{\partial u} = \sqrt{2} e^{\frac{1}{2}(\phi - \psi)}, \tag{58}$$

$$\frac{\partial \phi}{\partial v} - \frac{\partial \psi}{\partial v} = \sqrt{2} e^{\frac{1}{2}(\phi + \psi)}, \tag{59}$$

which constitute a Bäcklund transformation between $\phi$ and $\psi$. By differentiating (58) with respect to $v$, (59) with respect to $u$, eliminating first derivatives on the RHS using the "other" of these, and adding and subtracting the results we recover (57) together with the simpler equation

$$\frac{\partial^2 \psi}{\partial u \partial v} = 0. \tag{60}$$

So far we have proved that (58), (59) imply (57), (60), but not the converse. Our scheme is to solve (60) and substitute the result into (58), (59); if these together can be solved consistently for $\phi$ a solution of (57) is generated. The general solution of (60) is

$$\psi = F(u) + G(v) \tag{61}$$

for arbitrary functions $F$, $G$. Substitution into (58) gives

$$e^F \frac{\partial}{\partial u}(\phi + F - G) = \sqrt{2} e^{\frac{1}{2}(\phi + F - G)} \tag{62}$$

with solution

$$-\sqrt{2}e^{\frac{1}{2}(\phi+F-G)} = U(u) + g(v) \tag{63}$$

where

$$U(u) = \int^u du' e^{-F(u')} \tag{64}$$

and $g$ is an arbitrary function. Likewise, (59) gives

$$-\sqrt{2}e^{-\frac{1}{2}(\phi+F+G)} = V(v) + f(u), \tag{65}$$

where

$$V(v) = \int^v dv' e^{G(v')} \tag{66}$$

and $f$ is an arbitrary function. (63) and (65) are compatible if $f = U$, $g = V$. On eliminating $F$ and $G$ in favour of $U$ and $V$, either now gives the general solution of (57) as

$$\phi(u,v) = \ln\left[\frac{2U'(u)V'(v)}{\{U(u) + V(v)\}^2}\right] \tag{67}$$

where a prime denotes differentiation with respect to argument, and $U$ and $V$ are arbitrary functions.

If we demand that the solution be real for real $x$ and $y$, so that $v = u^*$, the function $V$ is the conjugate of the function $U$. The solution still contains two arbitrary functions, but these are now real.

Because the boundary conditions are often more obvious for a point charge than for a diffuse plasma charge, it is easier to specify the Green's integral form of the PB equation than it is to specify the differential form plus the boundary conditions. But we know more about solving nonlinear differential than integral equations. The best of both worlds is had by substituting general solutions of the differential equation into the integral version to give a simpler problem for the functions of integration. Write the PB equation temporarily as $\nabla^2\phi = -\rho_i(r) - \nu(r)F[\phi(r)]$, and recall that Green's function satisfies $\nabla^2 G(r,r') = -\delta(r - r')$. Then

$$\phi(r) = \epsilon_0^{-1}\int_V dr' G(r,r')\rho_i(r') + \int_{V_p} dr' G(r,r')F[\phi(r')], \tag{68}$$

where $V = V_p + \overline{V}_p$ denotes all space. Write the LHS of this as

$$\phi(r) = \int_{V_p} dr'\phi(r')\delta(r - r'), \quad r \in V_p \tag{69}$$

$$= -\int_{V_p} dr'\phi(r')\nabla_{r'}^2 G(r,r') \tag{70}$$

$$= -\int_{V_p} d\mathbf{r}' G(\mathbf{r},\mathbf{r}')\nabla_{\mathbf{r}'}^2\phi(\mathbf{r}')-$$

$$-\int_{\partial V_p} d\mathbf{S} \cdot [\phi(\mathbf{r}')\nabla_{\mathbf{r}'} G(\mathbf{r},\mathbf{r}') - G(\mathbf{r},\mathbf{r}')\nabla_{\mathbf{r}'}\phi(\mathbf{r}')] \tag{71}$$

on using the symmetry of $G(\mathbf{r},\mathbf{r}')$ and Green's vector identity. Upon equating this to the RHS of (68) the volume integrals cancel because $\phi(\mathbf{r}')$ satisfies the PB equation, leaving the boundary condition

$$\int_{\partial V_p} d\mathbf{S}_{\mathbf{r}'} \cdot [\phi(\mathbf{r}')\nabla_{\mathbf{r}'} G(\mathbf{r},\mathbf{r}') - G(\mathbf{r},\mathbf{r}')\nabla_{\mathbf{r}'}\phi(\mathbf{r}')]$$

$$= -\int_{\overline{V}_p} d\mathbf{r}' G(\mathbf{r},\mathbf{r}')\rho_i(\mathbf{r}'), \tag{72}$$

where the integral on the RHS is over fixed charge residing *outside* the plasma in the complement $\overline{V}_p$ of the plasma volume. (For a conducting boundary the RHS is zero.) This condition is linear in the potential, though not in the functions of integration.

This prescription may, for example, be combined with (67) to give the potential for a single species in arbitrary two-dimensional geometry, with Green's function $G(\mathbf{r},\mathbf{r}') = -(1/2\pi) \ln |\mathbf{r} - \mathbf{r}'|/D$ for the homogeneous boundary condition $\nabla G = 0$ at infinity: $D$ is any problem-defined length.

If dependence on $x$ and $y$ is presumed to depend only on the combination $x^2 + y^2 = r^2 \propto u^* u$, so that the RHS of (67) is a function of $u^* u$ alone, the resulting functional equation can be shown to have a unique solution corresponding to the cylindrically symmetrical solution (32).

Turn now to the two-species problem ($q_1 = -q_2 = q$) in two dimensions. In dimensionless characteristic form, and with a suitable shift of potential, the PB equation becomes

$$\frac{\partial^2 \phi}{\partial u \partial v} = \sinh \phi, \tag{73}$$

the "sinh-Gordon" equation [23]. The Bäcklund transformation pair

$$\frac{\partial}{\partial \binom{u}{v}}(\phi \pm \psi) = 2\lambda^{\pm 1} \sinh \tfrac{1}{2}(\phi \mp \psi) \tag{74}$$

unfortunately gives for $\psi$ an equation identical to (73). However the trivial solution $\psi = 0$ generates a non-trivial solution for $\phi$ which depends on a linear combination of $u$ and $v$: it is the one-dimensional solution (20), rotated. This can now be used in (74) to generate a further (and genuinely two-dimensional) solution for $\phi$, and so on, giving a hierarchy of soliton solutions [23] but never the general solution. The general solution has been obtained, at least for negative temperature, by Ting et al [24] by means of the inverse scattering transform and complex analysis. As usual with inverse scattering, the solution is left in implicit form: reduction to an associated linear problem is the key. That the solution remains implicit is predictable, for the cylindrically symmetrical case is Painlevé, and the full two-dimensional problem can be no simpler (though it may have the Painlevé property generalized to partial differential equations [25]).

## 7. THREE-DIMENSIONAL AND SPHERICALLY SYMMETRICAL PROBLEMS

In three dimensions, or any special symmetry of it, no solution in closed form is known. The three-dimensional problem for a single species, however, can be transformed to a functional problem. Stoyanov [26] has given a spinorial reduction which yields the general solution in four dimensions (easier to find than three, ironically); its form is reminiscent of the two-dimensional solution (67). Special cases can be sought in which the dependence is only on three Cartesian coordinates, thereby generating a functional problem yet to be explored.

Consider now spherical symmetry, when the PB equation is of general Emden type [27]. Examine first the single species case:

$$\frac{1}{r^2}\frac{d}{dr}\left(r^2\frac{d\phi}{dr}\right) = -(Nqn/\epsilon_0)e^{-q\phi/kT}, \tag{75}$$

where $r$ is now spherical not cylindrical. Once again this is invariant under a stretch of coordinates and a translation of $\phi$ which leaves the combination $r^2\exp(-q\phi/kT)$ invariant, so choose

$$R = (2NLn)^{1/2}r, \quad \xi = \ln R, \quad e^{\xi} = R^2 e^{-q\phi/kT} \tag{76}$$

to give the dimensionless form

$$\frac{d^2X}{d\xi^2} + \frac{dX}{d\xi} - 2 - e^{\xi} = 0. \tag{77}$$

A trivial unphysical solution $X = \ln(-2)$ exists; there is no other for $\phi$ of the form of an ascending power series in $r$ plus a term behaving at small $r$ as $\ln r$. On defining $\alpha \equiv dX/d\xi$, (77) reduces to the first order equation

$$\alpha\frac{d\alpha}{dX} + \alpha - 2 - e^{\xi} = 0. \tag{78}$$

If this can be solved for $\alpha(X)$, the solution of the PB equation (75) is given by quadrature:

$$\ln(2NLnr^2) = 2\int^{-(q\phi/kT)+\ln(2NLnr^2)} dX/\alpha(X) \tag{79}$$

and reversion.

(78) simplifies to an Abel equation, with algebraic form, upon taking $\exp(X)$ as dependent variable. This is not reducible to Riccati form and so does not have the Painlevé property [28]. Therefore not only is the single species spherical PB equation non-Painlevé, but also its generalizations to several species and to arbitrary three-dimensional geometry. Conjecture has it that the monodromy and inverse scattering approaches are inapplicable to non-Painlevé equations.

Though no general solution is available, a series solution of (75) can successfully be sought. With $\Phi = q\phi/kT$ and $R$ as in (76), and denoting differentiation with respect to $R$ by a prime, the PB equation becomes

$$\Phi'' + 2R^{-1}\Phi' + e^{-\Phi} = 0. \tag{80}$$

We look for a solution

$$\Phi = \sum_{m=0}^{\infty} a_m R^{m+s}, \quad a_0 \neq 0 \tag{81}$$

and assume *pro tem* that $s > 0$, so that $\Phi = 0$ at $R = 0$ and we can write

$$e^{-\Phi} \equiv F(R) = 1 + \sum_{m=1}^{\infty} f_m R^m. \tag{82}$$

Substitution of (81) and (82) into (80) yields an equation which is only consistent at small $r$ if $s = 2$, justifying our assumption. Choosing this value gives $a_0 = -\frac{1}{6}$, $a_1 = 0$ and the relation

$$(m+2)(m+3)a_m + f_m = 0, \quad m \geq 2. \tag{83}$$

But the $f$-series is the exponential of the $a$-series and we have $F' = -F\Phi'$, with a corresponding relation

$$(m+2)f_{m+2} = -\sum_{l=0}^{m}(m-l+2)f_l a_{m-l}; \quad f_1 = 0. \tag{84}$$

(83) can be substituted into (84) to give nonlinear recurrence relations for the $f_m$ (density) or the $a_m$ (potential). Tidied up, the result for the potential is [2]

$$\Phi = -\sum_{m=0}^{\infty} \frac{e_m}{2m+2} \left(\frac{R}{2\sqrt{3}}\right)^{2m+2} \tag{85}$$

where the $e_m$ are coefficients in an expansion of the electric field, satisfying

$$e_{m+1} = \frac{2(m+3)}{(m+1)(2m+5)} \sum_{l=0}^{m} e_l e_{m-l}, \quad e_0 = 1. \tag{86}$$

The density coefficients $f_m$ have been tabulated and numerical solutions given by Lampert and Martinelli [29].

The series solution contains no unspecified constants and cannot be fitted to arbitrary boundary conditions. It corresponds to $\Phi' = 0$ at $R = 0$, implying that there is no net point charge there, and to gauge $\Phi(0) = 0$. It is appropriate for the confinement of plasma within the sphere $r \leq b$. As usual $n$ is determined *a posteriori*, to be eliminated in favour of $b$.

For the two species $(q_1 = -q_2 = q)$ spherically symmetrical plasma, the PB equation is already proven to be non-Painlevé. Furthermore there is no series solution of the form (81). Recourse to numerical solution is necessary [17, 30, 31].

## 8. CHARGE CONDENSATION IN THREE DIMENSIONS

It is easy to show that a fixed point charge immersed in a Poisson-Boltzmann plasma attracts an equal and opposite charge arbitrarily close to itself (provided sufficient is present) and is, locally, precisely neutralized. This remarkable result was first proved by Lampert and Crandall [32] using a lengthier method. Here, we begin by demonstrating it for a plasma consisting of two species of opposite polarity. Suppose that there is a fixed point charge $Q_t$ at the point $\mathbf{r} = 0$ of the same polarity as $q_1$, and assume *pro tem* that there is no condensation of opposite plasma charge ($q_2$) onto it, or insufficient condensation to neutralize it. Then, sufficiently near the charge, the potential $\phi \sim O(r^{-1})$ with the same sign as $Q_t, q_1$, and $\exp(-q_2\phi/kT)$ behaves asymptotically as $\exp(+K/r)$ where $K$ is a positive constant. This form has a non-integrable singularity at $r = 0$, indicating that the density $\exp(-q_2\phi/kT)/\langle\exp(-q_2\phi/kT)\rangle$ is the delta function, and that *all* of the species $q_2$ has condensed on to the fixed charge — in contradiction to the original assumption of insufficient condensation or none. Inconsistency also arises on initially assuming over-condensation. The only possibility is precise neutralization.

If there is insufficient opposite plasma charge for neutralization all of it condenses, and the density of the species of like polarity behaves close by as $\exp(-K/r)$, $K > 0$, a very rapid fall-off indeed.

In view of this result the series solution (85) in fact allows for a neutralized fixed charge at the origin, provided that $n$ is calculated consistently. But by far the most important consequence of precise neutralization lies in revealing inapplicability of the linearized, Debye-Hückel theory. At high temperatures, or weak fields, the density factors $\exp(-q_s\phi/k) \approx 1 - q_s\phi/kT$ and the resulting linearised spherical equation has solution

$$\phi = kT\left(\sum_s N_s q_s n_s\right) \Big/ \left(\sum_s N_s q_s^2 n_s\right) + A r^{-1} e^{-\lambda r} + B r^{-1} e^{\lambda r} \tag{87}$$

where $\lambda^2 = (\epsilon_0 kT)^{-1} \sum_s N_s q_s^2 n_s$. The constant term is unimportant and $B$ is zero on physical grounds, leaving $\phi = A r^{-1} \exp(-\lambda r)$, an expression often quoted in textbooks wherein $A$ is established by comparison with the potential when no plasma is present: $A = Q_t/4\pi\epsilon_0$. But close to the charge the Debye-Hückel approximation is wildly inaccurate, and $A r^{-1} \exp(-\lambda r)$ should be matched to something different — which the neutralization argument shows is zero or a constant.

On writing the PB equation within the plasma for an assemblage of fixed point charges $Q_t$ at $\mathbf{r}_i$ as

$$\epsilon_0 \nabla^2 \phi = -\sum_i Q_i \delta(\mathbf{r} - \mathbf{r}_i) - \sum_s N_s q_s \frac{e^{-q_s\phi/kT}}{\langle e^{-q_s\phi/kT}\rangle}, \tag{88}$$

we see that there is a balance of singularities on the RHS, and that the $\nabla^2$ term is best viewed as a perturbation.

Suppose now that there are many species having opposite polarity to the fixed point charge. What are their proportions in the condensate? The result is that the species with the largest particle charge first condenses; then the next most highly charged; and so on until neutralization is reached. The previously heuristic result [12] is derived as follows. The PB equation (4) can be derived from an energy minimising variational principle $\delta I[\phi] = 0$, where

$$I[\phi] = \int d^3\mathbf{r}(\tfrac{1}{2}\epsilon_0|\nabla\phi|^2 - \rho_t\phi) - kT\sum_s N_s \ln(n_s[\phi]D^3) \tag{89}$$

and $D$ is any problem-defined characteristic length. Let the first $\ell$ species be those of the opposite polarity to the fixed charge $Q_t$, and denote the minimand, given this charge, by

$I[\phi; Q_t; N_1, \ldots, N_\ell]$. Denote the solution of the PB equation *without* the fixed charge but with the same boundary conditions as $\phi(\mathbf{r}; N_1, \ldots, N_\ell)$. Denote by $\gamma_i$ the number of particles of the $i$-th species in the condensate. Now minimise $I$ over all properties of the $\ell$ species making up the condensate:

$$\frac{d}{d\gamma_i} I[\phi(\mathbf{r}; N_1, \ldots, N_\ell); 0; N_1 - \gamma_1, \ldots, N_\ell - \gamma_\ell] = 0, \quad \sum_{i=1}^{\ell} \gamma_i q_i = -Q_t. \tag{90}$$

But

$$\frac{dI}{d\gamma_i} = \frac{\partial I}{\partial \phi} \frac{\partial \phi}{\partial \gamma_i} + \frac{\partial I}{\partial \gamma_i}, \tag{91}$$

and $\partial I / \partial \phi$ vanishes by construction, leaving

$$\frac{dI}{d\gamma_i} = kT \ln(n_i[\phi] D^3). \tag{92}$$

Upon writing this in full, and incorporating the neutralization constraint in (90) by means of the Lagrange multiplier $w$, we have

$$-kT \ln\left[V_p^{-1} \int_{V_p} d\mathbf{r}\, e^{-q_i \phi/kT}\right] = w q_i = 0, \quad i = 1, \ldots, \ell. \tag{93}$$

Dependence on the $\gamma_i$'s is implicit, through the potential. Rewrite this equation as

$$V_p^{-1} \int_{V_p} d\mathbf{r}\, e^{-q_i \phi/kT} = e^{-q_i \phi/kT}, \quad i = 1, \ldots, \ell. \tag{94}$$

This can be satisfied only if $\phi(\mathbf{r})$ is a constant within $V_p$ and $w$ takes that value. To prove this, order the species such that $|q_1| \leq |q_2| \leq \ldots \leq |q_\ell|$ and put $q_i = m_i q_1$, so that $m_1 = 1$ and the other $m_i$ exceed unity. Write $\exp(-q_i \phi/kT) \equiv f(\mathbf{r})$ and re-express (94) as

$$\left[V_p^{-1} \int_{V_p} d\mathbf{r}\, f(\mathbf{r})^{m_i}\right]^{1/m_i} = e^{-q_i w/kT}, \quad i = 1, \ldots, \ell. \tag{95}$$

But the LHS is shown by differentiation to be a non-decreasing function of $m_i$, and is stationary only for constant $f(\mathbf{r})$; the requirement that $\phi(\mathbf{r})$ be constant now follows.

But $\phi(\mathbf{r})$ satisfies the PB equation and is not in general constant within $V_p$, so that no solution of (94) exists and there is no smooth minimum of $I$ on the hyperplane $\sum_i q_i \gamma_i = -Q_t$ in $\gamma$-space. We therefore seek the smallest value of $I$ on this hyperplane where it intersects the surface of the hypercube $0 \leq \gamma_i \leq N_i$. This corresponds to $\gamma_i = N_i$ — complete condensation — for some species. The notion that $\exp(-q_i \phi_t/kT)/\langle \exp(-q_i \phi_t/kT) \rangle$ becomes "more singular" as $|q_i|$ increases implies that these species shall be those with greatest particle charge, completing the proof.

The result of introducing a fixed point dipole or multipole into the plasma is more bizarre still. Construct the dipole, of moment $M$, out of two fixed point charges of magnitude $\pm M/\sigma$ and separation $\sigma$, and let $\sigma \to 0$. As the magnitude of the charges increases these attract all of the plasma charge, of one polarity or the other, onto themselves. In the limit the entire plasma condenses! (Quadrupoles are obtained by a like process with dipoles, and so on.) This result is consistent with the greater singularity $r^{-2}$ of the radial dependence of the dipole potential than of a point charge ($r^{-1}$). Its manifest absurdity stems fom the fact that real dipoles are not point entities and from breakdown of the approximation implicit in the PB equation, discussed next. Results concerning charge condensation have been summarised by Garrett [33].

## 9. CONCLUSIONS

The Poisson-Boltzmann equation purports to describe the electrostatic potential within plasmas in thermal equilibrium. Nonlinearity renders it difficult to solve exactly, and very few solutions are known in closed form, although inverse scattering has solved a particular further case. Much information can nevertheless be extracted from numerical and series solutions, differential equation theory, and from looking at weak and strong field limits. In particular, the equation implies that a line charge immersed in plasma may attract a non-zero amount of opposite charge arbitrarily close to itself and undergo partial neutralization. Point charges are precisely neutralized in this manner, while a point dipole causes the entire plasma to condense on to either side of it. In such circumstances the linear Debye-Hückel equation fails dramatically.

We finish by indicating the limits of Poisson-Boltzmann analysis. It is inexact, for point charges are not in practice perfectly screened. Discrepancy is due partly to the potential itself being a macroscopic average, but above all to the neglect of interparticle correlations. The PB equation is derived by Maximum Entropy in the phase space of a single particle under a nonlinear energy constraint instead of in the full $6N$-dimensional phase space with a linear constraint; making it easier to find a simple result. The equation is a continuum approximation corresponding to smearing out of charged particles, such that the smeared parts of each particle are correlated with each other but not with elements of other smeared particles. Onsager [34] has indicated the inconsistency of this scheme, though its predictions should still be accurate for small electrostatic:kinetic energy density ratio. Correction can be made [35, 36] but the theory loses its simplicity: the full phase space is used and the assumptions concerning correlations, leading after lengthy analysis to Poisson-Boltzmann, are removed.

Provided that the user is aware of its limitations, the Poisson-Boltzmann equation provides a luminous application of Maximum Entropy analysis. The equation is an invaluable tool in equilibrium plasma theory, and also — with attractive inter-particle forces — in gravitational astrophysics [37].

## REFERENCES

[1] Garrett, A.J.M. 'Maximum Entropy with Nonlinear Constraints'; these Proceedings.

[2] Garrett, A.J.M. & Poladian, L. (1988) 'Refined Derivation, Exact Solutions and Singular Limits of the Poisson-Boltzmann Equation', *Ann. Phys.* **188**, 386-435.

[3] MacGillivray, A.D. (1972) 'Lower Bounds on Solutions of the Poisson-Boltzmann Equation near the Limit of Infinite Dilution', *J. Chem. Phys.* **57**, 4071-4075.

[4] Friedman, A. & Tintarev, K. (1987) 'Boundary Asymptotics for Solutions of the Poisson-Boltzmann Equation', *J. Differential Eqns.* **69**, 15-38.

[5] Tintarev, K. (1987) 'Fundamental Solution of the Poisson-Boltzmann Equation', in: I.W. Knowles & Y. Saito (eds), *J. Differential Equations and Mathematical Physics* (Proceedings, Birmingham, Alabama, 1986), Lecture Notes in Mathematics **1285**, Springer, Berlin, pp. 480-485.

[6] Berestycki, H. & Lions, P.-L. (1983) 'Nonlinear Scalar Field Equations, I: Existence of a Ground State, and II: Existence of Infinitely Many Solutions', *Arch. Rat. Mech. Anal.* **82**, 313-346 and 347-375.

[7] Atkinson, F.V. & Peletier, L.A. (1986) 'Ground States of $-\Delta u = f(u)$ and the Emden-Fowler Equation', *Arch. Rat. Mech. Anal.* **93**, 103-127.

[8] Debye, P. & Hückel, E. (1923) 'Zur Theorie der Elektrolyte, I: Gefrierpunktserniedrigung und verwandte Erscheinungen', *Physik. Zeits.* **24**, 185-206.

[9] Ninham, B.W. & Parsegian, V.A. (1971) 'Electrostatic Potential between Surfaces Bearing Ionisable Groups in Ionic Equilibrium with Physiologic Saline Solution', *J. Theoret. Biol.* **31**, 405-428.

[10] Fuoss, R.M., Katchalsky, A. & Lifson, S. (1951) 'The Potential of an Infinite Rod-Like Molecule and the Distribution of the Counter-Ions', *Proc. Nat. Acad. Sci. USA: Chemistry* **37**, 579-589.

[11] Alfey, T., Berg, P.W. & Morawetz, H. (1951) 'The Counterion Distribution in Solutions of Rod-Shaped Polyelectrolytes', *J. Polymer Sci.* **7**, 543-547.

[12] Rubinstein, I. (1986) 'Counterion Condensation as an Exact Limiting Property of Solutions of the Poisson-Boltzmann Equation', *SIAM J. Appl. Math.* **46**, 1024-1038.

[13] McCaskill, J.S. & Fackerell, E.D. (1988) 'Painlevé Solution of the Poisson-Boltzmann Equation for a Cylindrical Polyelectrolyte in Excess Salt Solution', *J. Chem. Soc. Faraday Trans.2*, **84**, 161-179.

[14] Ince, E.L. (1956) *Ordinary Differential Equations*, Dover, New York. Reprint of the original 1926 edition.

[15] Davis, H.T. (1962) *Introduction to Nonlinear Differential and Integral Equations*, Dover, New York.

[16] MacGillivray, A.D. & Winkleman, J.J. (1966) 'On an Asymptotic Solution of the Poisson-Boltzmann Equation – The Moderately Charged Cylinder', *J. Chem. Phys.* **45**, 2184-2188.

[17] Lampert, M.A. & Martinelli, R.U. (1985) 'Buffering of Charge in Non-Linear Poisson-Boltzmann Theory', *Chem. Phys. Lett.* **121**, 121-123.

[18] Alexandrowicz, I. & Katchalsky, A. (1963) 'Colligative Properties of Polyelectrolyte Solutions in Excess of Salt', *J. Polymer Sci.* **A 1**, 3231-3260.

[19] MacGillivray, A.D. (1972) 'Upper Bounds on Solutions of the Poisson-Boltzmann Equation near the Limit of Infinite Dilution', *J. Chem. Phys.* **56**, 80-83.

[20] MacGillivray, A.D. (1972) 'Bounds on Solutions of the Poisson-Boltzmann Equation near Infinite Dilution – The Moderately Charged Case', *J. Chem. Phys.* **57**, 4075-4078.

[21] Lampert, M.A. & Crandall, R.S. (1980) 'Non-Linear Poisson-Boltzmann Theory for Polyelectrolyte Solutions: The Counterion Condensate Around a Line Charge as a $\delta$-Function', *Chem. Phys. Lett.* **72**, 481-486.

[22] Liouville, J. (1853) 'Sur l'equation aux differences partielles $d^2 \log \lambda / du dv \pm \lambda / 2a^2 = 0$', *J. Math. Pures. Appl.* **18**, 71-72.

[23] Drazin, P.G. (1984) *Solitons,,* first corrected edition, Cambridge University Press, Cambridge, UK. Section 7.2.

[24] Ting, A.C., Chen, H.H. & Lee, Y.C. (1987) 'Exact Solutions of a Nonlinear Boundary Value Problem: the Vortices of the Two-Dimensional Sinh-Poisson Equation', *Physica D.* **26**, 37-66.

[25] Weiss, J., Tabor, M. & Carnevale, G. (1983) 'The Painlevé property for partial differential equations', *J. Math, Phys.* **24**, 522-526.

[26] Stoyanov, D.T. (1986) 'On the Classical Solutions of the Liouville Equation in a Four-Dimensional Euclidean Space', *Lett. Math. Phys.* **12**, 93-96.

[27] reference [15], Chapter 12, section 7.

[28] reference [15], Chapter 3, section 8.

[29] Lampert, M.A. & Martinelli, R.U. (1984) 'Solution of the Non-Linear Poisson-Boltzmann Equation in the Interior of Charged, Spherical and Cylindrical Vesicles. I: The High Charge Limit', *Chem. Phys.* **88**, 399-413.

[30] Lampert, M.A. & Crandall, R.S. (1979) 'Spherical, Non-Linear Poisson-Boltzmann Theory and its Debye-Hückel Linearization', *Chem. Phys. Lett.* **68**, 473-476.

[31] Martinov, N., Ouroushev, D. & Chelebiev, E. (1986) 'New types of polarization following from the non-linear spherical radial Poisson-Boltzmann Theory Equation, *J. Phys.* **A 19**, 1327-1332.

[32] Lampert, M.A. & Crandall, R.S. (1980) 'Fate of the Coulomb singularity in nonlinear Poisson-Boltzmann theory: The point charge as an electrically invisible object', *Phys. Rev.* **A 21**, 362-366.

[33] Garrett, A.J.M. (1988) 'Screening of point charges in two and three dimensions', *Phys. Rev.* **A 37**, 4354-4357.

[34] Onsager, L. (1988) 'Theories of Concentrated Electrolytes', *Chem. Rev.* **13**, 73-89.

[35] Fixman, M. (1979) 'The Poisson-Boltzmann equation and its application to polyelectrolytes', *J. Chem. Phys.* **70**, 4995-5005.

[36] Attard, P., Mitchell, D.J. & Ninham, B.W. (1988) 'Beyond Poisson-Boltzmann: Images and correlations in the electric double layer. I: Counterions only', *J. Chem. Phys.* **88**, 4987-4996. ' — II: Symmetric electrolyte', *J. Chem. Phys.* **89**, 4358-4367.

[37] Kiessling, M.K.-H. (1989) 'On the Equilibrium Statistical Mechanics of Isothermal Classical Self-Gravitating Matter', *J. Stat. Phys.* **55**, 203-257.

# LINEAR INVERSION BY THE MAXIMUM ENTROPY METHOD WITH SPECIFIC NON-TRIVIAL PRIOR INFORMATION

V. A. MACAULAY and B. BUCK
*Department of Theoretical Physics*
*1, Keble Road*
*Oxford*
*OX1 3NP*
*United Kingdom*

ABSTRACT. Here we present the MaxEnt solution of an inverse problem in scattering physics, the recovery of a nuclear charge density from noisy and incomplete measurements of its Fourier transform. Prior information on the charge density is used to motivate a Fourier-Bessel expansion and in addition to restrict the space of feasible reconstructions sufficiently to produce a convergent error estimate.

## 1. Introduction

To probe a nuclear charge distribution, high-energy electrons are scattered in the coulomb potential generated by the charge. In the case of light nuclei, it can be assumed that no more than a single scattering occurs at a given nucleus; this is the first Born approximation. The differential cross-section for the process, when divided by the cross-section for scattering from a point charge, gives $|F(\mathbf{q})|^2$, the squared form factor, conventionally expressed as a function of the momentum transfer $\mathbf{q}$. Then

$$F(\mathbf{q}) = \int e^{i\mathbf{q}\cdot\mathbf{r}} \rho(\mathbf{r}) \mathrm{d}^3\mathbf{r}, \qquad (1.1)$$

where $\rho(\mathbf{r})$ is the nuclear charge distribution, the volume integral of which is scaled to unity. To retrieve $\rho(\mathbf{r})$ from the data, this equation has to be inverted. Specializing to spherical nuclei gives

$$F(q) = 4\pi \int r^2 j_0(qr) \rho(r) \mathrm{d}r. \qquad (1.2)$$

Here $j_0(x)$ is the zeroth-order spherical Bessel function of the first kind, $\sin(x)/x$. The nucleus $^4$He appears to satisfy the conditions required for the validity of eqn(1.2) and this is the case we will treat here. The phase ambiguity of $F(q)$, arising because it is $|F|^2$ that is measured, reduces to one of sign and is resolved by the plausible identification of the diffraction minima of $|F|$ as points where the sign of $F$ changes.

The data come from [1, 2] and are massaged into the form of $M$ triplets $(q_m, \bar{F}_m, \sigma_m)$ where $q_m$ is a momentum transfer, assumed known with negligible error, $\bar{F}_m$ is an estimate

P. F. Fougère (ed.), Maximum Entropy and Bayesian Methods, 273–280.

of the form factor at $q_m$ and $\sigma_m$ is an estimate of the root mean square error of $\bar{F}_m$. For this analysis we assume ourselves to be ignorant of how $\bar{F}_m$ and $\sigma_m$ are obtained (for instance, whether there are averages over several runs or single estimates).

## 2. MaxEnt Inversion

Direct inversion of eqn(1.2) is, of course, impossible. Many charge densities are compatible with the data and so the inversion must necessarily be probabilistic.

The problem as it stands does not have sufficient structure for an analysis with Bayes' theorem to be possible. In particular we have no sampling distribution or, equivalently, likelihood. We are in what Jaynes has called [3] the exploratory phase of the problem and so must look to the MaxEnt principle. Then only an hypothesis space (e.g., a set of propositions along the lines of $\{A_i : \text{'The charge density is within an infinitesimal distance of } \rho_i(r)\text{'}\}$—this will be made precise later) and testable constraints on $\text{prob}(A_i)$ (the measured data) are required.

The probabilistic entity with which it is necessary to deal in this problem is a probability functional of the charge density given some information $\mathcal{X}$, $p[\rho(r)|\mathcal{X}]$. In addition, a generalized volume element in the function space $D(\rho)$ must be supplied. Then $\int_{\rho \in \Sigma} p[\rho(r)|\mathcal{X}]D(\rho)$ is the probability of a density in a sub-region $\Sigma$ of function space. The entropy of this probability functional is

$$S = -\int p[\rho(r)] \ln\left(\frac{p[\rho(r)]}{p_0[\rho(r)]}\right) D(\rho), \qquad (2.1)$$

where $p_0$ is a prior functional in which is encoded any knowledge of $\rho$ existing before the data are taken.

The constraints on $p[\rho]$ are:

$$\bar{F}_m = \int F_m[\rho]p[\rho]D(\rho), \qquad (2.2a)$$

$$\sigma_m^2 = \int \left(F_m[\rho] - \bar{F}_m\right)^2 p[\rho]D(\rho), \qquad (2.2b)$$

$$1 = \int p[\rho]D(\rho), \qquad (2.2c)$$

where $F_m[\rho]$ is the RHS of eqn(1.2) evaluated at $q = q_m$. In words, the expected value of the form factor is constrained to be equal to the measured form factor, the expected deviation to the measured error and the distribution is to be normalized.

The entropy of $p[\rho]$ is maximized subject to these constraints by Lagrange's method of undetermined multipliers and the result is

$$p[\rho(r)|\mathcal{D} \cdot \mathcal{I}] = Z^{-1}p_0[\rho(r)|\mathcal{I}] \exp\left(-\sum_m \lambda_m F_m[\rho] - \sum_m \mu_m \left(F_m[\rho] - \bar{F}_m\right)^2\right), \qquad (2.3)$$

where $\mathcal{D}$ and $\mathcal{I}$ are propositions stating the data and the prior information respectively, and the $2M + 1$ multipliers are $\lambda_m, \mu_m (m = 1 \to M)$ and $\ln Z - 1$. The normalizing 'partition function' $Z$ satisfies

$$Z(\{\lambda_m, \mu_m\}) = \int p_0\left[\rho(r)|\mathcal{I}\right] \exp\left(-\sum_m \lambda_m F_m[\rho] - \sum_m \mu_m\left(F_m[\rho] - \bar{F}_m\right)^2\right) D(\rho), \quad (2.4)$$

and is useful for expressing the equations relating $\{\lambda_m, \mu_m\}$ to $\{\bar{F}_m, \sigma_m\}$:

$$-\frac{\partial \ln Z}{\partial \lambda_m} = \bar{F}_m \quad \text{and} \quad -\frac{\partial \ln Z}{\partial \mu_m} = \sigma_m^2. \quad (2.5)$$

It is illuminating to compare eqn(2.3) with the equivalent solution of a problem where a sampling distribution is available and Bayes' theorem can be used:

$$p\left[\rho(r)|\mathcal{D} \cdot \mathcal{I}\right] = \left(p[\mathcal{D}|\mathcal{I}]\right)^{-1} p\left[\rho(r)|\mathcal{I}\right] p\left[\mathcal{D}|\rho(r) \cdot \mathcal{I}\right]. \quad (2.6)$$

It is clear that MaxEnt has generated the same solution and has created as its sampling distribution a product of Gaussians.

To proceed, it is expedient to consider expanding $\rho$ in a set of orthonormal basis functions (if necessary with some weight function). Using Dirac's notation for vectors in an Hilbert space, we have

$$|\rho\rangle = \sum_n c_n |\phi_n\rangle, \quad (2.7)$$

where $\langle r|\rho\rangle$ gives $\rho(r)$ and $\{\langle r|\phi_n\rangle\}$ are the basis functions $\{\phi_n(r)\}$. The $\{c_n\}$ are expansion coefficients. Orthonormality immediately gives $c_n = \langle \phi_n|\rho\rangle$. In addition, the kernel $|m\rangle$ is defined so that its representation as a function of $r$, $\langle r|m\rangle$ is $4\pi j_0(q_m r)$. Then we introduce a weight function $r^2$ so that $\langle m|\rho\rangle = \int 4\pi j_0(q_m r)\rho(r)r^2 dr = F(q_m)$. Applying $\langle m|$ to eqn(2.7) gives $F_m = \sum_n c_n \langle m|\phi_n\rangle$.

In our earlier paper [4] we solved for the multipliers by introducing the volume element

$$D(\rho) = \prod_n dc_n, \quad (2.8)$$

and assigning the prior on the $\{c_n\}$ to be uniform within upper and lower bounds based on the assumed positivity of nuclear charge distributions and use of the mean-value theorem. This requires restriction of the dimensionality of the function space to that of the data space with no account of redundancy; we shall see that in fact about half of the $^4$He data are redundant. The resulting probability distribution is familiar (e.g., [5]):

$$p[\rho|\mathcal{D} \cdot \mathcal{I}]D(\rho) = Z^{-1} \exp\left(-\sum_{m=1}^{M} \frac{\left(F_m[\rho] - \bar{F}_m\right)^2}{2\sigma_m^2}\right) \prod_{m=1}^{M} dF_m. \quad (2.9)$$

In the restricted function space, it represents the transformation of the *noise* in the data into uncertainty in the charge density using an uninformative prior, but has taken no account of the *incompleteness* of the data.

Manipulating the argument of the exponential, we find the structure governed by the operator

$$\mathbf{B} \equiv \sum_m \frac{|m\rangle\langle m|}{\sigma_m^2}, \quad (2.10)$$

which we call the *data operator*. In [4], we used the eigenfunctions of this operator corresponding to non-zero eigenvalues as the basis, in which case $p[\rho]$ separates into a product of independent Gaussians in the $\{c_n\}$:

$$p[\rho|\mathcal{D}\cdot\mathcal{I}] = Z^{-1} \prod_{n=1}^{M} \exp\left(-\frac{(c_n - \bar{c}_n)^2}{2\Delta_n^2}\right), \tag{2.11}$$

with $\Delta_n^{-2} = n^{\text{th}}$ eigenvalue of $\mathbf{B}$ and $\bar{c}_n$ determined from completing the squares in eqn(2.9) when it has been expressed as a function of the $\{c_n\}$. As an estimate of $\rho$, we take $\bar{\rho} = \int \rho p[\rho]\mathrm{D}(\rho)$. Its squared error $(\Delta\rho)^2$ is $\int(\rho-\bar{\rho})^2 p[\rho]\mathrm{D}(\rho)$. Such a reconstruction is displayed in Fig. 1. Remember that the error contains only contributions from the noise estimates $\{\sigma_m\}$ and note that the errors at different $r$ are correlated.

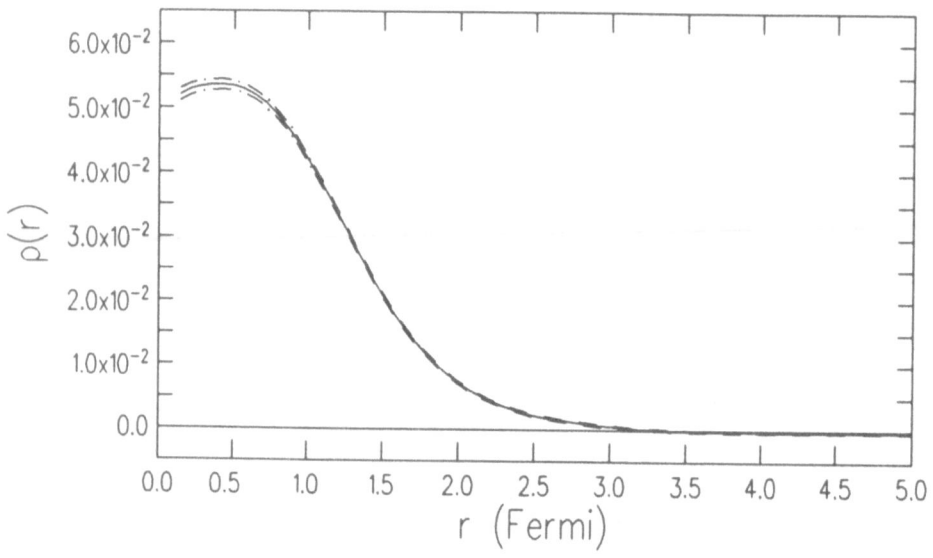

FIGURE 1. Reconstruction of $^4$He charge density in the sub-space spanned by the eigenfunctions of the data operator with non-zero eigenvalue. It is to this space that the error bands refer.

## 3. Redundancy in the Data

The theory of the singular value decomposition of $\mathbf{B}$ guarantees (except for pathological data sets) just $M$ positive non-zero eigenvalues, the sizes of which decrease steadily and correspond to coefficients which are decreasingly accurately determined. Experiments with varying numbers of data points lead one to suspect that there is redundancy in the data—given fewer points one could accurately predict the rest. Thus it would seem efficient to drive well-determined degrees of freedom into a small number of sharply defined coefficients (less than the number of data points) and to deal with these separately from the rest. However the above choice of basis functions yields eigenvalues and hence dispersions on the coefficients which show no obvious division between the well- and the badly-determined. See Fig. 2.

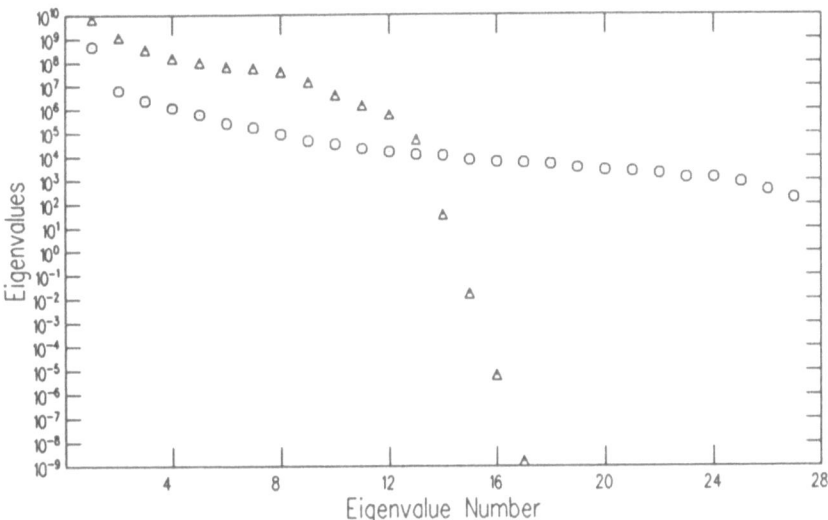

FIGURE 2. The eigenvalues of the data operator in the eigenfunction basis (circles) and in the Fourier-Bessel basis (triangles). Note the 'cliff' in the latter case indicating the redundancy in the data.

We are, however, free to diagonalize **B** in any basis we like. In the next section, a basis is described which does display such a division and also makes an assessment of the contribution to the error from incompleteness very simple.

## 3.1. MOTIVATION FOR A FOURIER-BESSEL EXPANSION

As noted in [6], a Fourier-Bessel (FB) expansion leads to a simple relation between the expansion coefficients and the form factor at certain values of $q$. For if

$$\langle r|\phi_n \rangle \propto \begin{cases} j_0(n\pi r/R), & \text{for } r \leq R \\ 0, & \text{for } r > R \end{cases}, \quad \text{then} \quad c_n \propto F(n\pi/R), \tag{3.1}$$

where $R$ is the radius beyond which the charge is assumed negligible (this is prior information e.g., from muonic atom data). It seems reasonable, bearing in mind eqn(3.1), that only those coefficients $c_n$ with measured data points near $q = n\pi/R$ can be well determined. With data distributed roughly evenly in $(0, q_{max})$, only the first $q_{max}R/\pi$ coefficients should be well determined from the data. As we shall see, this is the case. Prior information is needed to pin down the higher coefficients.

Another attractive feature of the FB expansion is a concentration property of its basis functions. It is clear from the data that $F(q)$ is very concentrated near $q = 0$. Let us firm up this statement by defining concentration (cf. [7]) to satisfy

$$(\text{Concentration})^{-1} = \frac{\int q^2 |F(\mathbf{q})|^2 d^3\mathbf{q}}{\int |F(\mathbf{q})|^2 d^3\mathbf{q}}. \tag{3.2}$$

In addition, we believe $\rho(r)$ to be negligible beyond some radius $R$. It would seem efficient to expand in functions the transforms of which are the most concentrated compatible with the latter condition. A simple variational calculation quickly reveals that these are the $j_0(n\pi r/R)$.

## 3.2 THE FOURIER-BESSEL EXPANSION

The eigenvalues of the data operator in the FB expansion are displayed in Fig. 2. Note the 'cliff' corresponding to a large increase in the dispersion of the expansion coefficients with the plunge beginning near the thirteenth eigenvalue. It is interesting to compare this to the estimate of §3.1 for the number of well-determined coefficients. With $q_{max} \sim 7F^{-1}$ and $R \sim 5F$, this is $q_{max}R/\pi \sim 11$ coefficients. The slightly smaller value might have been expected; one's intuition suggests that the measured data might allow one to predict the form factor for a small distance beyond $q_{max}$. These first 13 coefficients (with their associated eigenfunctions, which are mixtures of the original Bessel functions) are assigned the independent Gaussian probability distributions which the entropic analysis has shown to be appropriate. We shall return to the remaining $M - 13$ ill-determined coefficients in the next section, after we have considered the rest of the function space $(n > M)$.

## 4. Incompleteness in the Data

The more challenging uncertainty to estimate is that arising from the incompleteness of the data. Such an estimation would give error bands referring to the whole Hilbert space defined by the $\phi_n : n = 1 \to \infty$. Here we follow Dreher et al. [6]. In order to estimate $c_n : n = M + 1 \to \infty$ some information on the form factor in the region $q > q_{max}$, by definition, prior information, is required. The assumption of the positivity of $\rho$ is inadequate to reduce the feasible hyper-volume in the higher coefficient space sufficiently to avoid an infinite contribution to the uncertainty $(\Delta\rho)^2$. A condition which does lead to a finite error is the following smoothness constraint:

$$\left(r\rho(r)\right)'' \text{ has limited total fluctuation.} \tag{4.1}$$

Then it is easy to show, with the help of a theorem of Bessel functions [8], that, asymptotically, the form factor must vary like $O(q^{-4})$. Converting this to a statement about the coefficients gives

$$\lim_{n\to\infty} c_n = \frac{\text{constant}}{n^3}. \tag{4.2}$$

A further assumption which we make at this point is that the measured data extend to high enough $q$ that $q_{max}$ is essentially in the asymptotic region and the above constant can be determined by matching to the last measured extremum of $F(q)$. Then the $c_n$ for $n > M$ are assigned independent flat distributions between the limits $\pm$constant $\times n^{-3}$. There is no contribution to $\bar{\rho}$ but the contribution to $(\Delta \rho)^2$ is

$$(\Delta \rho)^2_{n>M} = \sum_{n=M+1}^{\infty} (\Delta c_n)^2 \phi_n^2(r) \qquad (4.3a)$$

$$\propto r^{-2} \sum_{n=M+1}^{\infty} n^{-6} \sin^2(n\pi r/R). \qquad (4.3b)$$

The sum can be reexpressed as $\sum_{n=1}^{\infty} - \sum_{n=1}^{M}$, the first term of which is proportional to $B_6 - B_6(r/R)$, where $B_6$ and $B_6(x)$ are the sixth Bernoulli number and polynomial respectively, and the second term can be computed numerically.

The last remaining coefficients to consider are the $M - 13$ ill-determined ones arising from diagonalizing $\mathbf{B}$. Now, not unexpectedly, the corresponding eigenfunctions are predominantly made up of the basis functions with $13 < n \le M$, that is those that would be determined by $F$ at $q > q_{\max}$. Thus, to a first approximation, we may treat them as part of the asymptotic region and the lower limit of the sum in eqn(4.3b) is replaced by 14.

The reconstruction of $\rho$ produced by such a treatment is displayed in Fig. 3. As a result of having only 13 basis functions contributing to $\bar{\rho}$, the data are not fitted exactly, which is desirable since they are, after all, noisy.

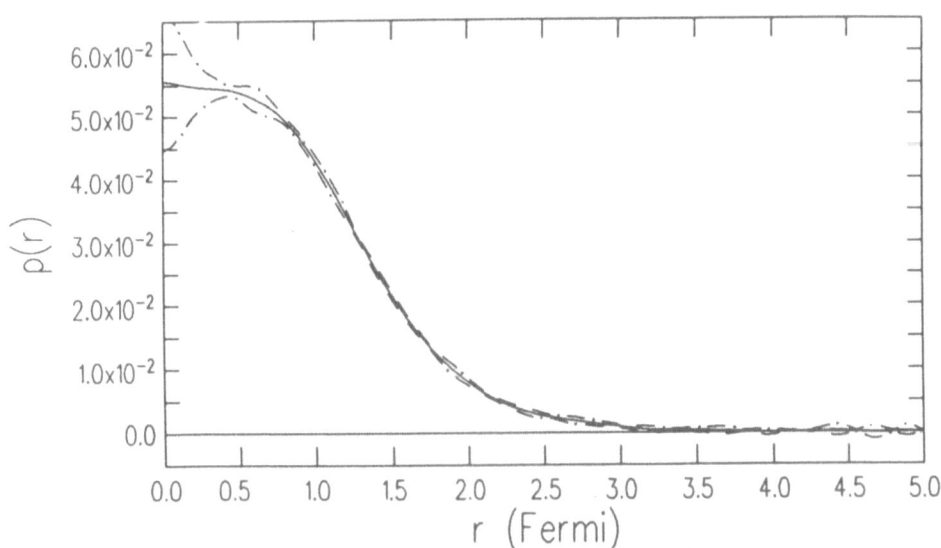

FIGURE 3. Reconstruction of $^4$He charge density using the FB expansion. The error bands refer to the full function space.

## 5. Conclusion

The MaxEnt recipe has been used in a problem where the necessary structure for a Bayesian analysis was not available. The use of an uninformative prior $p[\rho|\mathcal{I}]$ allows the Lagrange multipliers to be solved for and yields a familiar result for $p[\rho|\mathcal{D} \cdot \mathcal{I}]$. More informative priors can restrict the space of feasible reconstructions sufficiently to produce finite error bands on a reconstruction.

## Acknowledgement

One of us (V.A.M.) would like to acknowledge financial support from the Science and Engineering Research Council of Great Britain.

## References

[1] Frosch R F et al. 1967 Structure of the $^4$He Nucleus from Elastic Electron Scattering Phys. Rev. **160** 874–879

[2] McCarthy J S et al. 1977 Electromagnetic Structure of the Helium Isotopes Phys. Rev. C **15** 1396–1414

[3] Jaynes E T 1988 The Relation of Bayesian and Maximum Entropy Methods Maximum-Entropy and Bayesian Methods in Science and Engineering (Volume 1) Erickson G J and Smith C R (eds.) (Dordrecht: Kluwer) 25–29

[4] Macaulay V A and Buck B 1989 Linear Inversion by the Method of Maximum Entropy Inverse Problems (in press)

[5] Borysowicz J and Hetherington J H 1973 Errors on Charge Densities Determined from Electron Scattering Phys. Rev. C **7** 2293–2303

[6] Dreher B et al. 1974 The Determination of the Nuclear Ground State and Transition Charge Density from Measured Electron Scattering Data Nucl. Phys. A **235** 219–248

[7] Slepian D 1983 Some Comments on Fourier Analysis, Uncertainty and Modeling SIAM Review **25** 379–393

[8] Watson G 1944 A Treatise on the Theory of Bessel Functions (second edition) (Cambridge: Cambridge University Press) 595

# MINIMUM DISSIPATION AND MAXIMUM ENTROPY

David Montgomery
Department of Physics & Astronomy
Dartmouth College
Hanover, NH 03755

Lee Phillips
Code 4410
U.S. Naval Research Laboratory
Washington, DC 20375-5000

ABSTRACT. It is argued that for several driven, dissipative, fluid flows, maximum entropy states are the same as minimum energy dissipation states. The demonstration requires a definition of entropy for continuous solenoidal fields that has previously been employed in ideal magnetohydrodynamics.

## 1. Introduction

Variational principles have been of perennial interest for fluid mechanics. In fluid theory (Serrin, 1959; Finlayson, 1972), as elsewhere in mechanics (Lanczos, 1970; Yourgrau & Mandelstam, 1968), they provide perhaps the most economical statements possible about the behavior of the system.

Most such variational principles fall into one of two categories. (1) Some are logically equivalent to the dynamical equations of the system; for example, Lagrange's equations of particle dynamics are equivalent to Hamilton's principle or the principle of least action (Landau & Lifshitz, 1976). (2) Others are an assertion of some exact or approximate property of the system, one from among perhaps several possible consequences that might be extracted from the relevant dynamical equations, but not implying them or logically equivalent to them (Ting, Matthaeus & Montgomery, 1986; Taylor, 1974, 1986). In this second category one may put the hydrodynamic principle (Lamb, 1945) of "minimum energy dissipation rate," or its apparent offspring, the principle (Onsager, 1931; Prigogine, 1947; de Groot & Mazur, 1984; Keizer, 1987) of "minimum rate of entropy production." Both principles are closer to conjectures than to theorems, but seem to represent reality well enough in some important cases. The satisfactions to be obtained from the principles in the first category are largely aesthetic and, occasionally, computational. It is in the second category that most of the possibilities for new physical information lie. They are specific statements about less than the total behavior of the system.

The purpose of this article is to point out the likely equivalence of two variational principles for hydrodynamic systems that have apparently been considered as distinct. Both belong to category (2) above. The first principle can be called that of *maximum entropy* (Jaynes, 1957, 1979; Katz, 1967). Close in spirit to Jaynes's "information theory" approach to statistical mechanics, the method has previously been used (Montgomery, Turner

*P. F. Fougère (ed.), Maximum Entropy and Bayesian Methods*, 281–296.

& Vahala, 1979; Ambrosiano & Vahala, 1981) to infer "most probable" states of plasmas, fluids, and magnetofluids, by maximizing a suitably defined probability of the spatial distribution of the fields, subject to the constraints imposed by boundary conditions. The second is the principle (Lamb, 1945) of *minimum energy dissipation rate*, an older and relatively pedestrian predictor of hydrodynamic flow profiles that has found recent application (Montgomery and Phillips, 1988; Montgomery, Phillips & Theobald, 1989) in magnetohydrodynamics (MHD). The equivalence of the two principles may have been suspected by Jaynes (1980).

We shall argue that the minimum energy dissipation rate principle and the maximum entropy principle, properly interpreted for continuous fluids, are likely the same. We explore their equivalence in the context of several steady-state viscous flows for which the answers are already known (Batchelor, 1967), both from explicit solutions of the Navier-Stokes equations and from laboratory experience. (Heretofore, the principal applications (Montgomery, Turner & Vahala, 1979; Ambrosiano & Vahala, 1981) of the maximum-entropy method have been to ideal, non-dissipative systems.) Our purpose is less to gain new information about hydrodynamics than it is to make clear the equivalence of the two methods and to gain confidence in both for applications in the future to MHD. The plausibility of the maximum entropy method combined with the relative ease of implementation of the minimum dissipation method go part way toward substituting for a deductive proof of either one.

The essential problem in the maximum entropy method is a combinatorial one of deciding what statistical weight to assign to a particular configuration of streamlines of a solenoidal field such as vorticity, fluid velocity, or magnetic field. A reasonably good intuitive grasp of this problem results from first restricting attention, as we do here, to cases in which the fields can vary with no more than two spatial coordinates and have either vortex lines or streamlines which all intersect a single plane that can be ruled off in the two spatial coordinates. Subtleties having to do with full three-dimensional variations and multiply-linked vortex lines are deferred for future consideration, although they are briefly remarked upon at the end of the paper (of which an outline follows).

In Section 2, the maximum entropy method is used to calculate plane Poiseuille flow as a "most probable" state. The calculation of the same flow as a minimum dissipation rate state is quite simple. It is necessary to go into some detail about the limiting process in which a continuous flow field is achieved as a limit of a discretized line vortex description. After the most probable state is calculated in the discrete model, it is shown that the overall entropy that has been defined increases monotonically as the number of vortices becomes infinite and their individual strengths shrink to zero. (This reinforces a numerical conclusion reached by Ambrosiano and Vahala (1981).)

The limit should be understood as something slightly different from what are sometimes called "mean field" theories (Joyce & Montgomery, 1973; Lundgren & Pointin, 1977; Kraichnan & Montgomery, 1980; Ting, Chen & Lee, 1987), in which an underlying discrete particle mechanics is being approximated by a smoothed macroscopic average field. There are never any references to particle discreteness in this development. The results suggest passage to the limit at an early stage of the derivation, and this is done for an application to plane Couette flow, also calculable as a minimum-dissipation state.

In Section 3, the class of problems under discussion is treated at a greater level of generality, where it is argued that in the limit of an infinite number of vortices, maximization of the entropy integral becomes equivalent to the minimization of the viscous dissipation integral: not an obvious result. Two more applications, each with one or two *sui generis* features, are treated in Section 4: rotating Couette flow and pipe flow. Unanswered questions and possible future extensions are speculated upon in Section 5.

## 2. Plane Poiseuille and Couette Flow

We first define what is meant by the most probable state of a hydrodynamic flow and show that plane Poiseuille flow is such a state. We then also calculate plane Poiseuille flow as a state of minimum energy dissipation rate, which is straightforward. The equivalence of the two methods can then be made transparent and applied to other examples, such as plane Couette flow.

Two infinite, rigid, parallel plane boundaries may be put at $y = a$ and $y = -a$, say. The uniform-density fluid between them moves in the $x$-direction in response to a mean pressure gradient. The velocity field is well known to be $\mathbf{v} = (U_x(y), 0, 0)$ and the vorticity field to be $\omega = (0, 0, \omega)$. Here $U_x(y) = U_{max}(1 - y^2/a^2)$ and $\omega = -dU_x/dy = 2U_{max}y/a^2$. Either field, $\mathbf{v}$ or $\omega$, determines the other in the presence of no-slip boundary conditions. The average flow speed over the whole profile is $U_0 = 2U_{max}/3$. The flux across the channel width will be defined as $F = 2aU_0$ (per unit length in the spanwise direction). The system extends to infinity in the $\pm x$ and $\pm z$ directions. All variables are independent of $z$.

We now show how this plane Poiseuille profile can be calculated independently by assuming that the fluid assumes its "most probable" state, compatible with the constancy of $F$ and the boundary conditions that $\mathbf{v} = 0$ at $y = \pm a$. For convenience, periodic boundary conditions in $x$ will be assumed, with an arbitrarily long but finite period $L$. Since $L$ can be allowed to approach infinity, this is no restriction.

The first requirement is a definition of the probability of a fluid state. This may be obtained by a temporary discretization of the vorticity field (though this is not the only possibility). We consider all possible two-dimensional vorticity distributions $\omega(x, y)$, and defer consideration of the case in which spatial variation with $z$ is permitted, where the vortex lines may be other than straight and parallel.

The vorticity field will first be discretized by considering very large numbers of straight, parallel vortex lines of strength $\pm\kappa$, aligned perpendicularly to the $x, y$ plane. Boltzmann combinatorial statistics will be applied to the vortices to arrive at an expression for the probability of a state. After all the combinatorial steps are carried out, $\kappa$ may be allowed to approach zero and the number of vortices to become infinite. If the limiting process is carried out correctly, a finite, continuous $\omega$-field will result and will behave in a Navier-Stokes way.

The basic region of the $xy$ plane $(0 \leq x \leq L, -a \leq y \leq a)$ is divided into identical small cells, each of edge $\ell$ and area $\ell^2$. Let the discrete index $i$ label the cells in the $x$-direction and let $j$ label them in the $y$-direction. The $x, y$ coordinates of the center of the $ij$th cell are $(x_i, y_j)$. By $n_{ij}^+$ and $n_{ij}^-$, we shall mean the integer numbers of vortices of strength $+\kappa$ and $-\kappa$ which intersect the $ij$th cell, perpendicularly. By $N^+$ and $N^-$ we mean the total (integer) number of vortices of strength $\kappa$ and $-\kappa$ in the whole system. Ultimately, $\kappa$, $\ell$, and $(N^{\pm})^{-1}$ will approach zero, but they must do so together. Thus, if $\ell$ is selected as the basic small parameter, $\kappa$ will $\sim \ell^2$, and $N^{\pm}$ will $\sim \ell^{-3}$, if we treat $a$, $L$, $U_0$ and $F$ as $O(1)$ quantities which remain independent of $\ell$. It will turn out that $n_{ij}^{\pm} \sim \ell^{-1}$, which is important, since we will want to treat the $n_{ij}^{\pm}$ as $>> 1$, even though they are $<< N^{\pm}$. It will also be important that $n_{ij}^+ - n_{ij}^-$ be of $O(1)$, i.e., small compared to $n_{ij}^+$ and $n_{ij}^-$

separately.

A state of the system, under the assumption of independence of $z$ and before boundary conditions are imposed, can be completely specified by enumerating all the occupation numbers $n_{ij}^+$, $n_{ij}^-$ for all $i, j$. The vorticity field, in this representation, will become

$$\omega_{ij} \equiv (n_{ij}^+ - n_{ij}^-)(\kappa/\ell^2) \rightarrow \omega(x, y) \tag{2.1}$$

in the limit $\ell \rightarrow 0$. Clearly, many possible sets of $n_{ij}^+$ and $n_{ij}^-$ fulfill the boundary conditions and the constraint of constant $F$. It is necessary to select from these the set of $n_{ij}^\pm$ that are most probable, given only the constraint and the boundary conditions.

According to Boltzmann statistics (ter Haar, 1966), the probability to be assigned to any set of $n_{ij}^\pm$ occupation numbers is proportional to $W$, where

$$W \equiv \frac{(N^+)!(N^-)!}{\{\Pi_{ij}(n_{ij}^+)!\}\{\Pi_{ij}(n_{ij}^-)!\}}. \tag{2.2}$$

The products in the denominators run over all the cells.

Finding the $n_{ij}^\pm$ that will maximize $\ell n\, W$ is equivalent to finding the set that will maximize $W$, and easier. Stirling's approximation can be employed,

$$S \equiv \ell n\, W \cong N^+ \ell n\, N^+ + N^- \ell n\, N^- - \sum_{ij}(n_{ij}^+ \ell n\, n_{ij}^+ + n_{ij}^- \ell n\, n_{ij}^-), \tag{2.3}$$

as long as $n_{ij}^+$ and $n_{ij}^-$ are $\gg 1$.

Hereafter, $S$ will be called the "entropy" to be associated with a given configuration. The dependence upon $N^+$ and $N^-$ is carried along because we suspect, from the numerical discovery of Ambrosiano and Vahala (1981), that the dependence of $S$ on $N^+$ and $N^-$ will be of significance.

The most probable set of $n_{ij}^\pm$ can now be defined as the one which maximizes $S$, subject to whatever constraints apply. These are constant vortex numbers of both signs,

$$N^\pm = \sum_{ij} n_{ij}^\pm = \text{constant}, \tag{2.4}$$

and constant flux $F$. The constancy of $F$ may be expressed in an arithmetically convenient way by multiplying the definition of $\omega \equiv \partial v_y/\partial x - \partial v_x/\partial y$ by $a - y$ and integrating over the whole region, giving

$$\int_0^L dx \int_{-a}^a dy(a - y)\omega = -FL = -2U_0 aL. \tag{2.5}$$

(It is desirable to express $F$ as a two-dimensional integral.)

The discretized version of (2.5) is

$$2aU_0 L = FL = -\kappa\ell^2 \sum_{ij}(n_{ij}^+ - n_{ij}^-)(a - y_j)\ell^{-2}, \tag{2.6a}$$

since the correspondence between summation and integration is

$$\int dx \int dy \longleftrightarrow \ell^2 \sum_{ij}.$$

Stripped of superfluous constants, the constraint (2.6a) can be rewritten as

$$\sum_{ij}(n_{ij}^+ - n_{ij}^-)(a - y_j) = \text{constant} = -(2aU_0L/\kappa). \tag{2.6b}$$

The maximization proceeds via standard methods (Lanczos, 1970). We let each $n_{ij}^\pm \to n_{ij}^\pm + \delta n_{ij}^\pm$, and set, in $S$, the coefficients of the first-order terms in $\delta n_{ij}^\pm$ to zero. The constraints are taken account of by the method of Lagrange undetermined multipliers (Lanczos, 1970). Necessary conditions for a maximum in $S$ are then

$$\ell n\ n_{ij}^+ + \alpha^+ + \beta(a - y_j) = 0$$

$$\ell n\ n_{ij}^- + \alpha^- - \beta(a - y_j) = 0. \tag{2.7}$$

Here, $\alpha^\pm$ and $\beta$ are constants, to be determined by the requirements that the $n_{ij}^\pm$ which result from (2.7) shall satisfy (2.4) and (2.6). We first maximize $S$, treating $N^\pm$ as constants, then consider the dependence on $N^\pm$.

In this example, though not in some others to be considered, symmetries imply no net integrated vorticity. We are at liberty to pick $N^+ = N^- = N$, say. The content of (2.4) is then

$$N = \exp(-\alpha^\pm \mp \beta a)(L/\ell) \sum_{j} \exp(\pm\beta y_j). \tag{2.8}$$

Since $\sum_{j} \exp(+\beta y_j) = \sum_{j} \exp(-\beta y_j)$, (2.8) implies that $\alpha^+ + \beta a = \alpha^- - \beta a$, and therefore that

$$n_{ij}^+ - n_{ij}^- = \exp(-\alpha^+ - \beta a)2\sinh(\beta y_j) \tag{2.9a}$$

$$n_{ij}^+ + n_{ij}^- = \exp(-\alpha^+ - \beta a)2\cosh(\beta y_j). \tag{2.9b}$$

The relations so far are all exact but are unwieldy. Explicit solutions for $\alpha^\pm$ and $\beta$ can be extracted if we take advantage of the various small parameters. We anticipate, and then confirm by the consistency of the results, that $|n_{ij}^+ - n_{ij}^-|$ is $<< n_{ij}^+$ and $n_{ij}^-$ separately. From (2.9), this can only be true if $|\beta y_j| << 1$ for all $j$, or $|\beta a| << 1$. This yields

$$n_{ij}^+ - n_{ij}^- = 2\exp(-\alpha^+ - \beta a)\beta y_j, \tag{2.10}$$

plus terms two orders of magnitude higher in $\beta a$. Also up to terms of higher order in $\beta a$, (2.8) implies that

$$N = \exp(-\alpha^+ - \beta a)(2aL/\ell^2)$$

(to this order, the summation just counts the cells), so that

$$n_{ij}^+ - n_{ij}^- = (\ell^2/aL)N\beta y_j. \tag{2.11}$$

It remains to determine $\beta$. From (2.6),

$$F = 2aU_0L = \kappa(L/\ell)\sum_j y_j(N\beta y_j)(\ell^2/aL) = \frac{\kappa\beta\ell}{a}\sum_j y_j^2 N. \tag{2.12}$$

As $\ell \to 0$, $\sum_j y_j^2$ becomes approximated arbitrarily accurately by $\sum_j y_j^2 \cong \ell^{-1}\int_{-a}^{a} y^2 dy = 2a^3/3\ell$, giving

$$\beta a = \frac{3U_0L}{\kappa N}. \tag{2.13}$$

Note that since $\kappa \sim \ell^2$ and $N \sim \ell^{-3}$, $\beta a \sim \ell$ and is thus indeed $<< 1$, justifying the assertions made before eq. (2.10).

Eq. (2.1) now becomes, using this solution

$$\omega_{ij} = (n_{ij}^+ - n_{ij}^-)(\kappa\ell^{-2}) = (\ell^2/aL)(N\beta y_j)(\kappa\ell^{-2})$$

$$= \left(\frac{N\ell^2}{aL}\right)\left(\frac{3U_0L}{\kappa aN}\right)y_j(\kappa\ell^{-2}) = \frac{3U_0 y_j}{a^2} \to \frac{2U_{max}y}{a^2}, \tag{2.14}$$

as $\ell \to 0$.

Eq. (2.14) is the correct vorticity profile, and the solution is almost complete. A final necessary observation concerns the dependence of $S$ upon $N$, where $N \sim \ell^{-3}$ is now considered as a variable. Evaluating (2.3), using (2.7) and the results for $\alpha^\pm$ and $\beta a$, gives

$$S = 2N \ln N - \sum_{ij}\{n_{ij}^+(-\alpha^+ - \beta a + \beta y_j) + n_{ij}^-(-\alpha^- + \beta a - \beta y_j)\}$$

$$= 2N \ln N + 2(\alpha^+ + \beta a)N - \beta\sum_{ij}(n_{ij}^+ - n_{ij}^-)y_j \tag{2.15}$$

$$= 2N \ln N - 2N \ln (N\ell^2/2aL) - \beta 2aU_0L/\kappa$$

$$= -2N \ln (\ell^2/2aL) - 6U_0^2 L^2/(\kappa^2 N),$$

retaining only the most dominant terms. In view of the chosen scaling, the first term on the right of (2.15) is of $O(N \ln N^{2/3})$ and is positive, while the next term is negative and of $O(N^{1/3})$. Thus, both $S$ and $S/N$ (the "entropy per vortex") are maximized by allowing $N$ to increase without bound.

The limit is that of two continuous fields, each with an infinite flux through any finite area, but with a difference field having only a finite flux through any finite area. The solution for plane Poiseuille flow has been reproduced in every respect as this kind of "most probable" state. The limit is not, and should not be expected to be, a classical "thermodynamic limit" where $S/N$ approaches a constant as $N \to \infty$.

We now demonstrate the (nearly trivial) result that the plane Poiseuille profile is also one of minimum energy dissipation rate, given the boundary conditions and the constraint of constant $F$. For this purpose, it is algebraically slightly more convenient to note that the boundary conditions imply that

$$\int_0^L dx \int_{-a}^a dy\, \omega y = FL = 2U_0 a L \tag{2.16}$$

instead of (2.5). Seeking the stationary states of the dissipation rate for kinetic energy $R_\omega$,

$$R_\omega \equiv \rho_0 \nu \int \omega^2 dx dy, \tag{2.17}$$

where $\rho_0$ is the (uniform) mass density and $\nu$ is the kinematic viscosity, the method of undetermined multipliers leads us at once to the Euler-Lagrange equation,

$$\omega + \alpha y = 0. \tag{2.18}$$

The constant $\alpha$ may be immediately determined to be $-3U_0/a^2$ from (2.16), and the solution is recovered. The minimization of $R_\omega$ has led to the same profile as the maximization of $S$.

The question then arises as to whether there is an earlier stage at which minimizing (2.17) comes to the same thing as maximizing $S$. For the present problem, the answer can be given by noting that adding eqs. (2.7) gives at once that

$$n_{ij}^+ n_{ij}^- = \text{constant}, \quad \text{all } i,j. \tag{2.19}$$

The limiting processes that have been performed make clear that $n_{ij}^+$ and $n_{ij}^-$ are equal up to the terms of $O(1)$, and therefore may be written as

$$n_{ij}^\pm = n_{ij}^0 + \Delta n_{ij}^\pm, \tag{2.20}$$

where $n_{ij}^0$ is a very large constant which is much greater than $\Delta n_{ij}^\pm$. From (2.19) it follows that for the variable part of $n_{ij}^\pm$,

$$\Delta n_{ij}^+ = -\Delta n_{ij}^-, \tag{2.21}$$

up to terms of higher order in $\Delta n_{ij}^\pm / n_{ij}^0$. ($n_{ij}^0$ is fixed and $\sum_{ij} \Delta n_{ij}^+ = 0 = \sum_{ij} \Delta n_{ij}^-$.)

The vorticity fields associated with the positive and negative vortices separately are of much larger magnitude than their difference, and we may define them as the limit of

$$\omega_{ij}^\pm = n_{ij}^\pm (\kappa/\ell^2),$$

while (2.21) implies

$$\omega_{ij} = \omega_{ij}^+ - \omega_{ij}^- = (n_{ij}^+ - n_{ij}^-)(\kappa/\ell^2) = 2\Delta n_{ij}^+ \kappa/\ell^2 = -2\Delta n_{ij}^- \kappa/\ell^2.$$

Still in the discrete representation, the $R_\omega$ dissipation integral in discrete form is unimportant constants multiplied by

$$\ell^2 \sum_{ij} \omega_{ij}^2 = \frac{4\kappa^2}{\ell^2} \sum_{ij} (\Delta n_{ij}^+)^2 = \frac{4\kappa^2}{\ell^2} \sum_{ij} (\Delta n_{ij}^-)^2. \tag{2.22}$$

Keeping (2.22) in mind, we now consider $S$ and make use of the result that $|\Delta n_{ij}^\pm|$ is $<< n_{ij}^0$:

$$S = N^+ \ln N^+ + N^- \ln N^-$$

$$- \sum_{ij} n_{ij}^0 \left( 1 + \frac{\Delta n_{ij}^+}{n_{ij}^0} \right) \ln \left[ n_{ij}^0 \left( 1 + \frac{\Delta n_{ij}^+}{n_{ij}^0} \right) \right]$$

$$- \sum_{ij} n_{ij}^0 \left( 1 + \frac{\Delta n_{ij}^-}{n_{ij}^0} \right) \ln \left[ n_{ij}^0 \left( 1 + \frac{\Delta n_{ij}^-}{n_{ij}^0} \right) \right]$$

$$= N^+ \ln N^+ + N^- \ln N^-$$

$$- \sum_{ij} n_{ij}^0 \left[ 1 + \frac{\Delta n_{ij}^+}{n_{ij}^0} \right] \left[ \ln n_{ij}^0 + \frac{\Delta n_{ij}^+}{n_{ij}^0} - \frac{(\Delta n_{ij}^+)^2}{2(n_{ij}^0)^2} + \cdots \right]$$

$$- \sum_{ij} n_{ij}^0 \left[ 1 + \frac{\Delta n_{ij}^-}{n_{ij}^0} \right] \left[ \ln n_{ij}^0 + \frac{\Delta n_{ij}^-}{n_{ij}^0} - \frac{(\Delta n_{ij}^-)^2}{2(n_{ij}^0)^2} + \cdots \right]$$

$$= \text{constant} - \text{constant} \times \sum_{ij} [(\Delta n_{ij}^+)^2 + (\Delta n_{ij}^-)^2] + \cdots$$

$$= \text{constant} - \text{constant} \times \sum_{ij} (\Delta n_{ij}^+)^2 + \cdots, \tag{2.23}$$

the linear terms in $\Delta n_{ij}^\pm$ having cancelled. By comparing (2.23) with (2.22), we see that the minimization of $R_\omega$ is the same thing as the maximization of $S$, in the limit.

Having become familiar with the limiting process, $\ell \sim \kappa^{1/2} \sim (N^\pm)^{-1/3} \to 0$, with $n_{ij}^0 \sim \ell^{-1}$ and $\Delta n_{ij}^\pm$ of $O(1)$, we see that we can pass to the continuum limit at a much earlier stage. Let $\omega^\pm$ be the limits of $(\kappa/\ell^2) n_{ij}^\pm$, let $N^\pm$ be $\Phi^\pm/\kappa$, and set

$$\Phi^\pm \equiv \int \omega^\pm dx dy. \tag{2.24}$$

The physical vorticity field $\omega = \omega^+ - \omega^-$ is understood to be finite, but $\omega^+$ and $\omega^-$ become large without bound. The variational problem can be posed as one of varying the continuous fields $\omega^+$ and $\omega^-$ to maximize the continuous version of (2.3). The constraints, as well, need to be expressed in terms of $\omega^+$ and $\omega^-$.

We next illustrate this by a calculation of plane Couette flow as a most probable state, starting from the continuum formulation. This flow is, of course, also a minimum-dissipation state for its boundary conditions. The discussion will lead naturally to a discussion in Section III that demonstrates the equivalence of maximum entropy and minimum dissipation entirely from the continuum viewpoint.

## 2.1 PLANE COUETTE FLOW

Plane Couette flow also occurs between infinite parallel planes, but with a moving boundary and no pressure gradient. We put the planes at $y = a$ and $y = 0$, and require that the upper one move in the $x$-direction with a constant speed $U_0$. The velocity and vorticity profiles are well known to be $\mathbf{v} = (U_0 y/a, 0, 0)$ and $\omega = \omega \hat{e}_z = (0, 0, -U_0/a)$; either one determines the other, given no-slip boundary conditions.

We work hereafter with the entropy defined for the continuum, non-negative fields $\omega^+$ and $\omega^-$, as described at the end of the last subsection. The continuum version of eq. (2.3) is

$$S = (\Phi^+/\kappa) \, \ell n \, (\Phi^+/\kappa) + (\Phi^-/\kappa) \, \ell n \, (\Phi^-/\kappa)$$

$$- \int dx dy \left\{ (\omega^+/\kappa) \, \ell n \, (\omega^+ \ell^2/\kappa) + (\omega^-/\kappa) \, \ell n \, (\omega^- \ell^2/\kappa) \right\} = \text{constant} +$$

$$(-1/\kappa) \int dx dy \left\{ \omega^+ \, \ell n \, \omega^+ + \omega^- \, \ell n \, \omega^- \right\}. \tag{2.25}$$

For purposes of the variation of $\omega^+$ and $\omega^-$, (2.25) shows that it is adequate to maximize only

$$S \equiv - \int dx dy (\omega^+ \, \ell n \, \omega^+ + \omega^- \, \ell n \, \omega^-). \tag{2.26}$$

The fact that the state arrived at maximizes $S$ in the limiting process may be confirmed each time from considering (2.25) as $\ell \to 0$, $\kappa \to 0$, etc. after $\omega^\pm$ and $\omega = \omega^+ - \omega^-$ have been calculated.

The constraints under which $S$ is to be maximized are $\Phi^\pm \equiv \int \omega^\pm dx dy = \text{constant}$ (understood eventually to $\to \infty$), and for plane Couette flow, the constraint $\mathbf{v} = (U_0, 0, 0)$ at $y = a$ can be expressed as

$$U_0 L = \int dx dy (\omega^+ - \omega^-).$$

Again using the method of Lagrange multipliers, we are led immediately to

$$\omega^\pm = \exp(-\alpha^\pm \pm \beta) = \text{constant} = \Phi^\pm/La. \tag{2.27}$$

There is only one constant, in effect, and it is $\omega^+ - \omega^- = \omega = (\Phi^+ - \Phi^-)/La = U_0/a$. Verification that $S$ in (2.25) is maximized by allowing $\Phi^+ \to \infty$, $\Phi^- \to \infty$, and $\Phi^+ - \Phi^- = LU_0$ is straightforward, if we stay with the previous orderings.

### 3. Continuum Formulation; Equivalence of Maximum Entropy and Minimum Dissipation

If we had passed to the limit $\ell \to 0$ before the variational calculation, the remarks at the end of the last section show that we would in effect have been maximizing

$$S \equiv - \int dx dy (\omega^+ \ln \omega^+ + \omega^- \ln \omega^-) \tag{3.1}$$

subject to the constraints that $\Phi^\pm = \int \omega^\pm dx dy$ are constants which become infinite according to the limiting processes described. Other constraints, specific to the problem at hand, are what give individual cases their content. The constraints considered here are all of the class which may be written as

$$I = \int dx dy F(\omega) = \text{ constant.} \tag{3.2}$$

These are very general since $F(\omega)$ may contain $x$ and $y$, may perhaps depend nonlinearly upon $\omega$ and its derivatives, or even upon functionals of $\omega$.

In general, $I$ will have a variation

$$\delta I = \int \frac{dF(\omega)}{d\omega} \delta \omega dx dy = \int \frac{dF(\omega)}{d\omega} \left[ \delta \omega^+ - \delta \omega^- \right] dx dy. \tag{3.3}$$

(If $F(\omega)$ contains derivatives of $\omega$, it will be assumed that integrations-by-parts (Lanczos, 1970; Landau & Lifshitz, 1976) can transfer the derivatives to the part of the integral in (3.3) that multiplies $[\delta \omega^+ - \delta \omega^-]$.)

We may employ the property that $S$ is maximized by letting $\Phi^\pm \to \infty$ at an earlier stage of the calculation. We may write

$$\omega^\pm = \Omega_0 + \Delta \omega^\pm, \tag{3.4}$$

with the understanding that $\Omega_0$ will be a constant part of both fields that will $\to \infty$ and so is only superficially an important part of the formalism. The variational part of the problem comes from independent variations $\Delta \omega^+ \to \Delta \omega^+ + \delta(\Delta \omega^+)$, $\Delta \omega^- \to \Delta \omega^- + \delta(\Delta \omega^-)$, subject to the constraints that

$$\int \delta(\Delta \omega^+) dx dy = 0 = \int \delta(\Delta \omega^-) dx dy. \tag{3.5}$$

The variationally important part of $S$ can also be expressed entirely in terms of $\Delta \omega^\pm$; for, expanding $S$,

$$S = S_0 - \int dx dy \Bigg\{ ( \ln \Omega_0)(\Delta \omega^+ + \Delta \omega^-)$$

$$+(\Delta \omega^+ + \Delta \omega^-) + \frac{(\Delta \omega^+)^2 + (\Delta \omega^-)^2}{2\Omega_0} + \cdots \Bigg\}. \tag{3.6}$$

where $S_0$ is an infinite, additive constant.

In the maximization of $S$, the variation of the linear terms in (3.6) will contribute at most additive terms to the Lagrange multipliers that will go with (3.5). The only variation of any consequence in (3.6) will be the quadratic term. The maximization of $S$ involves only the maximization of $S_2$ where

$$S_2 \equiv -\frac{1}{2\Omega_0} \int \left[ (\Delta\omega^+)^2 + (\Delta\omega^-)^2 \right] dx dy. \tag{3.7}$$

Both the constraints (3.5) and (3.3), which becomes

$$\delta I = \int \frac{dF(\omega)}{d\omega} \left[ \delta(\Delta\omega^+) - \delta(\Delta\omega^-) \right] dx dy, \tag{3.8}$$

can now be taken into account by Lagrange multipliers when maximizing (3.7). The resulting Euler-Lagrange equations are immediate:

$$\Delta\omega^+ + \alpha^+ + \beta \frac{dF(\omega)}{d\omega} = 0$$

$$\Delta\omega^- + \alpha^- - \beta \frac{dF(\omega)}{d\omega} = 0. \tag{3.9}$$

$\alpha^\pm$ and $\beta$ are constants in (3.9); and $\omega$, wherever it appears, is to be expressed as $\omega = \Delta\omega^+ - \Delta\omega^-$. Specific examples of $F(\omega)$ have already been produced in the two illustrations of Section 2, but fancier applications, in which (3.9) represents, for example, differential equations, can be produced. These have been important in MHD generalizations (Montgomery, Turner & Vahala, 1979; Ambrosiano & Vahala, 1981).

Adding the relations (3.9), we see at once that

$$\Delta\omega^- = -\Delta\omega^+ - \alpha^+ - \alpha^-, \tag{3.10}$$

a result which follows whether or not eqs. (3.9) are easy to solve. The useful fact about (3.10) is that, introducing it into $S_2$ reduces $S_2$ at once to

$$S_2 = \text{constant} - \frac{1}{\Omega_0} \int dx dy (\Delta\omega^+ - \Delta\omega^-)^2 =$$

$$\text{constant} - \text{constant} \times \int dx dy \omega^2, \tag{3.11}$$

once the constancy of $\int \Delta\omega^+ dx dy$ and $\int \Delta\omega^- dx dy$ is taken into account.

Eq. (3.11) expresses the somewhat remarkable result that, with this definition of entropy and with infinite fluxes of the independent fields $\omega^+$ and $\omega^-$, maximizing $S$ is equivalent to minimizing $S_2$, which is proportional to the viscous dissipation integral (2.17). This statement is independent of the detailed nature of the constraint (3.2), which is quite general and may stand for multiple constraints.

A remark may be in order as to the intuitive basis of "maximum entropy equals minimum dissipation." In the absence of information (i.e., constraints) to the contrary, one would expect the most probable state to be one of maximum smoothness allowable by the boundary conditions: there is no reason for detailed structure not implied by the conditions of the problem. Similarly, dissipation rates in fluids and magnetofluids are direct, mean-square measures of the spatial variability of the fields, and minimum dissipation states also represent a kind of "maximum allowable smoothness."

## 4. Rotating Couette Flow and Pipe Flow

Flow between rotating concentric cylinders is easy to obtain as a maximum entropy or minimum dissipation state, but pipe flow introduces a few novelties. We solve first the former and then the latter.

Assume two infinite, concentric, rotating cylinders at radii $r = a_1$ and $r = a_2 > a_1$. Let the first rotate at angular velocity $\Omega_1$ and the second with angular velocity $\Omega_2$. The only serious constraint beyond the constancy of the fluxes $\Phi^\pm = \int \omega^\pm r dr d\theta$ is that the integrated vorticity between the boundaries account for the imposed difference in circulation around them:

$$\oint_{r=a_2} \mathbf{v} \cdot d\ell - \oint_{r=a_1} \mathbf{v} \cdot d\ell = \Phi^+ - \Phi^- = \text{constant}, \tag{4.1}$$

where the line integrals run azimuthally around the cylinders. The Euler-Lagrange condition resulting from the maximization of $S$ subject to (4.1) is

$$\ell n\, \omega^+ + \alpha^+ = 0$$

$$\ell n\, \omega^- + \alpha^- = 0 \tag{4.2}$$

or

$$\omega = \text{constant} = \omega_0, \text{ say.} \tag{4.3}$$

Integrating $\omega_0 = r^{-1} d(rv_\theta)/dr$ and imposing boundary conditions to find $v_\theta$ gives the usual rotating Couette flow profile

$$v_\theta = (\Omega_2 a_2^2 - \Omega_1 a_1^2)(a_2^2 - a_1^2)^{-1} r + (\Omega_1 - \Omega_2) a_1^2 a_2^2 (a_2^2 - a_1^2)^{-1} r^{-1}. \tag{4.4}$$

Minimizing $R_\omega$ subject to the same constraints leads immediately to $\omega =$ constant, and imposing the boundary conditions on $\mathbf{v}$ leads again to (4.4).

Pipe flow is a more interesting case and is an example in which the vortex lines, which now run azimuthally around the axis, are curved, and we must take into account a location dependence that was not present in the three previous cases, wherein all spatial cells were of volume $\ell^2$ times a length that was arbitrary but did not need to be mentioned because it was the same for all $i, j$.

We consider first the axisymmetric case $\omega = \omega_\varphi(r, z)\hat{e}_\varphi$ only. We rule the region $0 \leq r < a$, $0 \leq z < L$ (cylindrical coordinates) off into cells centered at the discrete locations $(z_{ij}, r_{ij})$. However, if we are to assign the cells equal statistical weight in an expression like (2.2), these cells must not be of equal area (ter Haar, 1966). If a vortex ring is transported with the fluid to a larger radius and the fluid is incompressible, then the volume must stay the same, so that the ring area must shrink by a factor of the ratio of the initial to the final radius. The easiest way to implement this mathematically is to carry out the discrete calculations of Sections 2 and 3 as before, with cells of equal statistical weight, but when it comes time to convert the $\sum_{ij}$ to integrals over the $r, z$ plane, we do it slightly differently; namely, let

$$\sum_{ij} \rightarrow \int \frac{dzdr}{\ell^2} \left( \frac{r}{r_0} \right), \tag{4.5}$$

where $r_0$ is some reference length such as the radius $a$ of the pipe, or perhaps a factor times $a$. (Only the normalization is affected by the choice of $r_0$.)

The definitions of $\omega^+$ and $\omega^-$ are unaffected, but the entropy changes slightly. For example, (2.26) becomes

$$S = - \int dzdr \left( \frac{r}{r_0} \right) (\omega^+ \, \ell n \, \omega^+ + \omega^- \, \ell n \, \omega^-) \tag{4.6}$$

Developments like those leading to (2.23) and (3.11) go through as before except that now there is an extra factor of $(r/r_0)$ in the integrands. Eq. (3.11), for example, becomes

$$S_2 = \text{constant} \quad - \quad \text{constant} \times \int dzdr \left( \frac{r}{r_0} \right) \omega^2. \tag{4.7}$$

Since the dissipation integral $R_\omega$ is always

$$R_\omega = \rho_0 \nu \int rdrd\varphi dz\omega^2, \tag{4.8}$$

it is again clear that maximizing (4.7) is equivalent to minimizing (4.8). The constraint of constant axial flux may be written as

$$\int dzdr r^2 \omega = \text{constant}, \tag{4.9}$$

for no-slip boundary conditions at $r = a$. Minimizing dissipation or maximizing entropy subject to (4.9) leads to a linear dependence of $\omega$ on $r$. Solving for $v_z(r)$ subject to no-slip boundary conditions at $r = a$ gives the familiar pipe-flow profile

$$v_z(r) = V_0(a^2 - r^2) \tag{4.10}$$

where $V_0$ is the maximum speed and can easily be expressed in terms of the flux down the pipe.

## 5. Discussion and Closing Remarks

The question naturally arises as to how far the notion of the entropy of a continuous, solenoidal, vector field can be pushed. We do not answer the question here. It seems reasonable, in an axisymmetric field in which $\omega = \omega_\varphi(r, z)\hat{e}_\varphi + \omega_z(r, z)\hat{e}_z$ and the vortex lines do not intersect the cylinder wall, to seek to calculate helical flows by maximizing an expression such as

$$S^* = - \int \left[ \omega_\varphi^+ \, \ell n \, \omega_\varphi^+ + \omega_\varphi^- \, \ell n \, \omega_\varphi^- + \omega_z^+ \, \ell n \, \omega_z^+ + \omega_z^- \, \ell n \, \omega_z^- \right] rdrdz, \tag{5.1}$$

subject to constraints.

The reason we have not pushed further with the maximization of such expressions as (5.1), which effectively treats $\omega_z^{\pm}$ and $\omega_\varphi^{\pm}$ as independent fields, is that the variational fields are constrained in a non-standard way by the requirement that $\nabla \cdot \omega = 0$. We do not doubt that techniques will emerge for defining the statistical weight for a solenoidal field of a completely general character, but the results may come slowly.

Other than the method presented here, there are two known techniques for assigning probabilities to continuous flow fields. First, the fields may be decomposed into a family of orthogonal functions such as Fourier series, and a phase space defined in terms of the expansion coefficients. There is some basis for this for ideal fluids; but for viscous (non-ideal) ones, we have little understanding of how to weight the various possible states. The other method is to regard the fundamental basis of the fluid as molecular and define probability distributions in the phase space of the molecules. This method, too, seems limited by the absence of useful ensembles to describe fluid systems in which organized fluid motion is being steadily driven and dissipated into heat. It seems worthwhile to experiment with other discretizations such as this one based upon field lines, particularly if the calculation of specific flows, rather than just expectations or ensemble averages, is of interest.

We believe also that there is a powerful intuitive appeal in the idea that the "most probable" state of a low Reynolds number dissipative fluid subject to constraints is also the least dissipative one; both conditions represent in a sense the fluid's best efforts to reach the (unconstrained) *most* probable state of thermal equilibrium. We do not see any clear basis at this point for merging these considerations with those of "minimum entropy production rate," if by this it is meant molecular entropy in the standard kinetic theory sense. Note that no use has been made of the fact that $R_\omega$ represents physically an energy dissipation rate. Minimizing $R_\omega/\nu$ would have done just as well. This is perhaps the reason why more discontinuous differences have not shown up in the formulation of the method for ideal and non-ideal fluids. These difficult discontinuities have been standard in other discretizations.

We do believe that there is compelling reason now to believe that maximum entropy and minimum dissipation can be shown to be the same in a non-trivial sense, for a considerable variety of fluid and magnetofluid systems. Our intent is to try to take these considerations further for MHD, in particular.

Another slightly anomalous feature of the present method needs mentioning. In the standard textbook treatment (Batchelor, 1967) of fluid-mechanical energy dissipation, $R_\omega$ as given by Eq. (2.17) is only called the energy dissipation for the case of stationary, no-slip boundaries. For the general case of moving boundaries in incompressible, uniform-density flow, the total viscous contribution to the energy balance consists of two terms: $R_\omega$ and a surface integral over the boundary, which is

$$\rho_0 \nu \int_{S_B} d\sigma \cdot (\omega \times \mathbf{v}).$$

Here, $S_B$ represents the bounding surface, and $d\sigma$ is the vector element of area on it. By "minimizing the dissipation," in our usage, we mean always the minimization of the volume integral $R_\omega$, only. At this point, the method appears not quite identical with the result quoted (for example, by Batchelor (1967), p. 227) that the flow with negligible inertial forces (i.e., $\mathbf{v} \cdot \nabla \mathbf{v} = 0$) is the one which minimizes the energy dissipation subject to a given value of $\mathbf{v}$ over the boundary. In that context, the dissipation being minimized is not necessarily $R_\omega$. However, it may readily be shown that the first variations of the energy dissipation rate per unit volume and the mean square vorticity are proportional, even for the case of moving boundaries. Minimizing the one is equivalent to minimizing the other, even if they are unequal.

## Acknowledgements

This work was supported in part by NASA Grant NAG-W-710 and U.S. Department of Energy Grant DE-FG02-85ER53194.

## References

Ambrosiano, J. and Vahala, G. (1981) Most-probable magnetohydrodynamic tokamak and reversed-field pinch equilibria, *Phys. Fluids* **24**, 2253.

Batchelor, G.K. (1967) *An Introduction to Fluid Dynamics*, Cambridge University Press, Cambridge, U.K.

de Groot, S.R. and Mazur, P. (1984) *Non-Equilibrium Thermodynamics*, Dover, New York.

Finlayson, B.A. (1972) Existence of variational principles for Navier-Stokes equation, *Phys. Fluids* **15**, 963.

Jaynes, E.T. (1957) Information theory and statistical mechanics: I and II, *Phys. Rev.* **106**, 620 and **108** 171 (1957).

Jaynes, E.T. (1979) "Where Do We Stand on Maximum Entropy?" in *The Maximum Entropy Formalism*, ed. by R.D. Levine and M. Tribus, (MIT Press, Cambridge, MA), pp. 15-118.

Jaynes, E.T. (1980) The minimum entropy production principle, *Ann. Rev. Phys. Chem.* **31**, 579.

Joyce, G. and Montgomery, D. (1973) Negative temperature states for the two-dimensional guiding-centre plasma, *J. Plasma Phys.* **10**, 107.

Kraichnan, R.H. and Montgomery, D. (1980) Two-dimensional turbulence, *Rep. Prog. Phys.* **43**, 547.

Katz, A. (1967) *Principles of Statistical Mechanics: The Information Theory Approach*, Freeman, San Francisco.

Keizer, J. (1987) *Statistical Thermodynamics of Nonequilibrium Processes*, Springer-Verlag, New York.

Lamb, H. (1945) *Hydrodynamics*, 6th ed., Dover, New York, pp. 617-619.

Lanczos, C. (1970) *The Variational Principles of Mechanics*, 4th ed., Dover, New York.

Landau, L.D. and Lifshitz, E.M. (1976) *Mechanics*, 3rd ed., Pergamon Press, Oxford, Chapters I and VII.

Lundgren, T.S. and Pointin, Y.B. (1977) Non-Gaussian probability distributions for a vortex fluid, *Phys. Fluids* **20**, 356.

Montgomery, D., Turner, L. and Vahala, G. (1979) Most probable states in magnetohydrodynamics, *J. Plasma Phys.* **21**, 239.

Montgomery, D. and Phillips, L. (1988) Minimum dissipation rates in magnetohydrodynamics, *Phys. Rev.* **A38**, 2953.

Montgomery, D., Phillips, L. and Theobald, M.L. (1989) Helical, dissipative, magnetohydrodynamic states with flow, *Phys. Rev.* **A40**, 1515.

Onsager, L. (1931) Reciprocal relations in irreversible processes: I and II, *Phys. Rev.* **37**, 405 and **38**, 2265.

Prigogine, I. (1947) *Etude Thermodynamique des Phénomènes Irréversible*, Dunod, Paris.

Serrin, J. (1959) in *Fluid Dynamics I*, Vol. 8 of *Handbuch der Physik*, ed. by S. Flügge (Springer-Verlag, Berlin, 1959), Pt. 1, p. 125 ff.

Taylor, J.B. (1974) Relaxation of toroidal plasmas and generation of reverse magnetic fields, *Phys. Rev. Lett.*, **33**, 1139.

Taylor, J.B. (1986) Relaxation and magnetic reconnection in plasmas, *Revs. Mod. Phys.* **58**, 741.

ter Haar, D. (1966) *Elements of Thermostatistics*, 2nd ed., Holt, Rinehart and Winston, New York, 1966.

Ting, A.C., Matthaeus, W.H. and Montgomery, D. (1986) Turbulent relaxation processes in magnetohydrodynamics, *Phys. Fluids,* **29**, 3261.

Ting, A.C., Chen, H.H. and Lee, Y.C. (1987) Exact solutions of a nonlinear boundary value problem: the vortices of the two-dimensional sinh-Poisson equation, *Physica* **26D**, 37.

Yourgrau, W. and Mandelstam, S. (1968) *Variational Principles in Dynamics and Quantum Theory*, 3rd ed., Saunders, Philadelphia.

# MAXIMUM ENTROPY AND EQUATIONS OF STATE FOR RANDOM CELLULAR STRUCTURES

N. RIVIER[*]
*Materials Science Division*
*Argonne National Laboratory*
*Argonne, IL 60439, USA*

ABSTRACT. Random, space-filling cellular structures (biological tissues, metallurgical grain aggregates, foams, etc) are investigated. Maximum entropy inference under a few constraints yields structural equations of state, relating the size of cells to their topological shape. These relations are known empirically as Lewis's law in Botany, or Desch's relation in Metallurgy. Here, the functional form of the constraints is not known a priori, and one takes advantage of this arbitrariness to increase the entropy further. The resulting structural equations of state are independent of priors, they are measurable experimentally and constitute therefore a direct test for the applicability of MaxEnt inference (given that the structure is in statistical equilibrium, a fact which can be tested by another simple relation (Aboav's law)).

## 1. Introduction

We shall discuss the structure of soap froth, tissues, metallurgical grain mosaics, in short, of random space-filling cellular networks. These structures are at first glance all indistinguishable (apple, feather tissues look like soap froth, as do metallurgical ceramics), even though the local forces responsible for their architecture are very different (Weaire and Rivier 1984). These forces are therefore less relevant in molding the cellular structure than the inescapable, mathematical constraints of filling a topological space. This "universality", and the irrelevance of local, specific forces, are only possible if the structure is random. Why is it random? We shall see that randomness and space-filing imply remarkable (and observable) correlations in the cells. The main correlations are expressed in the structural equations of states (1) and (2), which are the focus of our discussion.

Maximum Entropy inference is the tool used. The present application of MaxEnt relies entirely on information expressible as constraints. It is independent of prior probabilities (which are here non-trivial measures as befits a continuous, geometrical problem) and constitutes therefore a direct test of the predicting power of MaxEnt methods.

Specifically, we will exploit the ability of MaxEnt to detect bias (in the colloquial sense). Bias manifests itself here through unexpected correlations between the size of cells and their topological shape (the number of sides). Roughly, larger cells have more sides. But a precise relation was discovered by Lewis (1928) in two-dimensional botanical tissues,

$$A_n = \alpha(n-n_0) \tag{1}$$

and - a different correlation - by Desch (1919) in metallurgical grain mosaics,

---

[*]Work supported by the U.S. Department of Energy, BES-Materials Sciences, under contract W-31-109-ENG-38.

*P. F. Fougère (ed.), Maximum Entropy and Bayesian Methods, 297–308.*

$$\Pi_n = \alpha'(n-n_0') \tag{2}$$

Here $A_n$ and $\Pi_n$ denote the average area and perimeter (or radius) of n-sided cells.

Correlations (1) and (2) can only appear as constraints in MaxEnt formalism, since metallurgical and botanical mosaics, similar looking random space-filling cellular networks, should have the same a priori measures for the size and the shape of their cells. Then, the correlations are physical laws or unavoidable mathematical restrictions. Our aims are to understand their origin, and why grains differ from tissues.

Most of the results of this paper have already appeared in print. The guiding principle in explaining and establishing significant correlations in random cellular structure, especially their equation of state, has been Maximum Entropy. It is a privilege, in this forum, to return the compliment and use empirical structural correlations to illustrate the power and versatility of MaxEnt inference, as "a method of reasoning which ensures that no unconscious arbitrary assumptions have been introduced". (Jaynes 1957, p.16).

## 2. Elementary Topological Transformations and Detailed Balance

Tissues, froths, etc. are cellular graphs with interfaces (edges, and faces in 3D) meeting at vertices and forming polygons in 2D, polyhedra in 3D. Randomness has two, complementary manifestations: Interfaces do not have fixed length, but fluctuate, and the cells' number of sides n is a random variable. The second manifestation is a consequence of the first, as we shall see.

Fluctuations of its interfaces cause elementary, local topological transformations of the cellular network. In two dimensions (2D), there are only two types of transformations, neighbor exchange (T1) or cell disappearance (T2) (and its inverse). (Fig. 1). Cellular division or mitosis, the essential topological transformation in the growth of biological tissues, is an inverse T2, usually combined with a few T1.

These elementary topological transformations have two direct consequences. i) Only vertices of coordination 3 (4 in 3D) are structurally stable. A "Four Corner Boundary", the critical point of T1 transformation, is tipped either way by infinitesimal fluctuation. It occurs with negligible probability in random cellular structures. ii)

$$<n> = 6, \tag{3}$$

for a 2D froth containing a very large number of cells. (A weaker relation holds in 3D) [Eq. (3) follows from Euler's relation, C-E+V = 1, between numbers of cells (C), edges (E) and vertices (V) of any large 2D (Euclidean) network, a topological conservation law (its left-hand side is conserved under elementary transformations), and from vertex coordination 3.]

a)                                                           b)

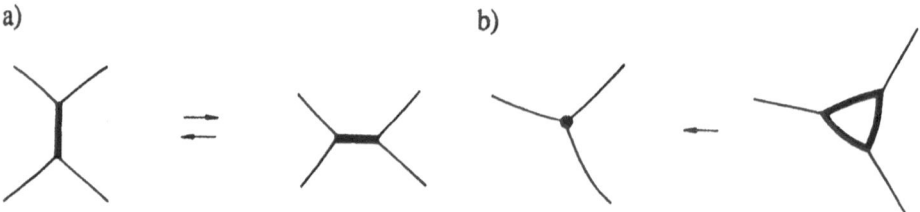

Figure 1. Elementary topological transformations in 2D cellular networks. a) Neighbor exchange (T1). b) Cell disappearance (T2).

Elementary topological transformations play the same role as collisions in kinetic theory of gases. They establish statistical equilibrium of the structure. [Indeed, concentrate on topological correlations m(n), where m is the average number of sides of any cell neighboring a n-sided cell. Unlike the constant <n>, m(n) is controlled by elementary topological transformations, which impose recursion relation between m(n) and m(n±1). Correlation m(n) is the solution of the same recursion relation for T1, T2, and their inverses. Elementary topological transformations can therefore occur independently, anywhere in the structure, and establish the observable correlation m(n), an empirical law due to Aboav (1971) which is universally obeyed by all known random cellular structures (cf. Weaire and Rivier (1984), Mombach *et al.* (1989))]. Aboav's relation indicates that the structure is in detailed balance under elementary topological transformations, a prerequisite for statistical equilibrium, whose characterization is our next step.

## 3. Statistical Equilibrium

Statistical crystallography (the description of cellular structures in statistical equilibrium) is based on the proposition that an assembly of an enormous number of cells will take up one of the most probable configurations subject to a few constraints like space-filling. The most probable configurations in the ensemble share the same equation of state, a relation between measurable, macroscopic parameters of the structure, and the same probability distribution function for the microscopic parameters. In kinetic theory of gases (Table 1), the equation of state is the ideal gas law (pV=NkT), and the distribution function Boltzmann's or Maxwell's distribution, with pressure p, temperature T and volume V as macroscopic parameters, and an atom's velocity or energy, its microscopic parameters {i}. The gas is in microscopic equilibrium as it satisfies detailed balance under collisions which are local, elementary transformations of the microscopic parameters. In liquids where atoms interact, a non-ideal equation of state like van der Waals's, reflecting their interaction, replaces the ideal gas law. Both equation of state and distribution function are obtained by maximizing the entropy, which takes the Gibbs form,

$$S = \Sigma_i \, p_i \ln p_i \,. \tag{4}$$

The first triumph of MaxEnt formalism was to reformulate these physical results in information terms, ridding them of the their unnecessary overcoat of mechanical hypotheses like ergodicity (Jaynes 1957). Gibbs's entropy is identical to Shannon's measure of information, and MaxEnt yields the least biased or maximally non-committal distribution subject to available knowledge encoded in the constraints. But, if the distribution has been analyzed extensively in MaxEnt literature, the equation of state between macroscopic parameters has always been relegated as a somewhat trivial consequence, of interest to experimental physicists only. This is, emphatically, an oversight. While the distribution contains detailed information on the system, it also depends on the "priors" - the measure in phase (or event) space - information which cannot be encoded in constraints, and which is therefore not controllable by physical means. (Statistical mechanics, for which quantum mechanics (Nernst theorem) imposes that all cells in phase space are equally probable a priori is an exception.) By contrast, the equation of state gives a rougher account of the system (it relates macroscopic or average parameters), but it is independent of the priors, as we shall show, and only indicates how many, and which physical (chemical, mathematical or biological) constraints are relevant, a very desirable tool for the scientist: "If it can be shown that the class of phenomena predictable by maximum entropy inference differs in any way from the class of experimentally reproducible phenomena, that fact would demonstrate the existence of new law of physics, not presently known" (Jaynes 1957, p.20). For "class of phenomena", read "equations of state", and you have the message and the content of this paper.

We shall see that froths, tissues, etc., are the least biased partitions of space, subject to a few, inevitable mathematical constraints pertaining to filling a topological space, and possibly, to an energy constraint. The surprise is that there is a particular way of filling topological space which increases the entropy, and therefore decreases the bias further, and this purely mathematical interplay between constraints is indeed reflected in botanical tissues: This is the content of their equation of state, Lewis's law (1). Additional constraints modify the structural equation of state (for example, from (1) into (2), see § 6), which becomes the simplest diagnostic tool for structural pathology.

## 4. Equation of State for Ideal Tissues. Lewis's Law

A cell in 2 dimensions is described by two microscopic parameters, its area A and the number n of its sides. We can marginalize the metric parameter A in the distribution function $P(n,A)$, and concentrate on the shape distribution $p_n$

$$p_n = \int dA\, P(n,A) \tag{5}$$

The topological parameter $n=3,4,...,N$ is an integer, so $\{p_n\}$ can be regarded as a vector in a N-2-dimensional space.

We want to find the most probable, or least biased distribution $\{p_n\}$, without going through the effort of computing the entropy (see Rivier (1985) for a complete solution). $\{p_n\}$ is subject to the inescapable constraints,

$$\begin{aligned}
&\Sigma_n\, p_n = 1, &&\text{normalization}\\
&\Sigma_n\, n\, p_n = 6, &&\text{topology} \qquad\qquad\qquad (6)\\
&\Sigma_n\, A_n\, p_n = A_{tot}/C, &&\text{space-filling}.
\end{aligned}$$

Here, $A_n$ is the average area of n-sided cells, $p_n A_n = \int dA\, A\, P(n,A) = p_n \int dA\, A\, P(A|n)$, $A_{tot}$ is the total area available and C is the number of cells. The topological constraint is familiar (3). The constraints are linear in $\{p_n\}$, and occupy a space of dimension d ($\leq 3$).

The solution $\{p_n\}$ lives in a space of dimension

$$D = N - 2 - d, \tag{7}$$

the dimension of $\{p_n\}$ less that of constraint space. The larger the dimension D of the solution space, the more probable that solution, so that the most probable solution is obtained when the dimension d of the space of constraints is lowest. (Calculation of the entropy fully confines this argument of linear algebra (Rivier (1985), see also § 7).

Now, d will be minimal (=2) if one constraint can be made redundant, that is if the constraints (6) are linearly dependant. Using arbitrary coefficients $\lambda_i$, $0 = 1 - \lambda_1 n - \lambda_2 A_n = 1 - \lambda_1 6 - \lambda_2(A_{tot}/C)$, one obtains,

$$A_n = (A_{tot}/C)\,\lambda\,[n-(6-1/\lambda)], \tag{1'}$$

a relation between average sizes and shapes of cells, discovered empirically by Lewis (1928). It has slope $<A> \lambda$ and intercept $n_0$,

$$(A_{tot}/C) \lambda = <A> \lambda , \tag{8}$$

$$n_0 = 6-1/\lambda , \tag{9}$$

and implies that the average cell area $<A>$ is equal to that of hexagons,

$$<A> = A_{tot}/C = A_6 . \tag{10}$$

Lewis's relation (1) was found originally in cucumber epidermis, human amnion (with an interesting comment (Lewis 1931) on the history of the sample in relation with its morphology) and pigmented epithelium of the retina; later in the epidermis of iris, begonia, and peas (Smoljaninov 1980, Fig 18-19), in iris stomata, axial fibroblasts and sections of the cerebellum of a mouse, cat and man (ibid, Fig. 26-27); also in onion, garlic, agave and anthurium (Mombach et al. 1989), and in the Voronoi cells of Poisson-distributed seeds ((Crain (1978); Two columns of his table are interchanged (Boots 1987)). All these tissues have intercept $n_0 \approx 1$-2, except for the Voronoi, the onion and the stomata $n_0 \approx 0$, and for the cerebella $n_0 \approx -2$. The intercept $n_0$ and slope of Lewis's law are given in term of the indeterminate multiplier $\lambda$ (which serves indeed as a Lagrange multiplier in maximizing the entropy). We shall see in § 5 that $\lambda$ is the time - it measures the ageing of the structure, at least for soap froth where a complete physical description of growth (von Neumann's law) is available. Thus, the larger the intercept $n_0$, the coarser and older (riper) is the structure. Fig. 2 reproduces Lewis's data (1931) on cucumber epidermis, and human amnion. If the average area $A_n$ is linear in n, the range in areas of individual cells is large ($\approx 0.6$-1 $A_n$).

Lewis's law is a structural equation of state, obtained because the tissue is free to adjust the arbitrary functional dependence of the constraint $A_n$ in order to increase the entropy. The derivation is due to Rivier and Lissovski (1983). It raises several questions: 1) What is the parameter $\lambda$ ? (Lagrange multipliers in physics are physical quantities, temperature, chemical potential, force of constraint, etc.). 2) Is Lewis's law universal ? Soap froths will be discussed presently, and I feel that they do obey Lewis's law despite their high $n_0$ (and $\lambda$). Metallurgical grain mosaics do not. They follow Desch's law (2) instead. Why ? (Note that $A_n$ and $\Pi_n$ are both averages, so that $A_n \neq \Pi_n^2$ in general.)

In soap froths (Glazier et al. 1989) as in the one photographic emulsion studied by Lewis (1931), the intercept $n_0 > 3$, so that the occurrence of a few three-sided cells badly spoils the linearity of Lewis's plot. However, these few cells are not statistically significant and the large, systematic departure from the linearity of Lewis's law for n = 3 is only a reflection of the impossibility of negative areas. Nevertheless, a definite conclusion on Lewis's law in soap froths must await statistical analysis of the data, which has yet to be made. In the froths, the average cell area $A_{tot}/C$ equals from that of the hexagon, $A_6$, in agreement with Lewis's law (10). This is not the case in the emulsion.

## 5. Evolution, von Neumann's law and the meaning of $\lambda$

Soap froths, metallurgical grain mosaics and biological tissues evolve in time, through topological transformations induced by the fact that air diffuses out of small bubbles into the large ones, larger botanical cells are more likely to divide, and larger grains grow at the expense of smaller ones. The time scale for evolution is long enough for the froth to

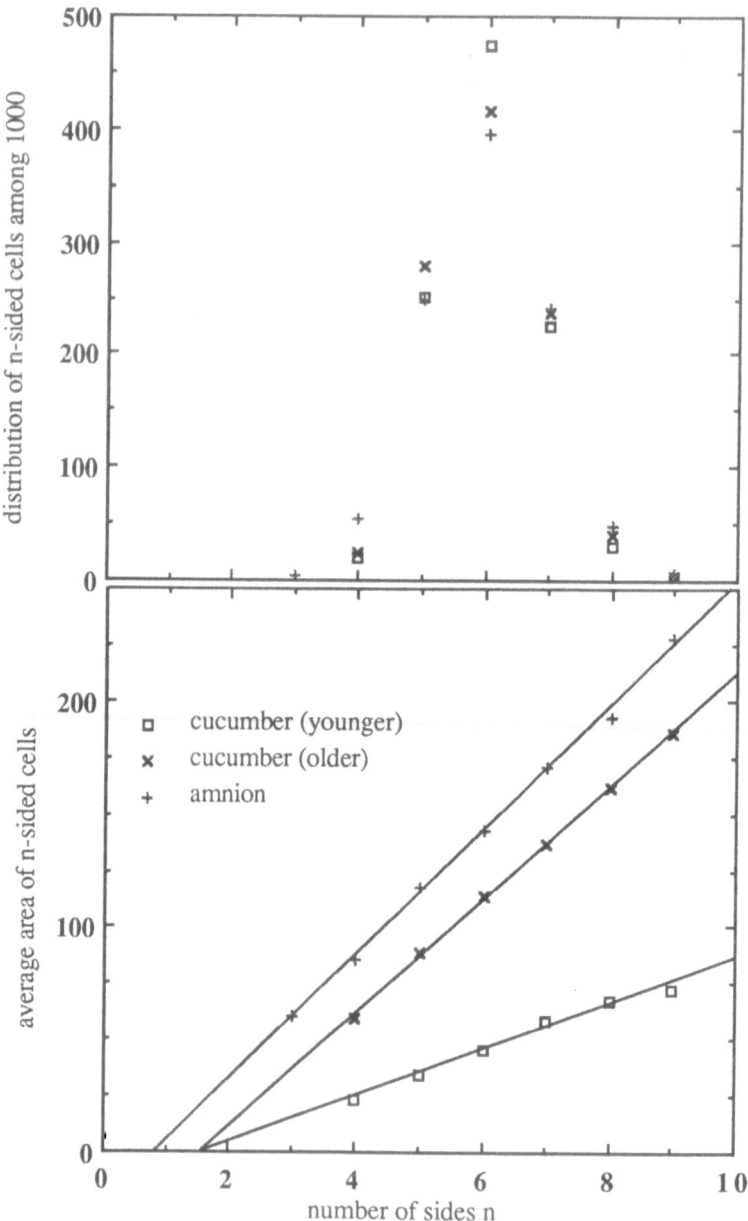

Figure 2. Lewis's law (1') and distribution of cell shape for cucumber and amnion. (Data from Lewis 1931). $<A> = A_6$.

remain in statistical equilibrium and to satisfy its equation of state at all times. In thermodynamics, the evolution is then called quasistatic.

*5.1 Coarsening.* For an ideal tissue or froth, Lewis's law is obeyed at all times. The simplest case is a soap froth, where $A_{tot}$ and C remain constant for a while (at least for young froths without triangular bubbles, so that no T2 transformations occur). Then, differentiation of Lewis's law (1') yields

$$(d/dt)A_n = (d/dt) [(A_{tot}/C) \lambda (n-6) + (A_{tot}/C)] = \gamma (n-6) , \qquad (11)$$

a relation obtained by von Neumann (1952) for 2D soap froths. Eq. (11) implies that the froth coarsens, a fact obvious to anyone washing the dishes. Von Neumann's derivation is a gem: A few, simple, physical laws conspire to produce a purely topological (and almost dimensionless) result (11). Any (not the average) pentagonal cell loses area at the same rate as any heptagon grows, at half the rate of any octogon, etc. The present derivation (Rivier 1983a) is slightly weaker because it involves averages, but like Lewis's or Desch's law, it can be generalized to 3 dimensions, and to all space-filling random structures in statistical equilibrium. [Von Neumann's independent derivation also shows that a froth need not obey Lewis's law to evolve according to (11) (as is the case for metallurgical aggregates)].

Von Neumann obtains $\gamma = 2\pi\sigma\delta/3 > 0$, where $\sigma$ is the surface tension and $\delta$ the gas diffusivity across the interfaces. Here,

$$\gamma = (A_{tot}/C) (d\lambda/dt) > 0 , \qquad (12)$$

hence,

$$\lambda = (C/A_{tot}) \gamma t + cst , \qquad (13)$$

so the indeterminate multiplier $\lambda$ is simply the time. It measures the ageing of the structure. The intercept of Lewis's law $n_0 = 6 - 1/\lambda$ (9) is therefore larger for older froths, which are also coarser and less uniform (the slope of Lewis's law, proportional to $\lambda$, is larger), a phenomenon known in metallurgical circles as Ostwald ripening.

If C does not remain constant (bubbles or grains disappear (T2)),

$$dA_n/dt = \gamma (n-n_1) \qquad (14)$$

$$n_1 = 6 - [(d/dt) (A_{tot}/C)] / \gamma < 6 \qquad (15)$$

$$\gamma = (d/dt) (\lambda A_{tot}/C) > 0 , \qquad \lambda = (C/A_{tot}) \int \gamma \, dt \qquad (16)$$

and the coarsening is qualitatively similar.

Why should metallurgical grains (whose evolution is driven by surface (grain boundary) tension) evolve like soap froths (which coarsen by diffusion of gas across interfaces from the smaller cells with high pressure into the larger ones) ? This is because the resultant of the surface tensions of three interfaces concurrent at a vertex is directed towards the cell with inner angle $< 2\pi/3$. Accordingly, hexagons with 6 angles equal to $2\pi/3$ are stationary, pentagons tend to shrink and heptagons grow. Recent simulation and modelisation of grain growth by Telley (1989) has made this result quantitative.

*5.2 Botanical tissues. Steady state growth.* In constrast, to froths or grains, biological tissues evolve by combination of growth [$(d/dt) A_{tot} > 0$] and cellular division [$(d/dt) C > 0$], while remaining structurally in a steady state. Cellular division (mitosis) is a combination of elementary topological transformations (inverse T2, together with one or several T1).

The steady state of vitally active botanical tissues has <u>constant</u> $\lambda$ (which does not measure time any longer), thus constant intercept of Lewis's law, but increasing $A_{tot}$ (continuous growth) and number of cells C (cell division). Strict steady state (topological invariance of the structure) corresponds to constant $A_{tot}/C$ over times longer than the period of the mitotic cycle, and is achieved through interplay between continuous cellular growth and discrete mitoses. It is not quite realized in cucumber epidermis, in which $A_{tot}/C = A_6$ and Lewis's slope both increase with age, while the intercept $n_0 \approx 2$, thus $\lambda$, remains constant. (Fig. 2). Cucumber epithelia coarsen a little.

Topological steady state has been discussed elsewhere (Rivier 1988). It is interesting because it yields the average shape of cells just about to divide (n=7) and that of their two daughters (5.5), as well as the period of the mitotic cycle. An importance consequence of Lewis's law (already hinted at in the original paper of Lewis (1931)) is that if detailed balance holds for the topological variable n (and it does, as shown by Aboav's law), it automatically holds for the metric variable A, since

$$\langle A_n \rangle = A_{\langle n \rangle} . \tag{17}$$

Notably, the average growth rate of a cell is given by

$$d\langle A \rangle / dt = \sum_n p_n (dA_n/dt) , \tag{18}$$

since

$$\sum_n A_n (dp_n/dt) = 0 \tag{19}$$

follows from topological and normalization constraints (6) and Lewis's law (1). Note that Eq. (19) does <u>not</u> imply topological steady state ($dp_n/dt = 0$).

The conclusion of the last two sections is a quotation lifted from Lewis (1931) paper: "Diese scheinbare Regellosigkeit sehr wohl geregelt ist" (Strasburger, 1866).

## 6. Metallurgical Aggregates, Desch's Law

In metallurgical aggregates, it is not the cell's area, but its perimeter $\Pi_n$ (or radius) which is proportional to n, a relation (2) discovered also empirically and even earlier on by Desch (1919), and confirmed by computer simulations (Srolovitz et al. 1984).

MaxEnt inference, and the inevitability of all the constraints (6) used in deriving Lewis's law, immediately provide the explanation for this non-ideal behavior, namely the presence of an additional physical constraint. Clearly, there is energy concentrated in the interfaces or grain boundaries between cells, known as surface tension $\sigma$. Accordingly, the cell has an additional parameter, its perimeter $\Pi$, and the mosaic has an additional constraint,

$$\sum_n p_n \Pi_n = 2\sigma E/C \qquad \text{energy} , \tag{20}$$

besides the inevitable ones (6). Here $p_n = \int dA d\Pi \, P(A,\Pi,n)$ and $p_n \Pi_n = \int dA d\Pi \, \Pi \, P(A,\Pi,n)$. Maximization of entropy proceeds as in § 4. (The Lagrange multiplier enforcing the energy constraint (20) is the inverse temperature $1/kT$, and we are on the solid grounds of classical thermodynamics.)

The functional dependence of $\Pi_n$, like that of $A_n$, is not imposed a priori. It can be adjusted to decrease the dimension of the space of constraints, and increase further the entropy. It turns out that there are only two alternatives (Rivier 1985), either $A_n$ and n, or $\Pi_n$ and n are linearly dependent, but not both. Moreover, the latter alternative has a slightly higher entropy $S_{ME}$ than the former, but this fact does not rule out Lewis's law even in the presence of surface tension, because the two alternatives are disjoint in events space. At any rate, the presence of surface tension, in addition to the inevitable constraints (6), offers Desch's law,

$$\Pi_n = (2\sigma E/C)\lambda\,[n-(6-1/\lambda)], \qquad <\Pi> = \Pi_6\,, \qquad (2')$$

as an alternative to Lewis's (1',10), as is indeed observed in metallurgical aggregates. Note that the slope, intercept and the average perimeter of Desch's law duplicate exactly the corresponding parameters of Lewis's law (8-10). The evolution of metallurgical mosaics (Ostwald ripening) is therefore expected to follow a von Neumann's type of law (11), but involving the cell's perimeter instead of its area. Again, $\lambda$ is the time, and the mosaic coarsens,

$$(d/dt)\,(\lambda 2\sigma E/C) > 0 \qquad (21)$$

but there is no simple expression for this positive rate in terms of physical parameters, as in soap froths. We have already justified in § 5 von Neumann's evolution for mosaics driven by surface tension. Simple trigonometry shows that a hexagon in a mosaic, with 6 interfaces incident at $2\pi/3$ on its vertices, can swell or shrink freely without changing the interfacial length, i.e., at no cost in energy. The extent of this free breathing is only limited by topological transformations T1 and T2.

Von Neumann's proof of his law (11) refers to each cell individually, for which $A \propto \Pi^2$. Thus, $(d/dt)\Pi_n$, like $(d/dt)A_n$, is proportional to (n-6) if C remains constant, in agreement with evolution of Desch's law (2'). But the coefficient of proportionality $(1/2\Pi_n)$ depends on n.

Jaynes's scenario quoted in §3, has been performed in this section by metallurgical grains. The fact that the "new law of physics" (2) was actually known by a few metallurgists since 1919 should not detract from this demonstration of the predictive and diagnostic power of MaxEnt, and of the role in understanding and clarifying physical phenomena by identifying the relevant constraints. It also shows that there was fundamental physics besides Einstein's in the late 1910's.

## 7.  Maximum Entropy and the Number of Constraints

In deriving Lewis's (1) or Desch's (2) laws, we have adjusted the unknown functional form of the space-filling or energy constraint to increase the entropy. This absence of information on, or arbitrariness in the form of the energy or space-filling law, represented as an additional degree of freedom in the system, is measured by an increase of the entropy, as we shall now demonstrate.

Specifically, two constraints were made linearly dependent. Equivalently, the dimension of the space of constraints was decreased by one, or one constraint was made redundant. Let us see how this increases the entropy (Rivier 1983b).

Consider s constraints, labelled by $\alpha = 1...s$,

$$<c_\alpha> = \Sigma_n \, p_n \, c_{n\alpha} = X_\alpha = - \, (\partial/\partial\lambda_\alpha) \ln Z \, , \tag{22}$$

imposed by the Lagrange multipliers $\lambda_\alpha$, with

$$Z = \Sigma_n \, q_n \, \exp \, (- \, \Sigma_\alpha \, \lambda_\alpha \, c_{n\alpha}) = Z \, (\{\lambda_\alpha\}) \, . \tag{23}$$

(The priors $q_n$, introduced for generality, are unimportant in our argument). The most probable, or maximum entropy $S_{ME}$ subject to the constraints, and the distribution $p_n$ are obtained by extremising the functional $-\Sigma_n \, p_n \ln p_n + \Sigma_\alpha \lambda_\alpha \, (\Sigma_n \, p_n \, c_{n\alpha})$,

$$S_{ME} = \ln Z + \Sigma_\alpha \, \lambda_\alpha \, X_\alpha = S_{ME}(\{X_\alpha\}) \, , \tag{24}$$

which is a function of $\{X_\alpha\}$ only, since

$$\partial S_{ME}/\partial\lambda_\alpha = 0 \tag{25}$$

is none other than the constraint equation (22). $S_{ME}(\{X_\alpha\})$ is therefore the Legendre transform of $\ln Z(\{\lambda_\alpha\})$, with

$$\partial S_{ME}/\partial X_\alpha = \lambda_\alpha \, , \tag{26}$$

and

$$\partial^2 S_{ME}/(\partial X_\alpha)^2 = \partial\lambda_\alpha/\partial X_\alpha = (\partial X_\alpha/\partial\lambda_\alpha)^{-1} = -[<(c_\alpha - <c_\alpha>)^2>]^{-1} \le 0 \, . \tag{27}$$

The last equality comes from (22), and the mathematically obvious inequality, the positivity of specific heats, constitutes one of the earliest postulates of thermodynamics (Truesdell (1980), p. 16). The entropy is a convex function of its coordinates $\{X_\alpha\}$.

Consider now constraint $\xi$, say. The entropy $S_{ME}$ is largest as a function of $X_\xi$ when $X_\xi$ is such that

$$\lambda_\xi = \partial S_{ME}/\partial X_\xi = 0 \tag{28}$$

because of inequality (27). But the vanishing Lagrange multiplier $\lambda_\xi = 0$ makes the corresponding constraint $\xi$ inoperative or redundant. This demonstration illustrates the discussion by Jaynes (1979) and Fougere (1988) of what counts as a degree of freedom in $Chi^2$ analysis.

## 8.  Conclusions

The principles governing the structure of random cellular networks and their evolution (statistical crystallography) are those of statistical mechanics. Accordingly, the development of statistical crystallography is parallel to that of classical statistical thermodynamics (Table 1). Its centerpiece is the equation of state, obtained by maximizing the entropy. Here, the cellular network is free to use the unspecified form of the constraints to increase its entropy, and does so, yielding equations of states (1) and (2) which were already known empirically. The ill-definition in the constraints is therefore a matter of physical indifference, a symmetry, rather than ignorance. A different equation of state would reveal the action of a new constraint, hitherto unidentified.

Slow evolution of the structure gives a physical meaning to the undetermined parameter in the equation of state. It is governed by an equation which is obtained almost automatically from the equation of state, but had been derived by von Neumann under specific physical assumptions. We are reminded of Jaynes' (1979, pp. 227, 232, and 236) comments as to how dynamics and ergodicity enter statistical mechanics: "Very efficiently".

TABLE 1. Methodology and equations of statistical thermodynamics and crystallography.

| Thermodynamics | Statistical | Crystallography |
|---|---|---|
| | Random variables: | |
| (position), velocity | | cell shape n, size A |
| $\Delta p \, \Delta x = h$ | | priors (geometrical measure) |
| | Detailed balance under: | |
| collisions | | elementary transformations |
| | | -> **Aboav law** |
| | | (shape correlations) |
| | Statistical equilibrium | |
| | (Maximum Entropy) | |
| | - equation of state (ideal) | |
| ideal gas law | | **Lewis law** |
| | - equation of state (non-ideal) | |
| van der Waals, etc. | | **Desch law**, etc. |
| | -most probable distributions | |
| Maxwell-Boltzmann | | P(A,n) |
| | | universality of cellular structures |
| | Slow evolution | |
| | (quasistatic transformations) | |
| irreversible thermodynamics | | **von Neumann law** (coarsening) |
| | | or **steady state** (growth, mitosis) |

## 9.   References

Aboav, D.A. (1970) 'The arrangement of grains in a polycrystal', Metallogr. 3, 383-390.
Boots, B.N (1987) 'Edge length properties of random Voronoi polygons', Metallogr.20, 231-236.
Crain, I.K. (1978) 'The Monte-Carlo generation of random polygons', Comput. and Geosc. 4, 131-134.
Desch, C.H. (1919) 'The solidification of metals from the liquid state', J. Inst. Metals 22, 241-276.
Fougere, P.F. (1988) 'Maximum Entropy calculations on a discrete probability space', in G.J. Erickson and C.R. Smith (eds.), Maximum-Entropy and Bayesian Methods in Science and Engineering (Vol. 1), 205-234.
Glazier, J.A., Gross, S.P., and Stavans, J. (1987) 'Dynamics of two-dimensional soap froths', Phys. Rev. A 36, 306-312.
Jaynes, E.T. (1957) 'Information theory and statistical mechanics', reprinted in R.D. Rosenkrantz (ed.), E.T. Jaynes: Papers on Probability Statistics and Statistical Physics, Reidel, Dordrecht (1983), 4-16.
Jaynes, E.T. (1978) 'Where do we stand on Maximum Entropy?', ibid., 210-314.
Jaynes, E.T. (1979) 'Concentration of distributions at entropy maxima', ibid., 315-336.
Lewis, F.T. (1928) 'The correlation between cell division and the shapes and sizes of prismatic cells in the epidermis of *cucumis*', Anat. Record 38, 341-376.
Lewis, F.T. (1931) 'A comparison between the mosaic of polygons in a film of artificial emulsion and the pattern of simple epithelium in surface view (cucumber epidermis and human amnion)', Anat. Record 50, 235-265.
Mombach, J.C.M., Vasconcellos, M.A.Z., and de Almeida, R.M.C. (1989) 'Arrangement of cells in vegetable tissues', J. Phys. D, to appear.
von Neumann, J. (1952) 'Discussion - Shape of metal grains', in Metal Interfaces, Amer. Soc. Metals, Cleveland, 108-110.
Rivier, N. (1983a) 'On the structure of random tissues or froths, and their evolution', Phil. Mag. B47, L45-49.
Rivier, N. (1983b) 'Topological structure of glasses', in V. Vitek (ed.), Amorphous Materials: Modeling of Structure and Properties, The Metall. Soc. of AIME, Warrendale, 81-97.
Rivier, N. (1985) 'Statistical crystallography. Structure of random cellular networks', Phil. Mag. B52, 795-819.
Rivier, N. (1988) 'Statistical geometry of tissues', in I. Lamprecht and A.I. Zotin (eds.), Thermodynamics and Pattern Formation in Biology, de Gruyter, Berlin, 415-446.
Rivier, N. and Lissowski, A. (1982) 'On the correlation between sizes and shapes of cells in epithelial mosaics', J. Phys. A15, L143-148.
Smoljaninov, V.V. (1980) Mathematical Models of Biological Tissues, Nauka, Moscow (in Russian).
Srolovitz, D.J., Anderson, M.P., Sahni, P.S., and Grest, G.S. (1984) 'Computer simulation of grain growth - II'. Acta Metall. 32, 793-802.
Telley, H. (1989) Modelisation et Simulation Bidimensionnelle de la Croissance des Polycristaux, PhD Thesis, EPFL, Lausanne.
Truesdell, C. (1980) The Tragicomical History of Thermodynamics 1822-1854, Springer, Berlin.
Weaire, D and Rivier, N. (1984) 'Soap, cells and statistics - Random patterns in two dimensions', Contemp. Physics. 25, 59-99.

# DELAY ESTIMATION USING MAXIMUM ENTROPY
# DERIVED PHASE INFORMATION

Jacqueline Yao
*The University Of Akron*
*Akron, Ohio 44325*
*USA*

Dr. Louis Roemer
*Professor Of*
*Electrical Engineering*
*Louisiana Tech University*
*Ruston, Louisiana 71272*
*USA*

Dr. Nathan Ida
*The University Of Akron*

Ke-Sheng Huo
*The University Of Akron*

ABSTRACT. The frequencies chosen for ultrasound reflectometry, based on attenuation and resolution constraints, often result in overlapping echo waveforms [1-3]. Phase information [4] can be used to estimate echo contributions in reflectometry using the Maximum Entropy Method [5]. The method of analysis is presented, along with experimental verification for a simple reflecting structure.

## 1. Introduction

The waveforms used in ultrasound reflectometry are typically several cycles, rapidly damped, at the resonant frequency of the transducer [6]. When low frequencies are chosen to obtain the low attenuation through the medium under test, then the time waveforms of reflections from adjacent obstacles will often overlap. This overlap of echos makes it difficult to interpret where the sources of the echos are located. The phase information, found through the use of the Maximum Entropy Method (MEM), can be used to identify such echos (which we demonstrate for a monolayer, though the method is not restricted to a monolayer).

The information content of the phase component of systems has been recognized [4,7] in different applications. In this work, the phase is obtained from the power spectral estimates of the even and odd constituents of the MEM power spectrum estimate (the terms even and odd originating from the respective time-domain sequences). We shall illustrate this by processing experimental signals obtained from a thin sheet obstacle.

The Maximum Entropy Method [8] is used to estimate the power spectra. The MEM spectral estimate provides a non-zero extension for missing data by making use of only the available data.

We must emphasize here that our aim is not to estimate the

*P. F. Fougère (ed.), Maximum Entropy and Bayesian Methods, 309–324.*
© 1990 *Kluwer Academic Publishers.*

power spectrum of the output signal nor is it to model the system as an all pole, minimum phase filter (although we, in fact, do both). We are simply utilizing the robustness and accuracy of the MEM power spectral estimation as a tool in our computations. The component of our calculation which makes use of the MEM spectrum estimation (which can be represented as a filter) allows us to consider superposition of signals or echos. Though the MEM uses a non-linear criterion of goodness, the predictive filter which is computed is a linear filter.

To demonstrate our technique we use a thin sheet of aluminum immersed in water; the resulting echoes overlap each other, resulting in a long, ill-defined reflected signal. The phase information obtained from the MEM spectra is further processed to yield an estimation of the time delay between a signal passing straight through the sheet and one reflecting back and forth within the sheet once before exiting (double the time delay to pass through the aluminum sheet).

Two examples are presented. The first example is a direct transmission of a pulse from one 3.5 MHz transducer to a similar receiving transducer (the case of removal of the obstacle from Figure 1a). No delayed pulse is observed, as spurious reflections have been reduced by echo absorbing material. The second example is a reflection from two nearby surfaces which cause overlapping echoes (created by the delay in passing through twice the reflecting sheet thickness). The signal path is shown in Figure 1b. The experimental apparatus corresponds to Figure 1a with the obstacle present.

All systems and processes are assumed real, unless otherwise specified.

## 2. Method: Obtaining Phase through the Use of Maximum Entropy Filters

The principal of maximum entropy, which involves autoregressive modeling, has been applied successfully to power spectral estimation [9-12]. The significance of the method compared to some traditional techniques [13] is higher frequency resolution, dependence only on the available data, and simple storage requirements owing to the infinite impulse response (IIR) structure of the resulting all pole filter. The all pole filter yields a smoothly changing phase estimate, which does not appear to have the experimental difficulties reported with other methods [14,15]. The model is linear, but overall, the processing is not linear.

The data $x(n)$ are used to compute a predictor polynomial, $H(z)$, which predicts the next data point, $x(n+1)$. The order

of the predictive filter chosen will set the number of poles
of the predictive filter, $Y(z)$. $Y(z)$ will be a minimum phase
filter, where

$$Y(z) = \frac{\sigma}{H(z)} \tag{1}$$

The unit sample response $y(n)$ of $Y(z)$ is not directly
related to $x(n)$, however their respective autocorrelation
values R.(n) are related by

$$R_{xx}(n) = R_{yy}(n) \quad n = 0, 1, \ldots, N \tag{2}$$

where N is the order of the filter. Hence $R_{yy}(n)$, for $n>N$,
provides a non-zero extension to $R_{xx}(n)$. Since the power
spectrum $s.(e^{ij\omega})$ is the Fourier transform of its autocorrelation sequence; then

$$|Y(e^{j\omega})|^2 = S_{yy}(e^{j\omega}) = |\frac{\sigma^2}{H(z)H(1/z)}|_{z=e^{j\omega}} \tag{3}$$

is an estimate of the power spectrum of $x(n)$. It is clear
that the estimation is independent of the phase of $x(e^{ij\omega})$.
The error noise power of the predictive filter is $\sigma^2$.

If the phase of the signal, composed of outgoing and
reflected waves, were only known, then the phase would yield
an estimate of the delay of the reflected wave compared to
the outgoing wave. In a simplistic model, a signal $x(t)$
consisting of an original unit amplitude cosine and a
reflected fraction, $a_0$, delayed by time $t_O$ would be

$$x(t) = \cos \omega t + a_o \cos \omega (t - t_o)$$

This can be written as

$$x(t) = \sqrt{\left(1 + a_o^2 + 2a_o \cos \omega t_o\right)} \cos(\omega t - \theta)$$

where

$$\theta = Arctan\left(\frac{a_o \sin \omega t_o}{1 + a_o \cos \omega t_o}\right)$$

For small reflection coefficients, $a_o$, and for $a_o \cos \omega t_o \ll 1$

$$\theta \doteq Arctan\left(a_o \sin \omega t_o\right)$$

$$\theta \doteq a_o \sin \omega t_o$$

The periodicity of $\sin \omega t_o$ with respect to $\omega$ yields $t_o$ which equals $2d/v$, for a back-and-forth pair of reflections in a slab of thickness d and sound velocity $v$.

Unfortunately, we do not have the infinite range of $\omega$ to allow observing the phase, but only a small window in the frequency domain, due to the transducer bandwidth (and corresponding window in the time domain covering the overlapped pulse and echo).

We wished to avoid the FFT, as the windowed FFT provides the usual windowing problems, though its imaginary and real parts might allow computing phase (with some restrictions). The MEM, in contrast, yields a power spectral estimate which is real and non-negative. Thus it is necessary to use the even and odd components of the data to arrive at a phase estimate, a minor trick.

To estimate the phase of the process x from the known data samples x(0), x(1),..., x(M) and still be consistent with the principles of maximum entropy, we write the time series x(k) as the sum of an even series, e(k), and an odd series o(k), where the point of symmetry is chosen at

$$N_s = M/2 \tag{4}$$

and M, the data length, is an even integer.

The power spectra of these three sequences are related by

$$S_{xx}\left(e^{j\omega}\right) = S_{ee}\left(e^{j\omega}\right) + S_{oo}\left(e^{j\omega}\right) \tag{5}$$

We denote by

$$X\left(e^{j\omega}\right), E\left(e^{j\omega}\right)e^{-j\omega T_s}, jO\left(e^{j\omega}\right)e^{-j\omega T_s} \tag{6}$$

the Fourier Transforms, respectively, of x(k), e(k), and o(k).

$$E\left(e^{j\omega}\right) \text{ and } O\left(e^{j\omega}\right)$$

are real functions of $\omega$ and the factor $e^{-j\omega T_s}$ is due to the fact that the origin of symmetry is $T_s = N_s \Delta t$. $N_s$ is the number of samples to the center of the data block, $\Delta t$ is the sample spacing. The phase $\theta(\omega)$ is defined as

$$\theta(\omega) = Arctan \frac{O\left(e^{j\omega}\right)}{E\left(e^{j\omega}\right)} \tag{7}$$

Use the trigonometric identity

$$\cos[2\theta] = \cos^2\theta - \sin^2\theta \tag{8}$$

and replace $|E|^2$ with $S_{ee}$, from the Fourier transform property

$$E\left(e^{j\omega}\right) = Re\left\{X\left(e^{j\omega}\right)\right\}$$

and $|O|^2$ with $S_{oo}$, from the corresponding

$$O\left(e^{j\omega}\right) = Im\left\{X\left(e^{j\omega}\right)\right\}$$

The geometry is described by

$$Cos\,\theta = \frac{E(\omega)}{\sqrt{E^2(\omega) + O^2(\omega)}}$$

and

$$sin\,\theta = \frac{O(\omega)}{\sqrt{E^2(\omega) + O^2(\omega)}}$$

Use of the Cos(2θ) identity and using

$$Cos\,2\theta = \frac{|E(\omega)|^2 - |O(\omega)|^2}{|E(\omega)|^2 + |O(\omega)|^2}$$

allows us to write

$$Cos[2\theta(\omega)] = \frac{S_{ee}\left(e^{j\omega}\right) - S_{oo}\left(e^{j\omega}\right)}{S_{ee}\left(e^{j\omega}\right) + S_{oo}\left(e^{j\omega}\right)} \tag{9}$$

Both power spectral terms in (9) can be expressed in the form of (3) as

$$S_{ee}\left(e^{j\omega}\right) = \frac{P_e}{H_e\left(e^{j\omega}\right)H_e\left(e^{-j\omega}\right)} \tag{10}$$

$$S_{oo}\left(e^{j\omega}\right) = \frac{P_o}{H_o\left(e^{j\omega}\right)H_o\left(e^{-j\omega}\right)} \tag{11}$$

Substituting (10) and (11) in (9) yields

$$Cos[2\theta(\omega)] = \frac{P_e H_o\left(e^{j\omega}\right)H_o\left(e^{-j\omega}\right) - P_o H_e\left(e^{j\omega}\right)H_e\left(e^{-j\omega}\right)}{P_e H_o\left(e^{j\omega}\right)H_o\left(e^{-j\omega}\right) + P_o H_e\left(e^{j\omega}\right)H_e\left(e^{-j\omega}\right)} \tag{12}$$

Since maximum entropy filters are minimum phase, the right hand side of (12) converges for all values of $\omega$. The phase $\theta(\omega)$ can be calculated from (12).

## 3.  Phase Estimate applied to Range Detection

A pulse of energy traveling through a uniform lossless medium is reflected from a boundary and the echo is received at the point of origin of the signal. Let the incident signal be denoted x(t), the total traveled distance 2d, and the velocity of propagation $v$. The received signal y(t) is a function of the delay time, $t_d$, where, $t_d = 2d/v$. The aim is to determine the distance d to the boundary by computing the delay $t_d$. Preprocessing the echo signal according to the method discussed in the previous section yields

$$Cos\,2\theta(\omega) = Cos\left[2\omega t_d\right] \tag{13}$$

where the argument is proportional to the delay time of the echo; all data is referenced to the the origin of symmetry (midpoint of the data). The power spectrum of the $\cos 2\theta$ (the phase estimate of our original data) curve exhibits a peak at a time corresponding to $t_d$. This problem of echo location of the source of echos (while having overlapping echos) is encountered in radar, seismology, and echo cardiography. To illustrate the usefulness of the foregoing technique to this class of problems, a simple geometry will be used as an example.

## 4.  Experimental Confirmation

To test for robustness of the Maximum Entropy Phase Estimation (MEPE) method, we inserted a sheet of aluminum of 0.16 cm thickness into the ultrasound path described in Figure 1. The ultrasound transducers used were 3.5 MHz focused transducers, mounted on opposing sides of a plastic tank. The data were oversampled at 20 M samples/second, then averaged for an effective rate of 10M samples/second. We recognize that the transducer energy is centered around the 3.5 MHz resonance of the transducer; however, the information content in the signal phase (and amplitude) is distributed around the unit circle in the z plane. This is because the poles of the AR predictive filter convey their information as we traverse the unit circle, changing most rapidly in the vicinity of the poles. Though the signal energy is concentrated near 3.5 MHz, the computation of the AR coefficients, and therefore the pole locations, evidence this information on the whole unit circle. To use only the information on the unit circle adjacent to the poles is to window the result, thereby throwing away information.

To obtain a sufficiently large signal, a bijunction transistor was used to discharge a capacitor across the transmitter transducer. The received signal was sufficiently large to be digitized by a Tektronix model 2220 digital oscilloscope without prior amplification.

A signal without multiple reflections is shown in Figure 2a. The amplitude is the digitized value, which may be regarded as a relative value. The evaluation of equation (12), the cosine of twice the phase angle, is shown in Figure 2b. Taking the spectrum of the Figure 2b data yields the echo identification, Figure 2c (We have used the fact that there is no DC term, due to the odd symmetry of phase around the unit circle). A single large return with minimal artifacts is shown in Figure 2c. The relative time of arrival has been shifted to allow a full display in the drawing.

A contrasting signal with multiple reflections is shown in Figure 3a. Here, an extended echo appears without clear delineation between the different components of the multiple echo. The cosine of twice the phase angle is shown in Figure 3b; the oscillatory nature of the curve allows identifying the multiple echoes, shown in Figure 3c. The position of the first peak in Figure 3c corresponds to the time of transit (round trip, i.e. twice the 0.16 cm path) through the aluminum sheet that comprises the test obstacle. The sample calculation is

$$d = \frac{1}{2}[6420 m/s][0.5\mu s] \sim 1.6mm$$

The relative amplitudes of the multiple echoes are consistent, too.

The simple geometry illustrates the efficacy of the algorithm. The driving reason for using it, of course, is the need to identify multiple echoes in systems in which the choice of waveforms (duration and frequency) is constrained by the attenuation (and the limitations in signal sources). In a medical application, Erdol has simulated waveform responses for thin layers of tissue, which encourages the algorithm's application to echocardiography and similar applications [16]. (This paper has extended Erdol's method by using the Maximum Entropy Method for obtaining both the odd and even sequence spectra, as well as the spectra of the resulting phase estimate.)

## 5. Conclusions

The algorithm proved resilient for the data examined. The method is of more interest than just the data above, in that it attempts to estimate phase using power spectral estimates.

## 6. Acknowledgements

The generous support of Timken Research Center through Dr. Michael French is deeply appreciated. The loan of equipment, technical discussions, funding of Timken Fellowships, and encouragement are appreciated.

## REFERENCES

1. Kemerait, R.C. and Childers, D.G., "Signal Detection and Extraction by Cepstrum Techniques", IEEE Transactions on Information Theory, Vol IT-18, No. 6, pp. 745-749, November 1972.
2. Riad, S.M. and Nahman, N.S., "Application of the Homomorphic Deconvolution for the Separation of TDR Signals Occurring in Overlapping Time Windows", IEEE Transactions of Instrumentation and Measurement, Vol IM-25, No. 4, pp.

388-391, Dec. 1976.
3.  Blitz, J., Ultrasonics: Methods and Applications, Van Nostrand Reinhold Co., New York, 1971.
4.  Piersol, A.G. "Time Delay Estimation Using Phase Data", IEEE Transactions on Acoustics, Speech, and Signal Processing, Vol ASSP-29, No. 3, pp. 471-477, June 1981.
5.  Roemer, L.E. Chen, C.S. and Hostetler, M.S. "Cepstral Processing Using Spread Spectra for Cable Diagnostics", IEEE Transactions on Instrumentation and Measurement, pp. 31-37, March 1981.
6.  Wells, P.N.T., Ultrasonics in Clinical Diagnosis, Churchill Livingston, London, 1972.
7.  Quatieri, T.F. Jr. and Oppenheim, A.V., "Iterative Teachniques for Minimum Phase Signal Reconstruction from Phase or Magnitude, IEEE Transactions on Acoustics, Speech, and Signal Processing, Vol ASSP-29, No. 6, pp. 1187-1193, Dec. 1981.
8.  Papoulis, A., "Maximum Entropy and Spectral Estimation: A Review", IEEE Transactions on Acoustics, Speech, and Signal Processing, Vol ASSP-29, No. 6, Dec 1981.
9.  Ulrych, T.J. and Clayton, R.W. "Time Series Modelling and Maximum Entropy", Physics of the Earth and Planetary Interiors, Vol 12, pp. 188-200, 1976.
10.  Burg, J.P. "Maximum Entropy Spectral Analysis", Modern Spectrum Analysis, Edited by: Childers, D.G., IEEE Press, New York, pp. 34-41, 1978.
11.  Burg, J.P., "New Analysis Techniques for Time Series Data", Modern Spectrum Analysis, Edited by: Childers, D.G., IEEE Press, New York, pp. 42-48, 1978.
12.  Burg, J.P., "The Relationship Between Maximum Entropy Spectra and Maximum Likelihood Spectra", Modern Spectrum Analysis, Edited by: Childers, D.G., IEEE Press, New York, pp. 132-133, 1978.
13.  Oppenheim, A.V. and Schafer, R.W., Digital Signal Processing, Prentice-Hall, Englewood Cliffs, N.J., 1975.
14.  E. Poggiagliomi, A.J. Berkhout, and M.M. Boone, "Phase Unwrapping, Possibliities and Limitations", Geophysical Prospecting, No. 30, P281-291, 1982.
15.  R. Kuc, "Modeling Acoustic Attenuation of Soft Tissue with a Minimum Phase Filter", Ultrasonic Imaging, Nol 6, P24-36, 1984.
16.  N. Erdol, Use of the Maximum Entropy Method for Phase Estimation, Ph.D. Dissertation, The University of Akron, Akron, Ohio, 1982.

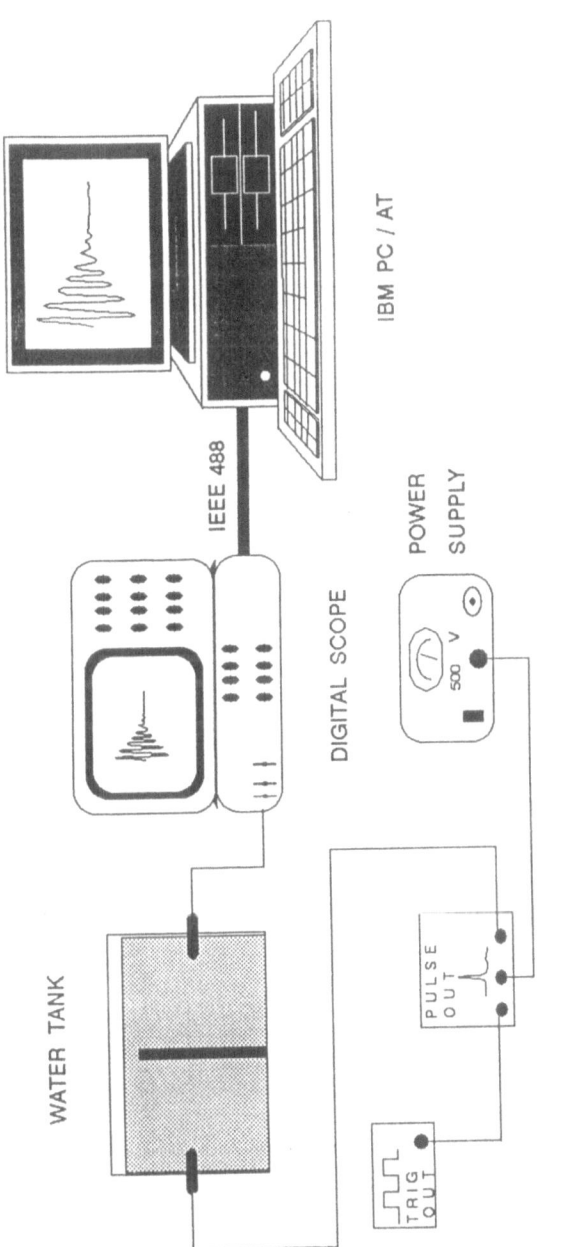

Figure 1a: Apparatus

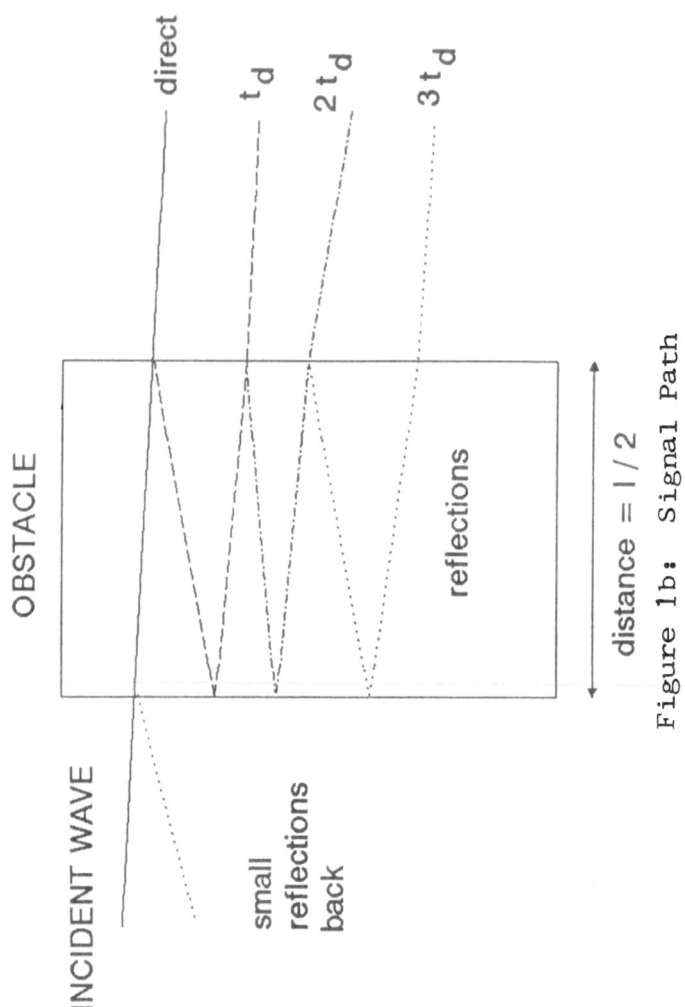

Figure 1b:  Signal Path

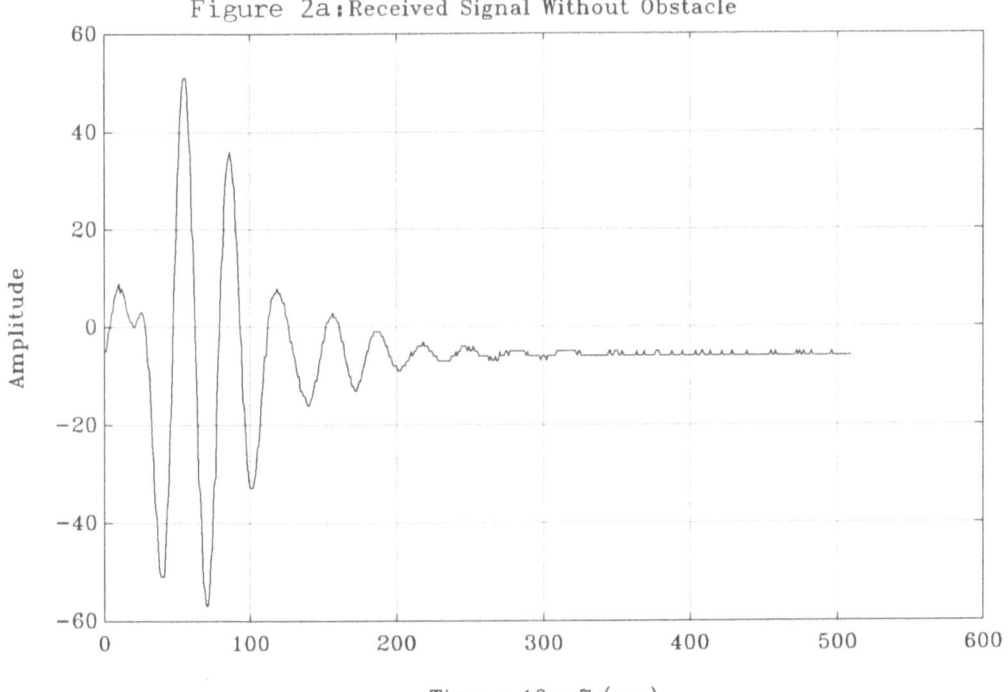

Figure 2a:Received Signal Without Obstacle

Time x 10^-7 (sec)

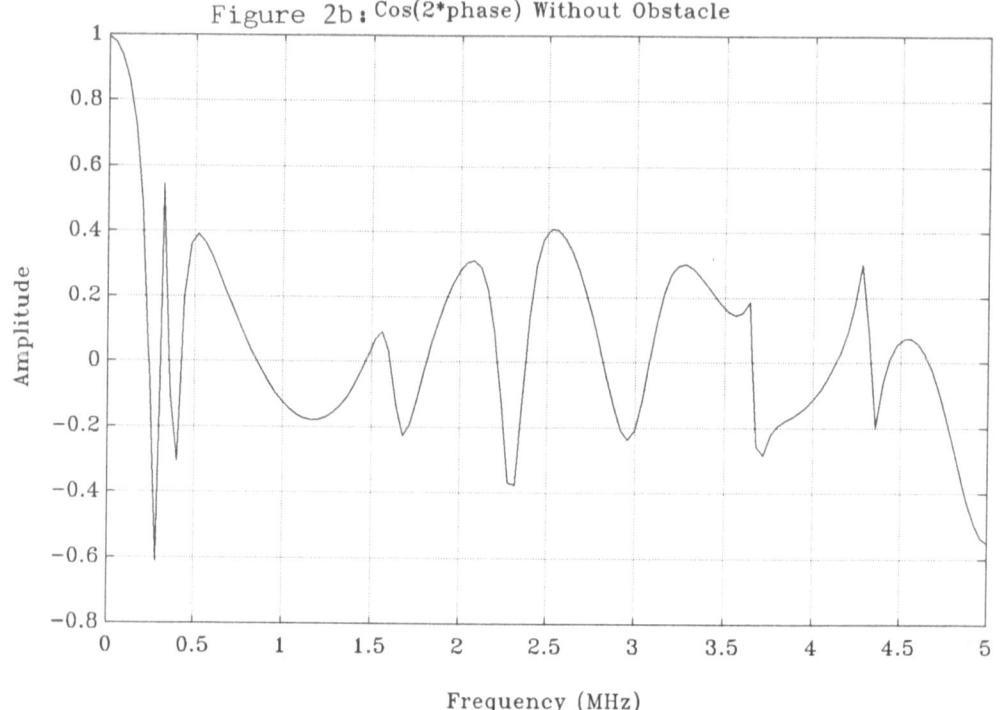

Figure 2b: Cos(2*phase) Without Obstacle

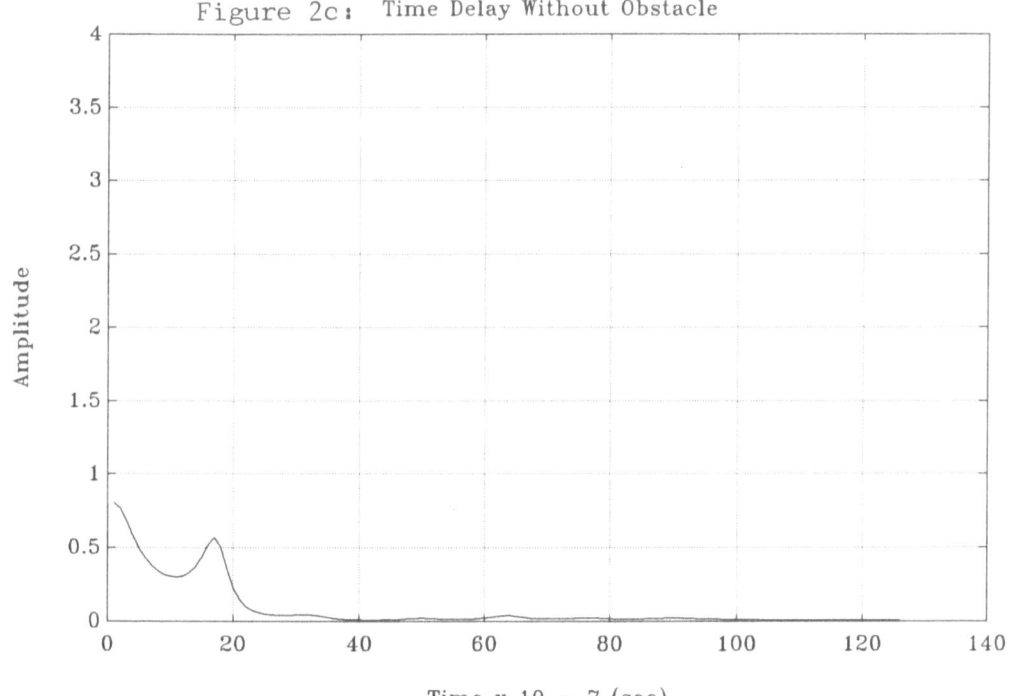

Figure 2c:  Time Delay Without Obstacle

Time x 10 ^-7 (sec)

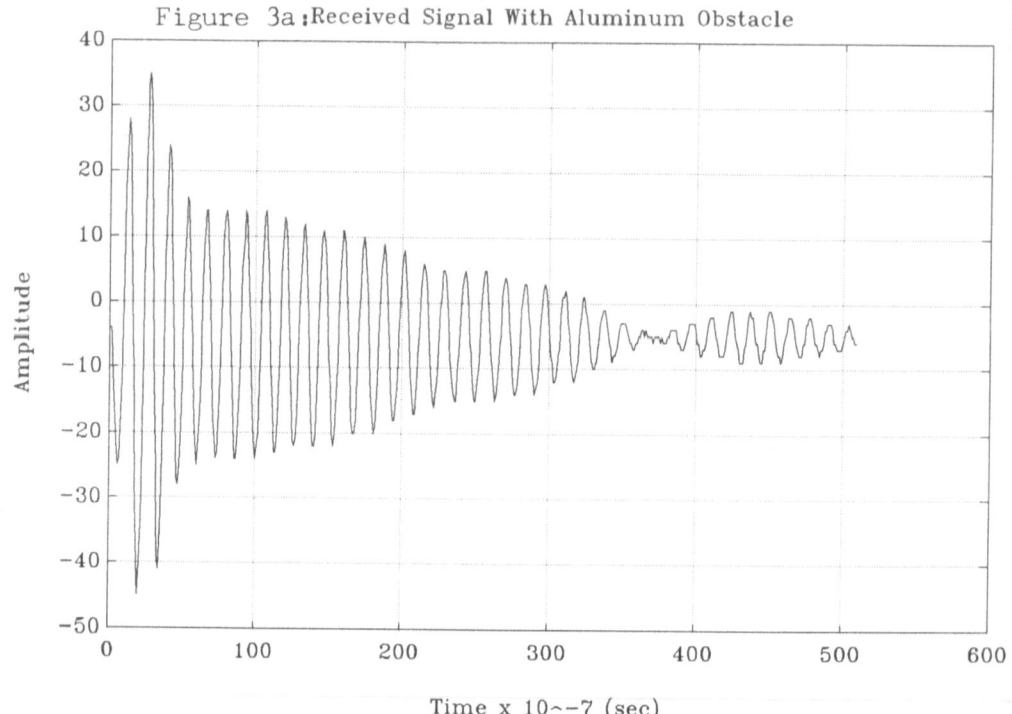

Figure 3a :Received Signal With Aluminum Obstacle

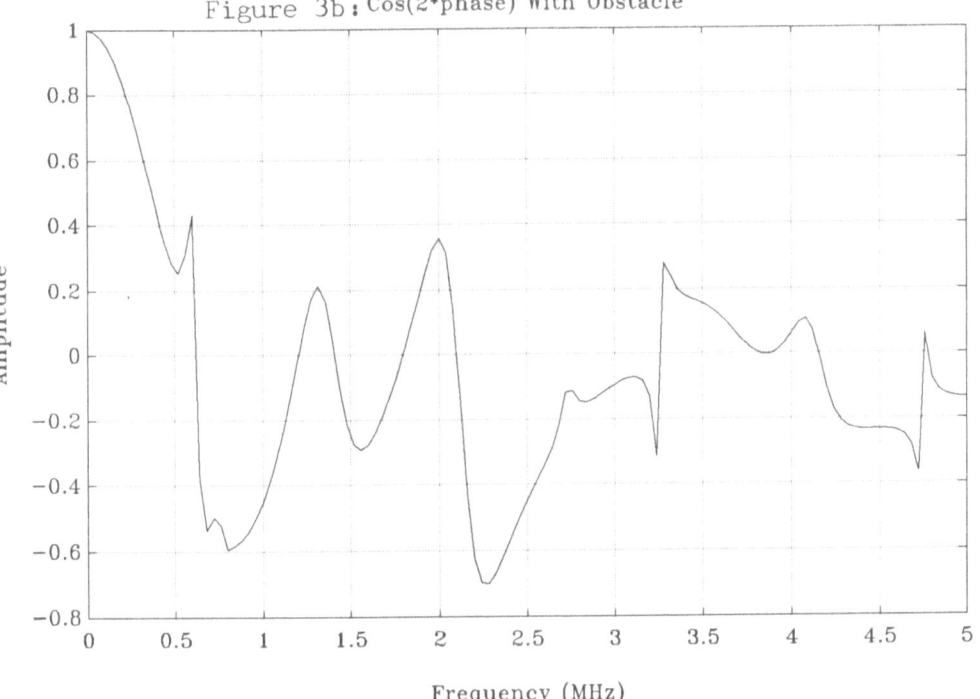

Figure 3b: Cos(2*phase) With Obstacle

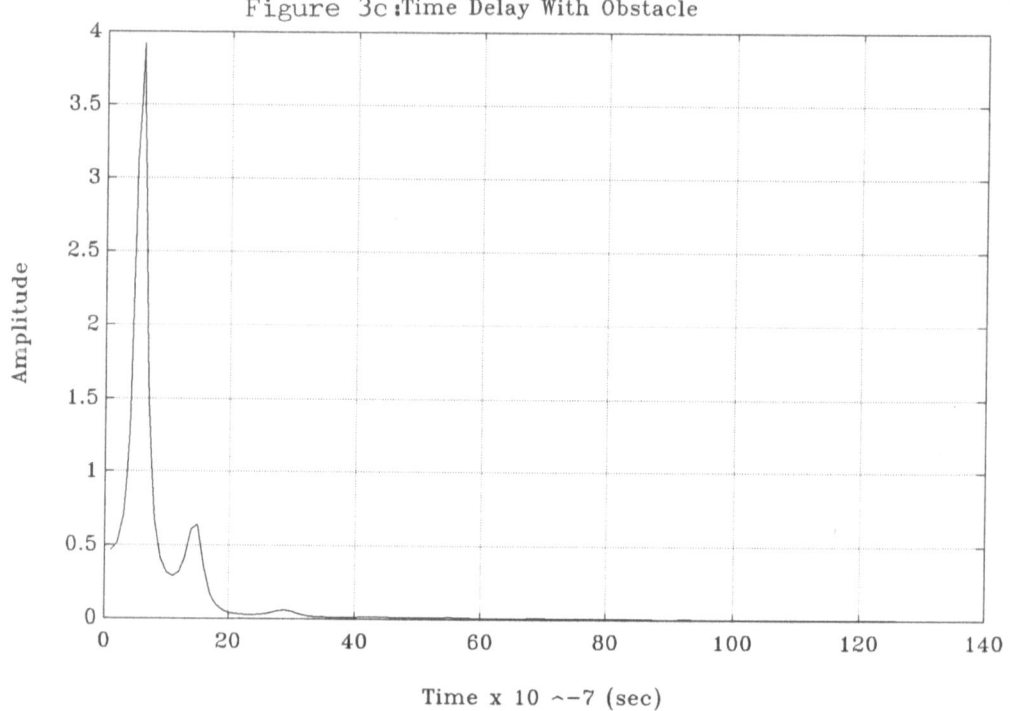

Figure 3c：Time Delay With Obstacle

Time x 10 ^−7 (sec)

# Drug absorption in man, and its measurement by MaxEnt

M.K. Charter

Mullard Radio Astronomy Observatory, Cavendish Laboratory,
Madingley Road, Cambridge, CB3 0HE, U.K.

### Abstract

In many cases the processes by which a drug is handled in .the body, once it has reached the bloodstream, are essentially linear at therapeutic doses. If so, then the concentration in blood after an oral dose may be considered as the convolution of the rate at which the drug reaches the bloodstream with the response of the body to an 'impulse' of the drug applied directly into the bloodstream. The input rate may be treated as a 'blurred' version of a positive additive distribution, where the 'blurring' reflects the diffusive processes which the drug must undergo during its transit from the point of dosing to the general circulation. The standard pharmacokinetic compartmental model of the body leads to an impulse response function which is of the form $\sum_i A_i e^{-\lambda_i t}$. One can extend this model, and express the function in terms of a continuous distribution of time constants $\lambda$, rather than just a small number of discrete values. Thus the input rate and impulse response can both be characterised using positive additive distributions, which should be reconstructed from experimental data by a process of Bayesian inference using Skilling's generalisation of the Shannon/Jaynes entropy.

## Introduction

Many drugs reach their site of action, in other words the place where they exert their pharmacological effect, via the bloodstream, in which they are carried around the body. The blood is also the means whereby drugs are transported to eliminating organs such as the liver or kidneys, which remove them from the body by metabolism or excretion. It therefore follows that modelling the processes which determine the concentration of the drug in the blood is of considerable importance in quantifying and predicting the action of the drug in the body. These kinetic processes can be divided into two parts: absorption, in which the drug travels from the point of dosing to the general circulation, and disposition, or the fate of the drug once it has reached the general circulation.

In a typical experiment, the subject is given the drug, and blood samples are taken from a vein in the forearm at various times after dosing. The concentration of the drug is determined in each blood sample, and these measurements form the raw data from which the absorption and/or disposition kinetics are to be inferred. The data are 'noisy' because of the errors in measuring very low concentrations of the drug in blood plasma or serum (often one part in $10^9$ or less), and incomplete in the sense that medical and ethical considerations limit the number of blood samples which can be taken.

*P. F. Fougère (ed.), Maximum Entropy and Bayesian Methods,* 325–339.
© 1990 *Kluwer Academic Publishers.*

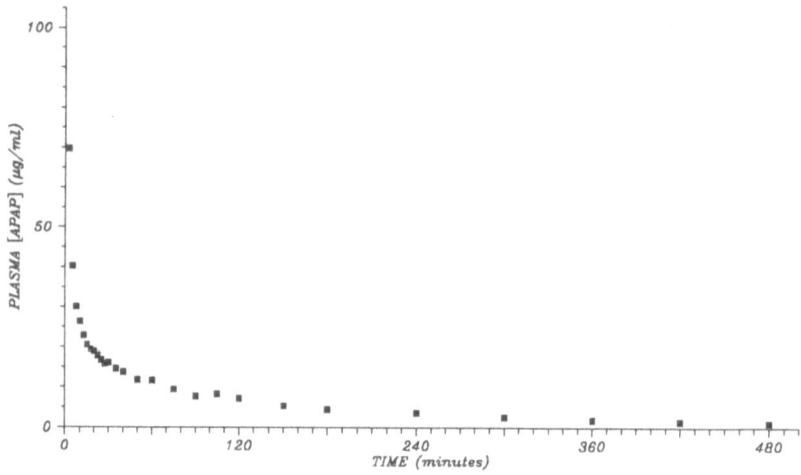

Figure 1: Plasma paracetamol concentrations after an intravenous dose.

## A simple model of the body

If a rapid intravenous dose of a drug is given, in other words a delta function input directly into the bloodstream, the data are typically as shown in Figure 1. A healthy volunteer was given 1000 mg of the common analgesic paracetamol (acetaminophen, APAP) by rapid intravenous injection. There is initially a rapid decline in the plasma paracetamol concentration, and then it decreases more slowly.

One of the models used most frequently to describe this behaviour is a compartmental system of the kind shown in Figure 2. The model consists of two interconnected compartments. One of them, labelled number 1, we may call the 'sampled compartment', since the concentration in it is the one which we actually measure. This compartment is supposed to have a volume $V$. The drug can move by a first-order transfer into the other compartment, labelled number 2, from which it can come back into the sampled compartment, again by a first-order process. The drug may also be irreversibly removed from the sampled compartment by a first-order process, characterised by the elimination rate constant $k_{10}$. This is plainly a gross simplification, and these two compartments cannot be associated with any particular anatomical features. Nonetheless, it does convey the basic principle that when a drug is introduced into the bloodstream it may distribute out of the blood into some other part of the body. It may then return to the bloodstream, from which it is eventually removed, by excretion or metabolic conversion to another chemical species.

The model can be described by the following system of differential equations

$$
\begin{aligned}
\dot{x}_1 &= -(k_{10} + k_{12})x_1 + k_{21}x_2 \\
\dot{x}_2 &= \qquad\quad k_{12}x_1 - k_{21}x_2
\end{aligned}
\tag{1}
$$

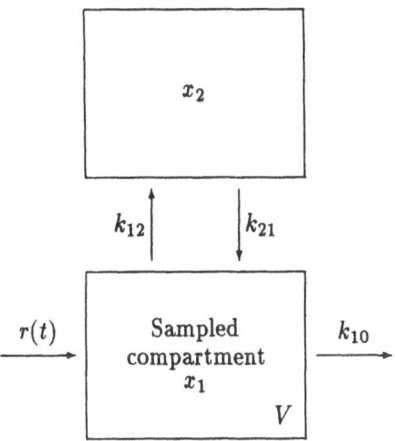

Figure 2: The two-compartment model

or

$$\dot{\mathbf{x}} = \mathbf{A}\mathbf{x}$$

where

$$\mathbf{A} = \begin{pmatrix} -(k_{10} + k_{12}) & k_{21} \\ k_{12} & -k_{21} \end{pmatrix}$$

for which the solution, expressed in terms of the concentration in the sampled compartment, is

$$R(t) = \frac{1}{V}\left(\frac{\lambda_1 - k_{21}}{\lambda_1 - \lambda_2}e^{-\lambda_1 t} + \frac{k_{21} - \lambda_2}{\lambda_1 - \lambda_2}e^{-\lambda_2 t}\right)$$

for a unit dose introduced at $t = 0$ into the sampled compartment, i.e., $r(t) = \delta(t)$, where $-\lambda_1$ and $-\lambda_2$ are the eigenvalues of $\mathbf{A}$. This system can be elaborated by the addition of more compartments, in which case the number of terms will, in general, increase. This type of model provides the motivation for using a simple sum of exponentials

$$R(t) = \sum_{i=1}^{N} A_i e^{-\lambda_i t} \tag{2}$$

to describe empirically the impulse response of the system. The fit of this function, with $N = 3$, to the intravenous paracetamol data shown in Figure 1, is shown in Figure 3. The data were weighted as $1/F_k^2$, where $F_k$ is the predicted concentration in the sample taken at time $t_k$ after dosing. The function clearly fits the data well. It is found that this type of model, with suitable time constants and coefficients, can be used for many drugs.

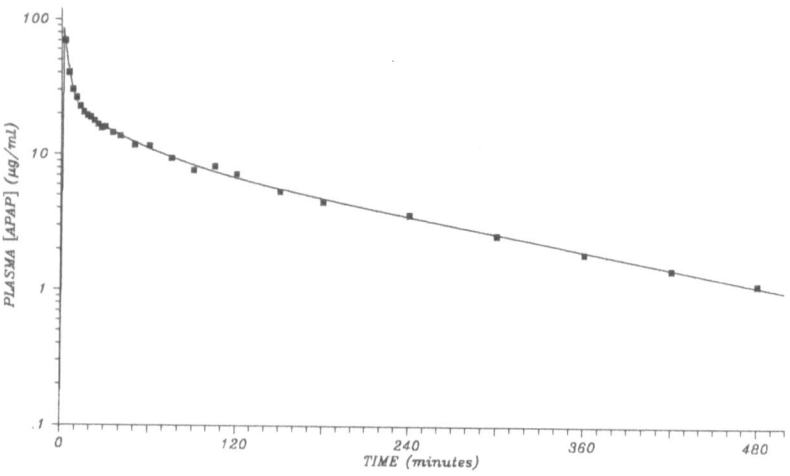

Figure 3: The intravenous paracetamol data of Figure 1, with $R(t)$ given by (2).

## Absorption kinetics

Having developed a model for the impulse response function, which describes the intravenous data, we now consider oral data—in other words the results of giving the drug by mouth. In this case the drug must be absorbed from the gastrointestinal tract before it appears in the bloodstream. The system (1) is linear, and so the concentration–time profile after an oral dose can be expressed as the convolution of the rate $r(t)$ at which the drug appears in the bloodstream with the response $R(t)$ to (an intravenous bolus of) the drug placed there. The determination of this input rate is therefore a deconvolution problem, with sparse and noisy data.

Before going on to discuss the application of MaxEnt to this problem, it is interesting to examine one of the standard methods commonly used to perform these deconvolutions. It illustrates the consequences of trying to pose this as an exercise in deductive logic, in other words trying to deduce the input rate from a small number of sparsely-sampled and noisy data, rather that recognising that it is a problem of inference. It is known as the Loo-Riegelman method [2]. We may imagine an extension of the compartmental model shown in Figure 2, to include $N - 1$ peripheral compartments, labelled from 2 to $N$. Transfers occur from the $i$th peripheral compartment to and from the sampled compartment, with rate constants $k_{i1}$ and $k_{1i}$ respectively, but there is no direct transfer between peripheral compartments. The system corresponding to (1) is

$$\dot{x}_1 = -\left(k_{10} + \sum_{i=2}^{N} k_{1i}\right)x_1 + \sum_{i=2}^{N} k_{i1}x_i + r(t)$$
$$\dot{x}_i = k_{1i}x_1 - k_{i1}x_i$$

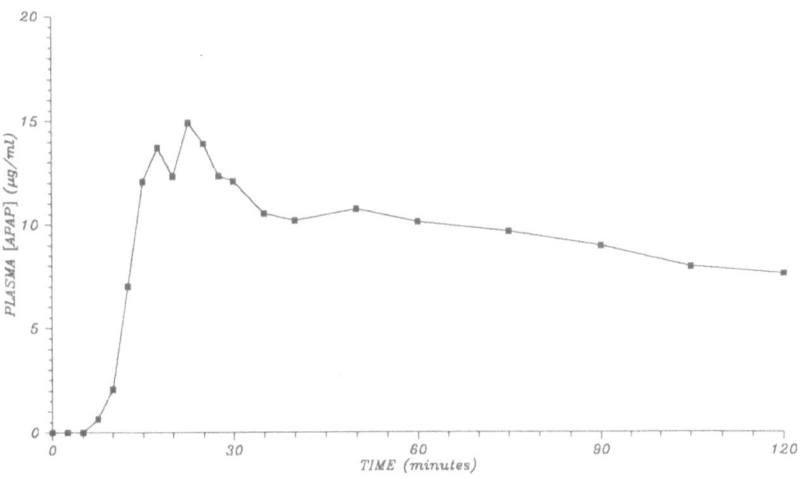

Figure 4: Oral paracetamol data, interpolated linearly.

whence

$$x_i = k_{1i} \int_0^t x_1(\tau) e^{-k_{i1}(t-\tau)} \, d\tau$$

and hence

$$
\begin{aligned}
r(t) &= \dot{x}_1 + \left(k_{10} + \sum_{i=2}^N k_{1i}\right) x_1 - \sum_{i=2}^N \left(k_{1i} k_{i1} \int_0^t x_1(\tau) e^{-k_{i1}(t-\tau)} \, d\tau\right) \\
&= V\left(\frac{dc_1}{dt} + \left(k_{10} + \sum_{i=2}^N k_{1i}\right) c_1 - \sum_{i=2}^N \left(k_{1i} k_{i1} \int_0^t c_1(\tau) e^{-k_{i1}(t-\tau)} \, d\tau\right)\right) \qquad (3)
\end{aligned}
$$

Equation (3) is a slight generalisation of Wagner's exact Loo-Riegelman equation [5]. The input rate is thus expressed in terms of the concentration $c_1(t)$ in the sampled compartment. This concentration function must be obtained by some arbitrary method of interpolation between the observations. Linear interpolation is generally used, although piecewise cubic polynomials have been proposed [6]. Linear interpolation produces an input rate with discontinuities at each sample time, because of the term $dc_1/dt$ in (3), which is hardly physiological. Moreover, no account is taken of possible noise in the data, so that small fluctuations in $c_1(t)$ can result in large swings in $r(t)$, to the extent that it may go negative even when this is physically impossible. This is an example of the noise amplification which is characteristic of 'deductive deconvolution'. Nonetheless, the Loo-Riegelman method is widely used. As an example, it is applied to paracetamol data from the same volunteer as before, after an oral dose of 1000 mg given in a hard gelatin capsule. Examination of the first two hours in detail, shown in Figures 4 and 5, shows the discontinuities in $r(t)$ at each sample time, and the wide swings.

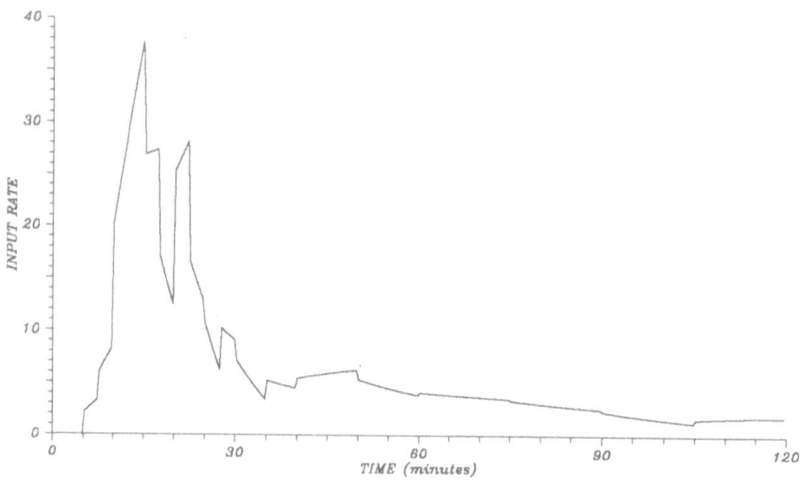

Figure 5: The input rate $r(t)$ given by (3), derived from the oral data in Figure 4.

We now turn to the application of MaxEnt to this problem. The input rate $r(t)$ has the attributes of a positive additive distribution (PAD)—indeed it has been noted previously that it can be considered as the distribution of transit times of drug molecules, from the point of dosing to the systemic circulation [1]. For computational purposes, it is convenient to express $r(t)$ as the vector $\mathbf{r}$, where $r_j$ is the proportion of the dose reaching the systemic circulation between times $(j-1)\Delta t$ and $j\Delta t$. Its entropy is defined using the generalised version of the Shannon/Jaynes entropy for a PAD $\mathbf{f}$ presented by Skilling [3]:

$$S(\mathbf{f}, \mathbf{m}) = \sum_j \left( f_j - m_j - f_j \log (f_j/m_j) \right)$$

where $\mathbf{m}$ represents the 'default', in other words our estimate of $\mathbf{f}$ before the experimental data $\mathbf{D}$ are known. In fact, $\mathbf{m}$ will be a low constant flat, since it will become apparent that sufficient structure can be incorporated into the model to give excellent reconstructions. For the moment the input rate $\mathbf{r}$ will be the same as $\mathbf{f}$. The prior distribution for $\mathbf{f}$ is proportional to $e^{\alpha S(\mathbf{f}, \mathbf{m})}$. The likelihood, in other words $\Pr(\mathbf{D} \mid \mathbf{f})$, will be essentially the usual $\chi^2$ statistic

$$\begin{aligned} L(\mathbf{f}) \quad &= \chi^2/2 \\ &= \tfrac{1}{2} (\mathbf{F} - \mathbf{D})^T \mathbf{V} (\mathbf{F} - \mathbf{D}) \end{aligned}$$

where $\mathbf{F} = \mathbf{F}(\mathbf{f})$ are the predicted data, obtained by convolving $\mathbf{r}$ with $R(t)$. It is assumed that the errors are Gaussian, with a variance-covariance matrix $\mathbf{V}$ known to within a scaling factor $\sigma^2$. We can then combine the prior and likelihood in the usual way to obtain the posterior distribution of $\mathbf{f}$, given the regularisation parameter $\alpha$ and error scaling $\sigma^2$. We choose the value $\hat{\mathbf{f}}$ which maximises this posterior probability, and determine $\alpha$ and $\sigma^2$ using the methods described by Gull [4]. These computations were performed using the

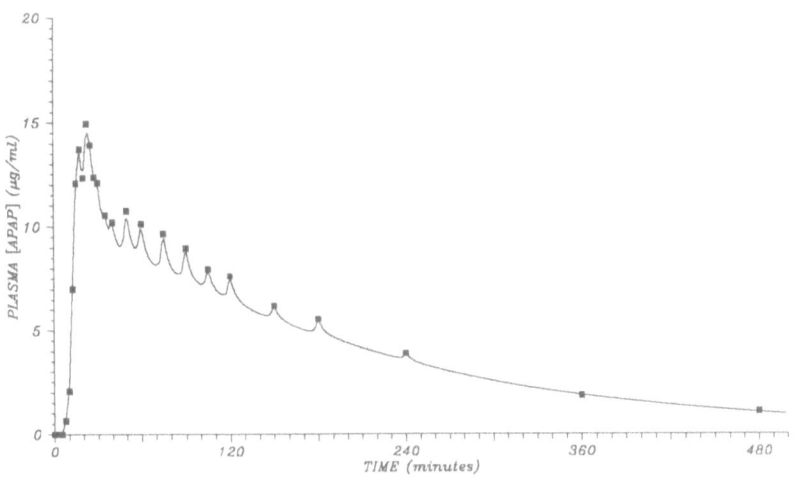

Figure 6: The oral data and predicted concentration function based on the input rate shown in Figure 7.

programme MemSys3 [7]. For the oral paracetamol data shown here, $r(t)$ was approximated by 96 'pixels' of width $\Delta t = 2.5\,\text{min}$.

The results shown in Figures 6 and 7, are, however, far from satisfactory. There is a spike in $r(t)$ at each of the later sample times, which is most unlikely unless it is supposed that the subject from whom the samples were taken underwent some profound metabolic or physiological change just before each blood sample was taken. It is useful to ask why these spikes seem to be so unrealistic, since the reconstructed input rate is positive, and the predicted concentration agrees well with the observations. The reason is that we know every molecule of the dose must undergo a number of diffusive processes during its transit from the point of dosing to the systemic circulation. For example, the dosage form must break up in the stomach or intestine. It will be emptied from the stomach into the intestine. The drug will have to dissolve in the gastric juice in order to pass through the gut wall by diffusion. It will then be transported to the liver in the bloodstream, which will also involve a mixture of diffusion and convection. Before it reaches the systemic circulation, it must pass through the liver, which is another process involving both diffusion and convection. We may therefore try modifying our simple model to incorporate this process. As with disposition kinetics, we cannot hope to develop a model which even approaches the full complexity of the various physiological processes involved, so this must again be an abstraction, aiming to capture the basic principle, but not the details.

The dose is considered to be distributed along a one-dimensional semi-infinite medium, through which it can move by a mixture of diffusion and convection. The rate at which it leaves the medium (at $x = 0$) is the input rate. We can write down a simple modification

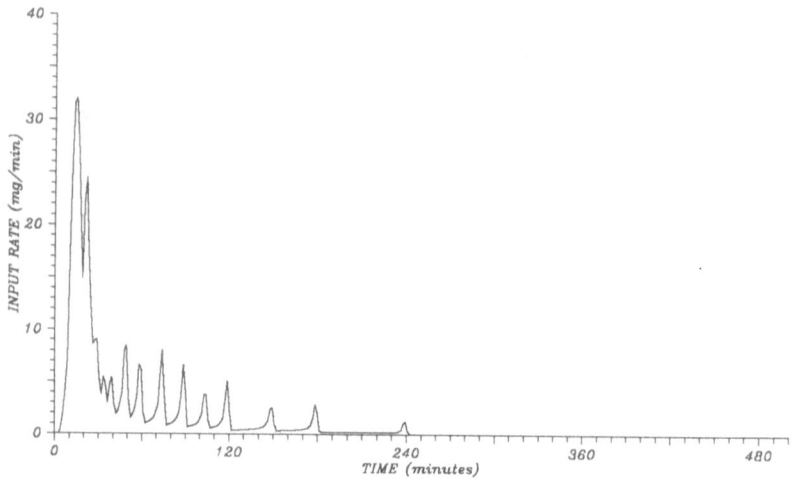

Figure 7: The input function $r(t)$ derived from the data shown in Figure 6.

of the diffusion equation to describe this process

$$\frac{\partial f}{\partial t} = D\frac{\partial^2 f}{\partial x^2} - v\frac{\partial f}{\partial x}$$

and this can be solved to give a 'pre-blur' operator **b**, where

$$b_{ij} = \frac{1}{2}\int_{(i-1)\Delta t}^{i\Delta t}\left(\frac{-(j-\frac{1}{2})\Delta x}{\tau} + v\right)\frac{1}{\sqrt{4\pi D\tau}}\exp\left(\frac{-\left(-(j-\frac{1}{2})\Delta x - v\tau\right)^2}{4D\tau}\right)d\tau$$

which acts on a vector **f**, representing the initial distribution of drug along the medium, to give the input rate **r** which is, in effect, a blurred version of **f** according to

$$\mathbf{r} = \mathbf{b}\,\mathbf{f} \tag{4}$$

The velocity $v$ gives merely the scaling between the hypothetical spatial distribution **f** in $x$, and the temporal distribution **r** in $t$, and so it may always be set to (minus) unity, without any loss of generality. The degree of 'pre-blur' is therefore controlled by a single parameter $D$. This process has analogies with the optical analogue of blurring an image of 'Susie' which has been described previously [4]. Some examples of this 'pre-blur' operator are shown in Figures 8 and 9. It can be seen that the width of the blur increases with time, and that it is asymmetric. These seem desirable features for the blurring process in this case, since it is describing a temporal distribution, rather than a spatial one in which a more symmetrical function, of constant width, might be appropriate.

The overall model of the system now consists of two stages. Firstly there is the absorption phase, in which the drug travels from the point of dosing to the systemic circulation

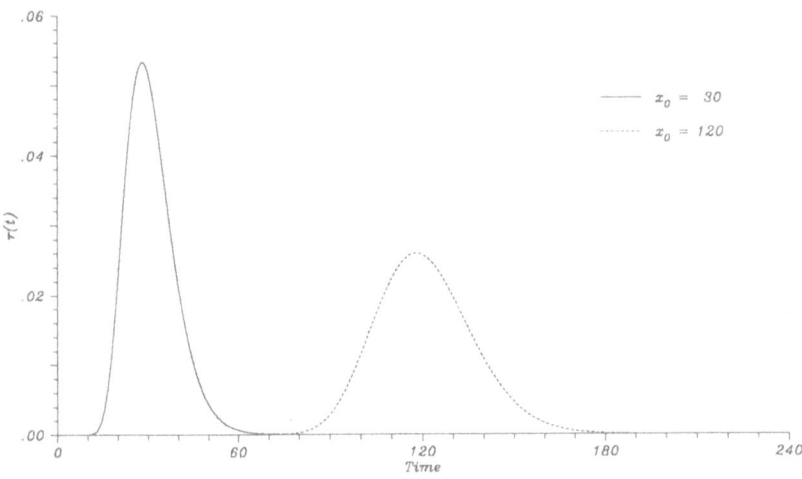

Figure 8: The 'pre-blur' operator $\mathbf{b}$ applied to delta functions at $x_0 = 30$ and $x_0 = 120$, with $D = 1$.

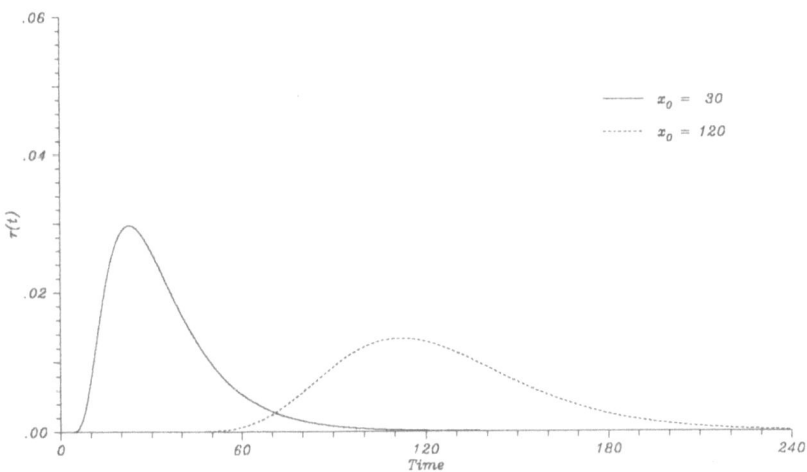

Figure 9: The 'pre-blur' operator $\mathbf{b}$ applied to delta functions at $x_0 = 30$ and $x_0 = 120$, with $D = 4$.

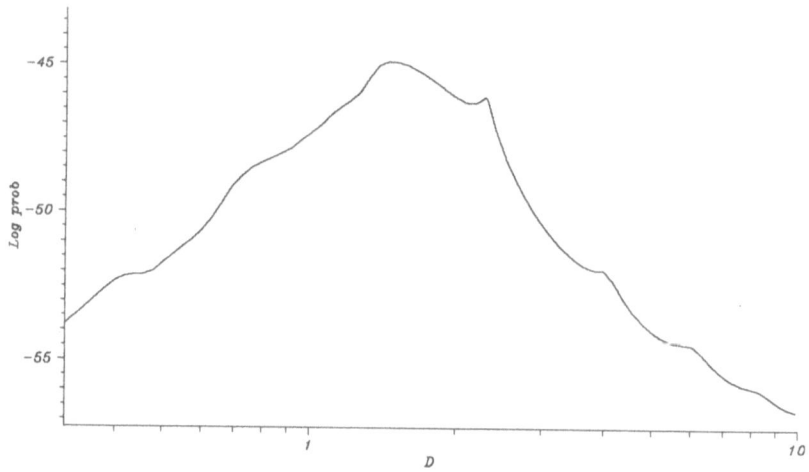

Figure 10: The posterior probability for the width $D$ of the 'pre-blur'.

by a mixture of diffusive and convective processes, modelled by the 'pre-blur' operator **b**. Secondly there is the disposition phase, in which the drug is transported around and out of the body, modelled by the impulse response function $R(t)$. It should be noted that the input rate **r** and the PAD **f** whose entropy is maximised are no longer the same, but are related by the blur operator, as in (4). Using this system, the excellent reconstruction shown in Figure 12 is obtained. It is smooth and physiologically realistic. Also, because it is inherent in this formulation of the problem that the data are noisy, the predicted concentration function, shown in Figure 11, does not pass through every datum, and there are no large oscillations in $r(t)$.

This leaves one parameter to be determined—the width $D$ of the 'pre-blur'. This can again be determined by the Bayesian argument presented previously [4], which gives the plot shown in Figure 10 for $\Pr(\mathbf{D} \mid D)$. If the prior distribution for $D$ is reasonably uniform over the region of interest, we may interpret this as a posterior distribution for $D$. There is a clear maximum, at a value of about 1.4, which is the value of $D$ used in the above reconstruction.

Thus it can be seen that MaxEnt gives physiologically realistic reconstructions of the input rate, even in the presence of sparsely-sampled and noisy data. No pre-processing of the data, such as smoothing, interpolation or extrapolation, is needed, and there are no arbitrarily-valued parameters in the analysis.

## Disposition kinetics revisited

Having considered how MaxEnt can be used to describe absorption kinetics, we now return to the question of disposition kinetics, and reconsider the compartmental model which was

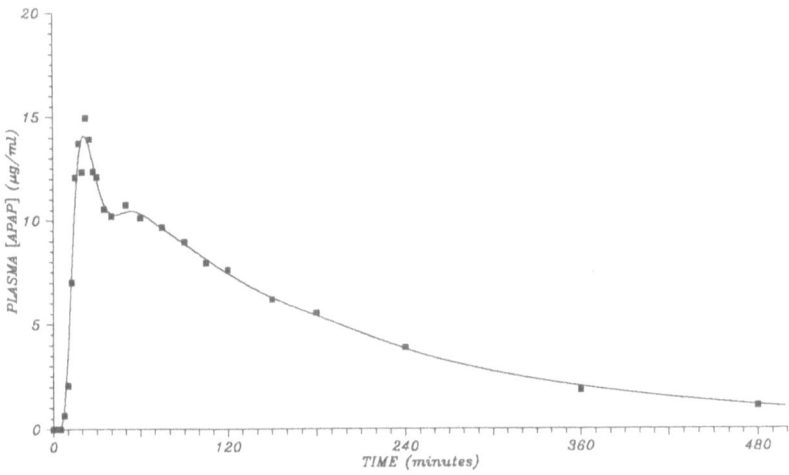

Figure 11: The data and predicted concentration function based on the input rate shown in Figure 12.

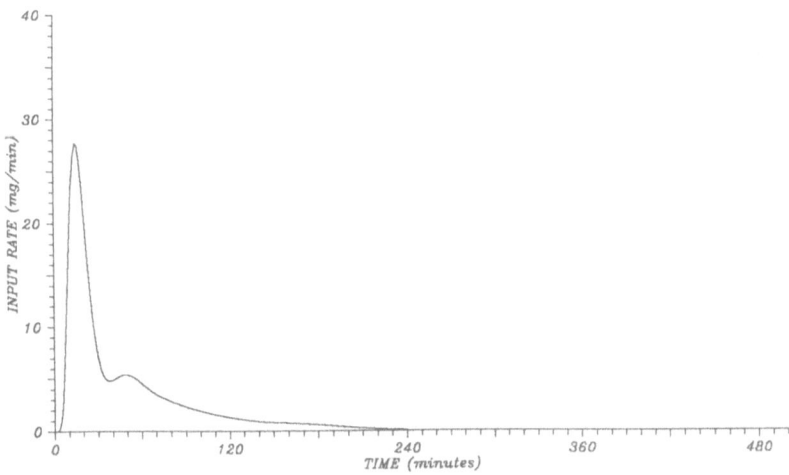

Figure 12: The input function $r(t)$ using 'pre-blur' with $D = 1.4$.

used earlier. The compartments do not represent any particular physiological or anatomical features, but the model contains some of the essential structure—that the drug can leave the sampled compartment, and then return back to it later on. One particular problem, however, is the choice of the number of peripheral compartments. In many ways it would seem closer to reality to imagine a very large number of compartments, which will result in a very large number of time constants $\lambda$, tending towards a continuous distribution as a function of $\lambda$, rather than just a small discrete number. In other words, instead of using a function of the form (2) with $N = 2$ or 3 for $R(t)$, we might replace the $\{A_i\}$ with a continuous distribution $f(\log \lambda)$, expressing the idea of a continuous distribution of compartments, with a continuous range of properties. It seems natural to consider the distribution $f$ to be a function of $\log \lambda$, since $\lambda$ is a scale parameter in the time domain. This gives an impulse response function

$$R(t) = \int_{-\infty}^{+\infty} f(\log \lambda) e^{-\lambda t} d\log \lambda \qquad (5)$$

This immediately removes the problem of how many compartments there should be, or a value for $N$. For a compartmental system of the type described, in which the impulse is applied to the sampled compartment, all the $\{A_i\}$ will be positive. It therefore seems reasonable to treat the continuum $f(\log \lambda)$ as a PAD, to be reconstructed by MaxEnt.

An example of this approach is shown for the intravenous paracetamol data discussed above. The fit to the observed data, shown in Figure 13, is, as expected, very good. Close comparison of Figures 3 and 13 also shows that in Figure 13 there is a slight curvature in the terminal phase of the profile plotted logarithmically, which is captured using (5) but not (2) for $R(t)$. This curvature is commonly seen, and in pronounced cases is difficult to model using conventional methods to characterise the terminal phase. One possible interpretation is that it represents small proportions of the dose in 'deep compartments', or areas of the body which they leave only slowly after distributing into them. The distribution $f(\log \lambda)$, shown in Figure 14, contains a series of peaks. The maxima in $f(\log \lambda)$ occur close to the estimates of the $\{\lambda_i\}$ obtained using (2) for $R(t)$.

The function $f(\log \lambda)$ obtained when more than one set of data is fitted simultaneously is shown in Figure 16. The single intravenous dose shown in Figures 1, 3 and 13 was actually one of a series in which the same volunteer was given 100, 200, 500, and 1000 mg paracetamol. The 1000 mg dose was given on three separate occasions. Observing the simultaneous fit to all six profiles, shown in Figure 15, gives an informal indication of the linearity of the system. While obviously not perfect, the assumption of linearity is probably a reasonable starting point. A series of doses ranging over an order of magnitude, with the resulting concentrations ranging over three orders of magnitude, is a fairly stringent test of the hypothesis. The greatest difference between the two distributions is in the fast time constants, representing the rapid initial distribution of the drug into the peripheral compartments. The slower time constants remain more constant.

Thus MaxEnt can be used to develop a model of drug disposition which seems closer to reality than a conventional sum of exponentials such as (2), and models some features not easily encompassed by it.

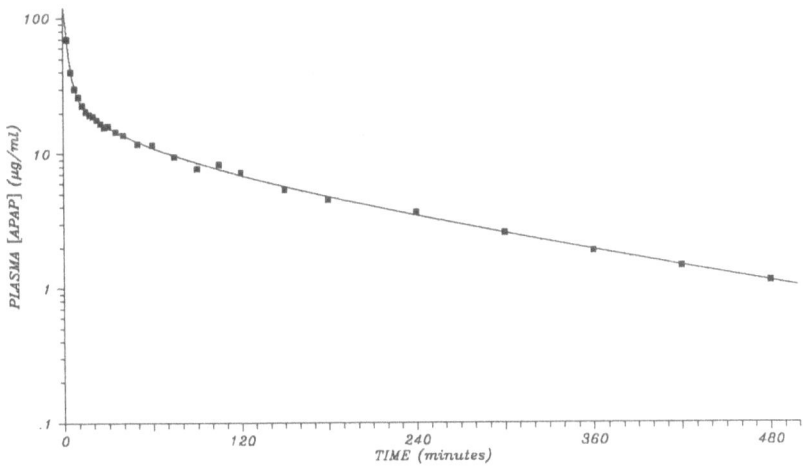

Figure 13: The intravenous paracetamol data with $R(t)$ given by (5).

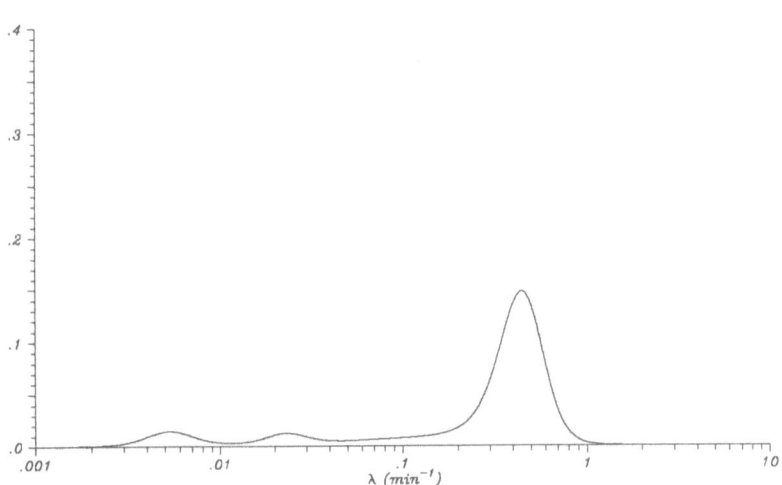

Figure 14: The function $f(\log \lambda)$ corresponding to Figure 13.

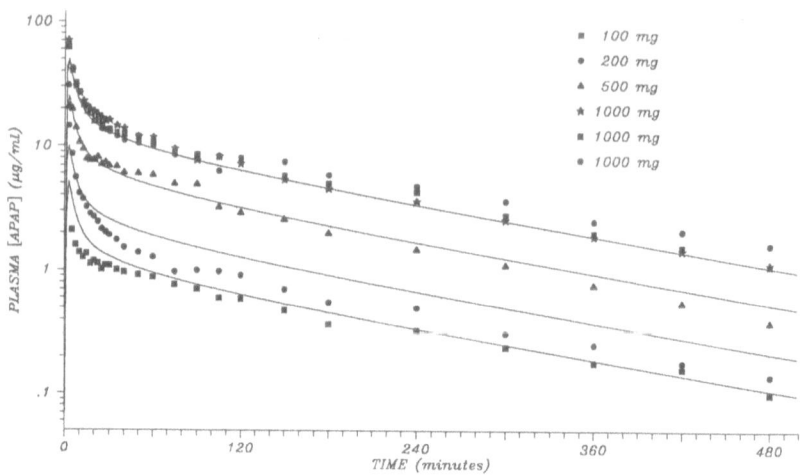

Figure 15: Six sets of intravenous paracetamol data fitted simultaneously.

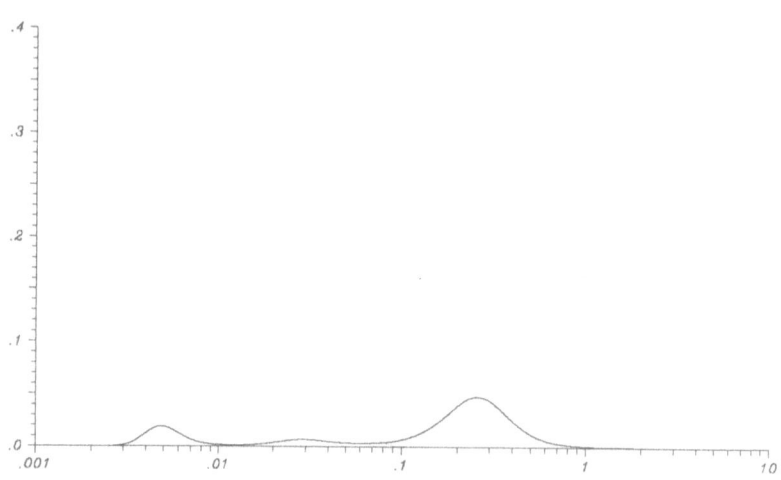

Figure 16: The function $f(\log \lambda)$ corresponding to Figure 15.

# References

[1] Riegelman S, Collier P. The application of statistical moment theory to the evaluation of *in vivo* dissolution time and absorption time. J Pharmacokinet Biopharm 1980;8(5):509–534.

[2] Loo JCK, Riegelman S. New method for calculating the intrinsic absorption rate of drugs. J Pharm Sci 1968;57(6):918–927.

[3] Skilling J. Classic maximum entropy. In: Skilling J, ed. Maximum entropy and Bayesian methods, Cambridge 1988. Dordrecht: Kluwer, 1989:45–52.

[4] Gull SF. Developments in maximum entropy data analysis. In: Skilling J, ed. Maximum entropy and Bayesian methods, Cambridge 1988. Dordrecht: Kluwer, 1989:53–71.

[5] Wagner JG. Pharmacokinetic absorption plots from oral data alone or oral/intravenous data and an exact Loo-Riegelman equation. J Pharm Sci 1983;73(7):838–842.

[6] Wagner JG. Fundamentals of clinical pharmacokinetics. Hamilton, Illinois: Drug Intelligence Publications Inc., 1975. p. 192.

[7] Gull SF, Skilling J. Quantified maximum entropy "MEMSYS 3" users' manual. Maximum Entropy Data Consultants Ltd., 33 North End, Meldreth, Royston, Herts, SG8 6NR, U.K.

# QUANTIFIED MAXIMUM ENTROPY

John Skilling
Department of Applied Mathematics and Theoretical Physics
Silver Street, Cambridge CB3 9EW, England

## Abstract
This tutorial paper discusses the theoretical basis of quantified maximum entropy, as a technique for obtaining probabilistic estimates of images and other positive additive distributions from noisy and incomplete data. The analysis is fully Bayesian, with estimates always being obtained as probability distributions from which appropriate error bars can be found. This supersedes earlier techniques, even those using maximum entropy, which aimed to produce a single optimal distribution.

## 1. Introduction

Many distributions of interest in science are positive and additive. For example, the intensity of light across an image is positive (obviously) and additive (meaning that the flux received across two non-overlapping patches is the sum of the individual fluxes). Likewise, a power spectrum is positive (obviously) and additive (meaning that the net power in two non-overlapping bands is the sum of the powers in the individual bands). More abstractly, a probability distribution is also both positive and additive, with the additional constraint of summing to unit total. We shall abbreviate "positive additive distribution" to "PAD". It will be our present task to investigate how we may use relevant data to learn about PADs. In so doing, we will gain some insight into probability theory, as well as into the craft of image reconstruction. Earlier papers containing much of this material are by Gull (1989) and Skilling (1988,1989).

Technically, if the domain D={x} of the PAD contains arbitrarily many points x, we should define f on the $\sigma$-algebra of sub-domains of D by

$$f(D_i) = \text{flux in sub-domain i.} \tag{1}$$

Positivity means

$$f(D_i) \geq 0 \tag{2}$$

and additivity means

$$f(D_i \cup D_j) = f(D_i) + f(D_j) - f(D_i \cap D_j) \tag{3}$$

A PAD f is correctly a distribution, or bra-vector $\langle f|$, from which scalar quantities such as fluxes in sub-domains or linear combinations thereof are obtainable by taking scalar products with appropriate ket-vectors. Thus

$$\rho = \langle f|P \rangle \tag{4}$$

where the ket-vector $|P \rangle$ is defined as an ordinary function P(x). This means

*P. F. Fougère (ed.), Maximum Entropy and Bayesian Methods, 341–350.*

that we should not think of our task as the estimation of f directly. Practical data are always macroscopic scalar products, never having infinite resolution. Accordingly, we should not hope to learn about f in all its infinitesimal detail. Instead, we can légitimately expect to use macroscopic data of the form $D = \langle f|R \rangle$ to learn about macroscopic structures of the form $\rho = \langle f|P \rangle$. Our tool for doing this will be Bayesian calculus.

Of course, in discrete domains (i.e. finite-dimensional f), we can ignore the bra-ket distinction and set

$$f_i = \text{flux in cell i} \quad , \quad \rho = \Sigma f_i P_i \quad \quad [5]$$

with the continuum limit commonly written as

$$f(x)dx = \text{flux in dx} \quad , \quad \rho = \int dx\ f(x)\ P(x) \quad \quad [6]$$

as if f could be specified pointwise as an ordinary function of x. Although the finite-dimensional case includes all practical computations, by casting our mathematics correctly we will ensure that our computer programs approach sensible limits as the continuum is approached.

## 2. Bayesian calculus

Scientific data analysis should aim to infer results from data in a logical manner. Each of the possible results A,B,C,... which might be inferred from an experiment corresponds to a logical proposition of the form "X is true", where X stands for any of A,B,C,..., so that scientific data analysis is a special application of our methods for dealing with logical propositions.

Let us suppose that there is a general calculus for dealing with such propositions. If there is, then obviously it must apply in special cases. Cox (1946) proved that any calculus which is consistent with the ordinary rules of Boolean algebra obeyed by propositions must be equivalent to Bayesian probability theory. Each proposition X carries a numerical code pr(X), obeying

$$pr(X,Y) = pr(X)\ pr(Y|X) \quad \quad [7]$$

$$pr(X) + pr(\sim X) = 1 \quad \quad [8]$$

where ",", means "and", "|" means "given", and "~" means "not". All such codes lie between 0 and 1, with the particular propositions "true" and "false" having the extreme values

$$pr(\text{false}) = 0 \quad , \quad pr(\text{true}) = 1 \ . \quad \quad [9]$$

These are the standard rules of probability calculus, and accordingly the codes are called "probabilities". The only freedom allowed is to rescale the Bayesian probabilities by some arbitrary monotonic function, rather as percentages are obtained by multiplying by 100. Whatever codes were first assigned to the propositions, these could be mapped uniquely back to the Bayesian probability values, so that we may as well adopt this standard convention.

This definition of "probability" is, of course, consistent with the commonly used definition as a limiting frequency ratio of "successes/trials". Frequency experiments are a compelling simple case for which any general calculus ought to give the obvious rules derived from multiplication and addition of (large) integers. Bayesian calculus passes this test. Moreover, the Cox definition of probability is more general than the frequentist definition, because it refers to arbitrary propositions without having to invent a "many-worlds" scenario involving all the events which might have happened but did not.

Conceivably, there may be no general language at all. One can imagine the existence of some other compelling simple case for which Bayesian calculus might give a demonstrably "wrong" result. We can not prove that such counter-examples do not exist. However, it is difficult to see how one might be constructed. Despite much effort, nobody has yet found one, and so we shall adopt a rigorous Bayesian approach.

In scientific data analysis, we wish to use our data D to make inferences about various possible results A,B,C,... We have no alternative but to code this inference in terms of conditional probabilities

$$pr(A|D), \ pr(B|D), \ pr(C|D), \ \ldots$$

Using "f" to represent any particular result A,B,C,..., we need the probability distribution $pr(f|D)$ as a function of f.

The data D do not give us this directly. Instead, the data give us the "likelihood" $pr(D|f)$ as a function of f. The particular algebraic form of the likelihood may involve delta functions for exact data, or Gaussian factors if the data have normally distributed noise, and may include various correlations and other peculiarities of the observing system which produced the data.

In order to reverse the conditioning from $pr(D|f)$ to the required $pr(f|D)$, we use Bayes' theorem. A useful trick for navigating through the morass of possible algebraic manipulations when dealing with multiple propositions is to start with the joint probability distribution of everything relevant, and expand it in various ways.

$$pr(f,D) = pr(f) \ pr(D|f) = pr(D) \ pr(f|D) \qquad [10]$$

This immediately gives <u>Bayes' theorem</u>

$$pr(f|D) = pr(f) \ pr(D|f) \ / \ pr(D) \qquad [11]$$

which is the basic tool of data analysis. As far as inferences about f are concerned, $pr(D)$ is just a constant which ensures that the probabilities of all the possible results add up to 1. We are given the likelihood $pr(D|f)$, but the factor $pr(f)$ remains.

This factor $pr(f)$ is the "prior probability" distribution of f. It codifies our prior expectations about possible results f <u>before</u> acquiring the new data D.

Probability theory tells us how to use data to <u>modulate</u> this prior expectation
into a posterior inference, but it does <u>not</u> tell us how to assign the prior
expectation in the first place. Actually, this is quite reasonable: a language
should not impose on what one wishes to say, but it leaves open the question of
assigning the prior. Before proceeding to this, we should stress that the
Bayesian formulation leads to a posterior <u>probability bubble</u> $pr(f|D)$ and <u>not</u> to
any single f, optimal or otherwise.

## 3. Parameterised prior

What should the prior be? In practical cases, a great deal of loosely-defined
prior information might be available, such as a general expectation of sharp
edges and corners in a photograph of a man-made scene, or of "lines" of some
particular shape in a power spectrum. For the present, we shall <u>not</u> attempt to
deal with these more difficult problems. We shall seek a simpler prior, more
general in its lesser knowledge, onto which such extra information might in
future be grafted. In particular, we shall not incorporate any complicated form
of spatial correlation into our prior.

Yet we should not be <u>too</u> ignorant about f. A priori, our prior for any
particular cell, $pr(f_i)$, should be uniform in the logarithm, because our
knowledge of it is scale-invariant under changes of units. Accordingly,
$pr(f_i) \propto f_i^{-1}$ . If we legislate that the cells be completely independent, we
reach

$$pr(f) \propto \Pi f_i^{-1} .$$

[12]

This "spiky prior" has horrible properties. The divergences at infinity
disappear when an upper limit is placed on the total flux $\Sigma f$ (and hence
implicitly on each $f_i$), but the divergences at zero remain. In the continuum
limit of many cells, this prior is (almost) 100% certain that almost all the
flux in a distribution is concentrated into almost none of the cells, so that
reconstructions using it shatter into violently oscillating ambiguity. It is
unusable in practice.

So the cells of our distribution ought not be completely independent. Indeed,
that accords with our intuition. If we had observed that the fluxes in the
first 99 cells all lay between, say, 8 and 18 units, then we might hazard a
guess that the flux in the 100th cell would lie somewhere in a similar range.
This will probably prove to be a more productive methodology than insisting on
perversely total ignorance. Introspection about this shows that we are
considering extra parameters, such as an average and a range, as well as the
fluxes themselves. Thus we are considering

    pr( f , parameters )

which factorises as

    $$pr(f, parameters) = pr(parameters) \, pr(f|parameters)$$

[13]

The Bayesian estimation of these parameters, which completes "classic maximum entropy", is addressed in the accompanying paper by Gull. Even without this analysis, there are circumstances in which the parameters might be known beforehand. Thus the distribution in question might be embedded in a larger ensemble - a grand canonical ensemble being the closest thermodynamic equivalent - on which extra data sufficed to fix the parameters. An astronomical photograph, for instance, might be (and in fact is) embedded in a much larger total sky of known mean brightness. In this paper, we take the parameters as given, so that we seek a "prior" for a PAD in the form pr(f|parameters) .

## 4. Quantified prior

In order to find the prior, we investigate a simple case for which the prior is known, and argue that any general theory must apply to this specific example. Our specific example is the traditional team of monkeys. Let the monkeys throw balls of quantum size q at r cells (i=1,2,...,r) independently and at random with Poisson expectations $\mu_i$ (possibly all equal). The probability of occupation numbers $n_i$ is known (from symmetry and straightforward counting of possible outcomes) to be

$$pr(n|\mu) = \prod_i \mu_i^{n_i} e^{-\mu_i} / n_i! \tag{14}$$

Define $f_i = n_i q$ and $m_i = \mu_i q$ to remain finite as the quantum size q is allowed to approach zero. This allows us to pass to continuous f. The "PAD-space" of f becomes constructed from microcells of volume $q^r$, each associated with one lattice-point of integers. Hence we have, for small q,

$$pr(f \in V|m,q) = \sum_{\text{lattice points in V}} pr(n|\mu,q)$$

$$= \int_V (d^r f /q^r) \prod_i \mu_i^{n_i} e^{-\mu_i} / n_i! \tag{15}$$

Because we are taking n large, we may use Stirling's formula

$$n_i! \approx (2\pi n_i)^{\frac{1}{2}} n_i^{n_i} e^{-n_i} \tag{16}$$

to obtain (accurately to within O(1/n))

$$pr(f \in V|m,q) = \int_V \frac{d^r f}{\prod_i (2\pi q f_i)^{\frac{1}{2}}} \exp \frac{\sum(f_i - m_i - f_i \log(f_i/m_i))}{q} \tag{17}$$

Our general prior must agree with this specific result in this specific example. However, our general prior should hold in the continuum limit of many small cells. Considering first the argument of $q^{-1}\sum(f_i - m_i - f_i \log(f_i/m_i))$ of the exponential, we see that the only continuum integral guaranteed to give this sum over a finite cellular division of x is $q^{-1}\int dx(f(x) - m(x) - f(x)\log(f(x)/m(x)))$. The denominator of [17], by contrast, can not be treated in this way as the exponential of a sum. Although

$$\prod_i f_i^{-\frac{1}{2}} = \exp -\tfrac{1}{2}\sum \log f_i , \tag{18}$$

neither $\int dx \log f(x)$ nor any variant of it can be guaranteed to give $\Sigma \log f_i$ over macrocells because $\log f$ is not an additive quantity. Accordingly, we assign the denominator $\Pi f_i^{-\frac{1}{2}}$ <u>not</u> to the prior of $f$, but to the <u>measure</u> $M(f)$ over f-space.

On rewriting $q=1/\alpha$ for later convenience, we have now arrived at our quantified prior

$$pr(f \in V|m,\alpha) = \int Df \; M(f) \; \exp(\alpha S(f,m)) \; /Z_S(m,\alpha) \qquad [19]$$

where

$Df \qquad = d^r f \qquad$ = volume element of f-space $\qquad [20]$

$M(f) \qquad = \Pi \; f_i^{-\frac{1}{2}}$ = measure on f-space $\qquad [21]$

$S(f,m) \qquad = \int dx( \; f(x) - m(x) - f(x) \; \log(f(x)/m(x)) \; )$
$\qquad\qquad\qquad = \underline{entropy}$ of $f$ relative to $m \qquad\qquad\qquad\qquad [22]$

$Z_S(m,\alpha)$ = normalising constant $\approx (2\pi/\alpha)^{r/2}$ . $\qquad\qquad\qquad [23]$

The entropy [22] generalises the normalised form $-\int dx \; f(x) \; \log(f(x)/m(x))$ usually defined on probability distributions, and discussed in that context by many authors, including Jaynes (1968), Shore and Johnson (1980) and Tikochinsky, Tishby and Levine (1984), from various different points of view. The most important single factor in the quantified entropic prior [19] is the pointwise probability

$$pr(f|m,\alpha) \propto \exp(\alpha S(f,m)) \qquad [24]$$

expressed as the exponential of the entropy of $f$. The "parameters" which we have needed are $m$ and $\alpha$. Of these, $m$ serves to define an expected magnitude for $f$, defined on each cell separately. Commonly, in the absence of any reason to favour one value of $x$ over another, $m$ is taken to be simply a constant. A priori, the most probable individual $f$, that which maximises the pointwise probability [24] at the entropy maximum, is $f=m$, so we think of $m$ as a "model" for $f$. The other parameter $\alpha$ serves as an inverse measure of the expected spread of values of $f$ about $m$. This parameter has inverse dimensions to $m$ and $f$, so that its numerical value in any particular problem depends on the units which have been chosen. Neither $m$ nor $\alpha$ can be a constant of nature, fixed by pure theory.

Actually, this particular decomposition of [19] into an entropic pointwise probability [22] on a measure [21] is not necessary to perform any of the subsequent calculations. We do it because it aids interpretation, with several authors (e.g. Skilling and Bryan 1984, Levine 1986, Rodriguez 1989) having commented on the remarkable result that the metric tensor $g_{ij}=\delta_{ij}/f_i$ equivalent to the measure [21] is just (minus) the entropy curvature:

$$g_{ij} = - \partial^2 S \; / \; \partial f_i \; \partial f_j \qquad , \qquad M(f) = (\det g)^{\frac{1}{2}} \qquad . \qquad [25]$$

## 5. Historic maximum entropy

In its simplest form, maximum entropy data analysis avoids the complexity of the correct Bayesian formulation by forcing the experimental data into a definitive "testable" form which rejects outright all but a "feasible" set of f, and then directly applying the Principle of Maximum Entropy by selecting that surviving feasible f which has greatest entropy S.

Although variants are allowable, most datasets comprise a list of N numbers linearly related to the required PAD f via a matrix R of response functions, and subject to noise of standard deviation $\sigma$, so that

$$D_k = \langle f | R_k \rangle \pm \sigma_k \qquad (k=1,2,\ldots,N) \qquad [26]$$

Assuming the noise to be Gaussian and uncorrelated, this gives the likelihood

$$pr(D|f) = \Pi(2\pi\sigma^2)^{-\frac{1}{2}} \exp(-\tfrac{1}{2}C(f,D)) \qquad [27]$$

where

$$C(f,D) = \sum_{k=1}^{N} ( D_k - \langle f|R_k \rangle )^2/\sigma_k^2 = (D-\langle f|R\rangle)^T[\sigma^{-2}](D-\langle f|R\rangle) \qquad [28]$$

is chisquared. Square brackets denote a diagonal matrix. On a frequentist view, chisquared might be expected to take a value not much greater than its formal expectation over possible data, namely

$$C(f,D) \leq N ,$$

which corresponds to the data being, on average, one standard deviation away from the reconstruction. This can be used as a testable constraint to define feasible distributions f.

In the event that f is a PAD, the "best" feasible f can be found by

"maximising S(f,m) over f under the constraint $C(f,D) \leq N$".

We call the "chisquared=N" technique "historic maximum entropy" after its use by Frieden (1972), Gull and Daniell (1978), Gull and Skilling (1984) and others. Although such reconstructions f are often much clearer and more informative than those obtained with cruder techniques, historic maximum entropy is not Bayesian, and hence imperfect. On theoretical grounds, the constraint C(f,D)=N is only an approximation to the true likelihood pr(D|f), no single selected f can fully represent the posterior probability pr(f|D) which theory demands, and it is difficult to define the number N of fully independent data in a suitably invariant manner. On practical grounds, accurate data tend to be fitted to within less than $\pm\sigma$, which is harmless and in fact useful, but then the less accurate data become relaxed beyond $\pm\sigma$, even though they too ought to be giving useful information.

## 6. Quantified maximum entropy

Quantified maximum entropy uses the proper Bayesian formulation to produce not just a single PAD, but the complete posterior <u>probability bubble</u> $pr(f|D)$.

The pointwise joint probability distribution of f and D is

$$pr(f,D|m,\alpha) = (\alpha/2\pi)^{r/2} \; \Pi(2\pi\sigma^2)^{-\frac{1}{2}} \; \exp(\; \alpha S(f,m) - \tfrac{1}{2}C(f,D)\;) \qquad [29]$$

and in the usual way this is proportional to the posterior

$$pr(f|m,\alpha,D) = pr(f,D|m,\alpha) \; / \; pr(D|m,\alpha) \qquad [30]$$

Now S is a convex function of f, with negative definite curvature [25], whereas $C(f)$ in [28] has positive curvature. Hence $pr(f|m,\alpha,D)$ has a unique maximum over f, say at $\hat{f}(m,\alpha,D)$, defined by

$$\alpha \; \partial S/\partial f - \tfrac{1}{2} \; \partial C/\partial f = 0 \quad \text{at} \quad f = \hat{f} \; . \qquad [31]$$

This $\hat{f}$ will be the <u>single most probable</u> PAD: if a single distribution must be selected as "the" reconstruction, it should presumably be this one. Although $\hat{f}$ can be obtained by maximising S over C, the relevant value of C is implicitly defined by $\alpha$, and is unlikely to equal the "historic" value of the number N of data.

In order to proceed further with quantification, we approximate $pr(f,D|m,\alpha)$ by a Gaussian about its maximum:

$$pr(f,D|m,\alpha) \approx (\alpha/2\pi)^{r/2} \; \Pi(2\pi\sigma^2)^{-\frac{1}{2}} \; \exp(\alpha S(\hat{f})-\tfrac{1}{2}C(\hat{f}))$$

$$\exp(\; - \tfrac{1}{2} \; (f-\hat{f})^T \; (\alpha[\hat{f}^{-1}] + R^T[\sigma^{-2}]R) \; (f-\hat{f}) \;) \qquad [32]$$

Our first use of this is to evaluate the normalising constant in [30].

$$pr(D|m,\alpha) = \int \mathbb{D}f \; M(f) \; pr(f,D|m,\alpha)$$

$$= \Pi(2\pi\sigma^2)^{-\frac{1}{2}} \; e^{\alpha S(\hat{f})-\tfrac{1}{2}C(\hat{f})} \; \det(\; I + \alpha^{-1} A \;)^{-\frac{1}{2}} \qquad [33]$$

where I is the identity matrix and, after a little algebra,

$$A = [\sigma^{-1}]R[\hat{f}]R^T[\sigma^{-1}] \quad , \text{ i.e. } \quad A_{jk} = \sigma_j^{-1} \; (\hat{f}|R_j R_k) \; \sigma_k^{-1} \qquad [34]$$

One slightly surprising application of [33] is in <u>parameter estimation</u>. Almost always in practical calculations there are parameters such as scaling constants, phase constants and the like, which may or may not be known perfectly in advance. The calculations will be conditional on the values x of these parameters, so that [33] gives the probability $pr(D|m,\alpha,x)$, conditioned also on x. Just as m and $\alpha$ can be estimated from [33] by locating their most probable posterior values, so can any other parameters x.

# 7. Inferences about the reconstruction

The posterior probability bubble pr(f|D) has the interesting and entirely correct property that as the number of cells $r \to \infty$ the quantity $f_j$ in any particular cell goes to zero as $O(r^{-1})$, whereas the corresponding standard deviation only goes to zero as $O(r^{-\frac{1}{2}})$. This means that the proportional error on individual components of f becomes arbitrarily large in the continuum limit. As foreseen in Section 1, we can not learn about microscopic structure from macroscopic data.

However, we <u>can</u> learn about macroscopic quantities by integrating the probability bubble. The evaluation of such integrals is greatly aided by the entropic measure [21]. With $\rho = \langle f|P \rangle$ as in [4], we have

$$pr(\rho|m,\alpha,D) = \int Df \; M(f) \; pr(\rho,f|m,\alpha,D)$$

$$= \int Df \; M(f) \; pr(f|m,\alpha,D) \; pr(\rho|f)$$

$$= \int Df \; M(f) \; pr(f|m,\alpha,D) \; \delta(\rho-\langle f|P\rangle) \qquad [35]$$

Taking f to be estimated by the Gaussian [32], it follows that our estimate for $\rho$ is also Gaussian with mean and variance

$$\hat{\rho} = \Sigma \; \hat{f}_j \; P_j \; = \; \langle \hat{f}|P \rangle \qquad [36]$$

$$(\delta\rho)^2 = P^T \; ( \; \alpha[\hat{f}^{-1}] + R^T[\sigma^{-2}]R \; )^{-1} \; P \; . \qquad [37]$$

We see that the single distribution $\hat{f}$ suffices to give the mean estimate of any feature $\rho$, but does not suffice to give the variance. For that, the experimental response functions R/$\sigma$ are also needed. A little algebra enables the variance to be recast as

$$(\delta\rho)^2 = P^T[\hat{f}]P \; /\alpha \; - \; P^T[\hat{f}]R^T \; ( \; \alpha[\sigma^{-2}] + R[\hat{f}]R^T \; )^{-1} \; R[\hat{f}]P \qquad [38]$$

which is well-defined in the continuum limit $r \to \infty$ because

$$P^T[\hat{f}]P = \langle f|P^2 \rangle \; , \quad (R[\hat{f}]P)_j = \langle f|R_jP \rangle \; , \quad (R[\hat{f}]R^T)_{jk} = \langle f|R_jR_k \rangle \qquad [39]$$

One use of these formulae is to set P(x) to 1 in a certain domain and 0 elsewhere, in which case $\rho$ estimates the amount of f in that domain. To investigate a difference between two regions, P might be set +1 in one region, -1 in the other, and 0 elsewhere. Clearly there are many possibilities.

# 8. Conclusions

Entropy plays a dual rôle in inferences about positive additive distributions. At one level, it can be used to choose a single PAD when a choice must be made. Presumably one should select the most probable individual PAD, and according to [22] that will be the one with greatest entropy. This is the Principle of Maximum Entropy (e.g. Jaynes 1978), commonly applied to the assignment of probability distributions, but also possessing wider validity. It can be used when the data form testable constraints, satisfied by some PADs but not others.

At the next level, the entropy defines a prior probability through its exponential $\exp(\alpha S)$. This form should be used when the data give conditioning information in the form of a likelihood $pr(D|f)$. It gives results in the form of a posterior probability bubble $pr(f|D)$, from which error bars can be extracted for any feature $(f|P)$ of the PAD. This is usually what is required in practical data analysis, and earlier methods which rely on finding single PADs are superseded by the fuller Bayesian analysis. In particular, the "historic" maximum entropy method is obsolete. As a general rule, the error bars which are found obey a form of uncertainty principle, in that errors are proportionally large on narrow features of the PAD, and proportionally small on wide features.

Quantified maximum entropy involves a sharpness parameter $\alpha$ as well as the entropic model m. The accompanying paper by Gull explains the Bayesian estimation of $\alpha$ known as classic maximum entropy, and develops more sophisticated use of the methods. The programming is entirely practical, as is shown by Sibisi in these proceedings.

## REFERENCES

Cox, R.P. (1946) "Probability, frequency and reasonable expectation",
    Am. J. Phys. 17, 1-13

Frieden, B.R. (1972) "Restoring with maximum likelihood and maximum entropy",
    J. Opt. Soc. Am., 62, 511-518

Gull, S.F. (1989) "Developments in maximum entropy data analysis" in
    Maximum Entropy and Bayesian Methods, ed. J. Skilling, 53-71. Kluwer.

Gull, S.F. & Daniell, G.J. (1978) "Image reconstruction from incomplete and
    noisy data", Nature, 272, 686-690

Gull, S.F. & Skilling, J. (1984) "Maximum entropy method in image processing"
    IEE Proc. 131(F), 646-659

Jaynes, E.T. (1968) "Prior probabilities?" Reprinted in
    E.T. Jaynes: Papers on Probability, Statistics, and Statistical Physics,
    (page 124) ed. R. Rosenkrantz, 1983. Reidel.

Jaynes, E.T. (1978) "Where do we stand on maximum entropy?" Reprinted in
    E.T. Jaynes: Papers on Probability, Statistics, and Statistical Physics,
    (page 240) ed. R. Rosenkrantz, 1983. Reidel.

Levine, R.D. (1986) "Geometry in classical statistical thermodynamics",
    J. Chem. Phys., 84, 910-916

Rodriguez, C. (1989) "The metrics induced by the Kullback number" in
    Maximum Entropy and Bayesian Methods, ed. J. Skilling, 415-422. Kluwer.

Shore, J.E. & Johnson, R.W. (1980) "Axiomatic derivation of the principle of
    maximum entropy and the principle of minimum cross-entropy"
    IEEE Trans. Info. Theory IT-26, 26-39 and IT-29, 942-943

Skilling, J. & Bryan, R.K. (1984) "The maximum entropy algorithm",
    Monthly Notices Royal Astronomical Soc., 211, 111-124

Skilling, J. (1988) "The axioms of maximum entropy" in
    Maximum Entropy and Bayesian Methods in Science and Engineering,
    vol 1. Foundations, ed. G.J. Erickson and C.R. Smith. Kluwer.

Skilling, J. (1989) "Classic maximum entropy" in
    Maximum Entropy and Bayesian Methods, ed. J. Skilling, 45-52. Kluwer.

Tikochinsky, Y., Tishby, N.Z. & Levine, R.D. (1984) "Consistent inference of
    probabilities for reproducible experiments" Phys Rev Lett, 52, 1357-1360

# QUANTIFIED MAXENT: AN NMR APPLICATION

Sibusiso Sibisi
Department of Applied Mathematics and Theoretical Physics
Silver Street
University of Cambridge
England CB3 9EW *

July 1989

### Abstract

'Classic MaxEnt' is a Bayesian derivation of the MaxEnt treatment of inverse problems leading to a posterior probability 'bubble' over the solution. This probability bubble—which is maximised at the optimal regularised solution—provides the framework for quantitative inferences about the solution. In particular, the framework allows the computation of fluxes and associated error bars over the solution. This is an important advance in the general theory of inverse problems which has thus far lacked a quantitative reliability treatment of the computed solution. This paper discusses this quantification procedure and applies it to practical NMR spectroscopy.

## 1   Introduction

The orthodox approach to inverse problems has one major inadequacy: given a single dataset, it is not possible to *quantitatively* assess features of the solution. For example, we may want to know to what extent we should believe a weak peak in a computed spectrum. In other words, we need to be able to compute the amplitude of the peak *with its associated error bar*. Such questions lie at the heart of data analysis, yet orthodox regularisation theory is not equipped to provide the answers; short, perhaps, of some Monte Carlo analysis on an ensemble of datasets. To truly come of age, the theory of inverse problems must be able to answer such questions from a single dataset. The Bayesian approach provides the necessary framework.

Adopting a Bayesian viewpoint, Skilling ([1]) presents a derivation of the entropic prior, leading to the 'Classic MaxEnt' method for inverse problems. Gull ([2]) completes the derivation by showing how to compute the associated regularisation parameter and further extends Classic MaxEnt to incorporate correlations in the solution. The posterior probability of the solution allows the reliability of the solution to be quantified—this paper shows that not only can the optimal solution be determined by maximising the posterior, but the associated error bars can also be computed.

We start with a brief discussion of Classic Maxent as developed in [1],[2] followed by a discussion of error bars. The formalism is then applied to the treatment of an NMR dataset.

*Supported by Mobil North Sea Limited

351

*P. F. Fougère (ed.), Maximum Entropy and Bayesian Methods*, 351–358.
© 1990 *Kluwer Academic Publishers*.

## 2    Classic MaxEnt

We are given $N$ data samples which are linearly related to the solution $f$ of dimension $M$ via an $N \times M$ matrix $R$ and subject to noise so that

$$D_k = \sum_{j=1}^{M} R_{kj} f_j \pm \sigma_k \qquad k = 1, \ldots, N$$

Assuming Gaussian uncorrelated noise, the data induce the likelihood

$$\Pr(D|f) = \prod_{k=1}^{N} \left(2\pi\sigma_k^2\right)^{-\frac{1}{2}} e^{-\frac{1}{2}C(f)}$$

where $C(f) = (D - Rf)^T \mathcal{E}(D - Rf)$, $\mathcal{E} = \mathrm{diag}\,(\sigma^{-2})$. Skilling ([1]) shows that the correct prior for positive, additive $f$ is the entropic prior

$$\Pr(f|\alpha) = \left(\frac{\alpha}{2\pi}\right)^{\frac{M}{2}} e^{\alpha S}$$

The scaling constant $\alpha$ is the 'regularising parameter' and $S$ is the entropy

$$S(f, m) = \sum_{j=1}^{M} (f_j - m_j - f_j \log(f_j/m_j))$$

The positivity condition on $f$ can be relaxed. This is achieved through introducing two subsidiary positive distributions $g$ and $h$ so that $f = g - h$, where $g$ and $h$ are determined simultaneously using the joint entropy $S(g, h) = S(g) + S(h)$. This extension has found application in NMR where spectra can consist of positive and negative peaks ([3]).

The global maximum of $S$—or, equivalently, that of the prior—is at $m$. Hence the assigned 'model' $m$ is the solution in the absence of data. Given data $D$, the joint probablity of $f$, $\alpha$ and $D$ is

$$\Pr(f, \alpha, D) = \Pr(\alpha) \Pr(f|\alpha) \Pr(D|f)$$

Since $\alpha$ is an unknown scale factor, the prior $\Pr(\alpha)$ is uniform (in $\log(\alpha)$, strictly speaking) over some sufficiently large range. Hence the joint distribution is a product of the known prior and likelihood. Also $\Pr(f, \alpha, D) \propto \Pr(f, \alpha|D)$—the posterior of $f$ and $\alpha$ given $D$—so

$$\begin{aligned}
\Pr(f, \alpha|D) &= \Pr(f|\alpha) \Pr(D|f) \\
&= \left(\frac{\alpha}{2\pi}\right)^{\frac{M}{2}} \prod_{k=1}^{N} \left(2\pi\sigma_k^2\right)^{-\frac{1}{2}} e^{(\alpha S - \frac{1}{2}C)}
\end{aligned}$$

(equality holds to within a constant which can be determined through normalisation). To determine $\alpha$ we must first marginalise over $f$. Here, we must incorporate the correct $f$-space measure $m(f) = \prod f^{-\frac{1}{2}}$ derived in [1]. The resulting integral is tractable if the exponent is of Gaussian form. To this end, let $\hat{f}$ maximise $\alpha S(f) - \frac{1}{2}C(f)$ at fixed $\alpha$. Noting that the Hessian of $\alpha S(f) - \frac{1}{2}C(f)$ is

$$-\frac{\partial^2}{\partial f \partial f}(\alpha S - C) = \alpha \mathcal{G} + R^T \mathcal{E} R$$

where $\mathcal{G}(f) = \mathrm{diag}(f^{-1})$, the Gaussian approximation of the posterior is

$$\Pr(f, \alpha|D) = \left(\frac{\alpha}{2\pi}\right)^{\frac{M}{2}} \prod_{k=1}^{N} \left(2\pi\sigma_k^2\right)^{-\frac{1}{2}} e^{(\alpha S(\hat{f}) - \frac{1}{2}C(\hat{f}))} e^{-\frac{1}{2}(f-\hat{f})^T (\alpha \hat{\mathcal{G}} + R^T \mathcal{E} R)(f-\hat{f})}$$

with $\hat{\mathcal{G}} = \mathcal{G}(\hat{f})$. The optimal scale parameter, $\hat{\alpha}$, is determined by marginalising over $f$, and then maximising with respect to $\alpha$. Using this optimal value of $\alpha$ finally leads to the Classic MaxEnt posterior probability bubble for $f$

$$\Pr(f|D) \propto e^{-\frac{1}{2}(f-\hat{f})^T(\hat{\alpha}\hat{\mathcal{G}} + R^T \mathcal{E} R)(f-\hat{f})}$$

centred at the optimal solution $\hat{f}(\hat{\alpha})$.

Orthodox MaxEnt regularisation consists in simply maximising $\alpha S(f) - \frac{1}{2}C(f)$ and using the discrepancy principle $C(\hat{f}) = N$ to determine the optimal $\alpha$. We refer to this as Historic Maxent.

# 3   Error Bars

The covariance matrix $\hat{\alpha}\hat{\mathcal{G}} + R^T \mathcal{E} R$ of the posterior bubble leads to inferences about errors on the optimal solution. However, there is a subtlety here. The *free-form* approach to inverse problems involves a choice of cell size in the solution. Hence the dimensionality, $M$, of the solution will vary according to the desired resolution although the total flux in a given range will remain roughly the same. Consequently, as $M \to \infty$ the flux in a given cell goes to zero as $\mathcal{O}(M^{-1})$ while its associated error goes to zero as $\mathcal{O}(M^{-1/2})$—the relative error becomes arbitrarily large in the continuum limit. One might expect this: macroscopic data ought not give accurate constraints on microscopic structure. However, the flux over a given range and its corresponding error will have the correct asymptotic behaviour. Technically, $f$ is a distribution defined over the $\sigma$- field of subsets of some total domain: it is only for convenience that we compute it as a function defined on discrete cells. Accordingly, instead of *pointwise* errors, we should seek errors on patches of the form

$$\rho = \sum_{j=1}^{M} f_j P_j = f^T P$$

where $P$ is a mask function which is non-zero over the region of interest. (The mask is a unit 'top-hat' if we seek the total flux in a given region but it may be a more general linear function.)

The posterior bubble for $f$ can be integrated to obtain a Gaussian distribution for $\rho$

$$\Pr(\rho|D) \propto e^{-\frac{1}{2}(\rho-\hat{\rho})^T P^T(\hat{\alpha}\hat{\mathcal{G}} + R^T \mathcal{E} R)P(\rho-\hat{\rho})}$$

where $\hat{\rho} = \hat{f}^T P$. Hence the error in $\rho$ is given by

$$(\delta\rho)^2 = P^T(\hat{\alpha}\hat{\mathcal{G}} + R^T \mathcal{E} R)^{-1} P$$

This result is an important advance in the theory of inverse problems: given a single dataset, we can quantitatively assess the significance of a feature $\hat{\rho}$ of the solution by computing its error bar $\delta\rho$. We shall refer to Classic MaxEnt with error bar quantification as Quantifed MaxEnt.

# 4   Application in NMR

Fourier transformation, with the aid of various filters, is the conventional processing method of NMR data. Quantification on the resultant spectra can be difficult: for instance, fluxes of overlapping lines cannot be readily inferred nor can a reasonable judgment be made on a line of low flux since there is no handle on the associated error. But these are the typical inferences that a chemist wishes to draw from the data. While Historic MaxEnt does allow line-sharpening without introducing gross artefacts and produces cleaner spectra ([5]), it too is not up to the task of proper quantification. Methods which fit a parametric model to the data have been developed in recent years. Notably,

Bretthorst's Bayesian approach provides a framework for determining the parameters and their associated errors ([4]).

Quantified MaxEnt has the flexibility of allowing the determination of fluxes and their error bars without parametrization of lines: the data do not have to conform to a particular model. However, if a given model is known to be appropriate, it is proper to fit the data to it. In such a case, we propose a hybrid scheme in which Quantified MaxEnt is used to determine the number of significant lines followed by a maximum likelihood (least squares) parametric fit to that number of lines using the MaxEnt frequencies as starting values. Any residual structure which does not conform to the chosen model (e.g. nonlinear baseline drifts) should then be further investigated through Quantified MaxEnt.

We feel that there may be the making of a general procedure here: Quantified MaxEnt should perform the Bayesian step of assessing the presence of 'parametric' features in $f$ (such as lines in a spectrum or ringing in an image). Such features may then be explicitly parametrized and, as appropriate, projected out of the data.

We note that, apart from noise, practical NMR data have various imperfections: detector mis-alignment, a time delay before data acquisition, baseline DC-offsets and corrupted early datapoints because of excitation pulse breakthrough. The first two imperfections constitute a two-parameter phase offset which must be determined and corrected for in order to obtain undistorted spectra. The two parameters are usually manually determined on the spectrometer. The tail-end of the data where the signal has decayed into the noise can be used to determine the noise-level and the DC-offset which is then subtracted from the data. The extent of pulse breakthrough is harder to determine *a priori*.

## 5   Historic MaxEnt on COSY33

We analyse a real 2-furoic acid dataset, COSY33. Fig. 1 shows the phased FT spectrum of the data. There are 12 slightly overlapping but clearly significant lines in the spectrum. Coupling between close lines changes their relative fluxes and widths, but the overall fluxes in the three isolated groups of lines should remain essentially the same. But there is also a narrow negative line in the middle of the spectrum and some nonuniformity in the baseline believed to be due to corrupted early datapoints.

We can sharpen the lines of the MaxEnt spectrum by deconvolving with a Lorentzian blurring function (the natural lineshape for exponentially decaying data). This amounts to replacing the Fourier response $\mathcal{F}$ between the spectrum and the data by

$$R_{kj} = e^{-\pi\lambda_0 k/M} \mathcal{F}_{kj}$$

where $\lambda_0$ is the blurring width. Thus $\lambda_0$ is a parameter in the likelihood and can, like $\alpha$, be determined by Bayesian means (in the framework of Classic MaxEnt). Other parameters in the likelihood, like phase parameters, can be determined in the same way if they are unknown. With the spectrum length $M = 2048$, the optimal Bayesian value of $\lambda_0$ is found to be 1.8 cells. Historic MaxEnt may then be applied using this optimal blur width leading to the positive spectrum of fig. 2 which is cleaner and has narrower lines than the FT spectrum of fig. 1. The spectrum also has a flat baseline: to achieve this, it suffices to ignore the first 4 datapoints in the computation by setting the corresponding entries of $\sigma^{-2}$ to zero in $C(f)$. However, quantitative questions about fluxes and associated errors remain unanswered.

## 6   Quantified MaxEnt on COSY33

Again, we use $\lambda_0 = 1.8$ and ignore the first 4 points. However, we relax the positivity condition because it biases a zero background toward positive flux and, in any case, we wish to quantify the

negative line. The resulting spectrum is shown in fig. 3; it may not be as clean as the Historic
MaxEnt spectrum but the important point is that we can quantify fluxes as shown in the table
below. We use top-hat masks of unit height and of width and position chosen so as to pick the
flux of individual lines, groups of lines or intermediate plateaus as described in the first column.
The lines are numbered in increasing order from left to right across the spectrum, as are the four
plateau regions.

| region | flux | error |
|---|---|---|
| plateau 1 | 1.59 | 1.57 |
| line 1 | 8.90 | .30 |
| line 2 | 9.50 | .32 |
| line 3 | 9.40 | .35 |
| line 4 | 9.96 | .43 |
| lines 1 - 4 | 37.75 | .76 |
| plateau 2 | -4.26 | 6.80 |
| (covers negative line) | | |
| negative line | -.34 | .28 |
| line 5 | 9.28 | .35 |
| line 6 | 8.92 | .34 |
| line 7 | 10.80 | .37 |
| line 8 | 10.33 | .35 |
| lines 5 - 8 | 39.87 | 1.04 |
| plateau 3 | -10.45 | 5.90 |
| line 9 | 10.62 | .42 |
| line 10 | 10.23 | .37 |
| line 11 | 8.82 | .37 |
| line 12 | 9.12 | .36 |
| lines 9 - 12 | 38.79 | 1.13 |
| plateau 4 | -.47 | 1.93 |
| total flux | 102.81 | 11.87 |
| (whole spectrum) | | |

The results are encouraging. The errors on the 12 lines are small, as expected. The sizes of
the errors on the first, second and fourth plateau regions indicates they do not contain significant
structure. The smaller relative error in the third plateau is due to a gentle but noticeable dip in
the baseline rather than structure due to individual lines. Nonetheless, a relative error of 0.56 is
quite substantial. The error on the negative line shows that it is also not significant.

# 7   Parametric Fitting

We use the data model

$$F_k = \sum_{j=1}^{r} a_j e^{-\pi \lambda_j k/M} \cos \left( \left( \frac{\pi k}{M} + \phi_1 \right) \nu_j + \phi_0 \right) \qquad k = 1 \ldots N$$

where $r$, the number of significant lines, is determined by Quantified MaxEnt. We must fit the
amplitudes (fluxes) $\{a_j\}$, the decays (line-widths) $\{\lambda_j\}$, the frequencies $\{\nu_j\}$ and the phases $\phi_0$ and
$\phi_1$. We fit these parameters iteratively; alternating between linear least squares for $\{a_j\}$ at fixed
$\{\lambda_j, \nu_j, \phi_0, \phi_1\}$ and nonlinear least squares for $\{\lambda_j, \nu_j, \phi_0, \phi_1\}$ at fixed $\{a_j\}$. The nonlinear parameters require starting values. The starting frequencies are simply read off the MaxEnt spectrum,
the starting phases are provided from the experiment and we start at some constant value (zero,

say) for all decays (decays are not conveniently determined from a spectrum). We can determine the errors on the computed parameters from the diagonal entries of the inverse of the covariance matrix in the Gaussian approximation of the likelihood.

We only fit to the significant signal of COSY33, which is in the first 500-odd points, and ignore the first 4 points as in MaxEnt. The experimental phases are $\phi_0 = 81°$ and $\phi_1 = 50°$. The results for the 12 lines of COSY33 are shown in the following table:

| line | $\nu$ | error | $\lambda$ | error | $a$ | error |
|------|-------|-------|-----------|-------|-----|-------|
| 1 | 256.33 | .059 | 2.53 | .056 | 9.24 | .13 |
| 2 | 268.73 | .060 | 2.68 | .057 | 9.78 | .13 |
| 3 | 281.84 | .060 | 2.85 | .057 | 10.45 | .14 |
| 4 | 294.33 | .058 | 2.64 | .055 | 9.92 | .13 |
| 5 | 1062.95 | .060 | 2.75 | .059 | 9.82 | .13 |
| 6 | 1075.60 | .061 | 2.82 | .059 | 9.86 | .14 |
| 7 | 1113.82 | .051 | 2.68 | .049 | 11.35 | .13 |
| 8 | 1126.47 | .054 | 2.87 | .052 | 11.44 | .14 |
| 9 | 1706.31 | .054 | 2.72 | .049 | 11.41 | .13 |
| 10 | 1732.01 | .055 | 2.71 | .051 | 11.04 | .13 |
| 11 | 1757.19 | .061 | 2.56 | .057 | 9.20 | .13 |
| 12 | 1782.72 | .066 | 2.94 | .062 | 9.97 | .14 |

| $\phi_0$ | error | $\phi_1$ | error |
|----------|-------|----------|-------|
| 82.2 | .50 | 47.5 | .92 |

The errors on frequencies and line-widths are much less than a tenth of a cell. The MaxEnt fluxes, although weaker by about one (MaxEnt) error, compare well with the displayed amplitudes. This is reasonable since the baseline in the Classic MaxEnt spectrum is slightly negative.

To examine residual structure, we subtract the dataset generated by these lines from the original dataset and Fourier transform to obtain the spectrum of fig. 4. This spectrum shows no evidence of residual structure.

# 8    Error Bars on a growing line

Finally, we would like to investigate the quantification of a line as a function of changing flux. We can do this by adding the dataset generated by a single line $(a_l, \nu_l, \lambda_l)$ to the COSY33 dataset for different values of amplitude $a_l$. We can thus quantify the line on datasets of the *same* noise realisation. The choice of $\nu_l = 719$, $\lambda_l = 1.8$ generates an isolated line in the corresponding spectra. The quantification on the line for different fluxes gives the following results:

| true flux | quantified flux | error |
|-----------|-----------------|-------|
| 0.0 | -.35 | .75 |
| .1 | -.27 | .75 |
| .2 | -.20 | .75 |
| .4 | -.02 | .75 |
| .8 | .38 | .75 |
| 1.0 | .58 | .75 |
| 1.6 | 1.13 | .76 |
| 2.0 | 1.53 | .76 |
| 3.2 | 2.73 | .77 |
| 4.0 | 3.61 | .78 |
| 8.0 | 7.37 | .80 |
| 16.0 | 15.24 | .87 |

These results are extremely plausible. The quantified flux bears a linear relationship to the true flux and the error remains quite uniform as the flux increases. This simple relationship makes it possible to identify the threshold where the quantified flux falls below the error and the line can no longer be considered significant. Again we see that the background is slightly negative, hence the quantified fluxes are reduced by about $(2/3) \times error$ compared to the true fluxes.

## 9   Conclusion

We have derived a procedure for computing quantitative errors on solutions of inverse problems. This is an important advance in the theory of inverse problems: it lays to rest the criticism that regularisation is simply a 'cosmetic' smoothing mechanism giving visually better solutions but with no handle on the quantitative reliability of the structure in the solution. Although we have demonstrated the procedure on a spectrum, the formalism readily generalises to higher dimensional problems and can thus be used to quantify images in many disciplines where the need has arisen but the tool has been lacking; frequently forcing practitioners to eschew the general free-form approach in favour of a more restrictive but quantifiable parametrisation of the image.

## References

[1] Skilling J., (1989). Classic Maximum Entropy. *Maximum Entropy and Bayesian Methods* (ed. J. Skilling), 45-52, Kluwer.

[2] Gull S.F., (1989). Developments in Maximum Entropy Data Analysis. *Maximum Entropy and Bayesian Methods* (ed. J. Skilling), 53-71, Kluwer.

[3] Laue E.D., Skilling J., Staunton J., (1985). Maximum Entropy Reconstruction of Spectra Containing Antiphase Peaks. *J. Mag. Res.*, **63**, 418- 424

[4] Bretthorst G.L., (1989) Bayesian Model Selection: Examples Relevant to NMR. *Maximum Entropy and Bayesian Methods* (ed. J. Skilling), 377-388, Kluwer.

[5] Sibisi S., Skilling J., Brereton R.G., Laue E.D., Staunton J., (1984). Maximum Entropy Signal Processing in Practical NMR Spectroscopy. *Nature*, **311**, 446-447.

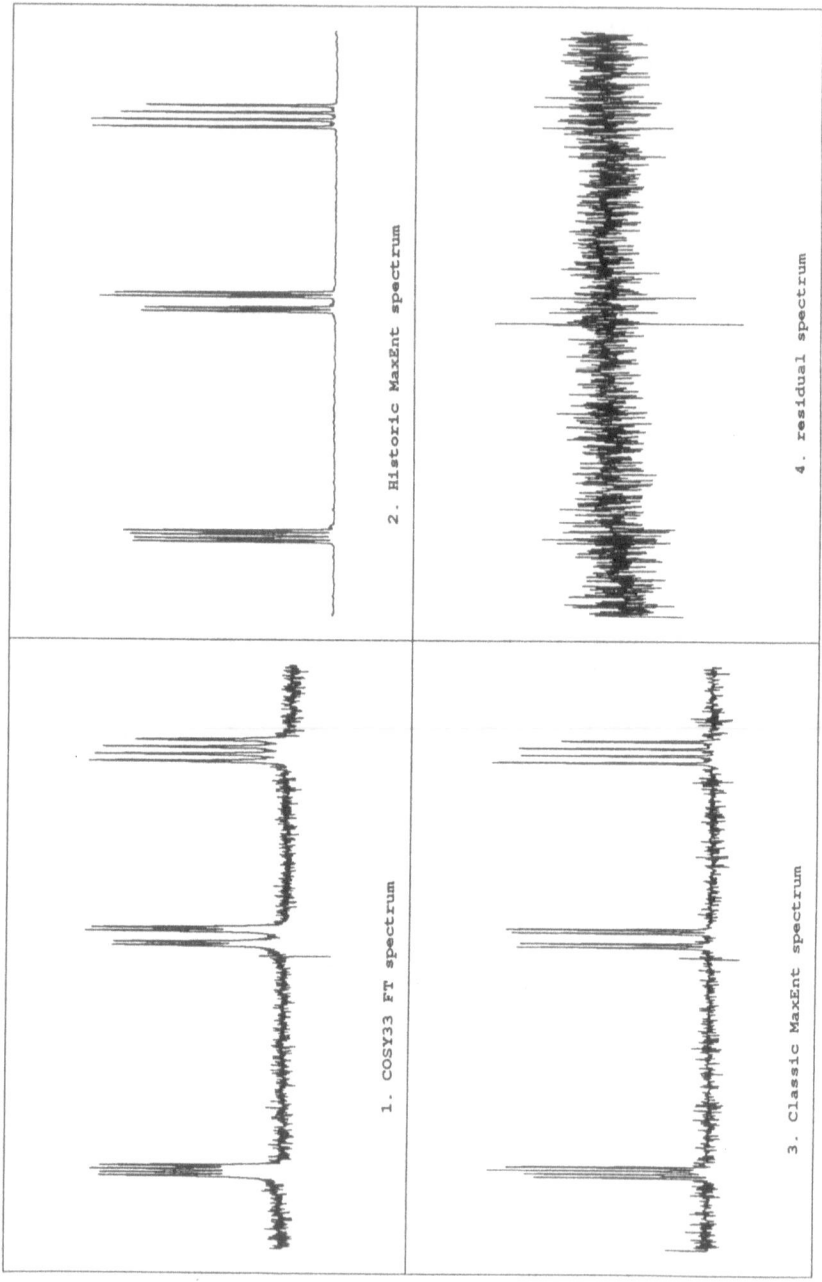

1. COSY33 FT spectrum

2. Historic MaxEnt spectrum

3. Classic MaxEnt spectrum

4. residual spectrum

# MAXIMUM ENTROPY RECONSTRUCTION OF RADAR CROSS-SECTION IMAGES

G. R. Heidbreder
Department of Electrical and Computer Engineering
University of California, Santa Barbara
Santa Barbara, CA 93117

F. van Roekeghem
Délégation Générale pour l'Armement, Direction des Engins,
Ministere de la Défense, Paris-Armées
FRANCE

ABSTRACT. We have applied the maximum entropy technique, using the Shannon-Jaynes entropy formulation and the Cambridge algorithm, to the reconstruction of two-dimensional high-resolution coherent radar images used in the study of radar cross-section diagnostics. In this application the image variable, namely the magnitude of a two-dimensional marginal distribution of radar reflectivity, is not linearly related to the measured data; rather the data are samples of the two-dimensional Fourier transform of a complex radar reflectivity function whose magnitude is desired. Adaptation of the algorithm is effected by formulating the problem as one of power spectral analysis of the data array using lagged product estimates of its autocorrelation.

## 1. Introduction

In maximum entropy (MAXENT) image reconstruction using the Cambridge algorithm [1] the image shape is specified by a real probability vector and its entropy is a concave function of that vector. Similarly, the cross-entropy of the image and its a priori estimate is a convex function of the image vector. The algorithm also requires a data dependent constraint which is a convex functional on the space of the image vector and unambiguous mapping operations between the image vector space and the space of measurement data. The usual constraint limits the chi-squared statistic

$$C(g) = \sum (G_k - D_k)^2 / \sigma_k^2 \tag{1}$$

obtained by summing normalized squared differences between measured data values $D_k$

and simulated values $G_k$ which would be obtained in the absence of measurement error if the image being observed were indeed represented by the vector $g$. A linear relationship

between the $G_k$ and $g$ assures that surfaces of constant $C(g)$ are convex ellipsoids in the image space, thus guaranteeing a unique result in the constrained maximization of entropy (or minimization of cross-entropy.)

P. F. Fougère (ed.), Maximum Entropy and Bayesian Methods, 359–368.
© 1990 Kluwer Academic Publishers.

We have applied the MAXENT technique, using the Shannon-Jaynes entropy formulation and the Cambridge algorithm, to the reconstruction of two-dimensional (2-D) high-resolution coherent radar images used in the study of radar cross-section (RCS) diagnostics [2]. In this application, the image variable, namely the magnitude of a 2-D marginal distribution of radar reflectivity, is not linearly related to the measured data; rather the data are samples of the 2-D Fourier transform of a complex radar reflectivity function whose magnitude is desired.

## 2. The Radar Data

The radar targets are turntable mounted in a benign (essentially clutterless) environment. High resolution is obtained in the range direction through time delay sorting of wideband signal returns and in the cross-range direction by inverse synthetic aperture (ISAR) processing. The model on which radar imaging is based is shown in Fig.1 The object to be imaged is described in object fixed polar coordinates $(r,\phi)$. The object is treated as planar with reflectivity distribution $g(r,\phi)$ which is in fact the marginal distribution obtained by projecting the actual three-dimensional reflectivity distribution on the plane containing the radar boresight axis and the normal to the axis of object rotation (the plane of the paper in Fig. 1.) The radar is shown in its initial position and shifted by an angle $\theta$, illustrating relative motion corresponding to target rotation through an angle $\theta$.

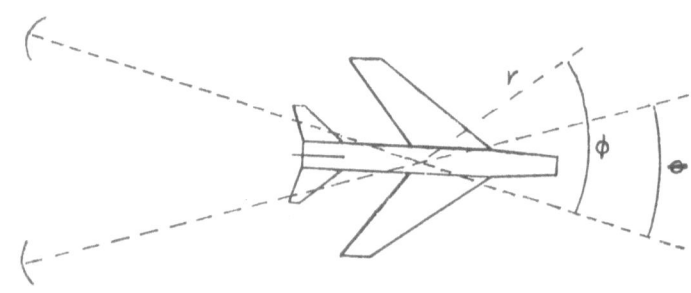

Figure 1. Radar-Target Geometry

Assuming the radar distance is sufficient for essentially plane wave illumination of the target and assuming constant velocity of propagation in the object and medium we can express the demodulated return signal for a continuous wave (CW) radar as [3]

$$G\left(\frac{2}{\lambda},\theta\right)=\int_0^\infty\int_0^{2\pi} g(r,\phi)\exp\left[-j\frac{4\pi r}{\lambda}\cos(\phi-\theta)\right]r\,dr\,d\phi \qquad (2)$$

where $\lambda$ is radar wavelength. We have neglected an unimportant complex constant dependent on the range to the center of rotation and the signal level. Substitution of

$$x = r\sin\phi, \quad y = r\cos\phi$$
$$f_x = \frac{2}{\lambda}\sin\theta, \quad f_y = \frac{2}{\lambda}\cos\theta$$

and $g_r(x,y) = g(r,\phi)$ places (2) in the more familiar form

$$G\left(\frac{2}{\lambda},\theta\right) = G_r(f_x,f_y) = \int_{-\infty}^{\infty} g_r(x,y)\exp[-j2\pi(f_x x + f_y y)]dxdy \tag{3}$$

Range-Doppler imaging exploits the Fourier transform relation in (2). Thus radar data is gathered at fixed $\theta$ over a band of spatial frequencies either by stepping the transmitter frequency or by sweeping it linearly over a band. The target is then rotated successively to other values of $\theta$ and the measurements are repeated. A 2-D set of samples is obtained with less than Nyquist spacing in spatial frequency. These samples are grouped in a keystone-shaped area in the spatial frequency domain. Typically the intervals of $\lambda$ and $\theta$ are small so that the keystone is well approximated by a rectangle and a 2-D fast Fourier transform is used to obtain estimates of $g_r(x,y)$ per the inverse of a discrete version of (3). Either the magnitudes or squared magnitudes of these estimates are taken as the image pixel values.

Provided that the target is spatially limited we could perfectly reconstruct its reflectivity function from the coefficients of its Fourier series expansion, i.e., samples of the Fourier transform $G_r(f_x,f_y)$. We define

$$G_{nm} = \frac{1}{4XY}\int_{-X}^{X}\int_{-Y}^{Y} g_r(x,y)\exp[-j2\pi(nx\,\Delta f_x + my\,\Delta f_y)]dxdy \tag{4}$$

where $2X = 1/\Delta f_x$ and $2Y = 1/\Delta f_y$ exceed maximum target extent. Then the reflectivity estimates are given by

$$\tilde{g}_r(x,y) = \sum_{n=-\infty}^{\infty}\sum_{m=-\infty}^{\infty} G_{nm}\exp[j2\pi(nx\,\Delta f_x + my\,\Delta f_y)] \tag{5}$$

$\tilde{g}_r(x,y)$ is just the periodic function associated with $g_r(x,y)$. The number of Fourier coefficients in (5) is, in view of the space limited reflectivity, unlimited. Hence, because we in fact have a limited number of Fourier coefficients (measured data), the reconstruction problem is ill posed. Conventional processing nevertheless truncates the Fourier series in (5) and uses windowing of the available coefficients to control the sidelobe artifacts resulting from the truncation.

## 3.　Adapting the Algorithm

The appropriate entropy is

$$S = -\iint |g_r(x,y)|^2 \log |g_r(x,y)|^2 \, dxdy \tag{6}$$

for a magnitude squared image or

$$S = -\iint |g_r(x,y)| \log |g_r(x,y)| \, dxdy \tag{7}$$

for a magnitude image.  Normalization of $|g_r(x,y)|^2$ and $g_r(x,y)$ is implied though the notation for the unnormalized functions is retained.  Introducing Cartesian and polar representations,

$$g_r(x,y) = g_x(x,y) + jg_y(x,y) = a(x,y)\exp[j\,\phi(x,y)] \tag{8}$$

we obtain the simpler expressions

$$\begin{aligned}
S &= -\iint \left[g_x^2(x,y) + g_y^2(x,y)\right] \log \left[g_x^2(x,y) + g_y^2(x,y)\right] dxdy \\
&= -\iint a^2(x,y) \log\left[a^2(x,y)\right] dxdy
\end{aligned} \tag{9}$$

and

$$\begin{aligned}
S &= -\iint a(x,y) \log a(x,y) dxdy \\
&= -\iint \sqrt{g_x^2(x,y) + g_y^2(x,y)} \, \log \sqrt{g_x^2(x,y) + g_y^2(x,y)} dxdy
\end{aligned} \tag{10}$$

Expressions (9) and (10) are concave functions of $a^2(x,y)$ and $a(x,y)$ respectively.  The (noisefree) data however is not linearly related to either variable.  On the other hand, $g_x(x,y)$ and $g_y(x,y)$ in (9) and (10) are linearly related to the data, but neither expression is a concave function of these variables.  It is nevertheless possible to formulate the problem in terms of an entropy which is a concave function of an image vector which maps linearly to data space.  We recognize the problem as one of estimating the power spectrum $|g_r(x,y)|^2$ of the complex field $G_r(f_x, f_y)$ .  As is so often customary in spectral analysis, we estimate samples of the autocorrelation of this field using lagged product sums from the data array. Assuming the measurements are given by (4) and that $g_r(x,y)$ is square integrable, the Fourier coefficients of $|g_r(x,y)|^2$ are

$$\frac{1}{4XY} \int_{-X}^{X} \int_{-Y}^{Y} |g_r(x,y)|^2 \exp\left[-j2\pi(nx\,\Delta f_x + my\,\Delta f_y)\right] dx\,dy =$$

$$\sum_{p=-\infty}^{\infty} \sum_{q=-\infty}^{\infty} G_{p,q} G_{p-n,q-m} \qquad (11)$$

This shows clearly that the knowledge of a limited number of Fourier coefficients of $g_r(x,y)$ is not sufficient to determine a limited number of Fourier coefficients of its squared magnitude.

Many targets of interest produce radar images exhibiting only a few strong scattering centers, i.e. peak responses whose widths are essentially those predicted by the bandwidth and synthetic aperture limits of the radar. This suggests modeling targets as a group of point scatterers of unknown location and unknown complex amplitude. With the resulting $\delta$-function representation for $g_r(x,y)$ , (4) becomes

$$G_{n,m} = \sum_{i=1}^{l} g_i \exp\left[-j2\pi(nx_i + my_i)\right] \qquad (12)$$

For simplicity we have assumed $\Delta f_x = \Delta f_y = 1$ in (12). A convolution of the data, modelled by (12), on a rectangular array yields

$$\sum_{p=n}^{N-1} \sum_{q=m}^{M-1} G_{p,q} G^*_{p-n,q-m} = \sum_{i=1}^{l} \sum_{k=1}^{l} g_i g_k^* \exp[-j2\pi(nx_k + my_k)] \times$$

$$\sum_{p=n}^{N-1} \sum_{q=m}^{M-1} \exp\{-j2\pi[p(x_i - x_k) + q(y_i - y_k)]\}$$

$$= (N-n)(M-m) \sum_{k=1}^{l} |g_k|^2 \exp\{-j2\pi(nx_k + my_k)\} + B_{n,m} \qquad (13)$$

where

$$B_{n,m} = \sum_{i=1}^{l} \sum_{k=1}^{l} g_i g_k^* \exp[-j2\pi(nx_i + my_i)] \ (1-\delta_{ik}) \times$$

$$\frac{[1-\exp\{-j2\pi(N-n)(x_i - x_k)\}][1-\exp\{-j2\pi(M-m)(y_i - y_k)\}]}{[1-\exp\{-j2\pi(x_i - x_k)\}][1-\exp\{-j2\pi(y_i - y_k)\}]} \qquad (14)$$

The first term of (13), but for the factor (N-n)(M-m), may be interpreted as the Fourier transform for the discretely modelled target. If we use the convolution in (13) to estimate the Fourier transform of the magnitude squared image function we obtain

$$\hat{R}(n,m) = \frac{1}{(N-n)(M-m)} \sum_{p=n}^{N-1} \sum_{q=m}^{M-1} G_{p,q} G^*_{p-n,q-m}$$

$$= R(n,m) + \frac{B_{n,m}}{(N-n)(M-m)} \tag{15}$$

where $R(n,m)$ is the desired Fourier transform value. We have neglected measurement noise in the above analysis. We assume that the radar measurements obtained in a controlled measurement facility are characterized by a high signal-to-noise ratio. A greater error in the estimates $\hat{R}(n,m)$ is deemed to result from the statistical uncertainty associated with the finiteness of the lagged product sum. Thus, we treat the second term of (15) containing $B_{n,m}$ as "noise" on the "measurements" of the $R(n,m)$ , which are appropriate linear constraints on the maximization of entropy (or on the minimization of cross-entropy.) Its value, which is zero for $l=1$ (a single point scatterer), depends on the phase of the scatterers and on their exact locations, all of which are unknown. In the absence of knowledge of the phase information, we assume scatterer phases to be random and independent, each uniformly distributed. Then

$$E\left[g_i g_k^*\right] = \begin{cases} 0 & \text{if } i \neq k \\ |g_i|^2 & \text{if } i = k \end{cases} \tag{16}$$

and the mean and variance of the error are

$$E\left[\hat{R}(n,m) - R(n,m)\right] = 0 \tag{17}$$

and

$$E\{|\hat{R}(n,m) - R(n,m)|^2\} = \frac{1}{(N-n)^2(M-m)^2} \sum_{i=1}^{l} \sum_{k=1}^{l} |g_i|^2 |g_k|^2 (1 - \delta_{ik}) \times$$

$$\frac{\sin^2[\pi(N-n)(x_i - x_k)]\sin^2[\pi(M-m)(y_i - y_k)]}{\sin^2[\pi(x_i - x_k)]\sin^2[\pi(y_i - y_k)]} \tag{18}$$

Each term of (18) has the familiar form of a 2-D grating lobe structure with integer main lobe separation in both the variables $(x_i - x_k)$ and $(y_i - y_k)$ . But the maximum possible scatterer separations consistent with the assumption $\Delta f_x = \Delta f_y = 1$ are $(x_i - x_k) = 1$ and $(y_i - y_k) = 1$. Thus we expect all terms of $B_{n,m}$ and of (18) to correspond to between grating lobe separations and to have small amplitude if either (N-n) or (M-m) is large. Accordingly we expect good estimates when one of the latter conditions is satisfied.

## 4. Results

Radar data were obtained from measurements on an aircraft-like model target. Both the longitudinal or roll axis and the wings of the model were in the plane defined by the synthetic aperture and radar boresight axis. Data corresponding to broadside and tail-on positions were chosen for processing.

The original data in each case was a 64 x 64 complex number array. To obtain the autocorrelation estimates in (19), a 128 x 128 array was formed by padding with zeros, transformed using a 2-D FFT, magnitude squared, and inverse transformed by a 2-D FFT.

The resulting 128 by 128 array of values of $\hat{R}(n,m)$ was then trimmed to a 65 x 65 array ($n$ and $m$ between -32 and 32) to eliminate the unreliable higher order lag estimates. In fact, because of the symmetry of the autocorrelation, only half of the latter array was introduced in the constraint function. A unique constraint

$$\frac{1}{(2N-1)M} \sum_{n=1-N}^{N-1} \sum_{m=0}^{M-1} \left( \hat{R}(n,m) - \right.$$
$$\left. \frac{1}{KL} \sum_{k=0}^{K-1} \sum_{l=0}^{L-1} A(k,l) \exp\left\{ -j2\pi \left( \frac{kn}{K} + \frac{lm}{L} \right) \right\} \right)^2 \leq C_{aim} \tag{19}$$

was constructed with $A(k,l)'s$ representing the reconstructed image pixel intensities. In (19) $K$ and $L$ determine the number of pixels, and $M$ and $N$ the number of lags for which autocorrelation estimates are used. We have used $M = N = 32$ and $K = L = 128$. $C_{aim}$ was set to $10^{-3} \hat{R}^2(0,0)$ .

The entropy

$$S = - \sum_{k=0}^{K-1} \sum_{l=0}^{L-1} A'(k,l) \log A'(k,l) \tag{20}$$

with $A'(k,l) = A(k,l) / \hat{R}(0,0)$, was maximized, subject to the constraint (19), by performing an iterative search using the Cambridge algorithm.

Figures 2 and 4 are the resultant maximum entropy images displayed as maps of equal-intensity contours. The dynamic range of the contours is indicated in the upper right of each figure. A plan view outline of the model target, drawn to scale, is superposed on each plot. Figures 3 and 5, respectively, are corresponding images obtained by conventional processing, i.e. by taking the squared magnitude of the inverse discrete Fourier transform of the cosine-square weighted data array. The comparisons are striking with the maximum entropy images showing improved resolution and freedom from artifacts at levels well below those of conventional image sidelobes. Although a flat a priori image was used in these reconstructions, it is clear that by minimizing a cross-entropy function in lieu of maximizing (20), the method provides a ready vehicle for utilizing a priori image estimates in the reconstruction process.

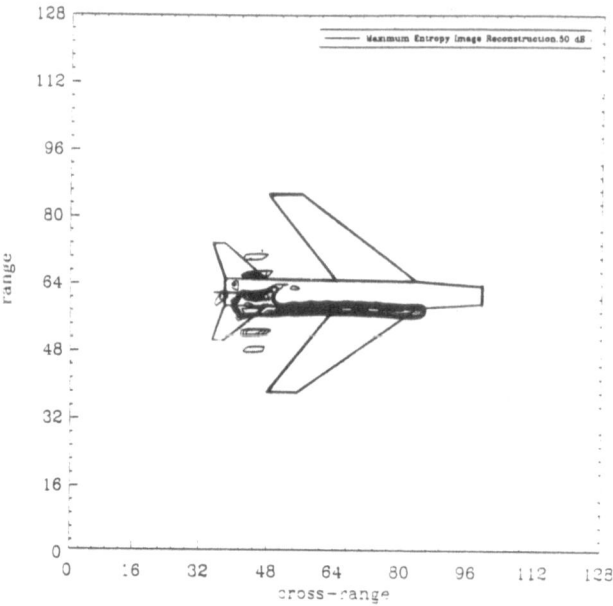

Figure 2.  Maximum Entropy Image of Aircraft Model at Broadside Aspect

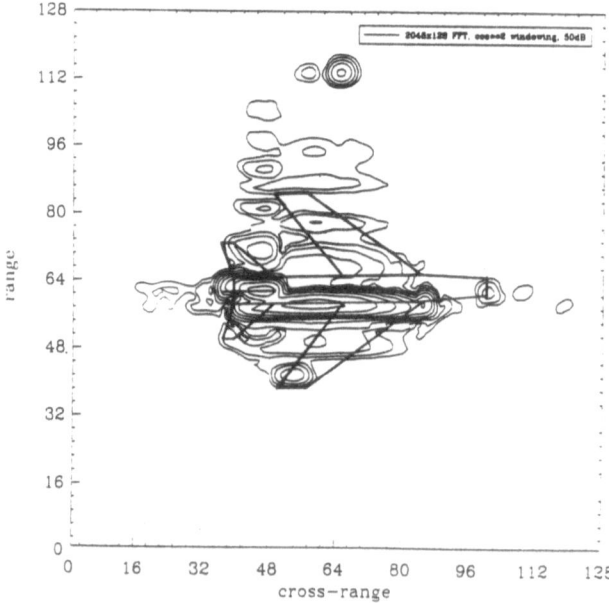

Figure 3.  Image of Aircraft Model at Broadside Aspect (Conventional
FFT Processing with Cosine Squared Weighting)

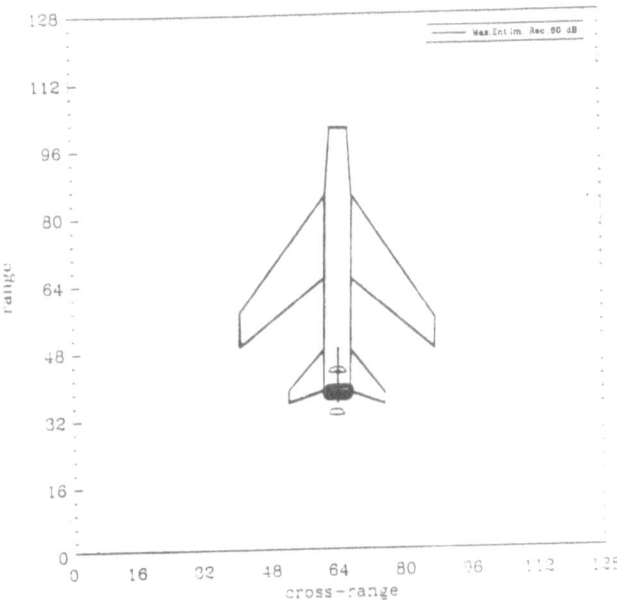

Figure 4.  Maximum Entropy Image of Aircraft Model at Tail-on Aspect

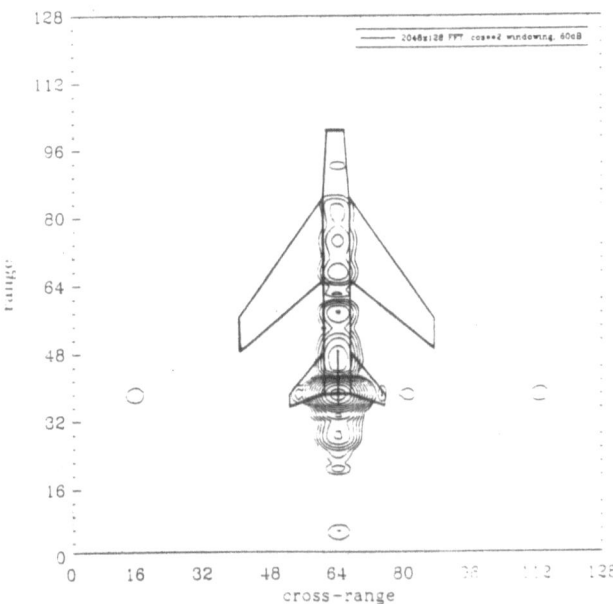

Figure 5.  Image of Aircraft Model at Tail-on Aspect (Conventional FFT
Processing with Cosine Squared Weighting)

## References

1.   Skilling, John and R. K. Bryan, "Maximum Entropy Image Reconstruction: General Algorithm, " <u>Mon. Not. R. Astr. Soc.</u>, (1984) 211, 111-124.  See also <u>Maximum Entropy and Bayesian Methods in Inverse Problems</u>, eds. C. Ray Smith and W.T. Grandy Jr., Reidel 1985, pp. 83-132.

2.   van Roekeghem, Frederic and G. Heidbreder, "Maximum Entropy Reconstruction of Radar Images of Rotating Targets," <u>Proc. IASTED Int. Symp. on Signal Processing and its Applications</u> (Brisbane, Australia, 1987,) pp. 486-490.

3.   Mensa, Dean, <u>High Resolution Radar Imaging</u>, Artech House, 1981.

FIER:  A Filtered Entropy Approach
to Maxium Entropy Image Restoration

R. A. Gonsalves, Tufts University
J. P. Kennealy and R. M. Korte, Mission Research Corp.
S. D. Price, A. F. Geophysics Laboratory

ABSTRACT

Object estimation by Maximum Entropy techniques are often slow-to-converge, iterative procedures which result in smooth, non-noisy estimates.  FIER is a new, robust, iterative procedure which employs a Wiener filter to generate an initial estimate and refines the estimate with a filtered entropy term.  It uses any point response function and converges in a controlled number of iterations.  It has been used extensively for estimation of astronomical objects measured in the 1983 IRAS (Infrared Astronomical Satellite) experiment.

1. Introduction

The Maximum Entropy (ME) method was introduced by Jaynes [1] to estimate probability functions and has been used by many authors, including Burg [2], Frieden [3] and Burch, Gull and Skilling [4], in a variety of applications.

The method balances two conflicting requirements: to maximize the entropy of an (estimated) object and to minimize the mean square error between the observed data and reconstructed data based on the object. The math leads to a nonlinear, implicit equation which the object must satisfy and the ME methods solve this equation, usually by iteration.

Most methods start with a flat featureless object, the unconstrained object with maximum entropy, and modify this estimate to acheive the desired balance between entropy and error.  In a series of papers, the authors and others [5,6,7] have used the Wiener filtered image as the starting estimate, an estimate which is modified iteratively to achieve a desired error while, at each iteration, the entropy is increased.  The search direction is a filtered entropy gradient; so the approach is called FIER, for Filtered Entropy Restoration.

P. F. Fougère (ed.), Maximum Entropy and Bayesian Methods, 369–382.

   The approach is shown graphically in Figure 1. The conventional
ME techniques start with a flat image, which implies large entropy, e,
and large mean square error, m. Iterations reduce the entropy and
error until the process converges on an object with the desired m,
namely $m_o$. The FIER technique starts with a Wiener filtered image,
which implies small m and small entropy. The iterations converge on
the same object but from the opposite direction.

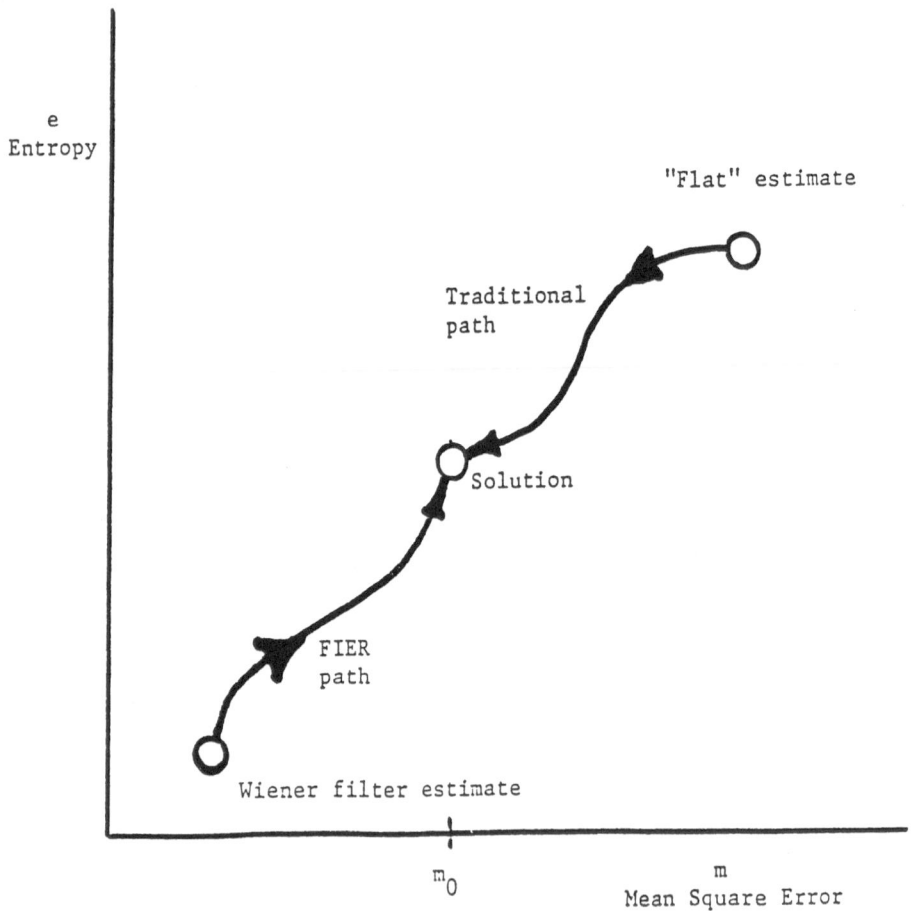

Figure 1.   Entropy-MSE path of Maximum Entropy Algorithms

## 2. Problem Statement

We adopt a vector/matrix notation which applies equally well to one or to multi-dimensional signals. Our model states that the object is distorted by a response matrix and that noise is added to form the observed data. Thus

$$D = H O + N , \qquad (1)$$

where the elements of (1) have the following meanings and dimensions:

D = data vector = K by 1

H = response matrix = K by J

O = object vector = J by 1

N = noise vector = K by 1.

Our problem is to estimate O based on D, with knowledge of H. Our estimate is F, a J by 1 vector.

The entropy of F is defined as

$$e = - \sum_{j=1}^{J} ( f(j) / c ) \ln( f(j) / c ) , \qquad (2)$$

where c is the area of F,

$$c = \sum_{j=1}^{J} f(j) . \qquad (3)$$

The mean square error, m, is found by constructing an estimate, G, of the data:

$$G = H F . \qquad (4)$$

Then m is

$$m = \sum_{k=1}^{K} ( d(k) - g(k) )^2 = |D - H F|^2 . \qquad (5)$$

The problem is to find an estimate, F, which maximizes the entropy, e, in (2) while constraining the mean square error, m, in (5) to be some fixed $m_o$ .

## 3. Gradients

To find a solution vector F we set up a metric q ,

$$q = e - a m , \qquad (6)$$

where a is a scalar weighting. Then F must have elements f(j) which satisfy

$$\frac{\partial q}{\partial f(j)} = 0 \qquad (7)$$

or, from (6)

$$\frac{\partial e}{\partial f(j)} = a \frac{\partial m}{\partial f(j)} , \text{ for all } j . \qquad (8)$$

In vector form, the entropy gradient, V, must be proportional to the mean square error gradient, W. Thus, (8) becomes

$$V = a W . \qquad (9)$$

To find the elements of V we differentiate e in (2) with respect to f(j). This yields, after considerable algebra,

$$v(j) \equiv \frac{\partial e}{\partial f(j)} = - (e + \ln ( f(j)/c))/c . \qquad (10)$$

The elements of W are found by differentiating m in (5) with respect to f(j). This yields

$$w(j) = \frac{\partial m}{\partial f(j)} = -2 \sum_{k=1}^{K} ( d(k) - g(k)) h(k,j) , \qquad (11)$$

or, for W, itself,

$$W = -2 H^T ( D - G )$$

$$= -2 H^T ( D - H F ) . \qquad (12)$$

When we put (12) into (9) we get

$$V = -2 \, a \, H^T \, ( \, D - H \, F \, ) \, , \qquad (13)$$

where the elements of V are given by (10). Equation (13) is an implicit equation for F, implicit since F appears on the right and ln f(j) appears on the left.

## 4. FIER Approach

Holding V fixed, we begin to solve (13) for F:

$$( \, H^T \, H \, ) \, F = H^T \, D + V \, / \, (2a) \, . \qquad (14)$$

If

$$R = H^T \, H \qquad (15)$$

has an inverse, we can solve (14) for F:

$$F = R^{-1} \, H^T \, D + R^{-1} \, V \, / \, (2a) \, . \qquad (16)$$

We define a new scalar multiplier

$$s \equiv 1 \, / \, (2a) \qquad (17)$$

and a vector B

$$B \equiv R^{-1} \, V \, . \qquad (18)$$

Thus, (16) becomes

$$F = R^{-1} \, H^T \, D + s \, B \, , \qquad (19)$$

our solution for F.

If R is not full rank, it will not invert. We guard against this by adding a small positive constant, Z to the diagonal of R. Thus

$$R \equiv H^T \, H + Z \, I \, , \qquad (20)$$

where I is the identity matrix. Equations (16) through (19) remain unchanged. Z is a noise control parameter which we discuss later.

The first term in (19) is a generalized Wiener filter. This becomes evident if we assume that the distortion matrix H is that of a linear, isoplanatic, finite impulse response filter. In this case the Wiener filter estimate for f(j) has Fourier Transform F(m) given by

$$F(m) = \frac{H^*(m) \ D(m)}{H^*(m) \ H(m) + Z} \tag{21}$$

This compares to the first term in (19)

$$R^{-1} H^T D = ( H^T H + Z I )^{-1} H^T D .$$

It also shows how one could calculate the first term if the distortion is, indeed, isoplanatic.

The second term in (19) has two parts: s, a scalar, and B, a vector. The vector B, given by (18), is the entropy gradient V, filtered by the inverse of R. Corresponding to (21), B(m) is

$$B(m) = \frac{V(m)}{H^*(m) \ H(m) + Z} . \tag{22}$$

Thus B is a filtered entropy gradient.

Since f(j) appears on both sides of (19), it is also an implicit equation for our solution vector F. To solve (19) for F we use an iterative approach. The $n^{th}$ iterate is

$$F^{(n)} = F^{(n-1)} + s B , \tag{23}$$

where

$$F^{(0)} = R^{-1} H^T D . \tag{24}$$

Thus we start with a Wiener filter estimate, calculate the filtered entropy gradient, B, for that estimate, and step a distance s in direction B get $F^{(1)}$. The iterations continue until the process converges on a solution vector F.

A block diagram of the algorithm is shown in Figure 2 and some details are given in the next section. The data elements d(j) are passed through a Wiener filter to get the initial estimate with elements $f^{(0)}(j)$. The $n^{th}$ iterate has elements $f^{(n)}(j)$, which follows from Equation (23). To get the direction elements b(j) we take the log of $f^{(n-1)}(j)$ as in Equation (10) and filter the result as in Equation (22). The step size, s, is calculated based on the desired MSE, m, as we explain next.

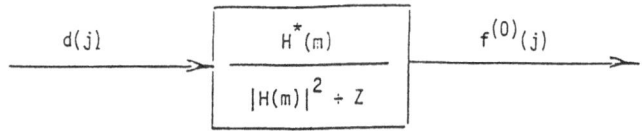

$$f^{(n)}(j) = f^{(n-1)}(j) + s\, b(j)$$

$$s = \text{step size}$$

$$b(j) = \text{filtered entropy gradient}$$

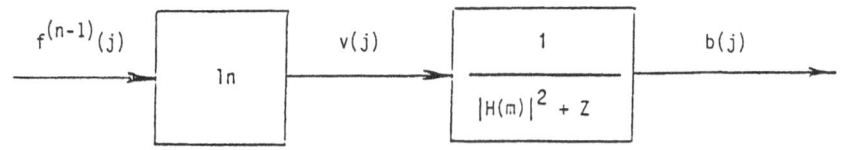

Figure 2    FIER Algorithm

5. Details of the Algorithm

We discuss the starting point, the step size, and control of the number of iterations.

The starting point is determined by $m_0$, the final mean square error that is to be acheived by the solution vector F. Refer to Figure 1. We want to start with m less than $m_0$, typically $m = m_0/2$. We can adjust the noise control parameter, Z, to achieve the desired level of m. In our algorithm we select Z to be equal to the largest diagonal element of $H^T H$, calculate m, and keep halving Z until m is just less than $m_0/2$.

Now we want to work our way back up an e-m curve to achieve the specified level of mean square error, $m_0$. But at each iteration, from (5),

$$m = \left| D - H F^{(n)} \right|^2 . \qquad (25)$$

where $F^{(n)}$ is given by (23). On the right side of (25) D and H are known; and, from (23) only s, the step size, is unknown. Thus, for a specified m we can solve (25), a quadratic equation for s, always choosing the solution which increases entropy. This is how s is chosen.

Finally, we discuss the number of iterations. This is entirely under our control. At each iteration we increase m from an initial $m \approx m_0/2$ to the final m. Thus if we have "time" for 10 iterations we can allow m to increase linearly from $m_0/2$ to m in 10 steps.

6. Examples

We show both one and two-dimensional examples. In Figure 3 we show a 15-sample object which has been distorted by a 3-sample, rectangular, isoplanatic response function and by additive noise with standard deriviation of 5.7. Since the peak value of the noiseless observed data is 109, the (peak) signal-to-noise ratio is about 20. The restoration, which used 28 iterations of the FIER algorithm, is also shown.

The negative values in our estimation point out a distinct departure of the FIER approach from conventional approaches to maximum entropy restoration. The FIER algorithm does not require a solution with only positive elements. This raises the problem of defining entropy for such signals. Our solution is to re-define the estimate at each iteration such that it stays positive. This requires addition of a bias, but only for purposes of entropy calculations. Also, at each iteration, we reduce the bias, as allowed, so that, at the final iteration the bias is nearly zero. The final estimate, itself, has no bias in it so it has negative values.

In Figure 4 we show the final gradients for entropy and for mean square error. As required, they are proportional.

Our second example is shown in Figure 5. The object, data, Wiener filter estimate, entropy gradient term and several iterates are shown. The final estimate used three iterations.

Our final example is shown in Figure 6. This is data from the IRAS experiment. The input data is shown in (a) and the restored image is shown in (b). Details of our use of FIER for IRAS data appear in Reference [7].

7. Summary

We have presented details and examples of the FIER approach to maximum entropy restoration of images from distorted, noisy data. The approach starts with a Wiener filter estimate and, iteratively, uses a filtered entropy gradient to increase the entropy while also increasing the mean square error. At each stage of the iteration we calculate a step size which will allow the mean square error to be increased in a controlled fashion.

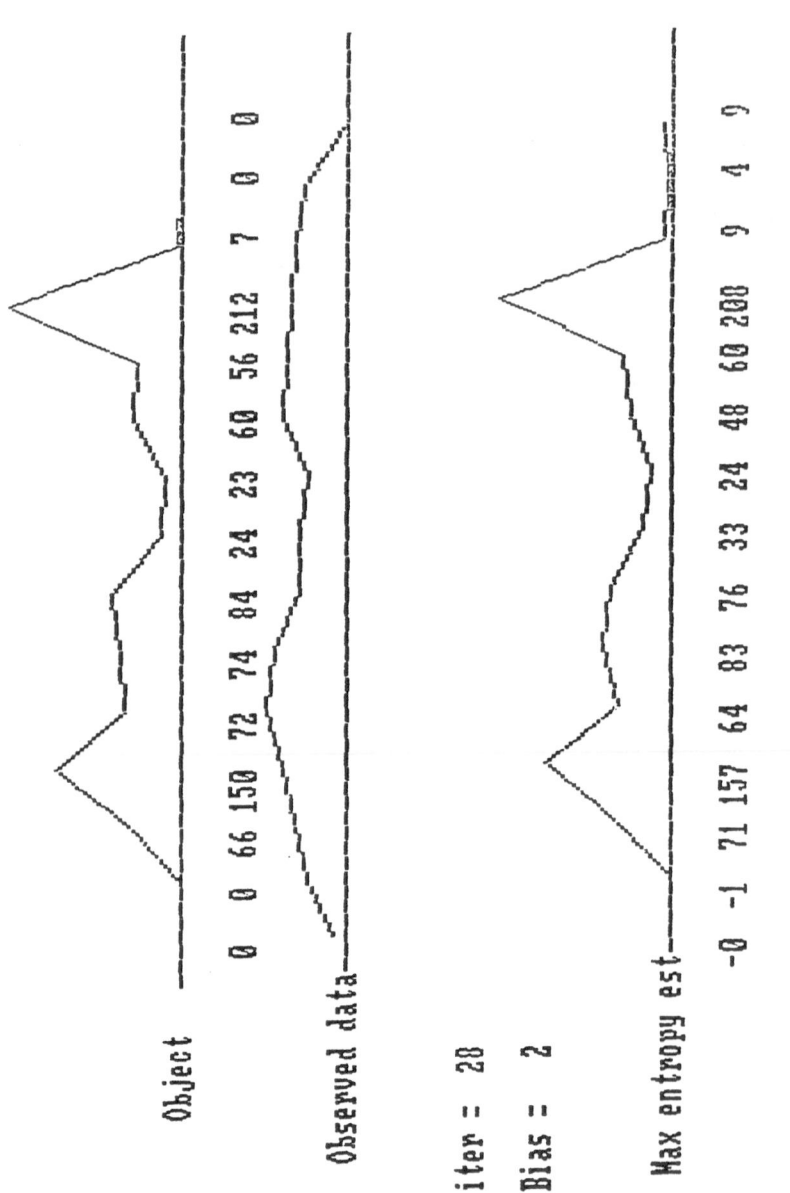

Figure 3. FIER ALGORITHM FOR A 1-D EXAMPLE.

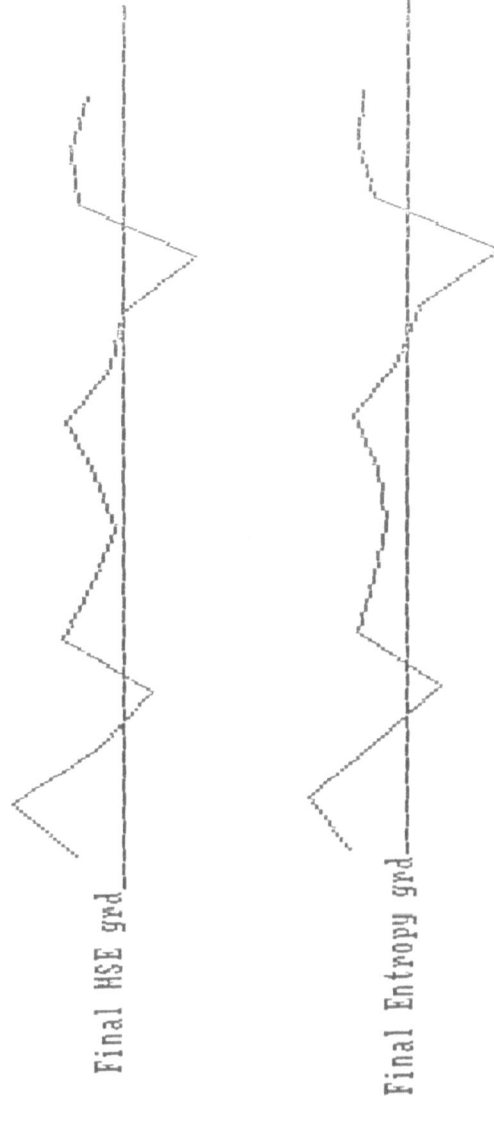

Figure 4. MSE AND ENTROPY GRADIENTS FOR THE SOLUTION VECTOR IN Figure 3.

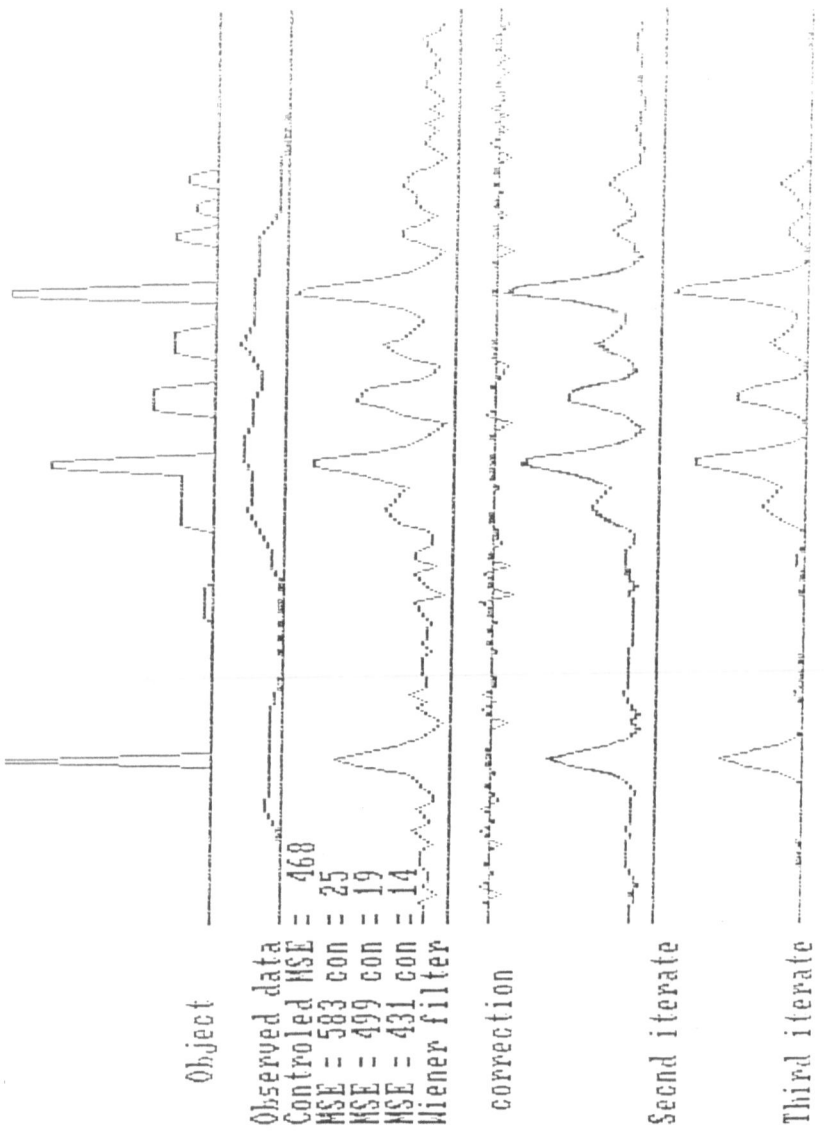

\imgs\NGC-2992.crd  NL10

\imgs\NGC-2992.rst  NL10

Figure 6. 2-D EXAMPLE OF FIER, APPLIED TO REGRIDDED IRAS DATA (LEFT).

## References

1. E. T. Jaynes, "Prior probability," IEEE Trans. SSC-4, 227-241 (1968).

2. J. P. Burg, "Maximum entropy spectral analysis," 37th Meeting, Soc. of Exploration Geophysicists (Oklahoma City, Okla., 1969).

3. B. R. Frieden, "Restoring with maximum likelihood and maximum entropy," J. Opt. Soc. Am. 62(4), 511-518 (1972)

4. S. F. Burch, S. R. Gull, and J. Skilling, "Image restoration by a powerful maximum entropy method," Comp. Vision, Graph., Image Proc 23, 113-128 (1983).

5. R. A. Gonsalves, H. Kao, "Entropy-based algorithm for reducing artifacts in image restoration," Optical Engineering, Vol. 26 No. 7, July 1987.

6. R. A. Gonsalves, R. Korte, J. P. Kennealy, "Entropy-based image restoration: modifications and additional results", Proc. SPIE Vol. 829, August 1987.

7. J. P. Kennealy, R. M. Korte, R. A. Gonsalves, T. D. Lyons, S. D. Price, P. D. LeVan, H. H. Aumann, "Infrared astronomical satellite (IRAS) image reconstruction and restoration", SPIE, Vol. 804, March 1987.

# REGULARISATION IN CODED-APERTURE SPECT

M. Barth, J. L. Denny, T. A. Gooley
*University of Arizona*
*Tucson, Arizona 85721*

ABSTRACT. In pinhole-coded-aperture medical imaging the linear system of equations to be inverted for an object reconstruction is often vastly underdetermined (up to 90%, say). Our interest is in a comparison of object reconstructions based on maximum a posteriori (MAP) estimates where the prior probability densities are members of exponential families. Preliminary results indicate that, for standard diagnostic tasks, a maximum entropy regulariser performs comparably to a positivity-norm regulariser.

## 1. Introduction

Emission tomography is used to determine the distribution of a radioactively labelled pharmaceutical in a part of the body such as the brain, heart or liver. We monitor physiological function in contrast to the morphological information obtained in X-ray tomography. In single-photon emission computed tomography (SPECT), as contrasted to positron emission tomography (PET), a single $\gamma$-ray is emitted in each radioactive decay. The mathematical descriptions of X-ray computed tomography (CT) and conventional SPECT methods are very similar if we measure the distribution of only those gamma rays whose trajectories are contained within a specified flat slice. The essential difference is that with X-rays we determine the cross-sectional distribution of linear attenuation coefficients, whereas with $\gamma$-rays we measure the distribution of radioactivity.

In conventional SPECT reconstruction algorithms the data are usually not sparse, and standard CT algorithms, like filtered backprojection, usually yield reasonable results. An intrinsic disadvantage of conventional SPECT systems is a parallel-hole collimator that restricts $\gamma$-rays reaching the detector plane to within a narrow angular range. The counting efficiency of such systems is usually low ($\Omega/4\pi \simeq 10^{-4}$).

An increase of detection efficiency may be obtained if the angular selectivity of the apertures between detector plane and object is relaxed. Systems that are currently under investigation consist of a number of pinholes in an aperture plane. With increasing density of pinholes the shadow images cast on the detector plane overlap more, which leads to an increasing information loss because of the uncertainty in the trajectories of a detected photon. The determination of an optimum arrangement of pinholes is a nontrivial task (e.g. Smith & Barrett (1988)).

The intention of coded-aperture SPECT systems is to increase the signal-to-noise ratio or, alternatively, reduce the exposure times. With no motion of the detectors or apertures, however, coded-aperture SPECT data are usually extremely sparse. A typical experimental setup is depicted in Fig. 1. One tries here to reconstruct an object defined on 39x39

383

*P. F. Fougère (ed.), Maximum Entropy and Bayesian Methods*, 383–389.

sampling points, say, using 2x64 data points.

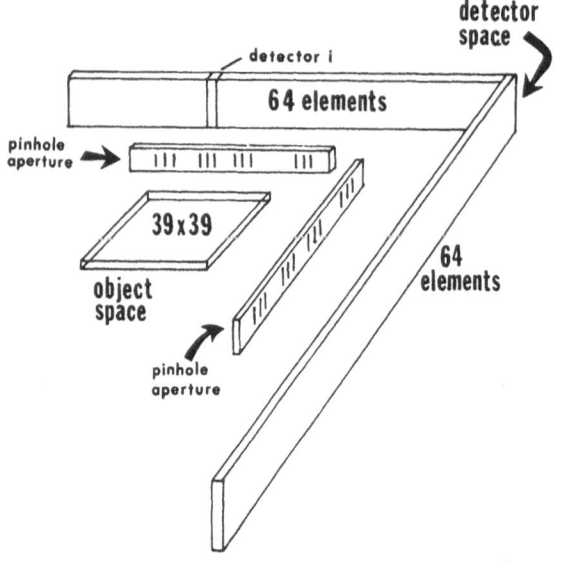

Fig. 1.
A schematic of the two-dimensional multiple-pinhole coded-aperture imaging system simulated for this study.

## 2. Description

For discretized object and data spaces, the imaging process may be described by a linear matrix equation,

$$g = Hf + n,$$

where $g$ is the data vector, $H$ is the mapping matrix, $f$ is the unknown object and $n$ describes noise or other errors in the data. If the mapping operator is modelled correctly, i.e., sampling effects and scattering contributions, etc. are properly incorporated, the only source of noise is the statistics of the emission process. It is instructive to consider the characteristics of a non-regularised solution of an ill-posed linear problem when the error term $n$ is represented by a Gaussian noise process. Because of the stochastic nature of the data, the reconstructed object, $\hat{f}$, is a random variable. We obtain the likelihood function

$$P(g|f) = \left[ \prod_i 2\pi(Hf)_i \right]^{-\frac{1}{2}} \exp( - (Hf - g))^t C^{-1}(Hf - g) ),$$

where $C = 2\mathrm{diag}\{(Hf)\}$ and t denotes a transpose operation. The maximum likelihood solution then is equivalent to the least squares solution,

$$\hat{f} = (H^t C^{-1} H)^{-1}(HC^{-1})^t g.$$

If $H^{-1}$ exists, we obtain for the expected value of the $L_2$ norm of $\hat{f}$,

$$E\{||\hat{f}||^2\} = ||f||^2 + tr[(H^tH)^{-1}C],$$

where tr denotes the trace of a matrix. The variance of the estimate is, if $H^{-1}$ exists,

$$\begin{aligned}Var(\hat{f}) &= E\{(\hat{f} - f)(\hat{f} - f)^t\} \\ &= ff^t + H^{-1}C(H^t)^{-1}.\end{aligned}$$

If $H^{-1}$ does not exist, we obtain similar formulas where $H^{-1}$ is replaced by a pseudoinverse, $H^+$. In medical imaging, the underlying noise process is often Poisson. For large counts, and particularly, for low-contrast objects, the Poisson distribution may be well approximated by a Gaussian distribution. In this approximation, a maximum likelihood solution with Poisson-distributed data is then expected to have similar characteristics to a solution based on Gaussian-distributed data. It is conspicuous that in the presence of small eigenvalues of $H^tH$, the length of the reconstructed object vector tends to be very much larger than the length of the original object vector. This property is responsible for the unsatisfactory quality of a maximum likelihood solution in many clinical applications. We realize further that the variance of the solution is poor if H is ill-conditioned. An appreciation of these features gave rise to early attempts of regularisation within the concept of ridge regression (Hoerl & Kennard (1970)).

Within a Bayesian framework for regularisation, an a posteriori probability density,

$$P(f|g) = \frac{P(f)}{P(g)} P(g|f),$$

is maximised. The usually controversial step is the specification of the prior probability density, $P(f)$. The formalism takes a simple form if $P(f)$ is a member of an exponential family, $P(f) = exp(\alpha S(f))$, where $S(f)$ is some smoothing function that does not depend on $\alpha$.

If one assumes no knowledge about the dependencies between neighbouring pixels in an object, certain arguments lead to a form (Skilling (1988), (1989)),

$$S_E(f) = \sum_i (f_i - m_i - f_i ln(f_i/m_i)),$$

where i denotes a pixel index and $m_i$ is a "model" to which one would like the reconstruction to be drawn in the absence of any data.

Irrespective of the various rationales leading to an entropy functional, we may be interested in the general characteristics of a reconstructed object. The entropy functional may be considered as a particular case of a Markov random field (MRF), where the potential term contains only self-interactions. In general, reconstructions based on an MRF cannot be given in closed form. A variational principle usually leads to intractable differential equations. If we admit no cross terms in the interaction potential of the regulariser, we may formulate a general variational principle which allows a semi-quantitative analysis of the properties of Maxent images. In a tomography problem, if noise is disregarded, the maximum entropy solution is a nonlinear function of the backprojection of a set of Lagrange multipliers. Nityananda & Narayan (1982) and Narayan & Nityananda (1986) derive from this observation a list of properties for maximum entropy reconstructions that need not always be propitious for some objects under consideration. It is therefore reasonable to look for more general approaches.

Often it appears reasonable to build into P(f) some dependencies between neighbouring pixels of an object. A useful smoothing function may be

$$S_G(f) = -\left\{ \sum_{[s,t]} \frac{1}{1 + \left[\frac{f_s - f_t}{\delta}\right]^2} + \frac{1}{\sqrt{2}} \sum_{\langle s,t \rangle} \frac{1}{1 + \left[\frac{f_s - f_t}{\delta}\right]^2} \right\},$$

where [s,t] indicates that s and t are nearest horizontal or vertical neighbours, and ⟨s,t⟩ indicates that s and t are nearest diagonal neighbours. $S_G(f)$ is minimised by configurations of constant intensity, therefore favouring smoothly varying reconstructions. The form of a Lorentzian is advantageous over $S(f) = \Sigma(f_s - f_t)^2$, e.g., since it penalizes large differences in $f_s - f_t$ merely moderately, thus allowing for the occurrence of edges (Geman & McClure (1985)).

In this study we investigated the characteristics of the three above suggested approaches. We consider the results preliminary, since the selection of the regularisation parameter was performed subjectively.

## 3. Experimental results

If a radiopharmaceutical is injected into the blood stream, its distribution allows an assessment of possible anomalies in e.g., the heart. If a heart suffers an infarction, the heart muscle often develops a local atrophy. In a contracting, or systolic stage it is expected that the dead heart tissue impedes locally the movement of the heart muscle. The reconstructed blood volume should therefore exhibit a protrusion at the site of a previous heart stroke.

Fig. 2 shows a simulated left ventricle featuring a protrusion at the top left part of a section through the volume. The homogeneous distribution of the radiopharmaceutical and its high contrast suggest a binary test object. We obtain for our object a total count of 45217 photons.

Fig. 3 is a positively restricted maximum-likelihood solution.

Fig. 4 is a constrained solution where the regularising functional was $S(f) = ||f/\Theta(f)||^2$ ($\Theta$ is the step function). The regularising parameter was subjectively adjusted such that the protrusion was well discernible in the reconstruction. We obtained in this case

$$\chi^2 = \sum_i ((Hf)_i - g_i)^2/(Hf)_i \simeq 20.$$ A reconstruction with $\chi^2 \simeq 128$, in accordance with an

early suggestion for the parameter choice (Skilling et al. (1979)), tended to oversmooth the reconstruction and to obscure the protrusion (Gooley et al. (1989)). For a criticism of the $\chi^2 = N$ rule see Gull (1989).

Fig. 5 is a maximum entropy reconstruction with $S_E(f) = \Sigma (f_i - m_i - f_i \ln(f_i/m_i))$. Since our test object consisted of a single connected binary object, the optimum $m_i$ is a model that has a constant value over some area and drops down to a low value outside of the support of the function describing our object. A simulation where the boundaries of the object is assumed to be known exactly may be considered unrealistic.

We created a series of 64 left ventricles, 32 showing protrusions at random positions, 32 corresponding to healthy hearts without protrusions. To create this set of images we sampled the perimeter of our original model as a function of angle. This function was expanded in a Fourier series and its components were randomly perturbed, giving rise to a set of reasonably randomized perimeters. We then created a mean object of this ensemble and identified m with its support. The value of $m_i$ (=constant) within the support of the mean object is of no particular importance as long as $m_k \ll m_l$ where 1 denotes pixels within the mean object's support and k denotes pixels within its complement. The reconstruction in Fig. 5 was created with $m_i \in (10^{-8}, 10^{-1})$ and is visually indistinguishable from a reconstruction for which $m_i \in (10^{-8}, 5)$. We note further that a reconstruction with a flat model, i.e., $m_i$ = constant for all pixels i, was virtually indistinguishable from the reconstruction given in Fig. 4.

The reconstruction in Fig. 5 appears somewhat ragged, much like a norm-regularised reconstruction. Both types of reconstruction, however, give clear evidence of the important feature in the object, viz. the protrusion. The choice of a support-type model may be compared to a finite-extent object constraint. In this application it yields only a mild improvement over a constant model since reconstructions with $m_i$ = constant feature values close to zero over extended regions outside the mean object's support.

In Fig. 6 we show the influence of the prior $\exp(\alpha S_G(f))$. Because of the nonconvexity of $S_G(f)$, a typical approach for its optimisation is the exploitation of a Monte Carlo method (Geman & Geman (1984)). Its intricacies are currently being investigated in our laboratory. Until the development of a robust algorithm, we take a resonable regularised maximum-likelihood solution, obtained e.g., by Maxent, as a starting configuration and search for the nearest local maximum in the energy surface of $S_G(f) + \log P(g|f)$ (Geman & McClure (1985)).

The parameters $\alpha$ and $\delta$ were determined empirically. In contrast to the results of Geman & McClure (1985), who used this prior for parallel-hole-collimator SPECT, our reconstructions are sensitive to the choice of the parameter $\delta$. We note, however, that the overall character of a solution using a $S_G(f)$ regularising functional with a reasonable starting point is smoother than a Maxent reconstruction, say. Yet some features may give rise to the detection of apparent protrusions that are not actually present.

With respect to a typical detection task (is a protrusion present or not?), the three approaches compare similarly in this example. A more complete assessment may be obtained by the study of receiver operating characteristic (ROC) curves on a series of objects (Metz (1986)). The choice of the regularisation parameter (Sibisi (1989), Gull (1989), Geman & McClure (1985)) and a reliable realization of a solution close to the global maximum of $S_G(f) + \log P(g|f)$ remain an active field of research.

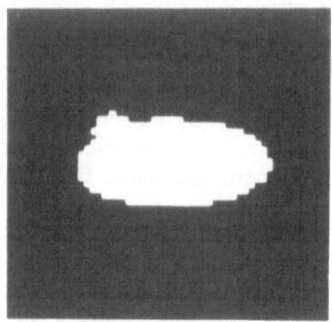

Fig. 2.
Original object to be reconstructed.

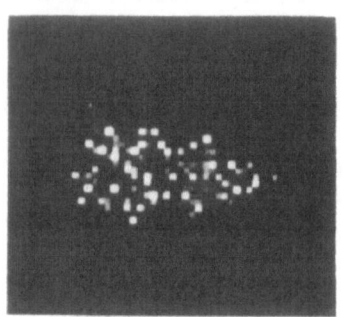

Fig. 3.
A maximum-likelihood reconstruction
with a positivity constraint.

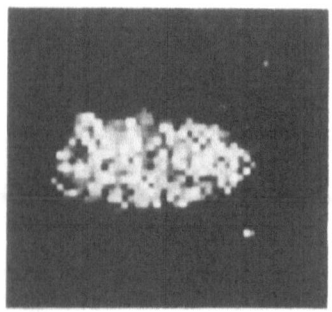

Fig. 4.
Reconstruction with a positivity-norm
regulariser, $S(f)=||f/\Theta(f)||^2$.

Fig. 5.
Reconstruction with an entropy
regulariser, $S_E(f)=\Sigma(f_i-m_i-f_i\ln(f_i/m_i))$.

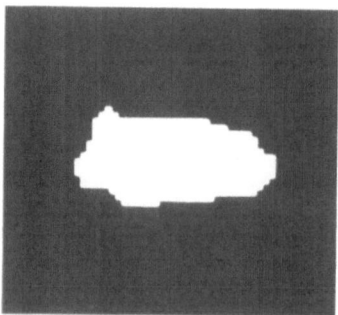

Fig. 6.
MAP          reconstruction          using
$Z^{-1}\exp(\alpha S_G(f))$ as a prior distribution,
where Z is a normalising constant.

## References

S. Geman and D. Geman (1984), IEEE PAMI-6, 721-741, "Stochastic relaxation: Gibbs distributions and the Bayesian restoration of images."

S. Geman and D. E. McClure (1985), Proc. Amer. Stat. Assoc., Statistical Computing Section, 12-18, "Bayesian image analysis: An application to single photon emission tomography."

T. A. Gooley, H. H. Barrett, M. Barth, J. L. Denny (1989), talk given at the OSA Topical Meeting on Signal Recovery and Synthesis, Cape Cod.

S. F. Gull (1989), "Developments in maximum entropy data analysis," In: Maximum Entropy and Bayesian Methods, ed. J. Skilling, pp. 389-396, Kluwer.

A. E. Hoerl and R. W. Kennard (1970), Tecnometrics 12, 55-67, "Ridge regression: Biased estimation for nonorthogonal problems."

C. F. Metz (1986), Invest. Rad. 21, 720-733, "ROC methodology in radiologic imaging."

R. Nityananda and R. Narayan (1982), J. Astrophys. Astr. 3, 419-450 (1982), "Maximum entropy image restoration - a practical non-information theoretic approach."

R. Narayan and R. Nityananda (1986), Ann. Rev. Astron. Astrophys. 24, 127-170, "Maximum entropy image restoration in astronomy."

S. Sibisi (1989), "Regularization and inverse problems," In: Maximum Entropy and Bayesian Methods, ed. J. Skilling, pp. 389-396, Kluwer.

J. Skilling, A. W. Strong, K. Bennett (1979), Mon. Not. R. astr. Soc. 187, 145-152, "Maximum-entropy image processing in gamma-ray astronomy."

J. Skilling (1988), "The axioms of maximum entropy," In: Maximum Entropy and Bayesian Methods in Science and Engineering, Vol. 1, ed. G.J. Erickson & C.R. Smith, p.p. 173-188, Kluwer.

J. Skilling (1989), "Classic maximum entropy," In: Maximum Entropy and Bayesian Methods, ed. J. Skilling, p.p. 45-52, Kluwer.

W. E. Smith and H. H. Barrett (1988), J. Opt. Soc. Am. A 5, 315-330, "Linear estimation theory applied to the evaluation of a priori information and system optimization in coded-aperture imaging."

BURG ALGORITHM APPLIED TO FOURIER TRANSFORM ION CYCLOTRON
RESONANCE MASS SPECTROMETRY

A. RAHBEE
Geophysics Laboratory/OPI
Hanscom AFB, MA  01731

ABSTRACT.  The method known as Burg's maximum entropy algorithm has been
applied to the time domain signals measured and recorded in a mass spectrometer known as Fourier transform ion cyclotron resonance (FTICR) mass
spectrometer.  Ordinarily  several thousands of data points are obtained
and analyzed by application of a fast Fourier transform algorithm to obtain the frequency domain spectra which are then converted to mass spectra.  In this paper we describe some results that we have obtained by
adapting Burg's algorithm to our FTICR, and will show that by employing
ahandful of data points, mass spectra can be gotten that are superior to
those obtained by the FFT techniques.

1.  INTRODUCTION

For the purpose of elucidating the subject matter of this paper a brief
description of the apparatus is necessary.  Prior to the advent of FTICR
mass spectrometers used in many laboratories had one quality in common.
Usually a number of parameters are  scanned to bring the ions of different charge-to-mass ratio onto the detector sequentially in time. Some of
these parameters are typically either electric or magnetic field or a
combination of the two parameters.  These are the well known scanning
type mass spectrometers.  In another type of instrument known as ion
cyclotron resonance (ICR) the ions are elevated into their cyclotron
orbits in the presence of a magnetic field and the power absorbed by the
ions is measured for the determination of their charge-to-mass ratios.
This brief  description, although not doing justice to the subject
matter, nevertheless is sufficient to lead us naturally into the next
stage of development that is the subject matter of this paper.  FTICR,
quite naturally developed as an extension of the ICR and the credit for
its development is due to the pioneering work of Comisarow and Marshall
in 1974[1].  Of the many unique features of the new instrument is its
ability to measure all the ion currents simultaneously.  Features such
as selective rejection of unwanted ionic species and availability of a
negative ion mode are also among the attractive features of the instr-
ments.  Investigators in many laboratories are currently engaged in

391

*P. F. Fougère (ed.), Maximum Entropy and Bayesian Methods*, 391–401.
© 1990 *Kluwer Academic Publishers.*

further refining the method and the instrument itself is also available
commercially from several manufacturers.

## 2. APPARATUS

Consider a positive ion of mass m and charge e in magnetic field of strength B and orbital velocity v in orbit of radius r. The angular frequency of the ion is w=v/r, and if an incremental change is denoted by
dw (corresponding to a change in m of dm) then following relationships
can be shown to hold

$$mw = Be \tag{1}$$

$$dw = -(Be/m^2)dm \tag{2}$$

$$(m/dm)_{1/2} = \text{const.}(BeT/m) \tag{3}$$

where equation 3 denotes the resolution at half maximum and T is observation time given by the number of data points times the sampling period. The method chosen by Comisarow and Marshall was to elevate the ions
formed in a cubical or rectangular cell(with six electrically insulated
sides) into their cyclotron orbits as in the old ICR.  This was done by
a fast rf chirp lasting under one millisecond and applied to two of the
parallel plates of the cell.  This rf pulse contained all the necessary
frequencies corresponding to the ion charge to mass ratios.  Rather than
measuring the power absorbed by the ions on the way to the cyclotron
orbits as in the older ICR, these investigators measured the image current induced by the circulating ions in the other set of parallel plates
of the cell after cyclotron orbits were reached.  This image current is
of course a sinusoid in case of one species and a composite of harmonic
currents in case of more than one species with frequencies that are related to ionic masses, as equation 2 above shows. This time domain signal is then subjected to an fast Fourier transform algorithm to get the
spectrum in frequency domain which is then  subsequently converted to a
mass spectrum.  In this fashion FTICR mass spectrometry becomes a matter of frequency measurement.  Figure 1 schematically depicts the principles involved in the FTICR apparatus.  The top panel of Figure 1
shows the ions in their cyclotron orbit inducing an image current by
setting up a field that sets electrons in motion with the same frequency
as the cyclotron frequency of the ions. The middle panel is a graphic
representation of an experiment in FTICR.   First the space within the
one  inch cubic cell is cleared of all charges by application of a
potential difference on the end plates.  Next an electron beam with
controlled energy enters the cell in the middle of one end plate along
the magnetic field, and exits in the opposite side onto  a collector
plate where the electron current is monitored.  Ions produced by
electron impact within the cell are trapped there by the magnetic field

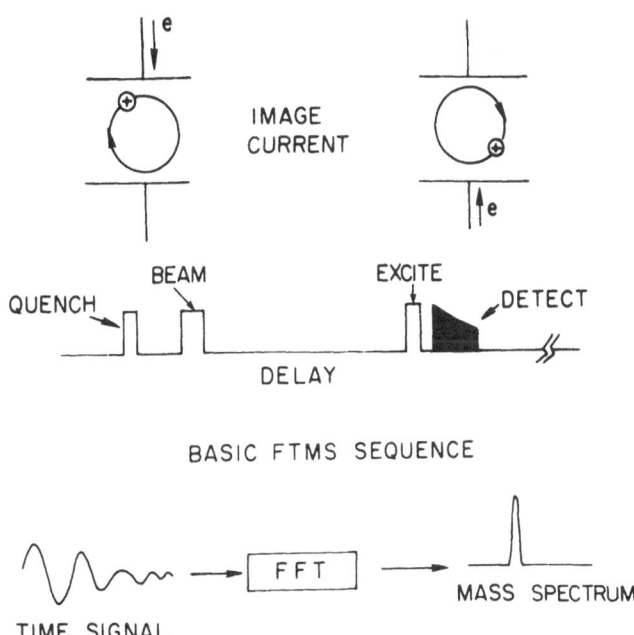

Figure 1. Schematics showing image current formation, a typical experimental sequence and fast Fourier transformation of the time domain
signal.

and low positive trapping voltage applied to both end plates.  After a
sufficiently long and variable delay for ion/molecule chemistry to
occur,  all ions (i.e. parent ions and ions produced due to chemistry)
are elevated into their cyclotron orbits and the image current is collected, digitized and stored on the computer disk for further analysis.
This time domain signal is then converted into a mass spectrum as shown
schematically in the bottom panel of Figure 1.
While in orbit, ions  undergo collision with the background gas.  These
collisions cause loss of phase and consequently the signal will eventually decay.  This decay will occur  faster at higher pressures and this
will in turn degrade the resolution of the apparatus as is known in
Fourier spectroscopy. Figure 2 is a measured time domain signal from
the mixture of two isotopes of carbon dioxide ($C_{12}O_2$ and $C_{13}O_2$)and
exhibits the transient nature of the signal. Figure 3 shows the mass
spectrum that results when the first four thousands points of the time
domain signal in Figure 2 are subjected to an FFT algorithm.  The sidelobes which are result of the FFT algorithm and  zero padding of the
signal are clearly discernible.
From the delay type experiment just described information regarding the
rate for ion/molecule reactions are derived. For further information on
the details of the apparatus itself the reader is referred to excellent
review articles in the literature [2,3].  The apparatus has been

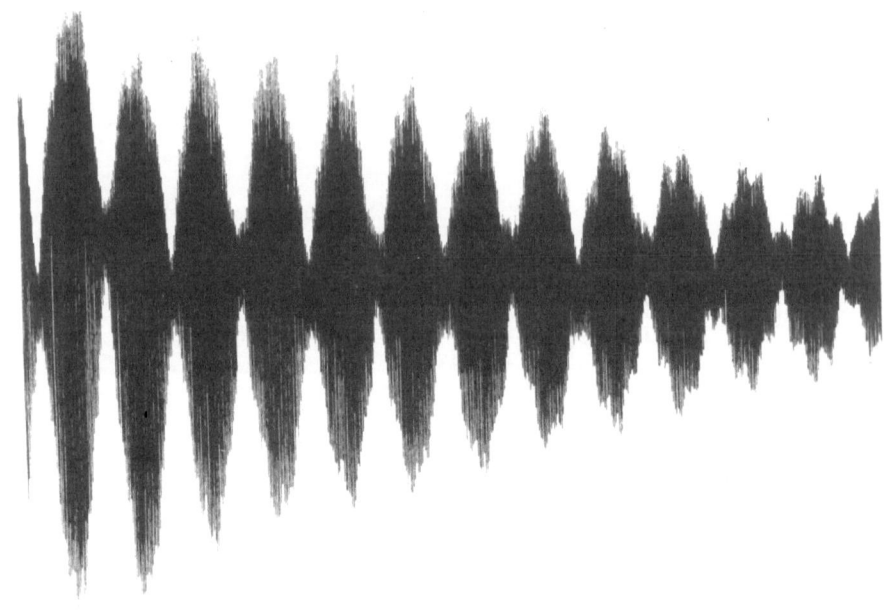

Figure 2. Time domain signal made up of two sinusoids (ions).

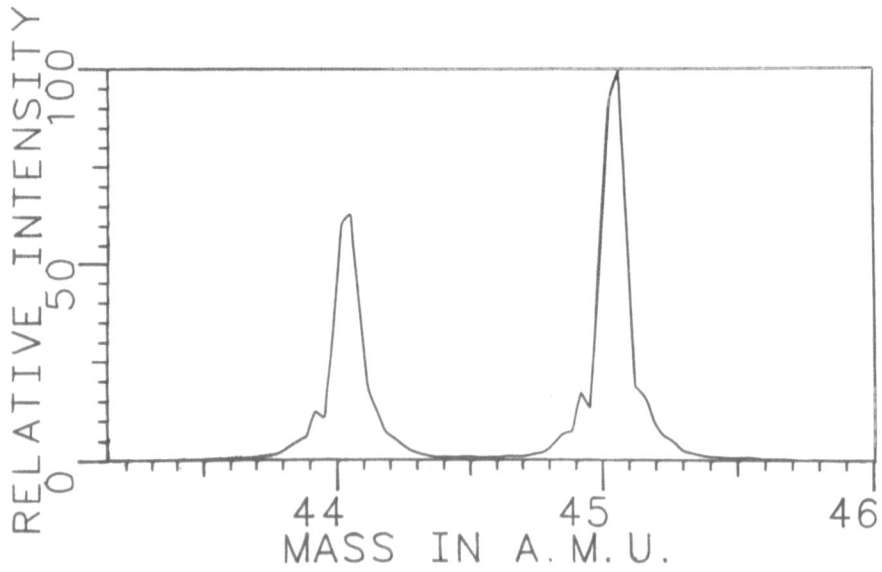

Figure 3. The FFT spectrum of the signal shown in Figure 2.

employed for studies in gas chromatography [4], collision induced dis-
sociation [5,6], and, more recently a new method of ion excitation in
place of rf excitation has been developed by Marshal et.al.[7].
At this point it is necessary to point out some aspects of the FTICR
mass spectrometer that will require caution in the interpretation of
the ion/molecule reaction rate studies and high resolution mass spectroscopy. Resolution of FTICR is proportional to magnetic field B and
observation time T and inversely related to the mass of the ion as seen
in equation 3 above. Magnetic fields of several Teslas  are usually
needed which require cryomagnets.  Even the very high resolutions
provided by FTICR has proved inadequate when very large organic molecules (with masses ranging into thousands of amu) are studied. In
addition,in high resolution work, the pressures must be lowered as much
as possible to resolve ions of the same nominal mass for example,where-
as measurable signals require just the opposite,i.e. higher pressures.
These are two conflicting requirements.
In addition to this we have found in this laboratory that, while in cyclotron orbit, different ions decay at different rates. We have seen

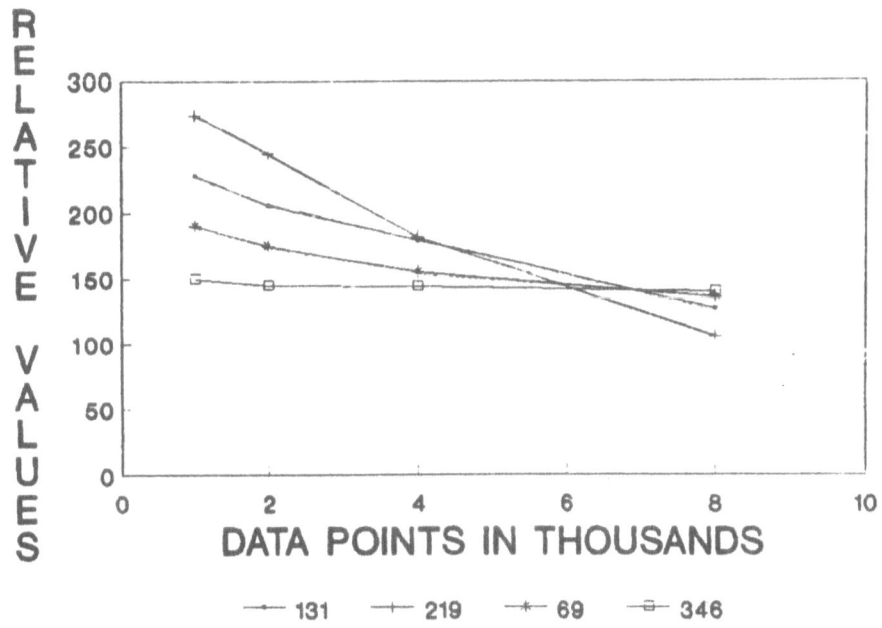

Figure 4. Decay of various ionic species as a function of time,i.e.
the number of data points.  Ionic masses in amu are also shown.

this in real ion signals.  Figure 4 shows experimental data obtained
from a time signal consisting of four sinusoids (ions). It clearly indicates that the ion  intensities as a function of number of data points
are decaying at different rates.   The implications of this in reaction
rate studies  are of serious concern since the FFT algorithm requiring,
at least 8k points of data and often 16k points for good quality spectra,  masks this effect and at times true ion intensity immediately after
their formation in chemical reactions are not known.  In addition to
this damping by collisions,  the resolution of the FTICR is affected by
problems shared by other systems employing fast Fourier transformation
These problems are inherent in the algorithm itself.  In particular the
zero padding required in the FFT algorithm can prove troublesome by
masking  the real effects, introducing the famous sidelobes in the frequency spectrum.   Although this effect can be reduced by "windowing"
techniques, nevertheless the sidelobes are always present and may inter-
fere with the task of proper identification of very closely spaced mass
doublets and isotopes.

## 3. BURG'S METHOD

To alleviate some of the problems we just enumerated and to develop new
ways of deducing the mass spectra from  time domain signals obtained in
FTICR apparatus, implementation of Burg's maximum entropy algorithm for
application to our apparatus was undertaken.  The fact that this method
based on Burg's original ideas proposed in 1967 [8],  required very few
data points was exremely attractive.  The well-known expression for the
power spectral density using Burg's method is

$$S(f) = 2 \, \Delta t \times P_M \left| \sum_{j=o}^{M} a_j \, \exp(-2\pi i \, f j \Delta t) \right|^{-2} \tag{4}$$

where $\Delta t$ is the sampling width, $P_M$ is the prediction error power for a
time domain filter of order M taken as the average of forward and backward predictions over the time series. The  $a_j$'s are the filter coefficients and f is the frequency which determines mass in our apparatus.
Expression 4 for the PSD was programmed in  Fortran and made  part of
the existing software in the Nicolet 1280 computer that controls the
apparatus.  Calculations were done in several stages.  The filter coefficients were calculated using a program  called  MEMPR [9] and then
S(f) was computed.  The programs were also made interactively and would
prompt the  operator for the starting point on the time signal,  number
of data points and the number of filter weights.  After the completion
of the calculation of the PSD,the graphics program already in the
software package was used to obtain the plots in various forms.

## 4. RESULTS

Figure 5 shows the result of using Burg's algorithm on the time domain signal shown above in Figure 2. This Burg spectrum was obtained using the first 200 data points and 100 coefficients and is to be compared to the FFT spectrum shown in Figure 3 that used 4000 points of data. The

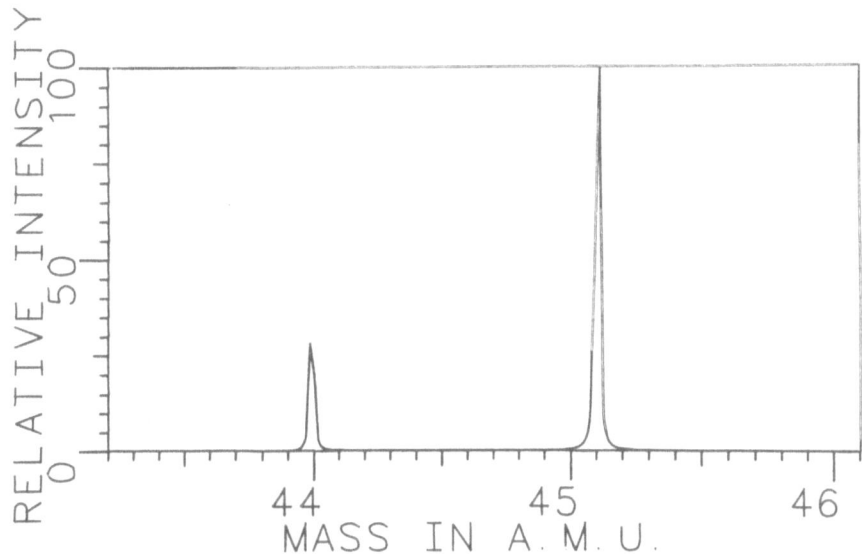

Figure 5. Burg spectrum of the time domain signal shown in Figure 2 and depicting the absence of the sidelobes and much narrower lines.

results of the application to FTICR have been published elsewhere in the literature[10,11]. In Figure 6 and 7 we show respectively spectrum of two ions of same nominal mass. Figure 6 is the FFT spectrum using 2000 points. It is clear the two ions are not totally resolved whereas in Figure 7 which is the Burg spectrum employing 80 points and 40 coefficients the two ions are clearly resolved. We have also extended our work to multi-ion time domain signals. Figure 8 shows the FFT result for a signal consisting of five sinusoids using 8000 data points and Figure 9 is the Burg spectrum of 300 data points and 120 coefficients. As seen from these results, the maximum entropy method employing Burg's method has produced results that are far superior to that obtained through the application of the FFT algorithm. Nevertheless we must point out that in a few instances we have encountered cases where shifting and splitting of the peaks in the spectrum were observed. For example using 17 points of a single sinusoid calculating 8 coefficients resulted in a very slight shift in the location of the spectral peak. This shift was only 1 part in one thousand and the shift could be avoided by going to more points, of the order of 50 to 100 points. Admittedly the shift

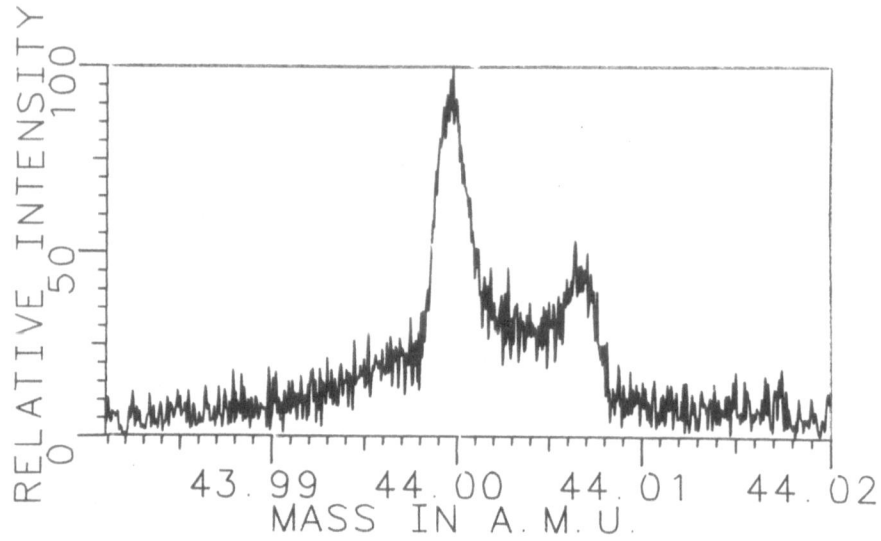

Figure 6. The FFT spectrum of two ions with the same nominal mass 44.

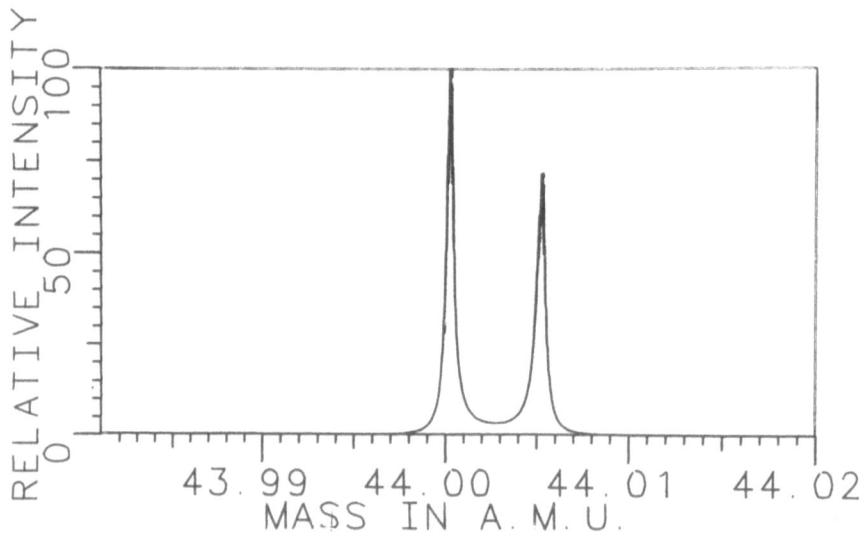

Figure 7. The Burg spectrum of the same two ions as in Figure 6.

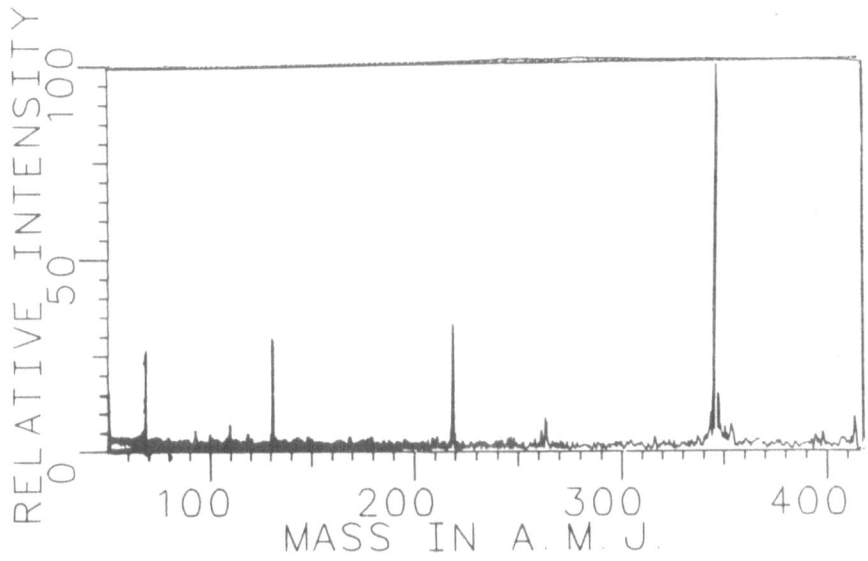

Figure 8. The FFT spectrum of five sinusoids produced by five ions.

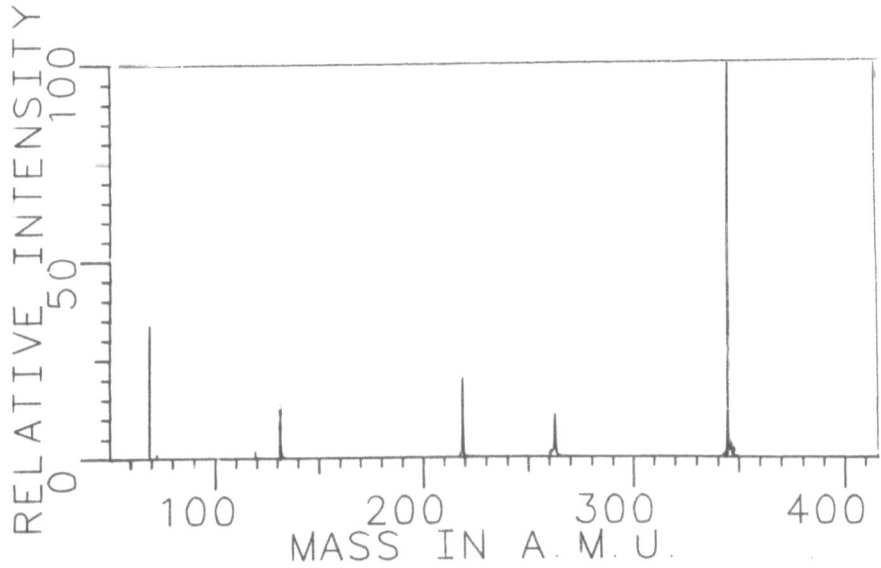

Figure 9. The Burg spectrum of the ions shown in Figure 8.

quoted here is not acceptable in many applications and consequently it must be guarded against. We have also observed some splitting of the spectral peaks which must also be looked for at this point until a cure is affected.  Generally when the signal to noise of the time domain signal is high, fewer and fewer data points are necessary.  In cases of moderate signal to noise ( on the order of 5 to 10 ) we have found by trial and error that the number of data points must be higher.  Some times after very good results are obtained, slight increase in the number of data points degraded the width of the line slightly but the results were still very acceptable. If the number of filter weights were more than 60% of the data points, the results were both shifted and split.  We have not yet encountered a case wherein a cure cannot be found but finding the right combination of data points  and filter coefficients to eliminate the problem faced in the few situations requires some experimentation with the number of data points, filter weights and other variables such as the start point on the data set. In the great majority of time the results are superior to FFT results and spectra are of extremely good quality.  Few situations where the Burg spectra manifest either shifting or splitting require further work. We are prsently working on adapting the nonlinear algorithm developed by Fougere [12] to FTICR software and some preliminary work an IBM PC has shown that elimination of the  shifting and splitting problem can be achieved,  but much work needs be done in that regard and it is hoped that results of further work can be presented in one of these workshops in the future.

Aknowledgement: We are indebted to Dr George Vanasse and Mr Floyd H. Cook of this laboratory for helpful discussions and encouragement. Programming help provided by Mr S. Davis of Real Time Engineering is also gratefully aknowledged.  Support for this research was provided by Air Force Office of Scientific Research under project 2310G4.

## 5. REFERENCES

[1]. Comisarow,M.B. and Marshall,A.G., Chem.Phys.Letters,25(1974)282.
[2]. Ghaderi,S. Kulkarni,P.S. Ledford,E.B. Wilkins,C.L. and Gross, M.
     L., Anal.Chem. 53(1981)428.
[3]. Gross,M.L. and Rempel,D.L., Science, 226(1984)261.
[4]. White,R.L. and Wilkins,C.L., Anal. Chem., 54(1982)2443.
[5]. Bricker,D.L., Adams,T.A. and Russel, D.H., Anal. Chem.,55(1983)
     2417.
[6]. Cody,R.B., Burnier,R.C.,Cassidy,C.L. and Freiser,B.S., Anal. Chem.
     54(1982)2225.
[7]. Chen, Ling and Marshall,A.G., Int.J.Mass Spectrom.Ion Processes,
     79(1987)115.
[8]. Burg,J.P.,in "Modern spectral analysis", IEEE Press, NY, P.42.

[9]. Ulrych,T.J. and Bishop,T.N., Rev. Geophys. Space Phys.13(1975)183
[10]. Rahbee, A., Chem. Phys. Letters, 117(1985)352.
[11]. Rahbee, A., Int.J. Mass Spectrom. Ion Processes, 72(1985)3.
[12]. Fougere, P.F., J. Geophys. Res.,90(1985)4355.

# MAXIMUM ENTROPY ANALYSIS FOR PATTERN RECOGNITION

C. H. CHEN
*Electrical and Computer Engineering Dept.*
*Southeastern Massachusetts University*
*N. Dartmouth, MA 02747 USA*

ABSTRACT. Feature extraction, classification, clustering and learning in pattern recognition are closely related to the maximum or minimum entropy principles. Such relationships are reviewed in this paper. The need for adaptive pattern recognition using neural networks is then emphasized. A comparison between neural networks and conventional statistical classifiers is also presented.

## 1. Introduction

Most of the progress in statistical pattern recognition was in 1960's and early 1970's [1], based mainly on the statistical decision theory [2] and to a small extent the information statistics [3]. Though the important work on statistical inference by Jaynes ([4], [5]) was not familiar to most researchers in pattern recognition at that time, the maximum entropy analysis has in fact been closely related to pattern recognition. In this paper such relationships are reviewed. The limited performance of conventional statistical classifiers has led to much interest in adaptive pattern recognition with the use of neural networks [6]. Indeed neural networks present a new approach to pattern recognition and statistical inference [7]. A comparison between neural networks and conventional statistical classifiers is also presented in the paper. In future development of pattern recognition, it is believed that Jaynes universal theory of statistical inference based on the twin principles of maximum entropy and Bayesian inference may lead to new directions or approaches that can overcome some limitations of the conventional pattern recognition.

## 2. Maximum Entropy Analysis In Feature Extraction

After thirty years of progress, the key problem to pattern recognition is still the extraction of effective features. The information or entropy measures have been used for many years in feature extraction. The idea is that best features are most informative and thus better able to discriminate among different pattern classes. Also the information measures such as the Shannon's entropy are indirectly related to the probability of misclassification (error probability), i.e. maximizing the information measure would tend to minimize the error probability. Let p(x) be the probability density of feature

*P. F. Fougère (ed.), Maximum Entropy and Bayesian Methods,* 403–408.

measurement x and $p(w_i|x)$ be the a posteriori probability of the ith class, $i=1,2, ..., m$. Effective features can be obtained by maximizing or minimizing entropy measures such as [8]

(1)      $$J_s = -\int \sum_{i=1}^{m} P(w_i|x) \log P(w_i|x) \ p(x) \ dx$$          (Shannon entropy)

(2)      $$J_q = \int \sum_{i=1}^{m} P^2(w_i|x) \ p(x) dx$$          (Quadratic entropy) or Bayesain distance)

(3)      $$J_k = \int \sum_{i=1}^{m} P(w_i|x) \ [1 - P^2(w_i|x)] \ p(x) dx$$          (cubic entropy)

Features based on entropy measures, from our experience, tend to be moderate to very effective. In texture image classification, entropy based texture features in the spatial domain has been shown experimentally to be comparable in performance as the gray level co-occurrence contrast features [9]. The latter requires much more computation.

The maximum entropy spectral analysis provides another approach to extract frequency domain features which are computationally efficient. Features determined from the power spectrum using maximum entropy analysis are more effective than those determined from FFT spectra. For example, two frequency ratio features computed from maximum entropy spectra of teleseismic events perform better than other features [10]. Similarly, the ring and wedge features of two-dimensional maximum entropy spectra of image segments perform better than similar features obtained from FFT spectra [11]. The ring and wedge features are texture features that measure directionality and coarseness of texture images. On the other hand, the autoregressive coefficients computed by the maximum entropy power spectrum are not more effective than other features in teleseismic discrimination [12]. Based on all the experimental results, it is reasonable to conclude that the maximum entropy analysis in general and maximum entropy spectral analysis in particular can provide good features in many pattern recognition problems though such features may not be the best in the recognition problem considered.

It is remarked that the statistical inference has not been able to solve the problem of selecting the best set of features. In general possible subsets of an n-element set of features total around $2^n$. An efficient search of the best subset would require the use of artificial intelligence algorithms [13].

### 3. Maximum Entropy Analysis in Pattern Clustering and Classification

The minimization of cross-entropy can be viewed as a refinement of a general classification method due to Kullback and can lead to a nearest neighbor classification rule or clustering method using a non-Euclidean information-theoretic distortion measure [14].

The major role of minimum cross entropy analysis is essential to determine cluster centroids for vector quantization or signal coding. Undoubtedly the minimum cross entropy analysis had led to a powerful and practical pattern clustering or signal coding method. As a classification method, there is no guarantee that minimizing the total distortion measure is equivalent to minimizing the error probability. For pattern classification, the minimum error probability is still the desired criterion. The difficulty with achieving the minimum error probability in practice is the lack of precise knowledge of probability densities. Our prior knowledge of the patterns, whether through sample estimates or pattern models, is very important. Such knowledge must be fully incorporated in the decision rules and must be updated to reflect the increased or improved knowledge from additional patterns or other sources. The Bayesian inference and maximum entropy methods are yet to be studied in the context of statistical pattern classification for possible improvement of the existing classification procedures.

## 4. Machine Learning and Adaptive Pattern Recognition

As discussed above, better pattern classification requires better use of prior knowledge about the patterns and the reduction of uncertainties in making decision. Learning and adaptive methods are much needed to reduce the uncertainties The term "learning" is more often used to estimating the parameters in probability densities or decision rules. Adaptive pattern recognition however refers to procedures to improve decision boundaries among the pattern classes considered. Essentially the two terms are equivalent and represent perhaps one of the most challenging problems in statistical inference. While many mathematically elegant learning and adaptive algorithms are available (e.g. [15]) with desirable convergence properties, the main problem has been the assumption of a large number of patterns which are not available in practice. Also there is always discrepancy between theory and practice. For example, the assumption of linear decision boundary may not be valid for the classification problem considered.

The emergence, or re-introduction, of the neural network technology is probably the most important development in pattern recognition in recent years. It can be considered as a practical and effective approach to utilize prior and acquired knowledge to reduce the uncertainties in complex recognition tasks. There is no need to make simplifying linear decision function assumption. Most neural networks provide computationally very efficient procedures for learning or adaptive pattern recognition. Generally speaking neural networks can provide rapid classification with comparable performance as the conventional methods and are more tolerant of noisy and limited data. However neural networks do not replace the existing pattern recognition techniques which make effective use of parametric models. Furthermore neural networks do not provide solutions to all statistical inference problems.

## 5. A Comparison between Neural Network and Conventional Classifiers

First the effects of finite number of training samples (or patterns) are considered experimentally. Both the Nestor's NDS-1000 neural network and the perceptron type neural network using back propagation algorithm are used in the experiment. A Fisher's

Iris data set is used which contains three pattern classes with 50 4-dimensional vectors per class. For the Nestor's network, Fig. 1 shows the percentage correct recognition versus the total number of training samples. Samples not selected for training are for testing use. For each training sample size, 5 trials are made and the results averaged. However the trial result which is more than two standard deviations below the average is deleted. Only one deletion is allowed. The curve shows that the percentage correct recognition stays at a fairly constant level of around 91.51%. In other words the performance is not sensitive to the training sample size. For the back propagation algorithm, similar conclusion can be made with an average percentage correct recognition of 96%. This shows that both neural networks can still be very effective in classification even when the number of training samples is small. For the conventional classifiers, however, the performance generally degrades significantly when the number of training samples becomes small. In this sense the neural networks are much better. As for the overall classification performance, the conclusion has been fairly consistent with that reported by Huang ad Lippmann [16], i.e. the neural networks perform about the same as the conventional nearest neighbor decision rule.

The features employed in the neural networks are the same as those for conventional classifiers. The neural network does nothing to solve the automatic feature extraction problem mentioned earlier. On the other hand, the neural networks, especially the Nestor's network, can evaluate feature subsets or weighted feature subsets quite efficiently. Also the fact that the networks rely highly on training makes them less dependent on the features used. The use of hidden layers allows for complex decision boundaries being established by training. This would make the feature extraction process less important in determining the classification performance.

### References

1.   C.H. Chen, "Statistical pattern recognition - early development and recent progress", International Journal of Pattern Recognition and Artificial Intelligence, vol. 1, no. 1, p. 43, April 1987.

2.   A. Wald, "Statistical Decision Functions", Wiley, New York, 1950.

3.   S. Kullback, "Information Theory and Statistics", Wiley, New York, 1959. Also published in paper back by Dover Publications, New York, 1968.

4.   E.T. Jaynes, "Information theory and statistical mechanics, Part I", Physics Review, vol. 106, p. 620; Part II, ibid, vol. 108, p. 171, 1957.

5.   E.T. Jaynes, "Prior probabilities", IEEE Trans. on Systems Science and Cybernetics, vol. SSC-4, no. 3, p. 227, September 1968.

6.   Y.H. Pao, "Adaptive Pattern Recognition and Neural Networks", Addison-Wesley, Reading, MA 1989.

7.  D. Hestenes, "Inductive inference by neural networks", Proc. of the 7th Annual Workshop on Maximum Entropy and Bayesian Methods in Applied Statistics, Reidel Publication Col, Dordrecht/Boston 1988.

8.  C.H. Chen, "On information and distance measures, error bounds, and feature selection", Information Science Journal, vol. 10, 1976.

9.  M.E. Jernigan and F.D'Astoud, "Entropy-based texture analysis in the spatial frequency domain", IEEE Trans. on Pattern Analysis and Machine Intelligence, vol. PAMI-6, no. 2, p. 237, March 1984.

10. C.H. Chen, "Nonlinear Maximum Entropy Spectral Analysis Methods for Signal Recognition", Research Studies Press, Wiley, Chichester, U.K., 1982.

11. C.H. Chen, "A study of texture classification using spectral features", Proc. of the 6th International Conference on Pattern Recognition, Munich, Germany, p. 1074, Oct. 1982.

12. I. Drydal, Teleseismic discrimination of earthquakes and nuclear detonations with features derived from MESA", International Journal of Pattern Recognition and Artificial Intelligence, vol. 1, no. 3, p. 323, 1987.

13. W. Siedlecki and J. Sklansky, "On automatic feature selection", vol. 2., no. 2, p. 221, 1988.

14. J. E. Shore and R.M. Gray, "Minimum cross-entropy pattern classification and cluster analysis", IEEE Trans. on Pattern Analysis and Machine Intelligence, vol. PAMI-4, no. 1, p. 11, January 1982.

15. R.O. Duda and P.E. Hart, "Pattern Classification and Scene Analysis", Wiley, New York, 1972.

16. W.Y. Huang and R.P. Lippmann, "Comparison between neural nets and conventional classifiers", Proc. of the First International Conference on Neural Networks, San Diego, vol. IV, p. 485, 1987.

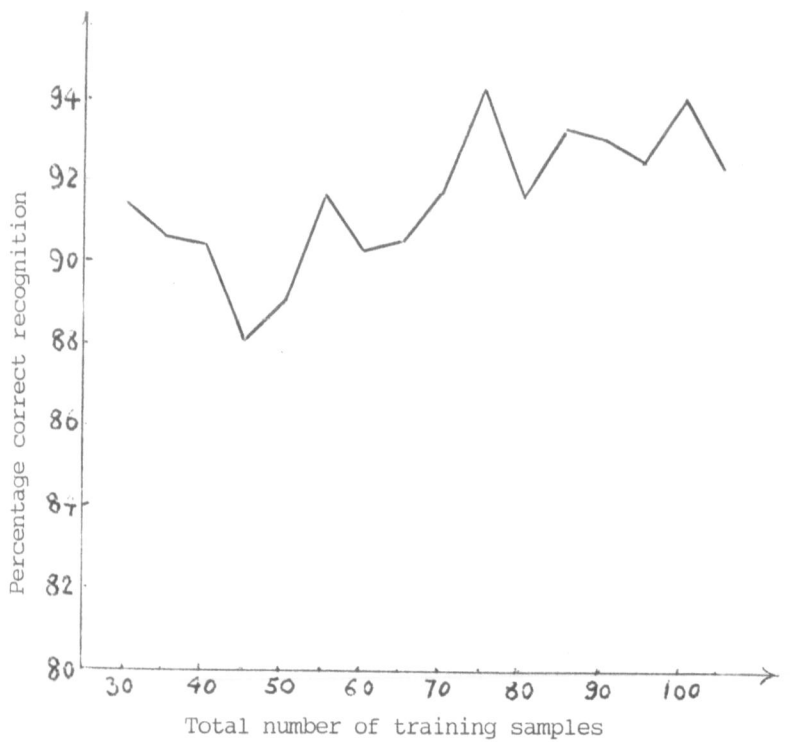

Figure 1.    Percentage correct recognition versus the number of training
samples for Nestor's network using the Iris data set.

# APPLICATION OF BURG'S SPECTRAL ANALYSIS TO HEAVE RESPONSE MODELING IN UNDERWATER APPLICATIONS

Ferial El-Hawary
Technical University of Nova Scotia
P. O. Box 1000, Halifax, Nova Scotia
B3J 2X4, CANADA

ABSTRACT. In numerous underwater acoustic operations, the received signals contain an undesirable component due to the dynamics of both source and receiver known as the heave phenomenon. We discuss a parametric modeling approach based on power spectral density estimation using Burg's maximum entropy methodology. Attention is focussed on the Auto Regressive model as an efficient tool for carrying out the estimation task. Results using field data are presented to illustrate the trade-offs involved this is to select models intended for heave component filtering in underwater remotely operated vehicles, seismic exploration, and float positioning applications.

## 1. Background

In numerous underwater acoustic operations, such as control and operations of autonomous remotely operated vehicles, underwater seismic exploration, and buoy wave data analysis, the received signals contain an undesirable component due to the dynamics of both source and receiver. This is commonly known as the heave phenomenon.

The vehicle's position is estimated both at the surface and at the vehicle in underwater remotely operated vehicles (see McFarlane et al. (1987)) . As detailed in Jackson, and Ferguson, J. (1984), a probing pulse is emitted from the surface, and the vehicle answers with a second pulse, thereafter the transponder array responds, and the time delays between the signals are measured separately at the surface and at the vehicle. Accuracy is essential in navigating the vehicle, and can be refined by compensating for motion effects. According to Smith (1987) , this aspect is also essential in controlling dynamic loads through the design of motion-compensated handling systems. The operation of manipulator controls in underwater robotics depend on compensation for motion effects ( for details see Collins (1987)). The application to floats of freely drifting acoustic sensors to measure signal propagation and ambient noise in the 1 to 20 Hz band is given in Culver and Hodgkiss (1988). The deployment of several freely drifting floats forms an array of sensors whose outputs can be combined after the experiment with a beam former. Float locations must be known to within one-tenth of a wavelength at the highest frequency of interest in order to effectively beamform their outputs. The floats generate and receive acoustic pulses and thereby measure float-to-surface and float-to-float travel times. Estimating float positions and float position uncertainties from these travel-time measurements is the main task in this application.

409

*P. F. Fougère (ed.), Maximum Entropy and Bayesian Methods, 409–417.*
© 1990 *Kluwer Academic Publishers. Printed in the Netherlands.*

Underwater seismic/acoustic exploration is adopted in evaluating the structure of the underwater layered media, and classification of sediments in terms of properties that aid in carrying out an exploration task. Identifying hydrocarbon formations is an important phase of the marine seismic process. Classical acoustic techniques have gained increasing use over the past several years for marine layer identification and classification of sediments (see Robinson and Durrani (1986), El-Hawary, and Vetter (1980 and 1982)).

A deep towed acoustic signal source and hydrophone receiver array are used. The source imparts energy to the water and underlying media. The source signal then undergoes multiple transmissions and reflections at the layers' boundaries. The ship's dynamics, coupled to the towed body (fish) through the towing cable, and the hydrodynamics of the towed body, cause vertical motions of the source and sensor. These components have the outcome of a varying acoustic wave travel path to the sea floor and to the sub-bottom reflectors between successive pings of the source. The motion's effects appear on the reflection records along the ship track as additional undulations of the sea floor and of the sub-bottom reflectors (the underlying media). Removal of the heave component is an important preprocessing task for improving displays of the raw and filtered reflection data, for extracting media parameters such as reflection coefficients and reflector depths. The design of the compensating filter facilitates reducing the residual heave effects, i.e., for delaying and advancing the recording trigger on successive firings so as to effect a smoothing or removal of such undulations. On the basis of the frequency response of the heave record, a model for the heave dynamics which is consistent with those found in the area of hydrodynamics can be postulated. This provides the basis for formulating the heave extraction problem as one of optimal linear estimation.

In all applications, filtering this component at the receiver side is required to enhance the quality of the received records for further processing to extract information. This demands an accurate model of the heave dynamics, and to identify its parameters based on short records. Spectral analysis can be instrumental in this regard.

A major part of signal processing is based on spectral analysis, typically for distinguishing and tracking signals of interest, and for extracting information from the relevant data. Given a finite number of noisy measurements of a discrete - time stochastic process, or its first few covariance lags, the classical spectral estimation problem involves estimating the shape of its continuous power spectrum. For some modern applications of signal processing, such as radar, sonar, and phased arrays, the spectrum of interest is a line spectrum, and the modern spectral estimation problem involves estimating the locations of these spectral lines.

The frequency resolution in conventional Fourier transform methods is roughly equal to the reciprocal of the data record length. As a result additional constraints (or prior information) are included to enhance the resolution capability. This is achieved in modern methods where the data are modeled as the output of a linear system driven by white noise. With a properly selected model, these methods will lead to enhanced performance. In underwater processing applications, the spectral estimates have to be based on short data records and yet low-bias, low-variance, high-resolution estimates are desired.

We treat source dynamic motion evaluation in underwater applications, and discuss a parametric modeling approach based on power spectral density estimation using Burg's maximum entropy methodology. Attention is focussed on the Auto Regressive model as an efficient tool for carrying out the estimation task. Results using field data are presented to illustrate the trade-offs involved this is to select models intended for heave component filtering in underwater remotely operated vehicles, seismic exploration, and float positioning applications.

## 2.  Parametric  PSD  Models

Power spectral density (PSD) is defined as the discrete-time Fourier transform of an infinite autocorrelation sequence (ACS). This transform relationship between the PSD and ACS may be considered as a non-parametric description of the second-order statistics of a random process (see Marple (1987), and Kay and Marple (1981)).

A parametric description of the second- order statistics may also be conceived by assuming a time series model of the random process. The PSD of the time-series model will then be a function of the model parameters rather than the ACS. A special case of models, driven by white noise processes and possessing rational system functions, is used in this approach.  This class includes the Auto Regressive (AR) process model, the Moving Average (MA) process model, and the Autoregressive - Moving Average (ARMA) process model. A major reference is Box and Jenkins (1970). The output processes of this class of models have power spectral densities that are totally described in terms of the model parameters and the variance of the white noise process.

The motive for parametric models of random processes is the capability to attain better PSD estimators based upon the model than produced by conventional spectral estimators. Better spectral resolution is one key advancement area. Both the periodogram and correlogram methods generate PSD estimates from a windowed set of data or ACS estimates. The unavailable data or unestimated ACS values outside the window are implicitly zero. This is as a rule an unrealistic assumption that leads to distortions in the spectral estimate. Some knowledge about the process from which the data samples are taken is often available. This information may be used to construct a model that approximates the process that generated the observed time sequence. Such models will make more realistic assumptions about the data outside the window other than the zero assumption. Therefore, the requirement for window functions can be avoided, along with their distorting effect. The degree of enhancement in resolution and spectral fidelity, if any, is determined by the appropriateness of the selected model and the ability to fit the measured data or the ACS (known or estimated from the data) with a few model parameters.

## 3.  Autoregressive  PSD  Estimation

The autoregressive (AR) time-series model approximates many discrete-time deterministic and stochastic processes. The sequence x[n] is assumed to be the output of a causal filter that models the observed data as the response to a driving sequence $e_K[n]$, which is a white noise process of zero mean and variance $\sigma_{oP}^2$. For a K-th order AR model we have:

$$x[n] = - \sum_{k=1}^{K} h_K[k] x[n-k] + e_{KK}[n] \qquad (1)$$

This can be written as:

$$e_K[n] = \sum_{k=0}^{K} h_K[k] x[n-k] \qquad (2)$$

Here we assume that $h_K[0] = 1$. Equation (2) reveals that $e_K[n]$ is obtained as the convolution of the filter sequence $h_K[n]$ and the input x[n]. As a result, we have using the Z-transform (see Oppenheim and Schafer (1975) and Oppenheim and Wilsky (1983)).

$$E_K(z) = H(z)X(z) \tag{3}$$

H(z) is the filter Z-transfer function. As a result the Z-transform of the correlation sequences is given by:

$$P_{e_K e_K}(z) = P_{xx}(z)H(z)H^*\left(\frac{1}{z^*}\right) \tag{4}$$

To obtain the auto-regressive power spectral estimate of x[n], we substitute $z = e^{j\omega T}$, to obtain :

$$P_{xx}(\omega) = \frac{\sigma_{eK}^2}{\epsilon_{h_K h_K}(\omega)} \tag{5}$$

Here we take T = 1. The denominator function is given by:

$$\epsilon_{h_K h_K}(\omega) = |\sum_{k=0}^{K} h_K[k]e^{-j\omega k}|^2 \tag{6}$$

It is clear from equation (6), that estimating the PSD is equivalent to estimating the AR model parameters $h_K[n]$. The autoregressive (AR) spectral estimation concept received considerable attention in the literature of time-series models, due to the following justifications:

1- AR modeling is equivalent to the minimum mean-square inverse filtering or the best linear prediction filtering approach.
2- Under the Gaussian assumption the infinite covariance extension to an M lag segment, provided by an AR (M) model, maximizes the entropy of the corresponding time series. In other words, among all possible extensions corresponding to the AR extension is the "whitest" and has the "flattest" spectrum. In fact, the AR method is equivalent to the maximum-entropy method.
3- The third justification is that autoregressive spectra tend to have sharp peaks, a nature often associated with high-resolution spectral estimates.
4- Estimates of the AR parameters can be obtained as solutions to linear equations. For example the AR parameters and the autocorrelation sequence are related by a set of linear equations. Estimates of MA and ARMA parameters, however, require the solution of nonlinear equations.

The parameters $h_K[K]$ of the AR (K) model can be found from the autocorrelation sequence for lags 0 to K by using the AR Yule-Walker normal equations or the discrete-time Wiener-Hopf equations given by

$$
\begin{bmatrix}
r_{xx}[0] & r_{xx}[-1] & \cdot & \cdot & \cdot & \cdot & r_{xx}[-K] \\
r_{xx}[1] & r_{xx}[0] & & \cdot & \cdot & \cdot & r_{xx}[-K+1] \\
\cdot & \cdot & & \cdot & \cdot & \cdot & \cdot \\
\cdot & \cdot & & \cdot & \cdot & \cdot & \cdot \\
\cdot & \cdot & & \cdot & \cdot & \cdot & \cdot \\
\cdot & \cdot & & \cdot & \cdot & \cdot & \cdot \\
r_{xx}[K] & r_{xx}[K-1] & \cdot & \cdot & \cdot & \cdot & r_{xx}[0]
\end{bmatrix}
\begin{bmatrix}
1 \\
h_K[1] \\
\cdot \\
\cdot \\
\cdot \\
\cdot \\
h_x(K)
\end{bmatrix}
=
\begin{bmatrix}
\sigma^2_{eK} \\
0 \\
0 \\
0 \\
0 \\
0 \\
0
\end{bmatrix}
\tag{7}
$$

Thus AR parameter estimation involves the solution of a Hermitian Toeplitz system, for $\sigma^2_{eK}$, $h_K[1]$,..., $h_K[p]$ , given the ACS of x [n] over 0 to K lags. This can be computed very efficiently using the Levinson algorithm. The linear prediction coefficients $h_1[1]$,..., $h_K[K]$ are often termed the reflection coefficients. A special symbol $k_K = a_K[K]$ is typically applied to distinguish these particular linear prediction coefficients from the remaining coefficients.

The Levinson recursive solution to the Yule-Walker equations relates the order K parameters to the order K-1 parameters as:

$$
h_K[n] = h_{K-1}[n] + k_K h^*_{K-1}[K-n]
\tag{8}
$$

for n = 1 to n = K-1. The reflection coefficient $K_K$ is obtained from the known autocorrelations for lags 0 to K-1

$$
k_K = h_K[K] = \frac{-\sum_{n=0}^{K-1} h_{K-1}[n] r_{xx}[K-n]}{\sigma^2_{e(K-1)}}
\tag{9}
$$

The recursions for the driving white noise variance are given by:

$$
\sigma^2_{eK} = \sigma^2_{e(K-1)}[1 - |k_K|^2]
\tag{10}
$$

where

$$
\sigma^2_0 = r_{xx}[0]
\tag{11}
$$

## 4. Application Results

We explored the performance of the autoregressive PSD estimation technique when applied to a series of heave records used earlier with Kalman filtering as benchmark. Figure (1) shows a sample heave record taken from a cruise off the coast of Newfoundland.

We concentrated on the influence of model order on the algorithm's results. Table (1) gives a summary of results listing the dominant peaks, their number, and the value of $\Sigma$. Figure (2) shows the power spectral estimates, for K = 20, 40, 80, and 100, and Figure (3) illustrates the corresponding error sequences. The impulse response sequences are given in Figure (4).

All models exhibited a peak at about 45 kHz. The model of order $K = 40$ appears to represent the spectrum more accurately from an error point of view. The dominant peak value is 14.76, with a $\Sigma$ of 118.09, the corresponding number of peaks is 6.

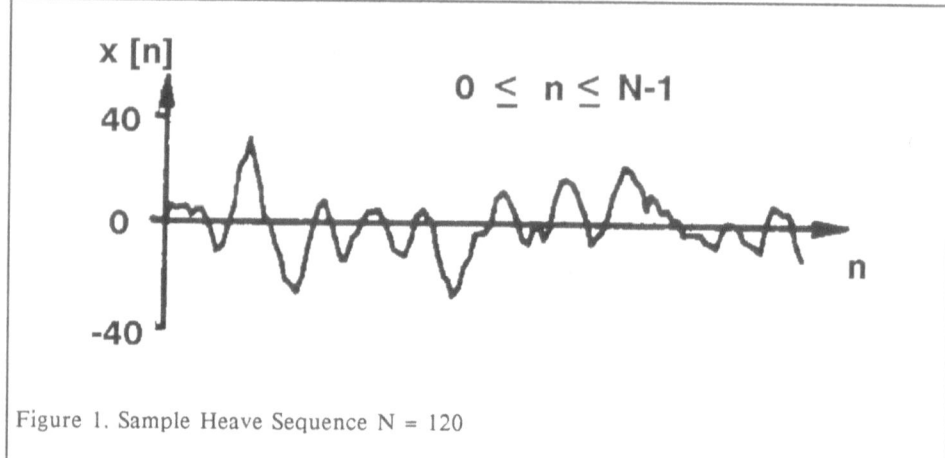

Figure 1. Sample Heave Sequence N = 120

## TABLE 1. Summary of results

| K | DOMINANT PEAK VALUE | $\Sigma$ | Number of Peaks |
|---|---|---|---|
| 10 | 1.82 | 117.28 | 1 |
| 20 | 1.66 | 116.96 | 2 |
| 30 | 2.58 | 116.95 | 4 |
| 40 | 14.76 | 118.09 | 6 |
| 50 | 10.20 | 115.99 | 6 |
| 60 | 19.51 | 112.32 | 5 |
| 70 | 7.43 | 43.87 | 12 |
| 80 | 8.52 | 45.64 | 12 |
| 90 | 5.60 | 34.20 | 11 |
| 100 | 5.33 | 29.74 | 18 |

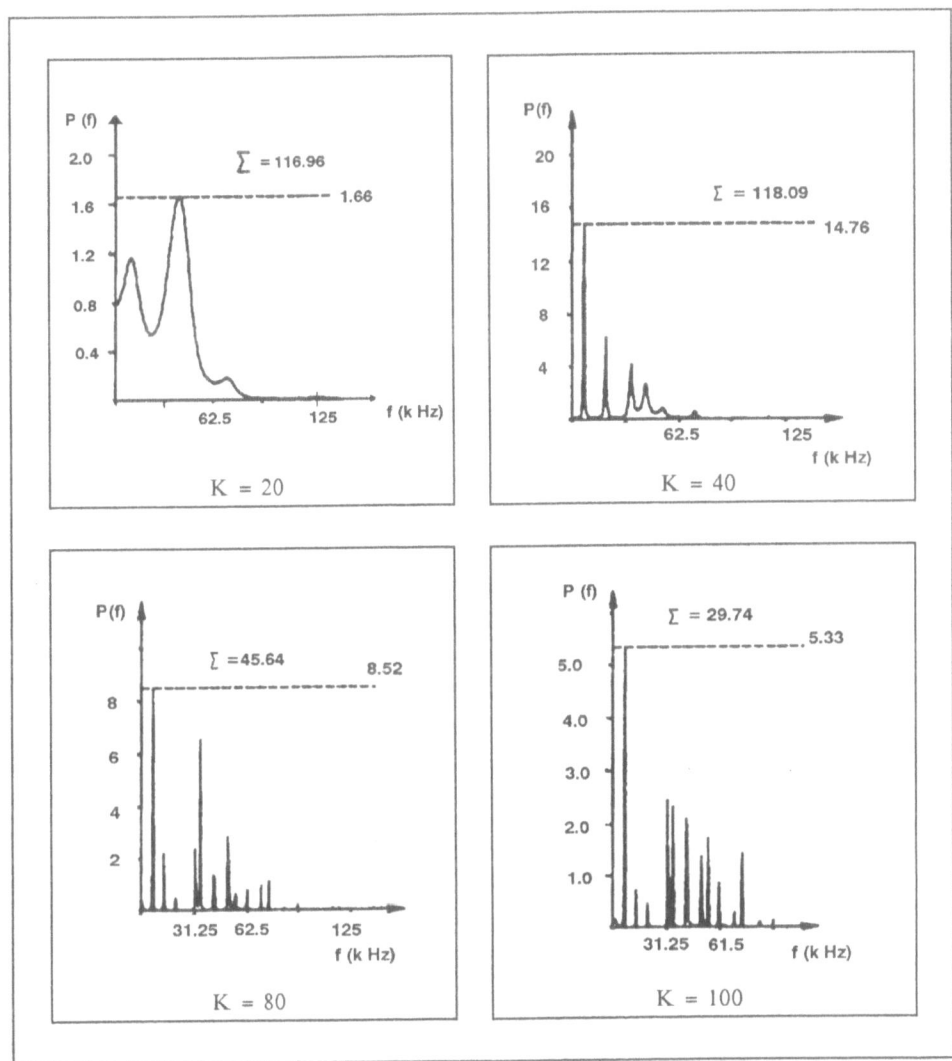

Figure 2.  Power spectral estimates

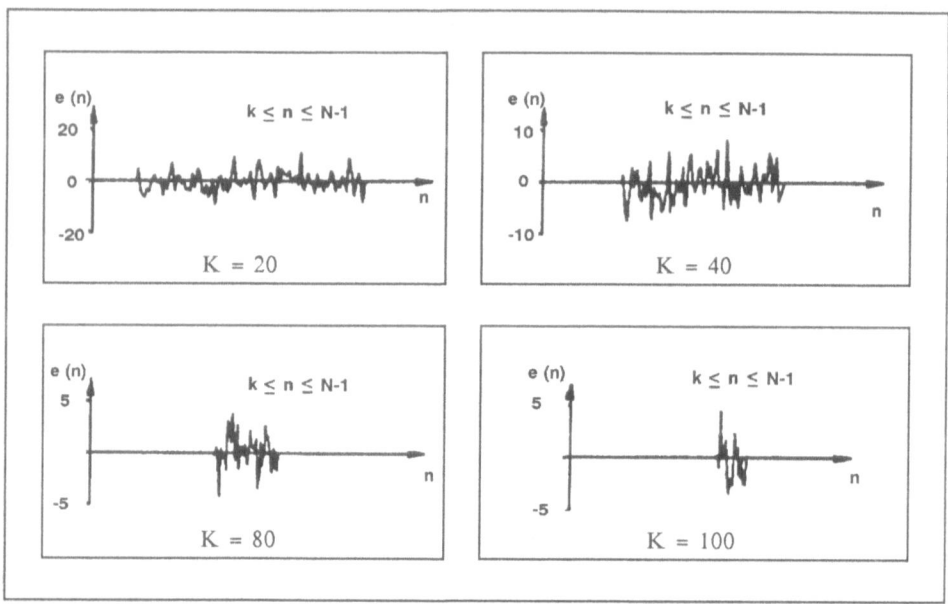

Figure 3.  Error sequences

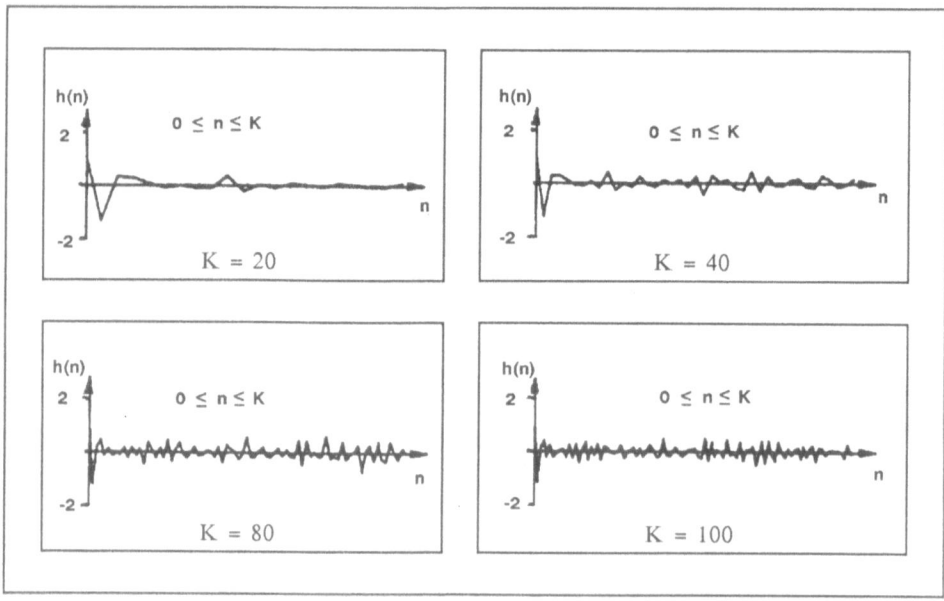

Figure 4.  Impulse response sequences

## 5. Conclusions

In this paper we treated the problem of source dynamic motion evaluation in underwater applications using the parametric approach to power spectral density estimation, and treated the Auto Regressive model for carrying out the estimation task.

The issue of compensating for underwater motion effects arises in a number of areas of current interest such as control and operations of autonomous remotely operated vehicles, underwater seismic exploration, and buoy wave data analysis. Earlier treatments of the problem relied on frequency response methods and Kalman filtering. We discussed the compensation problem and reported on the application of a spectral estimation technique to heave compensation. The model of order K = 40 appears to represent the spectrum more accurately from an error point of view. Further work remains to be done to attempt to verify the conclusions and refine the model order.

## 6. References

McFarlane, J.R., Frisbie, F.R., and Mullin, M. (1987) ' The Evolution of Deep ROV Technology in Canada', Oceans' 87 Proceedings, 1260-1266.

Jackson, E., and Ferguson, J. (1984) ' Design of ARC- Autonomous Remotely Controlled Submersible', Oceans' 84 Proceedings, 365- 368.

Smith, G.R. (1987) ' Development of a 5000 Meter Remote Operated Vehicle for Marine Research', Oceans' 87 Proceedings , 1254- 1259.

Collins, J.S. (1987) ' Advanced Marine Robotics as a Strategic Technology for Canada and Observations of a Related Large Scale Project in Japan', Oceans' 87 Proceedings , 1246- 1253.

Culver, R. L., and Hodgkiss, Jr., W.S. (1988) ' Comparison of Kalman and Least Squares Filters for Locating Autonomous Very Low Frequency Acoustic Sensors', IEEE Journal of Oceans Engineering, Vol. 13, No. 4, 282-290.

Robinson, E.A. and Durrani, T.S. (1986) 'Geophysical Signal Processing', Prentice-Hall, Englewood Cliffs, N.J.

El-Hawary, Ferial, and Vetter, W.J. (1980) 'Spatial Parameter Estimation for Ocean Subsurface Layered Media', Canadian Electrical Engineering Journal, Vol. 5, No. 1, 28-31.

El-Hawary, Ferial, and Vetter, W.J. (1982) 'Event Enhancement on Reflections from Subsurface Layered Media', IEEE Journal of Oceanic Engineering, Vol. OE-7, No. 1, 51-58.

Marple, S.L. (1987) ' Digital Spectral Analysis with Applications', Prentice-Hall Inc., Englewood Cliffs, N. J.

Kay, S.M. and Marple, S.L. (1981) ' Spectrum Analysis- A Modern Perspective' , Proceedings of the IEEE, Vol. 69, 1380 - 1418.

Box, G.E.P. and Jenkins, G.M. (1970) ' Time Series Analysis: Forecasting and Control ', Holden-Day, San Francisco, CA.

Oppenheim, A.V. and Schafer, R.W. (1975) 'Digital Signal Processing ', Prentice Hall Inc., Englewood Cliffs, N.J.

Oppenheim, A.V. and Wilsky, A.S. (1983) ' Signals and Systems' Prentice-Hall Inc., Englewood Cliffs, N.J.

# Maximum Entropy And Minimum Cross-Entropy Principles:
# Need For A Broader Perspective

H.K.Kesavan

Professor
University Of Waterloo, Canada

J.N.Kapur

Adjunct Professor
University of Waterloo, Canada.

## ABSTRACT

Jaynes' Maximum Entropy Principle (MaxEnt) has served as a uni-
fying principle in the study of a wide variety of probabilistic systems transcending all disciplinary boundaries. Here, we are concerned about the
two inverse problems of determining the most unbiased moment constraints, and the most unbiased entropy measure, when the remaining
two probabilistic entities are specified. The need for these inverse problems arises in application areas and has the same relevance as MaxEnt.
The problem of determining the most unbiased measure of entropy, however, takes us out of the framework of MaxEnt, where considerations of
Generalized Measures of Entropy become essential. Departure from the
well-established use of the Shannon measure has raised a variety of objections ranging all the way from the meaning of entropy to the mutilation of
its uniqueness properties. The Generalized Maximum Entropy
Principle(GMEP), which is the main contribution of this paper, takes cog-
nizance of these criticisms by giving justifications for the use of generalized measures. Acceptance of this new model will result in extending the
scope of MaxEnt to tackle problems which are at present beyond its
scope.

A very similar argument is advanced in favour of a Generalized
Minimum Cross-Entropy principle where measures other than Kullback-
Leibler measure are considered.

P. F. Fougère (ed.), Maximum Entropy and Bayesian Methods, 419–432.

## 1. Introduction

Jaynes' maximum entropy principle (MaxEnt) has met with great success in a wide variety of application areas in the realm of probabilistic systems. Apart from its practicality, MaxEnt goes to the very heart of the meaning of probability and inductive reasoning and consequently, provokes lively philosophical debates with other schools of thought on those subjects.

While MaxEnt covers a broad spectrum of problems, there are, however, application areas where the inverse problems arising from the principle assume paramount importance. These dwell on questions of determining the most unbiased set of moment constraints and the most unbiased measure of entropy when the remaining two probabilistic entities are specified. (Maximum entropy distribution is understood to have zero bias.) The first inverse problem concerning the determination of appropriate constraints does not raise fundamental questions regarding the legitimacy of such a benign inquiry. The feeling here is that this problem can still be accommodated within the framework of MaxEnt where Shannon entropy is sovereign. The only caution that is prescribed for proceeding with it is concerning the uniqueness of the solution thus obtained. But the second inverse problem, that of determining an unbiased measure of entropy generally meets with stiff opposition. The underlying reason for this resistance is based on objections to admit the use of generalized measures of entropy, which necessitates abandoning the exclusive use of the concept of Shannon entropy.

The research on generalized measures of entropy, however, is not new. It has gone on for a number of years in one context or another. There are several such measures which have found practical applications. The use of Burg entropy in spectral analysis is a case in point. Despite all the published literature on the subject, the use of of generalized measures of entropy in enunciating a principle underlying the second inverse problem is considered a major intrusion on MaxEnt. One is thus challenged to give convincing reasons before one would consider deviating from the well-established use of the Shannon measure which is pivotal to MaxEnt. There are concerns expressed about vitiating the physical meaning of entropy. Also, there are arguments related to the properties of uniqueness of the Shannon measure which generalized measures may not have. Finally, there are those who are convinced about the legitimacy of tackling the two inverse problems, but who have serious reservations about the nomenclature of generalized entropy. They would welcome an alternative name to connote the concept so that it may not be mistaken for the traditional notion of entropy.

In defence of our terminology, we can state that the concept of generalized entropy is already in vogue in the literature though it did not figure prominently in connection with MaxEnt. Furthermore, although one could theoretically define a number of generalized measures of entropy, we are not indulging in that futile exercise. Such proliferation of measures can only occur when one attempts to define them by relaxing one or more conditions from the list of axioms covering the Shannon measure, or by introducing some properties of parametrization. Rather, we are only concerned with a very small number of generalized measures which naturally arise as a result of the maximization procedure pertaining to the inverse problems. These measures will be admitted with the greatest of care only after getting convinced that Shannon entropy will not yield the desired results.

The constituents of GMEP are the three probabilistic entities: 1) generalized measure of entropy, which includes the most important measure of Shannon entropy; 2) a set of moment constraints; and 3) the a posteriori probability distribution. The principle addresses itself to the determination of any one of the three when the remaining two entities are specified. The global constraint which links the three mutually interacting probabilistic entities is the **Entropy Maximization Postulate(EMP).** It states that it is the maximum generalized entropy which is always the controlling quantity with respect to the states of the three probabilistic entities. GMEP then spells out three principles, and based on these, three deductive procedures for the determination of the unspecified entity when the other two are specified.

The principal differences of GMEP with MaxEnt are: 1) Generalized measures of entropy are allowed; 2) The entropy maximization principle which is central to MaxEnt is elevated to the Entropy Maximization Postulate; 3) The two inverse problems associated with determining the most unbiased set of moment constraints, and the most unbiased measure of entropy are also included.

The chief motivation for developing the above model arose in the context of overcoming hurdles that we faced in the formulation and solution of problems in a wide variety of areas, particularly those in the socio- economic disciplines. We were faced with the problem, for instance, of having to explain more than one model that exists in a discipline on the basis of the maximization principle. In such cases, we found out that uniqueness as guaranteed by the Shannon measure in MaxEnt was not a desirable property. The concept of generalized measures had the merit of overcoming such difficulties. Very frequently, the property of additivity could not be insisted upon because of the nature of the problem. Moreover, it is well-known that two distinct generalized measures, one satisfying the additivity property, and the second which does not, can lead to an identical probability distribution upon maximization with respect to the same set of constraints.

It is well-known that Kullback's minimum cross-entropy entropy principle is found very useful when apriori probability distributions are included.     From this vantage point, MaxEnt takes on the meaning of minimization of a probabilistic distance of a probability distribution from the uniform distribution. In fact, in this development, the concept of entropy itself takes on the meaning of a monotonic decreasing function of the cross-entropy measure $D(P:U)$. This concept of entropy becomes useful in gaining further insight into the concept of generalized measure of entropy.

Parallel to our development of GMEP, the generalization of the minimum cross-entropy principle(MinxEnt), deals with four probabilistic entities which are governed by the global constraint of the EMP. These are: 1) a generalized cross-entropy measure which need not be restricted to the Kullback-Leibler measure, 2) moment constraints, 3)a posteriori probability distribution, and 4) a priori probability distribution. The MinxEnt provides a method of determining a most unbiased probabilistic entity through a process of minimization when the rest of three are specified.

In this paper, we have first recapitulated the main results of MaxEnt before proceeding to enunciate the principles underlying the GMEP. The model is illustrated with some examples. The distributions of statistical mechanics are discussed from the fresh viewpoint not because of any novelty of results but on account of their historic importance in connection with MaxEnt. We give a few more examples taken from other areas. For reasons of brevity, we have avoided discussion of existence and uniqueness of the results of the inverse problems. The already published literature on GMEP [6,11,12], and a companion paper by the same authors in this volume will further clarify the results of this paper.

## The Maximum Entropy Principle (Jaynes' Formalism)

The problem is to maximize the Shannon's entropy function

$$- \sum_{i=1}^{n} p_i \ln p_i \tag{1}$$

subject to the constraints

$$\sum_{i=1}^{n} p_i = 1, \quad \sum_{i=1}^{n} p_i \, g_r(x_i) = a_r, \quad r = 1, 2, \cdots, m \tag{2}$$

The Lagrangian is

$$L = - \sum_{i=1}^{n} p_i \ln p_i - (\lambda_0 - 1) \left[ \sum_{i=1}^{n} p_i - 1 \right] - \sum_{r=1}^{m} \lambda_r \left[ \sum_{i=1}^{n} p_i \, g_r(x_i) - a_r \right] \tag{3}$$

Maximizing $L$, we get

$$\ln p_i + \lambda_0 + \lambda_1 \, g_1(x_i) + \cdots + \lambda_m \, g_m(x_i) = 0; \quad i = 1, 2, \cdots, n \tag{4}$$

or

$$p_i = \exp \left[ -(\lambda_0 + \lambda_1 \, g_1(x_i) + \cdots + \lambda_m \, g_m(x_i)) \right]; \quad i = 1, 2, \cdots, n \tag{5}$$

The Lagrangian multipliers are determined by substituting (2) into (5) to get

$$e^{\lambda_0} = \sum_{i=1}^{n} e^{-\lambda_1 \, g_1(x_i) - \lambda_2 \, g_2(x_i) - \cdots - \lambda_m \, g_m(x_i)} \tag{6}$$

and

$$a_r \, e^{\lambda_0} = \sum_{i=1}^{n} g_r(x_i) \exp \left[ -\lambda_1 \, g_1(x_i) - \cdots - \lambda_m \, g_m(x_i) \right], \quad r = 1, 2, \cdots, m \tag{7}$$

For the case of the continuous random variate, the maximum entropy principle requires: Maximize

$$- \int_a^b f(x) \ln f(x) dx \tag{8}$$

subject to

$$\int_a^b f(x) \, dx = 1, \quad \int_a^b f(x) \, g_r(x) dx = a_r, \quad r = 1, 2, \cdots, m \tag{9}$$

The results of the discrete version can be extended to this case.

### The Generalized Maximum Entropy Principle (GMEP)

Let $\phi(\cdot)$ be a convex function, and let

$$H(P) = - \sum_{i=1}^{n} \phi(p_i) \tag{10}$$

be the generalized measure of entropy. Let the constraint be

$$\sum_{i=1}^{n} p_i = 1 \quad \text{and} \quad \sum_{i=1}^{n} p_i \, g_r(x_i) = a_r, \quad r = 1, 2, \cdots, m \tag{11}$$

Using the method of Lagrange multipliers, we maximize (11) subject to the $(m + 1)$ constraints in (11), and get an expression for the first derivative of $\phi(p_i)$ as,

$$-\phi'(p_i) = \lambda_0 + \lambda_1 \, g_1(x_i) + \lambda_2 \, g_2(x_i) + \cdots + \lambda_m \, g_m(x_i) \tag{12}$$

### The direct principle:

Given the entropy measure $\sum_{i=1}^{n} \phi(p_i)$ and the constraints on mean values $g_1(x_i), g_2(x_i), \cdots, g_m(x_i)$, we wish to determine the probability distribution that maximizes the entropy measure.

Since $\phi(p_i)$ is a convex function, $\phi'(p_i)$ is a monotonically increasing function of $p_i$, so that corresponding to every value of $\phi'(p_i)$, we have a unique value of $p_i$. Finding these unique values for $p_1, p_2, \ldots, p_n$ from (12) and substituting into (11), we get $(m+1)$ equations to solve for the $(m + 1)$ Lagrange multipliers, which, in turn, yield the probabilities $p_i$.

### The first inverse problem (determination of constraints):

Given the generalized entropy measure $H(P)$ and the probability distribution for $p_i$, determine one or more probability constraints which yield the given probability distribution when the entropy measure is maximized subject to these constraints.

Since we know $\phi(p_i)$, we also know $\phi'(p_i)$, and hence the RHS of (12) can be determined. This will allow us to identify the values for $g_1(x_i), \cdots, g_m(x_i)$ by matching terms, and thus, a most unbiased set of constraints (11).

### The second inverse problem (determination of the generalized entropy measure):

Given the constraint functions $g_1(x_i), g_2(x_i), \cdots, g_m(x_i)$ and the probability distribution $p_i$, determine the most unbiased generalized entropy measure, which, when maximized subject to the given constraints, yields the given probability distribution.

We substitute the given values into (12) and get a differential equation that can be solved for $\phi(\cdot)$. Once $\phi(\cdot)$ is known, we can determine the generalized entropy measure (10)

$$H(P) = - \sum_{i=1}^{n} \phi(p_i) \tag{13}$$

In order to illustrate the applications of the GMEP, we consider Maxwell-Boltzmann, Bose-Einstein and Fermi-Dirac distributions of Statistical Mechanics. Kapur [4] and Forte and Sempi [2] have solved the direct problems of finding these as maximum entropy distributions where the measures of entropy and constraints are specified.

Here we consider the inverse problems where we assume the distributions are known either from observations or theory. We also assume the validity of our Entropy Maximization Postulate and proceed on the basis that these are maximum entropy distributions subject to some appropriate constraints and for some appropriate measure of entropy. Determination of these latter quantities subject to EMP will constitute the inverse problems.

The examples chosen are first from Statistical Mechanics because that is where Jaynes' MEP was first applied. We have also discussed other examples from transportation engineering, queuing theory, and autoregressive process.

## 2. The Direct Principle of GMEP (single-variate case)

Let the entropy measure of the probability distribution $p_1, p_2, \cdots, p_n$ be

$$H(P) = - \sum_{i=1}^{n} \phi(p_i) \tag{14}$$

where $\phi(\cdot)$ is a convex function, and let the constraints be

$$\sum_{i=1}^{n} p_i = 1, \quad \sum_{i=1}^{n} p_i \, g_{ri} = a_r, \quad r = 1, 2, \cdots, m; \quad p_i \geq 0 \tag{15}$$

Then maximizing (14) subject to (15) by Lagrange's method, we get,

$$- \phi'(p_i) = (\lambda_0 - 1) + \lambda_1 \, g_{1i} + \cdots + \lambda_m \, g_{mi} \tag{16}$$

when $\lambda_0, \lambda_1, \cdots, \lambda_m$ are obtained by using the $(m+1)$ equations (15). For the case of Shannon's measure of entropy, $H(P) = - \sum_{i=1}^{n} p_i \ln p_i$, (16) gives

$$p_i = \exp\left[- \lambda_0 - \lambda_1 \, g_{1i} - \lambda_2 \, g_{2i} - \cdots - \lambda_m \, g_{mi}\right] \tag{17}$$

For the continuous-variate case, equations (14) to (17) are replaced by

$$H = - \int_a^b f(x) \ln f(x) \, dx \tag{18}$$

$$\int_a^b f(x) dx = 1, \quad \int_a^b f(x) \, g_r(x) dx = a_r, \quad r = 1, 2, \cdots, m \tag{19}$$

$$- \phi'[f(x)] = (\lambda_0 - 1) + \lambda_1 \, g_1(x) + \cdots + \lambda_m \, g_m(x) \tag{20}$$

and,

$$f(x) = \exp\left[-\lambda_0 - \lambda_1 \, g_1(x) - \lambda_2 \, g_2(x) - \cdots - \lambda_m \, g_m(x)\right] \tag{21}$$

## Bivariate case of MEP

In this case, corresponding to equations (14) to (21), we have

$$H(P) = -\sum_{j=1}^{k} \sum_{i=1}^{n} \phi\left(p_{ij}\right) \tag{22}$$

$$\sum_{j=1}^{k} \sum_{i=1}^{n} p_{ij} = 1 \; ; \quad \sum_{j=1}^{k} p_{ij} \, g_{ij} = a_i \; ; \quad i = 1, 2, \cdots, n \tag{23}$$

$$\sum_{i=1}^{n} p_{ij} \, h_{ij} = b_j \; ; \quad j = 1, 2, \ldots, k$$

$$\sum_{j=1}^{k} \sum_{i=1}^{n} p_{ij} \, k_{ij} = c$$

$$-\phi'\left(p_{ij}\right) = (\lambda_0 - 1) + \lambda_i \, g_{ij} + \mu_j \, h_{ij} + \nu \, k_{ij} \tag{24}$$

$$p_{ij} = \exp\left[-\lambda_0 - \lambda_i \, g_{ij} - \mu_j \, h_{ij} - \nu \, k_{ij}\right] \tag{25}$$

and, the Shannon measure

$$S = -\int_{c}^{d} \int_{a}^{b} \phi\left[f(x, y)\right] dx \; dy \tag{26}$$

$$\int_{c}^{d} \int_{a}^{b} f(x, y) dx \; dy = 1; \quad \int_{c}^{d} g(x, y) \, f(x, y) dy = a(x)$$

$$\int_{a}^{b} h(x, y) \, f(x, y) dx = b(y)$$

$$\int_{c}^{d} \int_{a}^{b} k(x, y) \, f(x, y) dx \; dy = c \tag{27}$$

$$-\phi'[f(x, y)] = \lambda_0 - 1 + \lambda(y) \, g(x, y) + \mu(x) \, h(x, y) + \nu \, k(x, y) \tag{28}$$

$$f(x, y) = \exp\left[-\lambda_0 - \lambda(y) \, g(x, y) - \mu(x) \, h(x, y) - \nu \, k(x, y)\right] \tag{29}$$

## 3. Examples

**a) Maxwell-Boltzmann distribution:** In this example, we fix the two entities, namely, the probability distribution and the entropy measure which is taken as due to Shannon. The problem is to determine the constraint under which the distribution is a maximum entropy distribution. In such a case, according to GMEP, the entire set of the three probabilistic entities should constitute an unbiased set.

Let the probabilities of a particle being in the $n$ energy levels with energies $\epsilon_1, \epsilon_2, \cdots, \epsilon_n$ be

$$p_i = \frac{e^{-\mu \epsilon_i}}{\sum\limits_{i=1}^{n} e^{-\mu \epsilon_i}}, \quad i = 1, 2, \cdots, n \tag{30}$$

where $\mu$ is a constant.

We rewrite (30) as

$$p_i = e^{-\lambda_o - \lambda_1 g_{1i}} \tag{31}$$

where

$$\lambda_0 = \ln \sum\limits_{i=1}^{n} e^{-\mu \epsilon_i}, \quad g_{1i} = \epsilon_i, \quad \lambda_1 = \mu \tag{32}$$

Using (15), (17), (31) and (32), we get the constraints

$$\sum\limits_{i=1}^{n} p_i = 1 \quad \text{and} \quad \sum\limits_{i=1}^{n} p_i \, \epsilon_i = \text{constant} \tag{33}$$

Thus this distribution can be characterized as a maximum entropy distribution when the values of the mean energy is prescribed. Accordingly, it means that the knowledge of the mean energy of the system is enough to give us the Maxwell-Boltzmann distribution via the ME distribution.

**b) Bose-Einstein distribution:** In this example, we illustrate the procedure to determine most unbiased generalized entropy function, given the constraints and the distribution. The Bose-Einstein distribution is given as

$$p_i = \frac{\bar{n}_i}{N} = \frac{1}{N} \frac{1}{(e^{\lambda + \mu \epsilon_i} - 1)}, \quad i = 1, 2, \cdots, n \tag{34}$$

where $\lambda$ and $\mu$ are constraints to be determined from the equations

$$\sum\limits_{i=1}^{n} \bar{n}_i = N, \quad \sum\limits_{i=1}^{n} \bar{n}_i \, \epsilon_i = N \bar{\epsilon}. \tag{35}$$

$N$ is the expected number of particles in the system and $\bar{\epsilon}$ is the average energy per particle; $\bar{n}_i$ is the expected number of particles in the $i^{th}$ energy level:

$$\bar{n}_i = \frac{1}{(e^{\lambda + \mu \epsilon_i} - 1)}; \quad i = 1, 2, \cdots, n \tag{36}$$

First we note that (34) cannot be the maximum-entropy distribution if we use Shannon's measure of entropy, on the basis of comparison with (17). This takes it outside the scope of an inverse problem of Jaynes' MEP. Accordingly, we appeal to the GMEP, and by using (16), we get

$$\phi' \left[ \frac{1}{N} \frac{1}{e^{\lambda + \mu \epsilon_i} - 1} \right] = (\lambda_0 - 1) + \lambda_1 \epsilon_i \tag{37}$$

or

$$\phi'(x) = (\lambda_0 - 1) + \lambda_1 \left[ \frac{1}{\mu} (\ln \frac{1 + Nx}{Nx} - \lambda) \right] \tag{38}$$

or

$$\phi'(x) = a + b \ln \frac{1 + Nx}{Nx}, \tag{39}$$

where $a$ and $b$ are constants. Integrating

$$\phi(x) = ax + \frac{b}{N} (1 + Nx) \ln (1 + Nx) - b (x \ln x) + c \tag{40}$$

so that the unbiased entropy measure is [4, 5]

$$H(P) = - \sum_{i=1}^{n} p_i \ln p_i + \frac{1}{N} \sum_{i=1}^{n} (1 + N p_i) \ln (1 + N p_i) - \frac{1}{N} (1 + N) \ln (1 + N) \tag{41}$$

Had we insisted on using Shannon's measure only, a different set of constraints would have yielded the given distribution as the maximum entropy distribution. In subsection c, we shall illustrate this by finding the appropriate constraints for Fermi-Dirac distribution where use of the Shannon measure is insisted upon. In general, in such problems, when solutions exist, there is a trade-off between a generalized entropy measure along with some simple constraints on the one hand, and Shannon's measure and a more complicated set of moment constraints on the other.

c) **Fermi-Dirac distribution**: Here we are given that the expected number of particles in the $i^{th}$ energy level is

$$\bar{n}_i = \frac{1}{e^{\lambda + \mu \epsilon_i} + 1} , \quad i = 1, 2, \cdots , n \tag{42}$$

We are also given that the number of particles in each state can be 0 or 1 so that $j$ can vary from 0 to 1.

If $p_{ij}$ is the probability of there being $j$ particles in the $i^{th}$ energy level, we get

$$\sum_{j=0}^{1} p_{ij} = 1, \quad \sum_{j=0}^{1} j \, p_{ij} = \frac{1}{e^{\lambda + \mu \epsilon_i} + 1} ; \quad i = 1, 2, \cdots , n; \tag{43}$$

Using Shannon's measure of entropy, we get the maximum entropy distribution as

$$p_{ij} = c_i d_i^j, \quad \frac{d_i}{1 - d_i} = \frac{1}{e^{\lambda + \mu \epsilon_i} + 1}, \quad d_i = e^{-(\lambda + \mu \epsilon_i)} \tag{44}$$

$$c_i = \frac{1}{e^{-(\lambda + \mu \epsilon_i)}}; \quad p_{ij} = \frac{e^{-(\lambda + \mu \epsilon_i)j}}{1 - e^{-(\lambda + \mu \epsilon_i)}} \tag{45}$$

Comparing it with the probability distribution

$$p_{ij} = \exp\left(-\lambda_i'' - \lambda_j' - \mu' \, \epsilon_i j\right) \tag{46}$$

we get the constraints

$$\sum_{j=0}^{1} p_{ij} = 1, \quad i = 1, 2, \cdots, n; \quad \sum_{i=1}^{n} \sum_{j=0}^{1} j \, p_{ij} = N \tag{47}$$

$$\sum_{i=1}^{n} \epsilon_i \sum_{j=0}^{1} j \, p_{ij} = N \bar{\epsilon} \tag{48}$$

Thus the Fermi-Dirac distribution is characterized by the mean number of particles in the system and by the mean energy per particle if we use Shannon's measure of entropy. It will be characterized by the mean energy only if we use the entropy measure

$$H(P) = -\sum_{i=1}^{n} p_i \ln p_i - \frac{1}{N} \sum_{i=1}^{n} (1 - N p_i) \ln(1 - N p_i) + \frac{1}{N} (1 - N) \ln(1 - N) \tag{49}$$

where $0 < p_i < 1$. The measure (49) can of course be deduced in the same way also.

**d) Urban and Regional Planning** [10]: Let there be $m$ residential colonies and let $O_1, O_2, \cdots, O_m$ be the number of office workers living in these colonies. Let there be $n$ office blocks and let $D_1, D_2, \cdots, D_n$ be the number of workers in these offices so that

$$O_1 + O_2 + \cdots + O_m = D_1 + D_2 + \cdots + D_n \tag{50}$$

Let $T_{ij}$ be the number of trips made from the $i^{th}$ colony to the $j^{th}$ office block. The obvious constrains are

$$\sum_{j=1}^{n} T_{ij} = O_i, \quad \sum_{i=1}^{m} T_{ij} = D_j, \quad \sum_{j=1}^{n} \sum_{i=1}^{m} T_i = \sum_{i=1}^{m} O_i = \sum_{j=1}^{n} D_j = T \tag{51}$$

The problem of estimation of $T_{ij}$ has engaged the attention of transportation engineers for about fifty years. By a process of evolution, the successive models were refined till finally, the law

$$T_{ij} = A_i \, B_j \, O_i \, D_j \exp\left[-\nu \, c_{ij}\right] \tag{52}$$

fitted the observations reasonably well, where $c_{ij}$ is the cost of travel from the $i^{th}$ residential colony to the $j^{th}$ office block.

This result can be confirmed on the basis of GMEP. In this inverse principle, we fix Shannon's measure of entropy and $T_{ij}$. Assuming that (52) is a maximum entropy distribution, we rewrite (52) as:

$$T_{ij} \propto \exp\left[-\lambda_i - \mu_j - \nu \, c_{ij}\right] \tag{53}$$

which shows that the constraints are [refer to equations (14) to (25)]

$$\sum_{j=1}^{n} T_{ij} = O_i \quad i = 1, 2, \cdots, m$$

$$\sum_{i=1}^{m} T_{ij} = D_j \quad j = 1, 2, \cdots, n$$

$$\sum_{j=1}^{n} \sum_{i=1}^{m} T_{ij} c_{ij} = \text{constant} \tag{54}$$

Thus, by using the inverse principle, we conclude that the constraint must be on the mean, that is, the average cost of travel is a dominant consideration in the estimation of the number of trips. It can easily be shown on the basis of GMEP that the other sets of distributions which were suggested during the evolution of this transportation model do not satisfy the constraints in (51).

e) **Queue size distribution for an M/M/1 System** [4, 8]:

Assume Poisson arrival and exponential service-time distributions with mean arrival and service rates as $\lambda$ and $\mu$. For these distributions, it is well known that the probability that there are $n$ persons in the queue is given by

$$p_n = \frac{1 - \rho}{1 - \rho^{N+1}} \cdot \rho^n, \quad n = 0, 1, 2, \cdots, N \tag{55}$$

where $\rho = \dfrac{\lambda}{\mu}$ and $N$ is the maximum permitted size of the queue. We want to determine the constraint using an inverse principle of GMEP.

Accordingly, consider the distribution in (55) as a maximum entropy distribution pertaining to Shannon entropy and rewrite it as

$$p_n = e^{-\lambda_0 - \lambda_1 n}; \quad e^{-\lambda_0} = \frac{1 - \rho}{1 - \rho^{N+1}}, \quad e^{-\lambda_1} = \rho \tag{56}$$

We find that the most unbiased constraints, or the constraints under which (56) will be a maximum-entropy distribution, are

$$\sum_{n=0}^{N} p_n = 1, \quad \sum_{n=0}^{N} n\, p_n = \frac{1}{1 - \rho^{N+1}} \left[ \frac{\rho(1 - \rho^N)}{1 - \rho} - N\,\rho^{N+1} \right] \tag{57}$$

Thus, according to this theory, the M/M/1 queue size distribution is characterized by its mean queue length so that if the latter alone is known, the queue size distribution can be recovered via the maximum-entropy principle.

Though apparently, there are two parameters $\lambda$ and $\mu$, essentially, there is only one parameter of the distribution. It may be taken as either mean queue length or as the ratio $\dfrac{\lambda}{\mu}$. In the former case, it obviates the necessity of determining the arrival and service-time distributions. We can observe only the mean length of the queue and then derive the queue-size distribution by using the MEP.

**f) Auto-Regressive Process**:

This example will illustrate the second inverse principle of GMEP.

Let the generalized entropy measure by

$$\int_{-1/2}^{1/2} \phi[S(f)]df \tag{58}$$

where $\phi(\cdot)$ is a twice-differentiable concave function, and let

$$S(f) = \frac{1}{\lambda_0 + \sum_{k=-m}^{m} \lambda_k \exp(-2\pi i f k)} \tag{59}$$

The moment constraints are on the auto-covariance functions:

$$A_k = \int_{-1/2}^{1/2} S(f)\exp(-2\pi i f k)df \quad k=-m, \ldots, 0, \ldots, +m \tag{60}$$

Maximize (58) subject to (60):

$$\phi'[S(f)] = \lambda_0 + \sum_{k=-m}^{m} \lambda_k \exp(-2\pi i f k) \tag{61}$$

>From (59) and (61):

$$\phi'[S(f)] = \frac{h}{S(f)} \tag{62}$$

$$\phi[S(f)] = h \ln S(f) + constant \tag{63}$$

$\ln S(f)$ is a concave function of $S(f)$, so that (ignoring the additive and multiplicative constants) the required measure of entropy is

$$\int_{-1/2}^{1/2} \ln S(f)df \tag{64}$$

We recognize that the entropy measure in (64) was derived from entirely different assumptions by Burg in his well-known work on spectral analysis. Our purpose here was to demonstrate that his entropy measure, the moment constraints specified by auto-covariance functions in (60), and the spectral density function in (59) constitute an unbiased set according to the GMEP.

Similarly, we can also derive the results that the Havrda-Charvat entropy of the second order describes the moving average process, and the Kullback-Leibler measure describes the ARMA process. The latter is a good example to illustrate the Minimum Cross-Entropy principle.

The above examples serve to illustrate the scope of the principles comprising the GMEP. In complicated examples, computational algorithms become necessary to determine the most unbiased probabilistic entity.

## 4. Some Additional Observations on the GMEP

It is well known that almost all usually encountered statistical distributions can be obtained as maximum entropy distributions corresponding to some simple moment constraints. MaxEnt, in effect, links the three entities: Shannon's entropy function, the statistical distributions and the moment constraints. This interlocking is rendered possible by the simple and elegant nature of the Shannon's entropy function, and the fact that most statistical distributions can be expressed analytically in terms of exponential functions.

But we do encounter distributions in application areas which cannot be expressed in terms of exponential functions, in which case, they may very well elude the possibility of being treated as maximum-entropy distributions in the sense of Jaynes.

In the above context, the complicated distributions, which do not have an exponential form, are not necessarily linked either by Shannon's measure or by simple moment constraints. The GMEP provides the principles for addressing these questions.

Although the EMP is stated as a mathematical postulate, it also possesses the semblance of a theory. We hesitate to refer to it as a "physical law" since it also applies to many socio-economic systems which come under the purview of probabilistic laws. One could refer to it as a law of probabilistic systems which places the mutual interaction of the three probabilistic systems in vivid display. In fact, with a knowledge of EMP and GMEP, one could, in hindsight, view the problems tackled by Jaynes' MEP with a more comprehensive understanding.

We have not dealt with the questions of existence and uniqueness associated with the solutions of the GMEP. These have been published elsewhere [11]. Suffice it to say, they do not pose serious difficulties in the solution process.

## 5. Conclusions

The paper has presented arguments in support of using generalized measures of entropy in connection with the maximum entropy principle in those instances where the use of Shannon entropy is found inappropriate. On the basis of the Entropy Maximization Postulate, it has established the Generalized Maximum Entropy Principle where the mutual interaction of all the three probabilistic entities is viewed in a comprehensive way. Thus, it has rendered it possible to view the direct principle along with the two associated inverse principles, and their respective implementations within one single conceptual framework.

Departure from the exclusive use of Shannon entropy in the context of the GMEP has raised some serious objections. We have tried to answer some of these and provide our reasons for enunciating the new principle. There is one more objection which can be summarized like this: 'the flexibility of GMEP can be equated with the flexibility which is a characteristic feature of variational principles'. In answering this criticism we can only point out, that the 'choice function' which is the generalized measure of entropy, arises as a result of a deductive procedure based on the second inverse principle. It is not something that is identified from a trial and error procedure to fit the particular situation. Furthermore, the rationale for the second inverse principle is identical to the rationale for the first inverse principle which has to do with the determination of moment constraints. It is interesting to note that virtually no objection is raised to the latter principle which is also an integral part of GMEP.

In this paper, we have not discussed either the nature of solutions to the inverse problems of GMEP or their computational aspects. These were beyond the scope of the paper. Also, we have made only a statement of it. We found it necessary to allude to it, since it is an integral part of the thinking underlying the generalization. Furthermore, the concept of entropy arising from Kullback's minimum cross-entropy principle is more appropriate for the development of GMEP.

## 6. References

1.   El-Affendi, "A Maximum Entropy Analysis of the M/G/I and G/M/I Queuing Systems at Equilibrium", Acta Information 19, pp. 339-355, (1983).

2.   Forte, B. and C. Sempi, "Maximizing Conditional Entropy: A derivation of quantal statistics", Rendi Conla de Mathematics, Vol. 9, pp. 551-566, (1976).

3.   Jaynes E.T., "Information Theory and Statistical Mechanics", Physical Reviews, Vol. 106, pp. 620-630, (1957).

4.   Kapur J.N., "Measures of uncertainty, mathematical programming and physics", Jour. Ind. Soc. Agri. Stat., Vol. 24, pp. 47-66, (1972).

5.   Kapur J.N., "Four families of measure of entropy", Ind. Jour. Pure and App. Maths., Vol. 17, No. 4, pp. 429-449, (1986).

6.   Kapur J.N. and H.K. Kesavan, The Generalized Maximum Entropy Principle Sandford Education Press, Waterloo University, Waterloo, Canada, (1987).

7.   Kullback S., Information Theory and Statistics, John Wiley New York (1959).

8.   Shore J. E., "Derivation of Equilibrium and Time-dependent Solution for Queuing Systems using Entropy Maximization", Nat. Computer. Cong., Vol. 47, pp. 483-487, (1978).

9.   Tribus M., "Information Theory as the basis for Thermostatics and Thermodynamics", J. Appl. Mechanics, Vol. 28, p. 106, (1961).

10.  Wilson A.G., Entropy in Urban and Regional Modelling, Pion, London, (1970).

11.  Kesavan, H.K. and J.N.Kapur, "The Generalized Maximum Entropy Principle", IEEE SMC Vol. 19, no.5, Sept/Oct 1989.

12.  Kesavan, H.K. and J.N.Kapur, "On the Families of Solutions to Generalized Maximum Entropy and Minimum Cross-Entropy Problems," Int. J. General Systems, 1989.

# The Inverse MaxEnt and MinxEnt Principles and Their Applications

by

J. N. Kapur, Adjunct Professor,

and

H. K. Kesavan, Professor,

University of Waterloo, Canada

## Abstract

Two inverse MaxEnt (maximum entropy) and two inverse MinxEnt(minimum cross- entropy) principles are stated followed by some applications taken from diverse fields. A case is made out for the use of generalized measures of entropy and cross-entropy which naturally arise in such inverse principles.

## Key Words

Maximum Entropy/ Minimum Cross-Entropy/ Inverse Problems

## 1 Direct and Inverse Entropy Optimisation Principles

Jaynes [16] principle of maximum entropy states that given any constraints

$$\sum_{i=1}^{n} p_i = 1; \qquad \sum_{i=1}^{n} p_i g_r(x_i) = a_r, \qquad r = 1, 2, \cdots, m; \quad p_i \geq 0; \quad m + 1 \ll n \tag{1}$$

on a probability distribution $P = (p_1, p_2, \cdots, p_n)$, we should choose that probability distribution $P$ which satisfies the constraints (1) and minimizes Shannon's [37] entropy measure

$$S(P) = -\sum_{i=1}^{n} p_i \ln p_i \tag{2}$$

Kullback's [32] principle of minimum cross-entropy states that given the constraints (1) and given a priori probability distribution $Q = (q_1, q_2, \cdots, q_n)$,

*P. F. Fougère (ed.), Maximum Entropy and Bayesian Methods*, 433–450.
© 1990 *Kluwer Academic Publishers.*

we should choose that probability $P$ which satisfies (1) and minimizes Kullback Leibler [33] measure of cross-entropy

$$D(P:Q) = \sum_{i=1}^{n} p_i \ln \frac{p_i}{q_i} \qquad (3)$$

The generalised principle of maximum entropy states that we should choose that probability distribution $P$ which maximizes any specified measure of entropy for the given constraints (1) (Shannon [37], Renyi [36], Havrda-Charvat [14], Burg [8], Ferreri [10], Sharma-Mittal [40], Sharma-Taneja [39], Aczel-Daroczy [1], Behara-Chawla [6], Kapur [17, 18, 20, 22]).

Similarly the generalised principle of minimum cross-entropy states that given the constraints (1) and a priori probability distribution $Q$, we should choose that probability distribution $P$ which satisfies (1) and minimizes any specified measure of directed divergence or cross-entropy (Kullback-Leibler [33], Renyi [36], Havrda-Charvat [14], Csiszer [9], Hellinger [35], Bhattacharya [7], Kapur [21]).

If $Q$ is the uniform distribution, the generalised minimum cross-entropy reduces to the generalised maximum entropy principle.

All these four principles are **direct** principles. These are concerned with finding MaxEnt and MinxEnt probability distributions when the constraints and a priori probability distributions are given. In the **inverse** problems, a probability distribution is given and we have to find either (i) the constraints or (ii) the measure of entropy or cross-entropy or (iii) the a priori probability distribution, so that the given probability distribution is a MaxEnt or MinxEnt probability distribution.

The philosophy underlying these inverse problems is provided by the **Entropy Maximization/ Cross-Entropy Minimization Postulate and the Generalized Maximum Entropy/Generalized Minimum Cross-Entropy Principles** (Kapur and Kesavan [26, 27, 28], Kapur [23], Kesavan and Kapur [29, 30, 31]). According to these, the four entities *viz.*,

1. a priori probability distribution $Q$

2. the measure of entropy $H(P)$ or measure of cross-entropy $D(P:Q)$

3. the constraints set $C$

4. the a posteriori probability distribution $P$

are so interlinked that the entropy of $P$ (the cross-entropy of $P$ relative to $Q$) is maximum (minimum).

We now state our four inverse principles explicitly.

1. **First Inverse MaxEnt Principle:** Given a probability distribution $P$ and given a specific measure of entropy (say Shannon's) find the set of constraints for which $P$ is the maximum entropy probability distribution.

2. **Second Inverse MaxEnt Principle:** Given a probability distribution $P$ and given a set of constraints $C$, find a measure of entropy for which $P$ is a maximum-entropy probability distribution for the given set of constraints.

3. **First Inverse MinxEnt Principle:** Given a posteriori and a priori probability distributions $P$ and $Q$ and given a specific measure of cross-entropy (say Kullback-Leibler's), find the set of constraints for which $P$ is a minimum cross-entropy probability distribution.

4. **Second Inverse MinxEnt Principle:** Given a posteriori and a priori probability distributions and given a set of constraints $C$, find the measure of cross-entropy for which $P$ is a minimum cross-entropy probability distribution.

There are two other inverse principles concerned with finding the a priori probability distribution, but we shall not consider these here.

Sections 2 and 3 of the present paper will give some applications of the inverse MaxEnt and MinxEnt principles and section 4 will establish a case for the use of generalised measures of entropy and cross-entropy. Section 5 will discuss the implications of the inverse problems for the analysis of probabilistic systems.

# 2 Some Applications of the First Inverse MaxEnt and MinxEnt Principles

## 2.1 Characterisation of probability distributions as maximum-entropy probability distributions

If we maximize (2) subject to (1), we get, by using Jaynes' maximum-entropy principle,

$$p_i = \exp(-\lambda_0 - \lambda_1 g_1(x_i) - \cdots - \lambda_m g_m(x_i)) \tag{4}$$

where $\lambda_0, \lambda_1, \cdots, \lambda_m$ are obtained by using (1). Conversely, given the probability distribution (4), we can say that it is characterised by the moment
functions $g_1(x), g_2(x), \cdots, g_m(x)$ (or any $m$ linearly independent functions of
these) or we can say it is characterised by the moments $E(g_1(x)), E(g_2(x)), \cdots,$
$E(g_m(x))$. In other words, if (4) is given, then it can be obtained as a
maximum-entropy probability distribution when the values of some moments
$E(g_1(x)), E(g_2(x)), \cdots, E(g_m(x))$ are prescribed. Thus the knowledge of the
moments is in some sense equivalent to the knowledge of the probability dis-
tribution(P.D.), since the P.D. can be obtained from the knowledge of the
moments via MaxEnt.

For continuous variate distributions, we obtain the maximum entropy
P.D. by maximizing

$$- \int_a^b f(x) \ln f(x) dx \tag{5}$$

subject to

$$\int_a^b f(x) dx = 1, \quad \int_a^b f(x) g_r(x) dx = a_r, \quad r = 1, 2, \cdots, m, \quad f(x) \geq 0, \quad m+1 \ll n \tag{6}$$

to get

$$f(x) = \exp(-\lambda_0 - \lambda_1 g_1(x) - \cdots - \lambda_m g_m(x)) \tag{7}$$

The characterizing moments for some important distributions are given
in the following table: [23]

| Distribution | Moments | Distribution | Moments |
|---|---|---|---|
| Exponential | $E(x)$ | normal | $E(x), E(x^2)i$ |
| Gamma | $E(x), E(\ln x)$ | Beta(first kind) | $E(\ln x), E(\ln(1 - x))$ |
| Beta(second) | $E(\ln x), E(\ln(1 + x))$ | lognormal | $E(\ln x), E(\ln x^2)$ |
| multivariate normal | $E(x_i), E(x_i x_j)$ | Dirichlet | $E(\ln x)$ |

## 2.2 (b) Characterizing of Probability Distribution as Mini-mum Cross-Entropy Distribution

If the apriori probability distribution is $(q_1, q_2, \cdots, q_n)$ or if the apriori density function is $g(x)$, then (4) and (7) are replaced by:

$$p_i = q_i \exp(-\lambda_0 - \lambda_1 g_1(x_1) - \cdots - \lambda_m g_m x_m) \tag{8}$$

and,

$$f(x) = g(x) \exp(-\lambda_0 - \lambda_1 g_1(x) - \cdots - \lambda_m g_m x) \tag{9}$$

The probability distributions (8) and (9) are still characterized by the moment functions $g_1(x), \cdots, g_m(x)$ or the moment functions $E(g_1(x)), \cdots, E(g_m(x))$. The appropriate apriori probability distributions and moments for some standard distributions are given in the following table.

| Distribution | Apriori Distribution | Characterizing Moments |
|---|---|---|
| Binomial | $nc_r(\frac{1}{2})^n$ | mean |
| Poisson | $\frac{e^{-1}}{r!}$ | mean |
| Negative Binomial | $\frac{n}{n+r}(n + r_{c_r})$ | mean |
| log series | $\frac{1}{r}$ | mean |
| Multinomial | $\frac{N!}{r_1! r_2! \cdots r_n!}(\frac{1}{n})^N$ | $E(r_1), E(r_2), \cdots, E(r_n)$ |

## 2.3 (c) Pareto's Income Law Distribution and The Logarithmic Utility Functions

According to Pareto's law of income distribution[26],

$$f(x) = (\frac{x}{\theta})^{-(\theta+1)}, \qquad x \geq \theta \tag{10}$$

so that the characterizing moment is $E(\ln x)$. Thus the population in which Pareto's law holds is characterized by the expected value of the logarithm of the income. This is alternately expressed by saying that the utility function for this population is logarithmic. Thus we have deduced the utility function for the population by using the inverse maximum-entropy principle.

## 2.4 (d) Market Share of A Brand

Let $m_i$ be the market share and $p_i$ be the price of the ith brand of a product, $(i = 1, 2, \cdots, n.)$ Then if the purchase budget is fixed, we maximize the entropy $- \sum_{i=1}^{n} m_i \ln m_i$ subject to

$$\sum_{i=1}^{n} m_i = 1, \qquad \sum_{i=1}^{n} m_i p_i = B \tag{11}$$

to get

$$m_i = AC^{p_i} \tag{12}$$

If it is known that the market share of a brand is proportional to some power $\alpha$ of its attractiveness index, that is, if it is known that

$$m_i = k_i a_i^{\alpha}, \tag{13}$$

then from (12) and (13)

$$p_i \propto \ln a_i \tag{14}$$

Thus we have used the inverse maximum-entropy principle to show that if the market shares of different brands are proportional to some powers of their attractiveness indices, then prices of the brands are likely to be proportional to the logarithms of these indices [23].

## 2.5 (d) Product-Form Probability Distribution for Closed Queuing Network Model of A Flexible Manufacturing System

It is well known that probability of $n_1, n_2, \cdots, n_m$ jobs waiting in the m queues in front of m machines with normalized workloads $x_1, x_2, \cdots, x_m$ is given by

$$P(n_1, n_2, \cdots, n_m) = \frac{x_1^{n_1} x_2^{n_2} \cdots x_m^{n_m}}{\sum_{i=1}^{n} x_1^{n_1} x_2^{n_2} \cdots x_m^{n_m}} \tag{15}$$

where $\Sigma$ denotes summation over all non-negative integer solutions of

$$n_1 + n_2 + \cdots + n_m = N \tag{16}$$

and N is the total number of jobs in the closed network system. We can write (15) as

$$P(n_1, n_2, \cdots, n_m) = \exp(\lambda + n_1 \ln x_1 + n_2 \ln x_2 + \cdots n_m \ln x_m), \tag{17}$$

so that the product-form probability distribution (15) is the maximum-entropy probability distribution when the constraints are:

$$E(n_1) = a_1, E(n_2) = a_2, \cdots, E(n_m) = a_m, \tag{18}$$

so that the characterizing moments for (15) are the mean lengths of the queues [25].

The probability distribution (15) was first obtained on the assumption of exponential inter-arrival time distribution and Poisson service-time distribution. While (15) was found to be in excellent agreement with observations, the arrival and service distributions were different from those assumed. As such, efforts were made to find other arrival and service time distributions which could lead to (15) and a number of those were found. The distribution (15) was thus quite robust with respect to arrival and service time distributions and attempts were made to explain this fact. However, the real explanation was found in the inverse maximum entropy principle which showed that (15) was robust because it was characterized by the most natural constraints, namely, the mean lengths of the m queues.

## 2.6   (f) Earthquake-Frequency-Magnitude Relation

Purcaro [34] gave the empirical relationship

$$\ln N(x) = a - bx + \ln(c - x); \qquad x < c \tag{19}$$

where $N(x)$ is the number of earthquakes of intensity $\geq x$ in a certain region in an observed period of time, and $a, b, c, K$ are corresponding regional seismic parameters obtained empirically. If $T$ is the total number of earthquakes in this period in the region, then $\frac{N(x)}{T}$ can be regarded as the probability of an earthquake of intensity $\geq x$. If $f(x)$ is the probability density function, we get

$$f(x) = \exp(a - \ln(\frac{T}{b}) + (K - 1)\ln(c - x) + \ln(c + \frac{K}{b} - x) \tag{20}$$

This shows that $f(x)$ can be a maximum-entropy density function if the constraints are:

$$\int f(x)dx = 1, \int x f(x)dx = A, \int f(x)\ln(c - x)dx = \ln B, \text{and} \tag{21}$$
$$\int f(x)\ln(c + \frac{K}{b} - x)dx = \ln c$$

In other words, the seismic characteristics of the region are expressed in terms of three mean values, namely,

1. mean values of the earthquake intensity

2. geometric mean of the deficit of this intensity from a value c

3. geometric mean of the deficit of this intensity from another value $c + \frac{K}{b}$

The inverse maximum entropy principle suggests that we investigate the seismic interpretation of the parameters c and $c + \frac{K}{b}$ [26].

# 3  Applications of The Second Inverse MaxEnt And MinxEnt Principles

## 3.1  (a) Population Growth Models

Let $N(t)$ be the population at time t. If we assume the natural constraints that the total population size and the average age of the population are prescribed, we get the constraints [19]

$$\int_a^b N(t)dt = A, \qquad \int_a^b tN(t)dt + B \tag{22}$$

Let the entropy measure be

$$S = -\int_a^b \Phi N(t)dt \tag{23}$$

where $\Phi(.)$ is a twice-differentiable convex function; then given any population growth model,which we regard as a maximum-entropy model, we get, by using our second inverse maximum entropy principle a corresponding measure of entropy. In particular, we get:

$$\frac{dN}{dt} = kN \longleftrightarrow S = -\int_a^b N(t)\ln N(t)dt \tag{24}$$

$$\frac{dN}{dt} = N(c-dN) \longleftrightarrow S = -\int_a^b N(t)\ln N(t)dt + \frac{1}{d}\int_a^b (c-dN(t))\ln(c-dN(t))dt \tag{25}$$

$$\frac{dN}{dt} = (d+lN)^{-b} \longleftrightarrow S = \frac{1}{1-\alpha}\int_a^b (N^\alpha(t) - N(t))dt \tag{26}$$

$$\frac{dN}{dt} = \frac{1}{e^{(-d-ft)}}\pm 1 \longleftrightarrow -\int_a^b bN(t)\ln N(t)dt - \int_a^b (A\mp N(t))\ln(A\mp N(t))dt \tag{27}$$

$$\frac{dN}{dt} = (a_1 - b_1 N)\cdots(a_k - b_k N) \longleftrightarrow -\int_a^b bN(t)\ln N(t)dt$$

$$- \sum_{r=1}^k \frac{a_r}{b_r^2}\int_a^b (a_r - b_r N(t))\ln(a_r - b_r N(t))dt \tag{28}$$

Use of Shannon's measure will lead only to the Malthus law of population growth, while there are other population growth models, specially the logistic law [19], [24] which are more often applicable. This law leads to a certain limiting size of the population growth, and this characteristic is also shared by the corresponding entropy measure.

## 3.2   (b) Technological Innovation Diffusion Models

Let $f(t)$ be the proportion of potential adopters of a new technological innovation which has adopted the innovation till time t, and let the constraints
be:

$$\int_a^b f(t)dt = A, \qquad \int_a^b tf(t)dt = B \tag{29}$$

Then, corresponding to every innovation-diffusion model, we get a corresponding entropy measure. Thus we get: **Fisher-Pry Model** [1]

$$\frac{1}{c}\frac{df}{dt} = f(1-f) \tag{30}$$

given

$$S = -\int_a^b f \ln f dt - \int_a^b (1-f)\ln(1-f)dt \tag{31}$$

**Bass Model** [4]

$$\frac{df}{dt} = (p+qf)(1-f) \tag{32}$$

given

$$S = -\int_a^b (1-f)\ln(1-f)dt + \frac{1}{q}\int_a^b (p+qf)\ln(p+qf)dt \tag{33}$$

**Floyd Model** [12]

$$\frac{1}{c}\frac{df}{dt} = f(1-f)^2 \tag{34}$$

given

$$S = -\int_a^b f \ln f dt - \int_a^b (1-f)\ln(1-f)dt - \int_a^b \ln(1-f)dt \tag{35}$$

**Sharif-Kabir model** [39]

$$\frac{1}{c}\frac{df}{dt} = \frac{cf(1-f)^2}{1-f+\sigma f} \tag{36}$$

given

$$S = -\int_a^b f \ln f dt \int_a^b (1-f)dt - \sigma \int_a^b \ln(1-f)dt \tag{37}$$

Gompertz Model [13]

$$\frac{1}{c}\frac{df}{dt} = -f \ln f \tag{38}$$

given

$$S = -\int_a^b f(t) \ln f(t) df \tag{39}$$

Shannon's measure leads only to the Gompertz model. The most popular model due to Fisher and Pry requires a Fermi-Dirac type of measure of entropy.

## 3.3    (c) Spectral Density Function

We are given $(2m + 1)$ autocovariance functions

$$A_k = -\int_{-\frac{1}{2}}^{\frac{1}{2}} \exp(-2\pi i f k) S(f) df; \quad k = -m, \cdots, m \tag{40}$$

and we have to find the entropy measure such that by maximizing it subject to (40), we get a spectral density function. We get:

1. Autoregressive(AR) SDF

$$\frac{1}{\sum_{k=-m}^m \lambda_k \exp(-2\pi i f k)} \tag{41}$$

leads to the Burg [8] entropy measure

$$\int_{-\frac{1}{2}}^{\frac{1}{2}} \ln S(f) df \tag{42}$$

2. Moving Average(MA) SDF

$$\sum_{k=-m}^m \lambda_k \exp(-2\pi i f k) \tag{43}$$

leads to Havrda-Charvat [14] entropy measure

$$\int_{-\frac{1}{2}}^{\frac{1}{2}} (1 - S^2(f)) df \tag{44}$$

3. Autoregressive moving average (ARMA) SDF:

$$\frac{\sum_{k=-n}^{n} \mu_k \exp(-2\pi i f k)}{\sum_{k=-m}^{m} \lambda_k \exp(-2\pi i f k)} \tag{45}$$

leads to Kullback-Leibler [33] measure of cross-entropy of $T(f)$ from $S(f)$. (and not vice-versa):

$$-\int_{-\frac{1}{2}}^{\frac{1}{2}} T(f) \ln \frac{T(f)}{S(f)} df \tag{46}$$

4. Exponential SDF:

$$\exp\left[-\sum_{k=-m}^{k=m} \lambda_k \exp(-2\pi i f k)\right] \tag{47}$$

leads to the entropy measure

$$-\int_{-\frac{1}{2}}^{\frac{1}{2}} S(f) \ln S(f) df \tag{48}$$

Thus Shannon's measure leads to an exponential SDF while, in practice, we get other spectral density functions also. In fact the range of the measure

$$\frac{1}{\alpha(1-\alpha)} \int_{-\frac{1}{2}}^{\frac{1}{2}} (S^\alpha(f) - S(f)) df \tag{49}$$

varies from Burg's measure to Shannon's measure as $\alpha$ goes from 0 to 1, and by using this, we can get a continuous transition from Burg's to Shannon's spectral density functions.

## 3.4    (d) Most Likely Probability Distribution

Scientists and engineers often use MaxEnt and MinxEnt principles with the understanding that these lead to the most likely, or the most probable distribution. However, this need not be the case. If we consider teams of monkeys throwing balls in n boxes, and having multinomial, Poisson, binomial, negative binomial distributions, respectively, we get when the number of N balls is large:

$$\ln \frac{N!}{N p_1! N p_2! \cdots N p_n!} q_1^{N p_1} q_2^{N p_2} \cdots q_n^{N p_n} \approx -N \sum_{i=1}^{n} p_i \ln \frac{p_i}{q_i} \tag{50}$$

$$\ln \prod_{i=1}^{n} e^{-Nq_i} \frac{(Nq_i)^{Np_i}}{(Np_i)!} \approx -N \sum_{i=1}^{n} \left( \frac{p_i}{q_i} \ln \frac{p_i}{q_i} - \frac{p_i}{q_i} + 1 \right) q_i \tag{51}$$

$$\ln \prod_{i=1}^{n} N_{c_{N_{p_i}}} q_i^{Np_i} (1 - q_i)^{Np_i} \approx -N \sum_{i=1}^{n} \left[ p_i \ln \frac{p_i}{q_i} + (1 - p_i) \ln \frac{1 - p_i}{1 - q_i} \right] \tag{52}$$

$$\ln \prod_{i=1}^{n} \frac{(N + Np_i - 1)!}{(Np_i - 1)! Np_i!} \frac{q_i^{Np_i}}{(1 - q_i)^{N+Np_i}} \approx -N \sum_{i=1}^{n} \left( p_i \ln \frac{p_i}{q_i} + (1 + p_i) \ln \frac{1 + p_i}{1 + q_i} \right) \tag{53}$$

In the first two cases, maximizing Shannon's measure of entropy or minimizing Kullback-Leibler measure of cross-entropy subject to given constraints will give the most likely probability distribution, but in the other two cases, we have to maximize or minimize measures corresponding to Fermi-Dirac or Bose-Einstein distributions. If, in these latter two cases we use Shannon's entropy or Kullback-Leibler Cross-entropy, we expect to get probability distributions which will not be most likely.

# 4   The Case for Generalized Measures of Entropy and Cross-Entropy

1. Shannon and Kullback-Leibler measures can characterize exponential distributions as MaxEnt or MinxEnt probability distributions, but there is a large number of other probability distributions which do not belong to the exponential family. For example,

$$p_i = \frac{1}{e(\lambda + \mu\epsilon_i) + 1}, p_i = \frac{1}{e(\lambda + \mu\epsilon_i) - 1}, p_i = \frac{1}{e(\lambda + \mu\epsilon_i) + a}, p_i = \frac{1}{a + b\epsilon_i} \tag{54}$$

where

$$\sum_{i=1}^{n} p_i = 1, \qquad \sum_{i=1}^{n} p_i \epsilon_i = \hat{\epsilon} \tag{55}$$

The need for generalized measures of entropy and cross-entropy arises in order to bring this large number of non-exponential probability distributions within the purview of entropy optimization principles.

2. Shannon and Kullback-Leibler measures, because of their properties of uniqueness, can only explain specific models whereas there is a need for explaining several models which are current in certain disciplines. For this, we need generalized measures. There are population models

other than the Malthus model, there are innovation diffusion models other than the Gompertz model, there are spectral density functions other than exponential models, and similarly, there are a large number of phenomena in social, biological, and other sciences where we need the flexibility afforded by the generalized measures.

3. The parametric measures of entropy and cross-entropy do have advantages over the classical measures in certain contexts. For example, when Herniter[15] obtained switching probabilities for various brands by using MaxEnt, Bass[5] criticized his results on the ground that his solution gave the same results without regard to the product, and irrespective of whether the consumers were conservative or innovative. The anomaly was overcome by the use of parametric measures where the data also became a deciding factor in the final choice of the parameters.

4. The classical measures do not always lead to the most probable or most likely probability distribution. Attempts to arrive at such distributions constitute legitimate endeavors.

5. There are a large number of measures of cross-entropy which are already in vogue in diverse fields such as pattern recognition, signal processing, ecology [24], economics, and finance [41]. These were developed independently of the MaxEnt and MinxEnt principles, and cannot be wished out of existence.

6. Even if we confine ourselves to the Kullback-Leibler measure, we get two measures of cross-entropy:

$$\int f(x)\ln\frac{f(x)}{g(x)}dx \quad \text{and} \quad \int g(x)\ln\frac{g(x)}{f(x)}dx \qquad (56)$$

and two corresponding measures of entropy:

$$-\int f(x)\ln f(x)dx \quad \text{and} \quad \int \ln f(x)dx \qquad (57)$$

and both have been found to be useful.

7. Both Shannon and Kullback-Leibler measures have nice properties, and their use leads to elegant formalisms. They have been found so useful that invariably these are considered as the only measures of

entropy and cross-entropy. While these would continue to be used in a large number of cases, the availability of generalized measures will greatly enhance the scope and power of the entropy optimization principles. It would be erroneous to judge the efficacy of the concepts underlying generalized measures solely by their importance in applications to date.

# 5    Concluding Remarks

1. There are two approaches for determining a measure of entropy. The first is based on postulating some plausible axioms for a measure of uncertainty and determining a function satisfying these axioms. In this approach, while the discrete case is based on a solid foundation, the continuous case is somewhat weak. The second approach considers uncertainty as a monotonic decreasing function of the cross-entropy of P from the most uncertain (uniform) distribution. In this approach, the continuous case is considerably strengthened over the former interpretation while still retaining strength for the interpretation for the discrete case. The second approach also gives us a great deal of flexibility because of the wide latitude we would have in defining cross-entropy.

2. The measures of entropy and cross-entropy that we are proposing are not arbitrary choice-functions which are selected so as to get the desired results. A cross-entropy measure $D(P : Q)$ has to be a continuous function of $p_1, p_2, \cdots, p_n; q_1, q_2, \cdots, q_n$, has to be $\geq 0$, has to vanish if and only if $P = Q$, has not to change if pairs $(p_1, q_1), (p_2, q_2), \cdots, (p_n, q_n)$ are permuted among themselves and is to be a convex function of both $p_1, p_2, \cdots, p_n$ and $q_1, q_2, \cdots, q_n$. A measure of entropy $H(P)$ has to be a permutatively symmetric, continuous, concave function of MATH EXP which is maximum for the most uncertain distribution and has to be minimum and vanish for the most certain distributions.

3. The probability distribution obtained by maximizing Shannon measures may or may not be the most probable distribution, but quite a large proportion of the distributions lie in its neighborhood. This property will hold even if we use generalized measures of entropy. We shall also have greater flexibility to ensure that our probability distribution is also the most likely probability distribution.

4. If we use parametric measures of entropy and cross-entropy, we shall be able to ensure better fits to data than what we can achieve with non-parametric measures.

5. Most of the earlier generalized measures of entropy and cross-entropy were obtained due to the urge to generalize the Shannon and Kullback-Leibler measures. Aczel and Forte [3] have rightly proposed that these measures should be obtained to satisfy some practical considerations in applications. We endorse this idea, and that is precisely what is underlying our generalizations of MaxEnt and MinxEnt.

6. In normal statistical inference, we aim to find the best probability distribution from given data. In inverse problems, we start with the given probability distribution and ask,'what are the constraints which have given rise to this distribution?' We start by using Shannon's measure, and if we arrive at simple constraints, we stop right there. However, if Shannon's measure gives rise to very complicated constraints, we seek to determine a simple combination of a generalized measure of entropy and constraints which will still enable us to get the given probability distribution as a maximum entropy probability distribution.

# References

[1] Aczel, J. and Daroczy, Z. (1975) On Measures of Information and their Characterization, Academic Press, New York.

[2] Aczel, J., Forte, B., and Ng, C.T. (1974) 'Why Shannon and Hartley Entropies are Natural?', Adv. Appl. Prob., Vol. 6, 131–146.

[3] Aczel, J. and Forte, B. (1985) 'Generalized Entropies and MaxEnt Principle', in J.H. Justice (ed.), Maximum Entropy and Bayesian Methods in Applied Statistics, Cambridge University Press, pp. 95–100.

[4] Bass, F.M. (1969) 'A New Product Growth Model for Consumer Durables', Management Science, Vol.    , 215–227.

[5] Bass, F.M. (1974) 'Theory of stochastic preference and brand switching', J. Market. Res., Vol. 11, 1–20.

[6] Behara, M. and Chawla, J.S. (1974) 'Generalised Gamma Entropy', Selecta Statistica Canadiana, Vol. 2, 15–39.

[7] Bhattacharya, A. (1943) 'On a measure of Divergence between two Statistical Populations defined by their Probability Distributions', Bull. Cal. Math. Soc., Vol. 35, 99–109.

[8] Burg, J.P. (1972) 'The Relationship between Maximum Entropy Spectra and Maximum Likelihood Spectra', in D.G. Childers (ed.), Modern Spectral Analysis, IEEE Press, USA, pp. 130–131.

[9] Csiszer, I. (1972) 'A Class of Measures of Informativity of Observation Channels', Periodic. Math. Hungarica, Vol. 2, 191–213.

[10] Ferreri, C. (1980) 'Hypoentropy and Related Heterogeneity Divergence and Information Measures', Statistica (Bologna), Vol. 40, No. 2, 155–168.

[11] Fisher, J.C. and Pry, R.H. (1971) 'A Simple Substitution Model for Technological Change', Technological Forecasting and Social Change, Vol. 2, 75–88.

[12] Floyd, A. (1968) 'A Methodology for Trend Forecasting of Figures of Merit', in J. Bright (ed.), Technological Forecasting for Industry and Government: Methods and Applications, Prentice-Hall Inc., Englewood Cliffs, N.J., USA, pp. 95–109.

[13] Gompertz, B. (1825), Phil. Trans. Roy. Soc., Vol. 115, 513.

[14] Havrda, J.H. and Charvat, F. (1967) 'Quantification Methods of Classification Processes Concepts of Structural & Entropy', Kybernetica, Vol. 3, 30–35.

[15] Herniter, J.D. (1973) 'An Entropy Model of Brand Purchase Behaviour', J. Market. Res., Vol. 10, 361–373.

[16] Jaynes, E.T. (1957) 'Information Theory and Statistical Mechanics', Physical Reviews, Vol. 106, 620–630.

[17] Kapur, J.N. (1967) 'Generalised entropies of order $\alpha$ and type $\beta$', The Maths. Seminar, Vol. 4, 78–94.

[18] Kapur, J.N. (1972) 'Measures of uncertainty, mathematical programming and physics', J. Ind. Soc. Agri. Stat., Vol. 24, 47–66.

[19] Kapur, J.N. (1983) 'Derivation of logistic law of population growth from maximum entropy principle', Nat. Acad. Sci. Letters, Vol. 6, No. 12, 429–433.

[20] Kapur, J.N. (1983) 'Comparative assessment of various measures of entropy', J. Inf. and Opt. Sci., Vol. 4, No. 1, 207–232.

[21] Kapur, J.N. (1984) 'A Comparative assessment of various measures of directed divergence', Advances in Management Studies, Vol. 3, 1–16.

[22] Kapur, J.N. (1986) 'Four families of Measures of Entropy', Ind. J. Pure and App. Maths., Vol. 17, No. 4, 429–449.

[23] Kapur, J.N. (1988) Maximum-Entropy Models in Science and Engineering, Wiley Eastern Publishers, New Delhi, India.

[24] Kapur, J.N. (1985) Mathematical Models in Biology and Medicine, Affiliated East West Publishers, New Delhi, India.

[25] Kapur, J.N. and Kumar, V. (1985) 'A new derivation of product-form probability distributions for queuing theory networks', Ind. J. Management and Systems, Vol. 1, No. 3, 109–118.

[26] Kapur, J.N. and Kesavan, H.K. (1987) The Generalised Maximum Entropy Principle, Sandford Educational Press, Waterloo, Canada.

[27] Kapur, J.N. and Kesavan, H.K. (1988) 'A new approach to the study of probabilistic systems', Int. J. Management Systems, Vol. 4, No. 1, 67–70.

[28] Kapur, J.N. and Kesavan, H.K. (1989) 'Maximum Entropy Models', Proc. Symp. Math. Modelling, IIT, Madras, World Scientific Publishers, Singapore.

[29] Kesavan, H.K. and Kapur, J.N. (1989) 'Generalised Maximum Entropy Principle', IEEE Trans. Systems, Man and Cybernetics, (to appear).

[30] Kesavan, H.K. and Kapur, J.N. (1989) 'On the families of solution of generalised maximum entropy and minimum cross- entropy principle', Int. J. Gen. Syst., (to appear).

[31] Kesavan, H.K. and Kapur, J.N. (1989) 'Maximum Entropy And Minimum Cross-Entropy Principles: Need for a Broader Perspective' in

P. Fougere (ed.), Maximum Entropy and Bayesian Methods, Dartmouth College, Kluwer Academic Publishers, Dordecht, (to appear).

[32] Kullback, S. (1959) Information Theory and Statistics, John Wiley, New York, USA.

[33] Kullback, S. and Leibler, R.A. (1951) 'On Information and Sufficiency', Ann. Math. Stat., Vol. 22, 79–86.

[34] Purcaro, C. (1973) 'A new magnitude frequency relation for earthquakes and a classification of relation types', Geophy. J. Roy. Ast. Soc., Vol. 42, 61–79.

[35] Rao, C.R. (1973) Linear Statistical Inference and its Applications, Wiley Eastern, New Delhi, India.

[36] Renyi, A., (1961) On Measures Of Entropy and Informaton, Proc. 4th Berkeley Symposium Maths.Stat.Prob., vol1, pp. 547-561.

[37] Shannon, C.E. (1948) 'A Mathematical Theory of Communication', Bell System Tech. J., Vol. 27, pp. 379–423, and 623–659.

[38] Sharif, M.N. and Kabir, G.A. (1976) 'A generalised model for forecasting technological substitution', Technological Forecasting and Social Change, Vol. 8, 353–364.

[39] Sharma, B.D. and Taneja, I.J. (1974) 'Entropy of type $(\alpha, \beta)$ and other generalised measures of information', Matrica, Vol. 22, 205–214.

[40] Sharma, B.D. and Mittal, D.P. (1975) 'New non-additive measures of entropy for discrete probability distributions', J. Math. Sci., Vol. 10, 28–40.

[41] Theil, H. (1967) Economics and Information Theory, North Holland Publishers, Amsterdam.

[42] Tribus, M. (1966) Rational Descriptions, Decisions and Designs, Pergamon Press, Oxford.

# A COMPRATIVE ASSESSMENT OF ENTROPIC AND NON-ENTROPIC METHODS OF ESTIMATION

ANAND K. SETH
Smith Kline Beecham Pharmaceuticals
P. O. Box 1539
King of Prussia, PA 19406 U.S.A.

JAGAT N. KAPUR
System Design Engineering
University of Waterloo
Waterloo, Ontario N2L 3G1 Canada

ABSTRACT. The present paper gives a comparative assessment of entropic (maximum entropy and minimum cross-entropy) and non-entropic (orthodox and Bayesian) methods of statistical inference. It also investigates the conditions under which each approach is most appropriate. As an illustration, the general linear model is discussed from both entropic and non-entropic points of view.

## 1. Introduction

The present paper is inspired by the plea that Jaynes [4, 5, 6] has been making in almost all the MAXENT workshops for a deeper understanding of the interrelationship between MAXENT and Bayesian methods of statistical inference.

During the last three decades, after the explicit enunciation of the maximum entropy principle (MEP) by Jaynes [3], there have been remarkable achievements of this principle in a variety of fields (Kapur [8, 9, 10] and Kapur and Kesavan [11]). However, it is a fact that the powerful methods of entropic inference have not yet been fully integrated into courses and text books of statistical inference [9].

The entropic methods of statistical inference are themselves divided into two classes -- those based on principle of maximum entropy and those based on principle of minimum cross-entropy. Inspite of a great deal of commonality, the two methods are distinct conceptually. We will discuss the similarities and dissimilarities of the two approaches.

P. F. Fougère (ed.), Maximum Entropy and Bayesian Methods, 451–462.

The non-entropic methods (i.e., those which do not make an explicit use of the concept of entropy) are themselves classified into orthodox (or frequentist) and Bayesian methods, and there have been strong (and sometimes bitter) differences between the followers of the two approaches.

There have also been attempts at combining the entropic and non-entropic methods. Thus the maximum entropy principle is used to find the apriori probability distribution and then Bayes theorem is used to find the posteriori probability on the basis of the available information.

All the methods of statistical inference have had a fairly long history, spread over more than two centuries, at least.

The MEP was implicit in the works of Boltzmann and Gibbs, though it was stated explicitly by Jaynes. Kullback [12] stated the principle of minimum discrimination information (or minimum directed divergence or minimum cross-entropy) and applied it to a number of problems of statistical inference. Later, Johnson [7] placed it on a strong axiomatic foundation.

Bayesian methods have their origin in the theorem of Bayes, but important contributions have been made by Laplace, Jeffreys, Savage, Box, Lindley and others.

The orthodox methods include the method of maximum likelihood of Fisher [1], the method of minimum chi-squares of Neyman and Pearson [13], the method of moments of Pearson, the density estimation method of Gauss and the least squares method, also due to Gauss.

The organization of present paper is as follows: Section 2 is concerned with the derivation of the principle of maximum likelihood from the principles of maximum entropy and minimum cross-entropy. Section 3 is concerned with deriving the method of minimum chi-squares from the principle of minimum cross-entropy. Section 4 discusses the interrelationship between principles of maximum entropy and maximum likelihood on the one hand and methods of Gauss density estimation and methods of moments on the other. Section 5 discusses the solution of problems associated with the general linear model by Bayesian, minimum cross-entropy and least squares methods. The comparative assessment of minimum cross-entropy and Bayesian methods of statistical inference is given in Section 5. Finally, the conclusions are given in Section 6.

## 2. Derivation of Maximum Likelihood Principle from Maximum Entropy and Minimum Cross Entropy Principles

### 2.1 DERIVATION OF MLP FROM MEP

Let $f(x, \theta)$ be the density function of a population, where x and $\theta$ may be scalars or vectors. Given a random sample $x_1$, $x_2$, ..., $x_n$ from the population, our object is to find a good estimator for the parameter $\theta$ i.e., we have to find $\theta$, a function of $x_1$, $x_2$, ..., $x_n$, which in some sense, is as good an estimator for $\theta$ as possible. For this purpose, Fisher gave his method of maximum likelihood according to which we choose that $\theta$ which maximizes the likelihood function

$$L (x_1, x_2, \ldots, x_n, \theta) = f(x_1, \theta) \times f(x_2, \theta) \times \ldots \times f(x_n, \theta) \quad (1)$$

Fisher stated this almost as an axiom and the justification for the MLP was provided by the nice properties which maximum likelihood estimators were later shown to have.

Now according to the maximum entropy principle, we have to maximize the uncertainty that remains after all the available information has been used. In other words, we have to choose $\theta$ so as to minimize the information given by the random sample, so that we have to choose $\theta$ so as to minimize

$$- \int f(x, \theta) \ln f(x, \theta) dx \quad (2)$$

$$- \int \ln f(x, \theta) dF$$

$$= - \frac{1}{n} [\ln f(x_1, \theta) + \ln f(x_2, \theta) + \ldots + \ln f(x_n, \theta)]$$

$$= - \frac{1}{n} \ln L (x_1, x_2, \ldots, x_n, \theta)$$

i.e, we have to choose $\theta$ so as to maximize the likelihood function. Thus, we have deduced the MLP from the MEP. We cannot deduce the MEP from the MLP, since while MLP is concerned with estimating the parameter only, the MEP is concerned with estimating the density function itself. Accordingly we may say that the MEP is more basic then the MLP.

### 2.2. DERIVATION OF MLP FROM MINXENT

According to the minimum cross-entropy (MINXENT) principle, we should choose $\theta$, so that observed probability distribution of $g(x)$ is as close as possible to $f(x, \theta)$. The probability

distribution is obtained from the observed sample and its
distribution function is given by

$$G(x) = 0 \text{ when } x < x_1, \ G(x) = \frac{1}{n} \text{ when } x_1 \le x < x_2,$$

$$G(x) = \frac{2}{n} \text{ when } x_2 \le x < x_3, \ \ldots, \tag{3}$$

$$G(x) = 1 \text{ when } x \ge x_n$$

Here we have assumed that

$$x_1 \le x_2 \le x_3 \le \ldots \le x_n, \tag{4}$$

and that x is a scalar. We have to choose $\theta$ so as to minimize

$$\int g(x) \ln \frac{g(x)}{f(x, \theta)} dx \text{ or}$$

$$\int g(x) \ln g(x) dx - \int g(x) \ln f(x, \theta) dx$$

i.e, we have to choose $\theta$ so as to minimize

$$- \int \ln f(x, \theta) dG \tag{5}$$

$$= - \frac{1}{n} [\ln f(x_1, \theta) + \ln f(x_2, \theta) + \ldots + \ln f(x_n, \theta)]$$

$$= - \frac{1}{n} \ln L(x_1, x_2, \ldots, x_n, \theta), \tag{6}$$

so that again we have to choose $\theta$ so as to maximize the
likelihood function. We have thus deduced the MLP from the
minimum cross-entropy principle.

If x is a vector, we take

$$G(x) = \frac{1}{n} \delta(x-x_1) + \frac{1}{n} \delta(x-x_2) + \ldots + \frac{1}{n} \delta(x-x_n) \tag{7}$$

and again we can get the maximum likelihood principle.

Thus we have deduced the MLP from the MINXENT principle. Again
we cannot deduce the MINXENT principle from MLP as MINXENT
principle seeks to determine a density function, while MLP seeks
to determine only a parameter.

## 2.3. DISCUSSION

In statistical theory, MLP is taken as an axiom. Here it has
been deduced from the two more fundamental principles viz.
principle of maximum entropy and minimum cross-entropy. More

specifically, we use the principles that the uncertainty that remains after the knowledge of the sample has been used should be maximum or that the sample should be as close to the population as possible.

## 3. Derivation of Minimum Chisquare Method of Estimation From Principle of Minimum Cross-Entropy

Let $Np_1$, $Np_2$, ..., $Np_n$ be the expected frequencies in the n classes on the hypothesis that the parameter has the value $\theta$ and let $Nq_1$, $Nq_2$, ..., $Nq_n$ be the corresponding observed frequencies.

$$\text{Let } q_i = p_i (1 + \epsilon_i), \; i = 1, 2, \ldots, n \tag{8}$$

then

$$\sum_{i=1}^{n} p_i = 1, \; \sum_{i=1}^{n} q_i = 1 \Rightarrow \sum_{i=1}^{n} p_i \epsilon_i = 0 \tag{9}$$

and we get

$$\sum_{i=1}^{n} p_i \ln p_i/q_i = - \sum_{i=1}^{n} p_i \ln (1 + \epsilon_i)$$

$$= - \sum_{i=1}^{n} p_i \epsilon_i + \frac{1}{2} \sum_{i=1}^{n} p_i \epsilon_i^2 - \frac{1}{3} \sum_{i=1}^{n} p_i \epsilon_i^3 + \ldots$$

$$= \frac{1}{2} \sum_{i=1}^{n} p_i (q_i - p_i/p_i)^2 - \frac{1}{3} \sum_{i=1}^{n} p_i \epsilon_i^3 + \ldots \tag{10}$$

Assuming that $\epsilon_i$'s are sufficiently small so that quantities of the order $\epsilon_i^3$'s can be neglected, we get

$$D(P:Q) = \sum_{i=1}^{n} p_i \ln p_i/q_i \; \tilde{=} \; \frac{1}{2} \sum_{i=1}^{n} (q_i - p_i)^2/p_i$$

$$= \frac{1}{2N} \sum_{i=1}^{n} (Nq_i - Np_i)^2/Np_i$$

$$= \frac{1}{2N} \sum_{i=1}^{n} (o_i - e_i)^2/e_i \; = \; \frac{1}{2N} \chi_p^2 \tag{11}$$

where $e_i$ and $o_i$ are the expected and observed frequencies in the $i^{th}$ class and $\chi_p^2$ is Pearson's chisquare. According to the principle of minimum cross-entropy, we should choose $\theta$ so that $D(P:Q)$ is as small as possible i.e., $\chi_p^2$ is as small as possible. Thus we have deduced Pearson's method of minimum chi square from the principle of minimum cross-entropy. Similarly, if we put

$$P_i = q_i (1 + \epsilon_i^*), \quad i = 1, 2, \ldots, n \tag{12}$$

we get

$$\Sigma^n_{i=1} q_i \epsilon_i^* = 0$$

and

$$D(P:Q) = \Sigma^n_{i=1} P_i \ln P_i/q_i = \Sigma^n_{i=1} q_i (1 + \epsilon_i^*) \ln (1 + \epsilon_i^*)$$

or

$$D(P:Q) \overset{\sim}{=} \frac{1}{2} \Sigma^n_{i=1} q_i \epsilon_i^* + \frac{1}{2} \Sigma^n_{i=1} q_i (P_i - q_i)^2/q_i^2 + \ldots$$

$$= \frac{1}{2N} \Sigma^n_{i=1} (NP_i - Nq_i)^2/Nq_i$$

$$= \frac{1}{2N} \Sigma^n_{i=1} (e_i - o_i)^2/o_i = \frac{1}{2N} \chi_{Ne}^2 \tag{13}$$

where $\chi_{Ne}^2$ is Neyman's chisquare. Thus, we choose $\theta$ so as to minimize Neyman's chisquare. Again, Neyman's chisquare principle also follows from the principle of minimum cross-entropy.

## 4. Inter-Relationship Between Maximum-Entropy Principle, Maximum Likelihood Principle, Gauss Density Estimation Method and Pearson's Method of Moments

Let $f(x, \theta_1, \theta_2, \ldots, \theta_m)$ be the maximum-entropy density function, when $E(g_1(x))$, $E(g_2(x))$, $\ldots$, $E(g_m(x))$ are prescribed, so that $g_1(x)$, $g_2(x)$, $\ldots$, $g_m(x)$ are the characterizing moment functions and $E(g_1(x))$, $E(g_2(x))$, $\ldots$, $E(g_m(x))$ are the characterizing moments. Let

$$\int f(x, \theta_1, \theta_2, \ldots, \theta_m) g_r(x) \, dx = \phi_r(\theta_1, \theta_2, \ldots, \theta_m)$$
$$r = 1, 2, \ldots, m \tag{14}$$

then it can be shown that the maximum-likelihood estimators

$$\hat{\theta}_1, \hat{\theta}_2, \ldots, \hat{\theta}_m \text{ for } \theta_1, \theta_2, \ldots, \theta_m \text{ are obtained by solving}$$

$$\phi_r(\hat{\theta}_1, \hat{\theta}_2, \ldots, \hat{\theta}_m) = ((g_r(x_1) + g_r(x_2) + \ldots + g_r(x_n))/n$$
$$r = 1, 2, \ldots, m \tag{15}$$

where $x_1$, $x_2$,... ,$x_n$ is a random sample from the population. Thus, the maximum-likelihood estimators can easily be found provided we can get the given probability distribution as the maximum entropy probability distribution.

Pearson, in his method of moments, suggested that for estimating $\theta_1$, $\theta_2$, ... ,$\theta_m$, we should equate the first m algebraic moments of the sample and the population, i.e., we should use the equations

$$\int f(x, \theta_1, \theta_2, \ldots, \theta_m)x^r \, dx = (x_1^r + x_2^r + \ldots + x_n^r)/n,$$

$$r = 1, 2, \ldots, m. \qquad (16)$$

Except in some cases (like that of the normal distribution), Pearson's estimators were quite different from the maximum likelihood estimators which had very nice properties. As such, his method was vehemently criticized by Fisher [2].

If Pearson had used characterizing moments in place of algebraic moments, his estimators would have been the same as maximum likelihood estimators and there would have been no Fisher-Pearson controversy. However, neither Fisher nor Pearson was aware of the maximum-entropy principle and as such the controversy was not resolved in their times. In fact, even today many statisticians are not aware of the importance of maximum-entropy characterization of probability distributions.

## 5. General Linear Models

### 5.1 BAYESIAN APPROACH

Let the general linear model be

$$Y = X\beta + E, \qquad (17)$$

Y is the m x 1 observation vector,

X is the m x n known input or design matrix,

$\beta$ is the n x 1 vector of parameters to be estimated,

and E is the m x 1 normal N(0, R) error vector.

Let apriori probability distribution for $\beta$ be N($M_o$, $\Sigma_o$), then using Bayes' theorem, the posteriori probability distribution for $\beta$ is N($M_1$, $\Sigma_1$), where

$$\Sigma_1^{-1} = \Sigma_0^{-1} + X^T R^{-1} X \tag{18}$$

$$\Sigma_1^{-1} M_1 = \Sigma_0^{-1} M_0 + X^T R^{-1} Y \tag{19}$$

We make the following remarks:

(i) If $\Sigma_0$ and $M_0$ are given and $\Sigma_0$ is not singular, (18) and (19) will in general determine $\Sigma_1$ and $M_1$. If afterwards, we are given a new observation vector, a new known input matrix and a new error normal distribution, we can find $\Sigma_2$ and $M_2$ and in this way we can continuously update estimators for mean and covariance matrix of $\beta$

(ii) If $\Sigma_0^{-1}$ approaches zero matrix, we get

$$\Sigma_1^{-1} = X^T R^{-1} X, \quad \Sigma_1^{-1} M_1 = X^T R^{-1} Y \tag{20}$$

These are the classical results illustrating the fact that as the apriori distribution approaches the uniform distribution, Bayesian results approach the classical results.

(iii) If R is not of full column rank, $X^T R^{-1} X$ is singular and $\Sigma_1^{-1}$ does not exist. In this case the classical results do not exist, but if $\Sigma_0 \neq 0$, then the Bayesian results still exist.

## 5.2. MINIMUM CROSS-ENTROPY SOLUTION

Here we have to minimize

$$\int f(\beta) \ln \frac{f(\beta)}{g(\beta)} \, d\beta \tag{21}$$

subject to $E(Y - X\beta) = 0$ and $\tag{22}$

$$E(Y - X\beta)^T (Y - X\beta) = R \tag{23}$$

where

$$g(\beta) = 1/\left(2\pi\right)^{n/2} \; |\Sigma_o|^{1/2} \exp(-1/2(\beta - M_o)^T \Sigma_o^{-1} (\beta - M_o)), \qquad (24)$$

so that we get

$$f(\beta) = C \exp [-1/2 (\beta - M_o)^T \Sigma_o^{-1} (\beta - M_o)]$$
$$\times \exp (- \lambda_o - \lambda^T (Y - X\beta) - (Y - X\beta)^T D^{-1} (Y - X\beta)), \qquad (25)$$

where the scalar $\lambda_o$, the vector $\lambda^T$ and the matrix $D^{-1}$ are obtained by using the constraints (22) and (23).

## 5.3. LEAST SQUARES APPROACH

Here we have to minimize

$$(Y - X\beta)^T (Y - X\beta). \qquad (26)$$

This gives the usual normal equations

$$X^T X\beta = X^T Y$$

or

$$\beta = (X^T X)^{-1} X^T Y.$$

If $X^T X$ is singular, then

$$\beta = G X^T Y$$

where G is any generalized inverse defined by

$$X^T X G X^T X = X^T X.$$

## 5.4. DISCUSSION OF THREE METHODS

(i) In all the three methods, we have obtained estimates for $\beta$, but while in the first two cases, we obtained probability distributions for $\beta$, in the third case, we obtained simply an estimate for $\beta$.

(ii) In the Bayesian approach, we assumed the knowledge of the apriori distribution of $\beta$ and updated it in the light of knowledge of observations. In the minimum cross-entropy approach, we did not make use of Bayes' theorem, but instead made use of the minimum cross-entropy principle. Out of all the distributions consistent with the knowledge that $E = Y - \beta X$ is $N(0,R)$, we choose one which is closest to the given apriori distribution for $\beta$.

(iii) Since in the first two approaches, the principles used are different, the results are bound to be different. In the second approach, distribution of $\beta$ satisfies (22) and (23), while it need not satisfy these constraints in the first case.

(iv) Unlike the first two approaches, we do not assume any knowledge of apriori distributions of $\beta$ in the third approach. Also, here we use quite a different principle of estimation.

(v) For the same problem, different methods of estimation can lead, quite expectedly, to different results. Since statistical inference is inductive, there can be no perfect solution and there can be even differences of opinion as to which one is best.

## 5.5 Comparison of Bayesian and MINXENT Approaches

(i) In Bayesian approach, we assume apriori distribution on the parameter, while in MINXENT approach, we assume an apriori distribution on the density function itself.

(ii) In Bayesian approach, information is usually given in the form of a random sample, while in MINXENT approach, information is usually given in the form of values of some moments.

(iii) In Bayesian approach, information is complete and Bayes' theorem gives a unique posterior distribution. In MINXENT approach, there are many probability distributions consistent with the given values of moments and a unique distribution is obtained by choosing one of those ones which is closest to a given apriori distribution.

(iv) Bayesian approach uses Bayes' theorem whose truth follows from the theorems of total and compound probabilities, while in MINXENT we have to use a new axiom of statistical inference.

(v) In Bayesian approach, we continuously update the distribution of the parameters as more and more information becomes available, while in MINXENT approach, we continuously update the density function as values of more and more moments become available as new values of the same moments become available.

(vi) The results obtained from the Bayesian method approach the classical results as the apriori distribution approaches the uniform distribution, while the results of the MINXENT method approach the results of the MAXENT approach as the apriori density function approaches the uniform distribution.

## 6. Conclusion

In the present paper, a formal attempt has been made to relate the maximum entropy methods to the classical and Bayesian methods of statistical inference. This has been done by obtaining the maximum likelihood principle of statistical inference from the principle of maximum entropy and also from the principle of minimum cross-entropy. In addition, the interrelationship among maximum likelihood inference, Gauss density estimation method and Pearson's method of moments has been shown clearly. The other important aspect of this contribution is a demonstration of how Bayesian methods of statistical inference can be related to the minimum cross-entropy method. The connection between the two approaches is illustrated through a widely used general linear model. For this model, the similarities and dissimilarities of the Bayesian and the MAXENT solution are discussed.

It is hoped that by showing the relationship between entropic and non-entropic methods explicitly, the future development in statistical inference will be much broader and will include entropic method as one of the valid methods of inference.

## REFERENCES

1. Fisher, R.A. (1921) 'On the Mathematical Foundations of Theoretical Statistics', Phil. Trans. Roy. Soc., Vol. 222(A), 309-368.

2. Fisher, R. A. (1937), 'Professor Karl Pearson and the Methods of Moments', Annals of Eugenics, vii, 303-318.

3. Jaynes, E. T. (1957) 'Information Theory and Statistical Mechanics', Physical Reviews, Vol. 106, 620-630.

4. Jaynes, E. T. (1988) ' The Relationship of Bayesian and Maxmimum Entropy Methods, in G. J. Erickson and C. Ray Smith (eds.), Maximum-Entropy and Bayesian Methods in Science and Engineering, Vol I, Kluwer Academic Publishers, Boston, pp. 25-29.

5. Jaynes, E. T. (1989) 'Where Do We Stand on Maximum Entropy?, in R. D. Rosenkrantz (ed.) E. T. Jaynes: Papers on Probability, Statistics and Statistical Physics, Kluwer Academic Publisher, Boston, 210-314.

6. Jaynes, E. T. (1989) 'Clearing Up Mysteries - The Original Goal, ' in J. S. Skilling (ed.), Maximum Entropy and Bayesian Methods, Kluwer Academic Publishers, Boston, pp. 1-27.

7. Johnson, R.W. (1979) 'Axiomatic Characterization of the Directed Divergence and Their Linear Combinations', IEEE Trans. Inf. Th., Vol. II, 709-716.

8. Kapur, J.N. (1983) 'Twenty-five Years of Maximum Entropy', Journ. Math. Phy. Sci., Vol. 17, No. 2, 103-156.

9. Kapur, J.N. (1984) 'The Role of Maximum-Entropy and Minimum Discrimination Information Principles in Statistics', Jour. Ind. Soc. Agri. Stat., Vol. 36, No. 3, 12-55.

10. Kapur, J.N. (1989) Maximum-Entropy Models in Science and Engineering, Wiley Eastern, New Delhi.

11. Kapur, J.N. and H.K. Kesavan (1987) Generalized Maximum Entropy Principle (With Applications). Sandford Educational Press, University of Waterloo, Canada.

12. Kullback, S. (1959) Information Theory and Statistics, John Wiley, New York.

13. Neyman, J. (1949) 'Contribution to the Theory of $\chi^2$ test', Proc. First Berkeley Symp. Prob. Stat., 230-273.

# MAXENT APPLIED TO LINEAR REGRESSION

J.F. Cyranski
*The Heatherington*, #211
1421 Massachusetts Ave., NW
Washington, DC 20005

ABSTRACT. Given sparse, unreplicated data of poor instrumental resolution, we determine the probability of linear models using orthogonal least squares regression and MAXENT with an 'expert draftsman' constraint. An information bound condition enables MAXENT inference for the reliability of evidence determining the probability distribution for observations of a constant.

## 1. Introduction

There occur situations in which inferences must be based on observations that cannot be replicated, which are few in number, and which have intrinsic experimental resolution. In the Introductory Physics Laboratory, for example, one seeks to verify that the voltage across an inductor is proportional to the time rate of change of current. The combination of limited classtime, limited student experience, and limited CRO sensitivity, stray capacitance, etc., produce such a situation. In these circumstances, ordinary regression analysis is inadequate since there are 'error bars' in both variables; moreover, standard statistical methods cannot be justified for evaluating the reliability of the model.

To deal with such cases we employ orthogonal least squares regression, described in Section 2, modified to account for instrumental resolution. Reliability is expressed by the amount an 'expert draftsman' might 'wiggle' a straight line about the best fit consistent with the plotted data cells. This constraint is used with MAXENT to obtain a confidence distribution on the (linear) models. We illustrate the results in Section 3.

In Section 4 we consider the special case of discrete observations of a single continuous variable (determining the value of a constant). We present a novel

*P. F. Fougère (ed.), Maximum Entropy and Bayesian Methods*, 463–473.

information criterion as evidence to use in a MAXENT inference for the probability distribution on the data evidence. We also compare Bayesian and MAXENT approaches to induction.

## 2. Linear Regression

If one does not have arbitrarily high instrumental resolution, $N$ observations of paired variables $\eta, \xi$ define a set $D$ of $N$ data *interval* pairs:

$$\{(\xi_n - \Delta\xi_n, \xi_n + \Delta\xi_n) \times (\eta_n - \Delta\eta_n, \eta_n + \Delta\eta_n) \mid \quad n = 1,...,N\}. \tag{1}$$

In general, the endpoints of the intervals need to be sharply defined, nor need the 'instrument sensitivity' be uniform within the intervals. Data in this context are not points, but are *cells*.

Typically one seeks a relation between the variables which depends on $M$ parameters: namely, $F[(\eta, \xi), \theta] = 0$, with $\theta \subset R^M$. If we assume analyticity in (contravariant components) $\theta^m$ and that $F[(\eta, \xi), \theta] = -B$ (constant), then to *linear* order the relation becomes

$$B = \sum_{m=1}^{M} [\partial F(\eta, \xi) / \partial\theta_m \mid_{\theta=0}]\theta^m = \sum_{m=1}^{M} y_m(\eta, \xi)\theta^m. \tag{2a}$$

Note that this equation is intrinsically dimensionless, and corresponds to the equation of a hyperplane in the standard Euclidean metric space $R^M$,

$$R(\bar{\theta}, x) = \{y \in R^M \mid y \cdot \delta = x\}, \tag{2b}$$

where $\delta$ is a unit vector defining the normal to the plane and $x = B / \|\theta\|$ is its orthogonal distance from the origin.

Equation (2a) also may be recast in terms of *affine* variables by rescaling and introducing units via the maps $y_m \to f_m = y_m / \alpha(m)$ and $\theta^m \to a^m = \theta^m \alpha(m)$. If $a^1 \neq 0$, $f_1 = f_1(\eta)$, and $f_m = f_m(\xi)$, then (2a) becomes the familiar

$$f_1(\eta) = \sum_{m=2}^{M} (-a^m / a^1) f_m(\xi) + (x / a^1) \tag{2c}$$

$$= \sum_{m=2}^{M} A_m f_m(\xi) + C.$$

If one starts with (2c), by reversing the map one may always analyze in terms of hyperplanes in a conveniently rescaled and dimensionless Euclidean metric space.

Ignoring the 'error bars' for the moment, the hyperplane that best fits the data cells' 'centers of mass', $y_n$, is that from which the average *orthogonal distance* to the data points is minimum. Thus, we seek to minimize the 'distortion' of the 'data cloud' as described by the model

$$H(\bar{\theta}, x, \{y_n\}) = (1/N) \sum_{n=1}^{N} [\bar{\theta} \cdot y_n - x]^2 \tag{3}$$

$$= \bar{\theta} \cdot Y \cdot \bar{\theta} + [\delta \cdot <y> - x]^2,$$

where $y$ is the sample covariance matrix and $<y>$ is the sample average (center of mass of the 'data cloud'). The best fit hyperplane $(\bar{\theta}^*, x^*)$ is obviously defined by:

$$Y = \bar{\theta}^* = \lambda^* \bar{\theta}^*; \quad x^* = \bar{\theta}^* \cdot <y>, \tag{4}$$

where $\lambda^*$ is the minimum such eigenvalue. Only if $\lambda^*$ is degenerate will any ambiguity arise. The principal axes define the ellipsoidal representation of the data cloud having minimum distortion.

Finally, the $n$th datum's center of mass is determined by 'averaging' $y_n(\eta, \xi)$ over the $n$th cell; using the resolution function as weight. Thus, for example, if $f_2(\xi) = \xi^2$, and $(a - \Delta, a + \Delta)$ is observed with uniform sensitivity ('sharp resolution'), then this datum is defined by $\int_{a-\Delta}^{a+\Delta} d\xi \, \xi^2 / 2\Delta$.

We remark that ordinary least squares regression is *inconsistent* in that the estimates of the parameters depend on the form of the model. That is, if one performs $\eta$ vs. $\xi$ regression, the parameter values one obtains are not the same as if one did $\xi$ vs. $\eta$ regression. Orthogonal regression does not have this difficulty. All methods agree if the minimum eigenvalue is small, i.e. if the 'scatter' is not large.

## 3. Confidence Based on One Trial

By 'confidence' we mean the probability distribution of hyperplanes about the 'best fit' hyperplane. According to Jaynes, 'in a very fundamental sense no experiment can ever be repeated' [1]; hence each element of **D** represents a *single trial* (some values of the variables may recur, however). It follows that the probability we seek is a 'degree of belief' that cannot be determined by orthodox statistical arguments. Instead we employ MAXENT based on a 'prior' measure reflecting 'ignorance' before the experiment has been performed [2] and a *criterion* that bounds the expected 'distortion' measure $H(\bar{\theta}, x; \{y_n\})$ as one perturbs the hyperplane about the 'best fit' result (4), holding fixed the data centers of mass $\{y_n\}$.

Since $R^M$ is a Euclidean space, it is invariant under the group of rotations followed by translations. This implies the existence of an unique (up to a constant multiplier) measure on the Grassman manifold of the hyperplanes which is finite and invariant under the Euclidean group [3]. In case $M = 2$, the prior measure is simply $d\mu = Ad\phi dx$, where $\bar{\theta}_2 = \sin\phi$, $A > 0$ a constant. It can be shown that Jaynes' prior on straight lines [4] is precisely this [5].

The bound on the expectation of $H$ results by noting that if one holds the data cloud's CM($<y>$) fixed, variation of $H$ is equivalent to rotation of the model relative to the data set. A natural bound on such variation is the minimum eigenvalue. One may also translate the best fit plane parallel to itself by this same amount. The envelope of hyperplanes thus defined is equivalent to what an 'expert draftsman' would produce by 'wiggling' the lines consistent with the data plot. Additional variation is possible due to the uncertainty in the location of the CM (because the data 'points' are, in general, cells). This produces a bound $\Delta\lambda$, so that the ultimate constraint on the expected value of $H$ based on this 'evidence' $E$ is

$$\mathscr{E}(H|E) \leqslant 2\lambda^* + \Delta\lambda. \tag{5}$$

Note that (5) establishes a *criterion* for the maximum acceptable expected variation of the hyperplane, namely that the *extra* average squared orthogonal distance not exceed the minimum possible (plus its 'uncertainty'). We argue in the spirit of the Rayleigh criterion of optics, which defines the resolution of two points sources in terms of the distance between the central diffraction peak of the first point and the first diffraction minimum of the second. Analagous to the concept of the 'resolving power' of a telescope, we have the 'resolving power' of an experiment, defined by (5).

The application of MAXENT to orthogonal regression confidence levels is the essence of a laboratory analysis package developed for use by introductory physics students. The programs run on an APPLE/IIE (or IIGS) and are written in BASIC. The generic model has the form:

$$Y' = \sum_{m=1}^{'} A_m X^{r(m)} + B, \quad Y, X, r, \text{ and } r(m) \text{ are real.}$$

$Y$ and $X$ can be defined as arbitrary functions of $\eta$ and $\xi$ respectively. Resolution is assumed 'sharp'. Computations exploit the assumption that all but one of the eigenvalues are much greater than the minimum eigenvalue. Various confidence bounds have been derived. Details will be presented elsewhere [5].

As illustration, we present the results from a typical laboratory excercise, namely that described in Section 1.

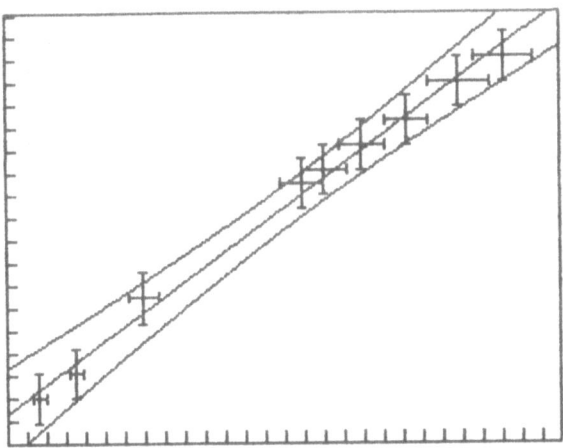

FIGURE 1. Voltage vs. current rate across an inductor.

Figure 1 shows the original data (with error bars), the best fit (central line), and (68%) contour envelope. Note that the envelope reasonably describes the variability a draftsman would permit in 'wiggling' lines. The 'error bars' are large primarily because the oscilloscope was operated at an extreme sensitivity range (the inductance was very low) producing thick traces. Bear in mind that these data are cells in which the measured value has uniform distribution.

If one suppresses the error-bars in computing the confidence region, the envelope is due entirely to data 'scatter'. From Figure 2 it should be clear that this results in unacceptably small 'wiggles' for the assumed uniform sensitivity. However, if the sensitivity behaved like a Gaussian probability distribution function (i.e., each error-bar represented a $\sigma$-deviation from the datum), then one might accept the envelope as reasonably consistent with such data.

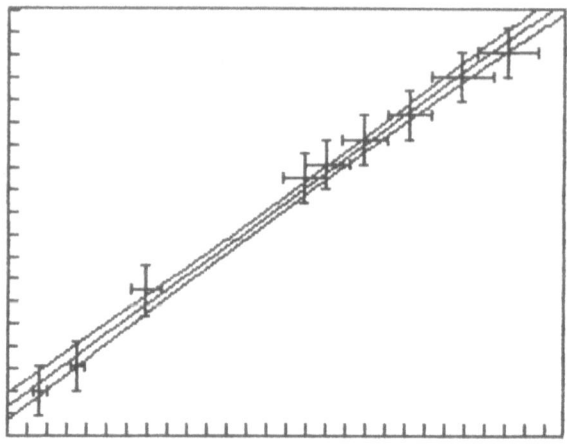

FIGURE 2. Same as Figure 1, but suppressing 'error bars' in calculating uncertainties.

## 4. Statistics for Measuring a Constant

Repeated measurement of a single 'constant' observable in order to obtain its 'true value' is a special case of $M$-dimensional inference where $M = 1$. One seeks to determine the parameter $B$ in the model $\eta = (0)\xi + B$ which best fits the $N \geqslant 1$ observations of $\eta$. Application of the MAXENT orthogonal least squares procedure to this case yields the normal distribution.

$$d p(\eta) / d\eta = N(a_0, \sigma_0), \tag{6a}$$

where the mean and variance are given by the sample estimates

$$x_0 = \;<\eta>, \quad \sigma_0^2 = \;<[|\eta - \eta_n| + \Delta\eta_n / 3^{1/2}]^2>. \tag{6b}$$

Note that (6) defines a continuum p.d.f. based on the discrete (finite resolution) observation(s). Thus, we have a simple solution to a classic problem [6].

Also, note that (6a) seems to indicate the standard assumption that the data have been independently 'drawn from a normal population'. Of course, in our case (6) may have been inferred from an unique, unrepeatable observation. Nevertheless, one usually proceeds to estimate the 'true' distribution of the parameters (mean and variance), typically using orthodox Bayesian analysis, which is based on the idea that repetitions of the experiment will yield a distribution of parameters '$a$' and '$\sigma$' [7]. From the viewpoint given at the outset, such repetition is not possible. The totality of data should be combined to achieve a new MAXENT distribution based on all the evidence.

Alternatively, Bayesian inference has been justified as the only 'common sense' procedure that determines the 'plausibility' of *propositions* of natural language based on other propositions. As such, the distributions obtained are not necessarily (Kolmogorov) measures on *sets of objects* that is, in general the 'propositions' need not correspond to measurable sets. MAXENT, on the other hand, specifies a true probability *measure*.

MAXENT also explicitly distinguishes between hypotheses (which define sets of objects for which certain attributes are true) and evidences about the truth of these hypotheses (which define sets of probability measures consistent with observed data). We have suggested that 'language' may be observed data). We have suggested that 'language' may be 'built up' from these disparate ingredients in a systematic way [8]. On the other hand, Bayesian calculus treats both types of 'propositions' on an interchangeable basis, treating such semantical distinctions as irrelevant.

In addition we note that Bayesians regard language as inherently Boolean. All statements have a binary truth value. Our experience with quantum 'logic', measurement theory, and the general problem of *defining* object classes suggests that it is often necessary to admit so-called 'fuzzy' truth. A priori (before any 'measurements' of attributes) it is necessary to define the objects in terms of observable attributes, but such definition may be subjective or inherently non-sharp. The subjective character of definition can be verified by asking various innocents whether a penguin is a bird? Or, is the fabulous Count Dracula a mythical 'bird' [9]?

It would seem, therefore, that 'fuzzy' truth is not a matter of lack of information. Alternatively, it may reflect an inherent inability to measure with perfect resolution. Indeed there often exists *non-uniform* instrumental sensitivity, which corresponds to a 'fuzzy' answer to the question 'what is the truth that the particle can be actually observed in this detector?'

Unfortunately, the resemblance of 'fuzzy' membership functions to conditional probability densities has led many to the belief that 'fuzziness' is simply an additional form of 'uncertainty' that can be treated in a Bayesian context. (The experimental question is rephrased. 'What is the 'probability' that the particle can be detected?') As in Quantum Theory, practical computations 'work' regardless of the interpretation. We merely remark that much of the confusion about the meaning of Quantum Mechanics could be dispelled if one were to accept QM as a classical probability theory about objects labelled by rays in a Hilbert space equipped with a 'fuzzy' membership function [10]! The Quantum 'state' is an object on which probabilities are asserted; it has no inherent meaning as a probability amplitude!

In any case, MAXENT admits suitable generalization to deal with this *prior* lack of definition. The general regression procedure described above clearly allows for this possibility.

Despite these caveats, we do not necessarily deny the legitimacy of Bayesian methods. Nevertheless, we are reassured when we can make inferences strictly within the MAXENT context. In the present case, we know from (6) that *if* another experiment were to be performed, the resulting data would yield a confidence probability $N(a, \sigma)$. However, this experiment has not been performed. The issue, then, is: What confidence do we have in the *reliability* of the experiment at hand? Put another way, what probability can we assign to the probability (6)? [8]

Clearly, some criterion is required, as we admit no repetition of the original experiment. Since a new experiment would generally provide a different amount of information, it seems natural to bound the expected increment of information produced by a new experiment over that supplied by the original one. Since the information of one distribution relative to a second (prior) is given (in 'nats') by

$$I(p,q) = \int dp \;\; \text{in} \;\; (dp/dq),  \tag{7}$$

the 'reliability' evidence is thus:

$$E^* = \{p^* \mid \int dp^*(a, \sigma) \mid [N(a, \sigma), N(a_0, \sigma_0)] \leqslant K\},  \tag{8}$$

where $dp^*$ represents the probability on the probability $dp(\eta)$, the latter being uniquely characterized by the first two moments. The appropriate 'prior' on $(a, \sigma)$ is $da d\sigma / \sigma$.

MAXENT predicts from (8) that in terms of the Lagrange parameter $\beta$ and the variable $\chi = \beta(\sigma/\sigma_0)^2$ the reliability distribution is given by

$$dp^*(a, \chi) / da\, d\chi = N[a_0, (\sigma_0^2 / \beta)^{1/2}]\chi_\beta[\chi],  \tag{9}$$

where $\chi_\beta$ has the form of a chi-squared distribution with $\beta$ degrees of freedom [11]. The function $K(\beta) = -\partial \ln Z(\beta) / \partial \beta$ defining the constraint condition is given by:

$$K(\beta) = (1/2)[1/\beta + \ln(\beta/2) - \psi(\beta/2)],  \tag{10}$$

where $\psi(z)$ is the digamma function.

Now the information bound ('$K$') should depend on the number of observations. That is, $K = K(N) = g(N)N$ should represent the maximum expected additional information supplied by the 'gedanken' experiment, where $g(N)$ is the number of nats per degree of freedom when there are $N$ measurements. One can verify that (10) is approximately satisfied with $g(N) = N$. In fact, for all $N > 2$ satisfaction is guaranteed to within 5% while for $N > 6$, the equation is satisfied to within 2% with improvement as $N$ increases. Thus, if the information rate $g(N)$ is roughly $N$ nats per degree of freedom when there are $N$ observations, the reliability level (9) for the experiment is just what one derives from the (untenable) assumption that each datum is independently distributed as $N(a_0, \sigma_0^2)$. Namely, that the means are effectively distributed according to a normal distribution centered at the observed mean which variance $\sigma_0^2 / N$ and the variances are distributed as the sum $\Sigma_{n=1} N(\sigma / \sigma_0)^2$, [11].

Combining both evidences, the appropriate distribution on the data values is given by the expectation value of the confidence level w.r.t. the reliability probability [8]

$$\mathcal{E}[dp(\eta \,|\, E) / d\eta \,|\, E^*] = \tag{11}$$

$$\int dp^*(a, \sigma) \exp(-[\eta - a]^2 / 2\sigma^2) / (2\pi\sigma^2)^{1/2}.$$

We remark that criteria are not 'absolute'. We feel that those we have proposed are general and natural, but they may be modified if one has reason to believe that they are inappropriate. Criteria are clearly part of the 'prior' information one assumes as part of the inference process.

## 5. Summary

Application of MAXENT to linear orthogonal regression permits meaningful inferences in cases where standard statistical arguments do not apply. This includes situations where trials are not repeated (in principle, always), where there are few data points, and where instrumental 'error bars' cannot be ignored. An 'expert draftsman' criterion is used to establish MAXENT evidence.

As a special case, one can infer a continuum probability distribution from discrete measurements of a 'constant'. In conjunction with an information criterion for MAXENT inference about the reliability of the original data set, one can obtain a measure of confidence for even a *single* observation! Orthodox predictions are recovered for $N > 6$ measurements.

We have tried to clarify our reasons for exclusive use of MAXENT inference as opposed to its use within a Bayesian context.

## Acknowledgements

I wish to acknowledge the special assistance of Ms. Robin McNemar in early stages of this work, and the patience of the many students whose lab experience included imperfect versions of the program. Also, I express appreciation to Dr. Paul Fougere for his heroic efforts on behalf of this workshop, and to several participants for their stimulating comments. Finally, I acknowledge the profound influence of Prof. Ed Jaynes.

## References

1.  E.T. Jaynes, Phys. Rev., **108**, 171 (1957).

2.  E.T. Jaynes, IEEE Trans, Systems Sci. Cybernetics, **SSC-4**, 227 (1968), Found. Phys., **3**, 477 (1973).

3.  C. Villegas, Ann. Math. Stat., **43**, 1767 (1972); L. Nachbin, *The Haar Integral* (Van Nostrand, Princeton, 1965).

4.  E.T. Jaynes, in W.L. Harper and C.A. Hooker, editors *Foundations and Philosophy of Statistical Inference* (Reidel, Dordrecht, Holland, 1976), 175-257.

5.  J.F. Cyranski, *Fitting Linear Models to Sparse, Unreplicated Data of Poor Resolution* (submitted for publication).

6.  J.F. Cyranski and N.S. Tzannes, Kybernetes, **12**, 187 (1983).

7.  W.T Eadie, D. Drijard, F.E. James, M. Roos and B. Badoulet, *Statistical Methods in Experimental Physics* (North-Holland, Amsterdam, 1971).

8.  J.F. Cyranski, in *Maximum Entropy and Bayesian Methods in Applied Statistics*, ed. J.H. Justice (Cambridge UP, Cambridge, 1986), 101.

9.  J.F. Cyranski, Inform. Contr., **41**, 275 (1979); Found. Phys., **9**, 641 (1979); J. Math. Phys., **22**, 1467 (1981).

10. J.F. Cyranski, J. Math. Phys., **23**, 1074 (1982).

11. A. Papoulis, *Probability Random Variables and Stochastic Processes* (McGraw-Hill, New York, 1965).

*Also of related Interest*

## Maximum-Entropy and Bayesian Methods in Inverse Problems
edited by C. Ray Smith and W.T. Grandy, Jr.
1985, 504 pp.                                  ISBN 90-277-2074-6
FUNDAMENTAL THEORIES OF PHYSICS 14

## Maximum-Entropy and Bayesian Spectral Analysis and Estimation Problems.
Proceedings of the Third Workshop on Maximum Entropy and Bayesian Methods
in Applied Statistics, Wyoming, USA, August 1–4, 1983,
edited by C. Ray Smith and Gary J. Erickson
1987, 332 pp.                                  ISBN 90-277-2579-9
FUNDAMENTAL THEORIES OF PHYSICS 21

## Maximum-Entropy and Bayesian Methods in Science and Engineering.
Volume 1: Foundations.
Volume 2: Applications.
edited by Gary J. Erickson and C. Ray Smith
1988, 1: 308 pp.; 2: 458 pp.,          Vol. 1. ISBN 90-277-2793-7
                                       Vol. 2. ISBN 90-277-2794-5
                                       Set ISBN 90-277-2792-9

FUNDAMENTAL THEORIES OF PHYSICS 31/32

## Maximum-Entropy and Bayesian Methods
## Cambridge, England, 1988
edited by J. Skilling
1989, 544 pp.                                  ISBN 0-7923-0224-9
FUNDAMENTAL THEORIES OF PHYSICS 36

## Papers on Probability, Statistics and Statistical Physics
by E. T. Jaines
edited by R. D. Rosenkrantz                     ISBN 0-7923-0213-3